*Early Sedimentary Evolution
of the Michigan Basin*

Edited by

Paul A. Catacosinos
Department of Geology
Delta College
University Center, Michigan 48710

Paul A. Daniels, Jr.
Petro-Hunt Corporation
1601 Elm Street
Dallas, Texas 75201

1991

© 1991 The Geological Society of America, Inc.
All rights reserved.

All materials subject to this copyright and included
in this volume may be photocopied for the noncommercial
purpose of scientific or educational advancement.

Copyright is not claimed on any material prepared
by government employees within the scope of their
employment.

Published by The Geological Society of America, Inc.
3300 Penrose Place, P.O. Box 9140, Boulder, Colorado 80301

Printed in U.S.A.

GSA Books Science Editor Richard A. Hoppin

**Library of Congress Cataloging-in-Publication Data**

Early sedimentary evolution of the Michigan Basin / edited by Paul A.
  Catacosinos, Paul A. Daniels, Jr.
      p.   cm. — (Special paper / Geological Society of America ;
  256)
    Papers from the 15th Annual Meeting of the Eastern Section of the
  American Association of Petroleum Geologists, held in 1986 in Ann
  Arbor, Mich.
    Includes bibliographical references and index.
    ISBN 0-8137-2256-X
    1. Geology—Michigan Basin (Mich. and Ont.)—Congresses.
  2. Geology, Stratigraphic—Congresses.   3. Sedimentation and
  deposition—Michigan Basin (Mich. and Ont.)—Congresses.
    I. Catacosinos, Paul A., 1933-      II. Daniels, Paul A.
  III. American Association of Petroleum Geologists. Eastern Section.
  Meeting (15th : 1986 : Ann Arbor, Mich.)   IV. Series: Special papers
  (Geological Society of America) ; 256.
QE78.E27    1991
557.74—dc20                              90-25692
                          CIP

**Cover Photo:** The color coded temperature image of the Great Lakes region was collected by NASA's Heat Capacity Mapping Mission (HCMM) satellite at 2:12 A.M. on August 21, 1978. Temperature was calculated from the intensity of thermal infrared radiation in the 10.5 to 12.5 micrometer wavelength region as measured by one channel of HCMM's multispectral scanner. The relation between color and temperature is shown on the color bar beneath the image. Lakes retaining the heat of the previous day appear red, urban areas are yellow to green, agricultural areas are green to aqua, forests are purple to black, and high-altitude clouds are black. Cold water is upwelling on the western shore of Lake Michigan. Photo supplied by John Herman, GeoSpectra Corporation, Ann Arbor, Michigan.

10   9   8   7   6   5   4   3   2

# Contents

*Preface* .................................................................................................. v

*Dedication* ............................................................................................. vii

*Thermal maturity of the Michigan Basin* ............................................................. 1
    K. R. Cercone and H. N. Pollack

*Economic geology and history of metallic minerals in the
    Northern Peninsula of Michigan* ..................................................................... 13
    Robert C. Reed

*Stratigraphy of Middle Proterozoic to Middle Ordovician
    formations of the Michigan Basin* ................................................................... 53
    Paul A. Catacosinos and Paul A. Daniels, Jr.

*Diagenetic history of the Trenton and Black River Formations
    in the Michigan Basin* .............................................................................. 73
    Joyce M. Budai and James L. Wilson

*Late Silurian pinnacle reefs of the Michigan Basin* ................................................. 89
    Gerald M. Friedman and David C. Kopaska-Merkel

*A history of study of Silurian reefs in the Michigan
    Basin environs* ..................................................................................... 101
    Robert H. Shaver

*The Salina evaporites in the Michigan Basin* ........................................................ 139
    Peter Sonnenfeld and Ihsan Al-Aasm

*Upper Devonian biostratigraphy of Michigan Basin* ................................................... 155
    Raymond C. Gutschick and Charles A. Sandberg

*Late Devonian history of Michigan Basin* ............................................................ 181
    Raymond C. Gutschick and Charles A. Sandberg

***Mississippian System of the Michigan Basin; Stratigraphy,
sedimentology, and economic geology*** ............................................. 203
    James A. Harrell, Craig B. Hatfield, and George R. Gunn

***Geological and geophysical evaluation of the region around
Saginaw Bay, Michigan, (central Michigan Basin) with
image processing techniques*** ....................................................... 221
    John D. Herman, Robert K. Vincent, and Ben Drake

***Index*** ............................................................................. 241

# *Preface*

In 1986 the 15th annual meeting of the Eastern Section of the American Association of Petroleum Geologists was held in Ann Arbor, Michigan. This meeting was co-hosted by the Department of Geological Sciences, University of Michigan and the Michigan Basin Geological Society, and coincided with the 50th anniversary of the MBGS. The inspiration for this volume, *The early sedimentary evolution of the Michigan Basin,* is taken from that meeting.

The main purposes of this volume are to recognize recent contributions to the understanding of Michigan Basin geology, and to provide an easier access to many of the voluminous, but scattered and sometimes obscure stratigraphic references. The chapters included are primarily stratigraphically oriented, with emphasis on the lower and middle Paleozoic sections. In some cases, the chapters provide critical summaries and reviews, while others present original research and/or new interpretations. The following is a brief summary of the chapters in the order in which they are presented in this volume.

Setting the stage are "Thermal maturity of the Michigan Basin" by Cercone and Pollack, and "Economic geology and history of metallic minerals in the Northern Peninsula of Michigan" by Reed. The chapter by Cercone and Pollack discusses the various heat-flow hypotheses that have been advanced for this basin and gives an alternative theory of the thermal evolution of the basin. It is a significant contribution and will be widely cited for years to come. The chapter by Reed provides insight into the probable lithologic setting for the Precambrian rocks that underlie the Michigan Basin and is an important aid to regional geophysical and geological interpretations. It also provides valuable historical perspectives and statistical information that are not readily available in any other source.

The third chapter, "Stratigraphy of middle Proterozoic to Middle Ordovician formations of the Michigan Basin" represents an attempt by the editors to expand on the idea that a "not-so-firm foundation" existed within the basin prior to the deposition of the lower Paleozoic succession. Discussion of the underlying Precambrian concentrates on sedimentary rocks that are interpreted as representing rift-fill sequences similar to those exposed within the adjacent Lake Superior Basin. It also provides a good summary and new regional interpretations of early Paleozoic stratigraphy and the associated basin evolution. We trust that it will generate discussion and comment and provide the impetus for future work.

"Diagenetic history of the Trenton and Black River Formations in the Michigan Basin" is written by Budai and Wilson. These rocks have consistently attracted considerable attention because of their importance as hydrocarbon reservoirs. Consequently, much research has been (and will continue to be) undertaken on the dolomitization and diagenesis of the formations. The chapter represents the leading edge of research. It also discusses the interrelation between hydrocarbon migration and mineralization in these rocks.

The inclusion of more than one chapter on Silurian pinnacle reefs was considered mandatory because of their economic importance and the part they play in interpretations concerning basin history. The first, "Late Silurian pinnacle reefs of the Michigan Basin," by Friedman and Kopaska-Merkel, gives an excellent summary of the current models of pinnacle-reef development. It also provides information on reef-reservoir characteristics, regional variations along the northern reef trend, and evaporite deposition. The second, "A history of study of Silurian reefs in the Michigan Basin," by Shaver, presents a stimulating perspective on reef relations. In addition, this chapter provides extensive historical and bibliographic information that will be invaluable to future scholars.

The "Salina evaporites in the Michigan Basin" by Sonnenfeld and Al-Aasm discusses the cyclical deposition of evaporites in the basin and how and why these cycles have varied through time. It contains bibliographic information (particularly older and/or more obscure references) that will be of great aid in future geochemical studies.

Because of the extensive information generated, two chapters on the Devonian systems were developed from one longer paper. These two are the "Upper Devonian biostratigraphy of the Michigan Basin" and the "Late Devonian history of Michigan Basin" by Gutschick and Sandberg. They deal with the stratigraphy, basin evolution, and the Upper Devonian formations within the basin, and their regional geologic relations. Further, new formational names are proposed in "Upper Devonian biostratigraphy."

"The Mississippian of the Michigan Basin: Stratigraphy, sedimentation, and economic geology" by Harrel, Hatfield, and Gunn presents a thorough summary of the system within the basin. It also addresses the sedimentation, tectonics, and the mineralogic and hydrocarbon importance of key formations in the system.

The final chapter of this volume is entitled "Geological and geophysical evaluation of the region around Saginaw Bay, Michigan (central Michigan Basin) with image-processing techniques" and is written by Herman, Vincent, and Drake. It is a good example of the synthesis and integration of diverse data sets that yield structural interpretations for exploration purposes. It also represents an excellent example of the use and interpretation of remote-sensing techniques as applied to the Michigan Basin.

As editors, we wanted to include coverage of the complete Michigan Basin stratigraphic column in this volume. Time and circumstance, however, conspired against us. This publication does form a natural trilogy with GSA Memoir 156, *Geology and tectonics of the Lake Superior Basin,* and GSA Memoir 160, *Early Proterozoic geology of the Great Lakes region,* and we trust it will encourage future efforts that will deal with other aspects of the basin's geology.

We thank all the authors for their patience and dedication to this project. We regret that all the manuscripts submitted could not be included, but appreciate the interest that was expressed in this effort. This volume could not have been completed if it were not for the encouragement of Campbell Craddock and Richard Hoppin in their roles as GSA Books Editors, and Lee Gladish, GSA Books Manager. In addition, many of our reviewers worked far beyond the ordinary obligations of critical readers, and to them we owe a great deal of gratitude. Our thanks also go to our employers, Delta College and Petro-Hunt Corporation, who were both understanding and generous in their support to us.

This has been a trying but rewarding labor for the both of us; we hope that this book will be of value for years to come.

Paul A. Catacosinos
Paul A. Daniels, Jr.

# *Dedication*

Louis I. Briggs, Jr.
April 24, 1921-May 15, 1979

Harold M. McClure, Jr.
January 13, 1921-December 27, 1977

Garland D. Ells
February 23, 1920-August 22, 1987

Since its inception, the Michigan Basin geological community has been closely knit, with the diverse interests of academia, industry, and government working hand in hand unravelling the geological puzzles of the basin. Although there are many past and present workers of the basin who could be singled out for their contributions, we feel that there are three individuals who epitomize the main areas mentioned above. These are: Dr. Louis I. Briggs, Jr., University of Michigan; Harold M. McClure, Jr., independent oil and gas producer; and Garland D. Ells, Michigan Geological Survey. Now deceased, these three dedicated geologists have left a legacy that continues to enrich our imaginations and minds to this day. It is to them that we dedicate this work.

Printed in U.S.A.

# Thermal maturity of the Michigan Basin

**K. R. Cercone**
*Geoscience Department, Indiana University of Pennsylvania, Indiana, Pennsylvania 15705-1087*
**H. N. Pollack**
*Deparment of Geological Sciences, University of Michigan, Ann Arbor, Michigan 48109-1063*

## ABSTRACT

Almost all Paleozoic strata in the Michigan Basin display elevated levels of organic maturity that cannot be explained by present-day burial depths, geothermal gradients, and heat flow. Likewise, higher heat flow from the basement in the past is unsatisfactory as an explanation of the elevated maturities, particularly for the younger sediments. Previous studies have concluded that a significant amount of Permo-Carboniferous overburden has been removed from this region by early Mesozoic regional uplift and erosion. If the missing overburden were sufficiently thick and thermally resistive, a "thermal blanket" effect would have caused elevated temperatures throughout the underlying stratigraphic section during late Paleozoic and early Mesozoic time. Models of organic maturation that take into account this "thermal blanket" effect as well as other variations in thermal conductivity attributable to lithologic differences can explain the anomalous maturity of most strata in the Michigan Basin without postulating any increase in ancient heat flow. An overburden thickness of 2,000 m with a gradient the same as that observed in the present-day Carboniferous section would provide an adequate explanation for the elevated maturities. Lesser thicknesses of overburden would require correspondingly higher geothermal gradients, a reasonable possibility if the missing overburden was a fluvio-deltaic sequence containing low-conductivity carbonaceous strata.

## INTRODUCTION

Over the past decade, the thermal evolution of the Michigan Basin has become the subject of controversy. Geophysical studies have related basin subsidence to thermal contraction following an Ordovician thermal event (Nunn and others, 1984; Nunn, 1986). These studies predict that only the basal Cambrian and Ordovician units of the basin should have been heated to the hydrocarbon generation stage (Nunn and others, 1984). Analyses of vitrinite reflectance, kerogen/spore coloration, and organic biomarkers, however, have led other workers to conclude that strata as young as Devonian are mature enough to generate petroleum (Moyer, 1982; Gardner and Bray, 1984; Rullkotter and others, 1990), although not all studies of thermal maturation indices agree (Hogarth and Sibley, 1985). Additional evidence for elevated maturity comes from geochemical analysis of crude oils in the Michigan Basin. Although early studies concluded that all the oils produced from Devonian reservoirs were derived from Ordovician source rocks (Vogler and others, 1981), more recent work suggests that at least the Traverse Group of Devonian oils was generated from mature Devonian source rocks (Pruitt, 1983; Illich and Grizzle, 1983, 1985; Rullkotter and others, 1986). As Devonian strata in the basin are presently buried at shallow depths (<1,500 m), oil generation in these beds reflects an anomalously high thermal maturity.

A previous study by one of the authors (Cercone, 1984) attributed the anomalous maturities of Michigan Basin strata both to the erosion of upper Paleozoic overburden from the basin and to the presence of a uniformly higher geothermal gradient of 35° to 45°C/km over the full stratigraphic section. The higher basement heat flow thus implied is not easily reconcilable with geophysical models of basin evolution. Several workers have

Cercone, K. R., and Pollack, H. N., 1991, Thermal maturity of the Michigan Basin, *in* Catacosinos, P. A., and Daniels, P. A., Jr., eds., Early sedimentary evolution of the Michigan Basin: Geological Society of America Special Paper 256.

concluded that such a thermal regime is probably unrealistic for the time interval of concern (Nunn and others, 1984; Rullkotter and others, 1986). In this chapter, we recognize that geothermal gradients in the basin primarily reflect vertical variations in sediment lithology, and to a much lesser extent, declining heat flow from the basement following the initiating thermal event. We explain the elevated organic maturity of strata in the Michigan Basin with an organic maturation model that uses gradients appropriate to the heat flow and individual formation lithologies of specific time intervals.

## DATA BASE

In order to allow other workers to test our conclusions, and to provide a basis for future modeling studies, we list in the Appendix all the organic maturity data for the Michigan Basin that we have been able to gather. Unfortunately, this data base has problems in terms of completeness and internal consistency. Because the majority of studies in the basin have been concerned with oil source beds, some units (e.g., the Upper Ordovician) have a disproportionate number of maturity analyses. Other units not generally sampled by oil wells (e.g., the Lower Cambrian) are underrepresented. Furthermore, since the individual sets of organic maturity data (listed separately in the Appendix) were generated at different institutions using different techniques, they are often difficult to compare. We have made no attempt to cast all the data in terms of a single consistent maturity parameter, as correlations between parameters are not agreed upon by all workers. Instead we have chosen to model selected subsets of vitrinite and sporopollen data (Fig. 1) individually, using appropriate stratigraphic constraints for each well.

A more serious problem for modeling concerns the reliability of different organic-maturity-measurement techniques at high levels of maturity. For Devonian and Carboniferous strata, which have an abundance of woody vitrinite material, maturity estimates are generally consistent. There appears to be two populations of vitrinite present in these samples: a low-reflectance indigenous population and a high-reflectance reworked population (Zempolich, personal communication, 1988). Note that reflectance values are reported for both populations in the Mobil data base whereas Shell reports only the low-reflectance values. Where two values have been reported, we have taken the lower value to represent the true thermal maturity of the sample. In Silurian and Ordovician samples, greater discrepancies appear in the data. Sporopollen coloration indices in these rocks, particularly in the deeply buried Ordovician units, seem to indicate extremely high maturities. Silurian and Ordovician rocks contain no vitrinite; any reflectance measurements reported for them have been performed on macrinitic or chitinous material. Reflectance values for Silurian and Ordovician rocks suggest somewhat lower levels of maturity. For example, in the Sun-Bradley #13186 well, Moyer (1982) reports a Sporopollen Coloration Index (SCI) value of 7.5 for the Utica Formation (1,818 m). This is equivalent to a Thermal Alteration Index (TAI) of 3.75 and a vitrinite

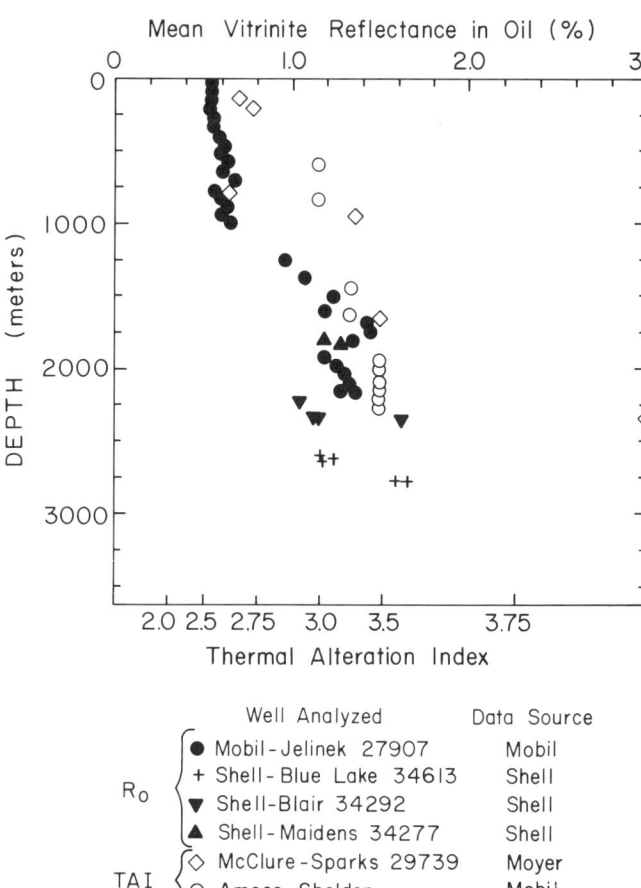

Figure 1. Organic maturity data from six wells selected from the Appendix for maturity modeling. Well locations are listed in Table 2. Note the discrepancies between thermal alteration indices (TAI) and vitrinite reflectance indices ($R_o$).

reflectance of 2.25 percent. For the same well, Shell reports a "vitrinite" reflectance of only 1.51 percent for the deeper Prairie du Chien Formation (2,021 m). Similarly, Gardner and Bray (1984) report a TAI of 4.0 for the Trenton Formation (1,930 m) of the Mobil-Jelinek #27907 well, equivalent to a vitrinite reflectance of 3.5 percent or more. Mobil, however, reports an actual reflectance of 1.19 percent for the Trenton at 1,920 m. The failure of any of our models (described below) to match high Ordovician sporopollen coloration values may indicate that the lower values are, in fact, more accurate. In order to avoid prejudicing the issue, however, we have used both sporopollen and reflectance data as constraints for our thermal models.

## ORGANIC MATURATION MODEL

Organic maturity can be modeled in one of two ways: by kinetic approximation of individual or collective organic reactions (Mackenzie and McKenzie, 1983; Tissot and others, 1980) or by the more general assumption of Arrhenius kinetics for all organic reactions (Lopatin, 1971; Waples, 1980). The latter approach yields somewhat higher estimates of maturity because it assumes maturation is a continuous process with a temperature-dependent rate, whereas the former approach incorporates temperature barriers that, unless exceeded, inhibit additional maturation (Tissot and others, 1987; Wood, 1988). Additionally, the Lopatin technique may overestimate the importance of time relative to temperature (Quigley and Mackenzie, 1988). Nevertheless, we use the Lopatin method because of its simplicity; given the uncertainty in our organic maturity data base, we believe that any errors introduced by the Lopatin method are subsumed in the larger uncertainties of the data themselves.

Our models have been constructed using the Lopatin method as incorporated in software written and marketed by Platte River Associates (*Lopatin,* v. 1.00). This program calculates burial-history curves from stratigraphic data, and isotherms from surface temperature and geothermal gradient data. Both surface temperature and geothermal gradients can be allowed to vary gradually through time. In addition, at any one time the geothermal gradient can comprise several different vertical segments, so that different lithologic units can be represented with appropriate gradients. These enhancements allow the generation of moderately complex thermal models. The thermal fields calculated by the Lopatin software are equilibrium fields, i.e., the period of time required for temperature adjustment associated with deposition and subsidence or uplift and erosion are ignored. However, it can easily be shown that for the rates of subsidence and uplift relevant to the history of the Michigan Basin, each on the order of 20 m/m.y., the assumption of equilibrium temperatures introduces no serious errors.

After the Lopatin program calculates burial history curves and isotherms, it generates a time-temperature index (TTI) value for the top and bottom of each stratigraphic unit according to the procedures outlined by Waples (1980). These TTI values provide estimates of the current organic maturity of strata and can be compared to actual maturity measurements, such as vitrinite reflectance or kerogen coloration index. The effect of sediment compaction on the depth of burial is not taken into account in this version of the Lopatin method. However, Dykstra (1987) has shown that when the effects of compaction on burial depth are taken into account in a Lopatin model, only the estimated timing of hydrocarbon generation changes; estimates of final organic maturity do not. The effects of sediment compaction, pore-water expulsion, and mineralogic changes on thermal conductivity and geothermal gradient were evaluated in our own study and found to be important only in the early depositional history under shallow-burial conditions. Because most organic maturation occurs under conditions of deeper burial, corrections to geothermal gradients for the above effects were not made.

Lopatin models can, through a series of forward numerical experiments, be used to constrain the history of a basin whose stratigraphic or thermal evolution is not known, but for which observations of the organic maturities of different strata exist. By varying the parameters of the model, a number of different stratigraphic/thermal histories can be constructed and used to predict current maturity levels in basin strata. The stratigraphic/thermal history whose predicted current maturities most closely match observed maturities is a satisfactory, albeit non-unique, model of the basin's actual evolution.

## ASSUMPTIONS

In order to model the thermal evolution of the Michigan Basin sediments, some preliminary assumptions must be made concerning the basin's stratigraphic history. First, following Nunn and others (1984), we assume that the principal subsidence of the basin began in the Ordovician. Individual Ordovician to Pennsylvanian unconformities are assumed to represent less than 100 m of erosion each and, therefore, are not included in the burial-history curve. However, a major erosional event is assumed to have occurred during Triassic and early Jurassic time, removing significant amounts of Pennsylvanian-Permian sedimentary cover from middle and eastern North America (Beaumont and others, 1988; Crowley and others, 1986a). Previous studies have variously estimated the amount of Michigan Basin overburden removed by this regional event at 0.3 to 0.5 km (Nunn and others, 1984; Beaumont and others, 1988); 1.0 km (Cercone, 1984; Vugrinovich, personal communication); and from other localities 1.5 to 2.0 km (Damberger, 1971); 2.0 to 3.0 km (Crowley and others, 1984; Crowley and others, 1986b), 3.5 to 4.0 km (Hower and Davis, 1981) or more (Friedman, 1987). As will be seen shortly, it is not the overburden thickness alone that in the controlling factor in the maturity, but rather the temperature at the base of the overburden. Any combination (product) of overburden thickness and geothermal gradient in the eroded sediments that yields the same temperature at the interface between the eroded overburden and the present-day preserved sediments will have the same effect on the thermal maturity of the preserved sediments. Finally, we assume that the Michigan Basin has undergone neither uplift nor subsidence between late Jurassic and Pleistocene time (Dorr and Eschman, 1971). Because the depositional and erosional events of the Pleistocene have occurred so recently, they have little significance for questions of organic maturation and can be safely ignored.

The generation of Lopatin models for the Michigan Basin requires specification of the surface temperature and subsurface geothermal gradients, both of which in principle can vary with time. We assume the mean annual surface temperature of the basin from Ordovician to Devonian time to have been 20°C, because in that time interval, Michigan was near the paleoequator and almost always submerged below epeiric seas (Droste and Shaver, 1983). From Permian to Holocene time, the surface temperature is assumed to have dropped continuously from 20 to 10°C, reflecting the drift of North America into higher latitudes.

Past geothermal gradients in the Michigan Basin were controlled both by sediment lithology and basement heat flow. Studies of bottom-hole temperature data in the Michigan Basin (Speece and others, 1985) have determined characteristic present-

**TABLE 1. AVERAGE GEOTHERMAL GRADIENTS USED FOR EACH SYSTEM CURRENTLY PRESERVED IN THE MICHIGAN BASIN***

| System | Present-day Gradient (°C / km) |
|---|---|
| Carboniferous | 26.2 |
| Devonian | 22.6 |
| Silurian | 14.1 |
| Ordovician | 21.2 |

*After Speece and others, 1985.

day geothermal gradients for each lithologic unit in the basin. We have calculated weighted averages of these lithologic gradients to obtain representative gradients for each of the Ordovician, Silurian, Devonian, and Carboniferous systems (Table 1). We allow changes in basement heat flow only in association with a thermal event, which may have initiated basin subsidence. The increased heat flow caused by such an event affects geothermal gradients in the basin principally during Ordovician time, with minor effects in the Silurian and no effects thereafter. However, Ordovician and Silurian strata were not deeply buried in Ordovician and Silurian time, and thermal maturity models that include increased gradients due to higher basement heat flow differ only slightly from those that use uncorrected modern gradients. Similar conclusions were reached by Nunn and others (1984).

The remaining free variables for the thermal evolution models are the thickness of and geothermal gradient through the eroded Permo-Carboniferous section. Taken together, these variables will yield a paleo-temperature at the surface of the preserved Carboniferous strata. In the following section we examine several cases that encompass all realistic possibilities of these variables.

## MODEL RESULTS

The stratigraphic and thermal assumptions discussed above are applied uniformly to several wells (see Table 2) for which complete stratigraphic and organic maturity data are available. A typical burial-history curve is shown for one of these wells (Fig. 2), both with and without the reconstructed section of Permo-Carboniferous overburden. For completeness we present a model calculation that has no overburden above the present-day preserved strata at any time in the basin history, and therefore no temperature increment to these strata due to an overlying "thermal blanket." Such a model is shown in Figure 3, curve A, and clearly it is inadequate to explain the observed thermal maturities.

We now turn to models that do include an assumption of Permo-Carboniferous overburden above the existing strata of the basin. We select as an initial reference model one with 1,000 m of overburden (see Cercone, 1984, for arguments in support of this thickness). If the present-day Carboniferous gradient is assumed for both the 1,000 m of eroded strata as well as for the modern sections, the thermal maturities of underlying strata are substantially underestimated (Fig. 3, curve B), thus suggesting a higher gradient through or a greater thickness of the eroded sediments. Gradients between 40° and 60°C/km through the 1,000 m of eroded strata yield a range of predicted maturity values (Fig. 3, curves C and D) that embraces most of the actual organic maturity measurements taken from underlying Ordovician through Carboniferous strata. Gradients in that range imply temperatures of 40° to 60°C at the base of the overburden (and at the surface of the preserved present-day sediments).

As noted earlier, any other combination (product) of thickness and gradient that yields the same temperature at the base of the former overburden would have the same effect on the thermal maturities of the preserved sediments. Thus, 500 m of overburden would require a gradient in the range of 80° to 120°C/km, whereas 2,000 m of overburden would require a gradient in the range of 20° to 30°C/km, similar to the present-day gradient at 26°C/km through the Carboniferous section. Overburden thicknesses greater than 2,000 m would require a gradient even less

**TABLE 2. WELL LOCATIONS FOR FIGURE 1**

| Well Name | Permit Number | Location | System Thickness |
|---|---|---|---|
| Mobil Jelinek Ferris 1 | 27907 | 5–5N–2E Shiawassee Co. | C: 470 D: 560 S: 610 O: 560 |
| Shell State Blue Lake 3-23 | 34613 | 23–28N–5W Kalkaska Co. | C: 194 D: 1010 S: 999 |
| Shell State Blair 2-24 | 34292 | 24–26N–11W Grand Traverse Co. | C: 162 D: 829 S: 969 O: 585 |
| Shell Maidens and others 5-25 | 34277 | 25–23N–15W Grand Traverse Co. | D: 763 S: 738 |
| McClure-Sparks and others 1-8 | 29739 | 8–10N–2W Gratiot Co. | C: 600 D: 630 S: 835 O: 790 |
| Amoco Irving Sheldon 1-21 | 29157 | 21–11N–14E Sanilac Co. | C: 382 D: 564 S: 795 |

than that which exists in the present-day preserved Carboniferous section, in order to avoid predicting maturities in excess of those observed. We believe that a lesser gradient, due either to higher conductivity sediments or a temporary drop in heat flow, is unlikely.

Sensitivity analyses were run to check the dependence of estimated Carboniferous gradients on other model parameters. The assumed timing of the overburden removal has the most significant effect on gradient estimation. Models which assume that erosion began immediately after the end of the Carboniferous, instead of at the beginning of the Mesozoic, require an estimated gradient for the eroded strata some 20 percent higher than would be necessary if the elevated temperatures due to the overburden persisted for the longer period of time.

## DISCUSSION

The principal results of the above model calculations are the determination of the required paleotemperatures (40° to 60°C) at the base of the former overburden, and the placing of constraints on the thickness of and geothermal gradient in the overburden that will yield such temperatures. The 2,000-m upper bound on overburden thickness, accompanied by a geothermal gradient similar to that found in the present-day Carboniferous section, is an acceptable solution, although we prefer an overburden thickness of only 1,000 m for reasons presented by Cercone (1984). It should be noted, however, that any thickness less than 2,000 m does require a compensatory increase in geothermal gradient to achieve the requisite temperatures at the base of the overburden. For a 1,000-m overburden thickness, gradients in the range of 40° to 60°C/km would be required (Fig. 3, curves C and D), and for lesser overburden thickness, even higher gradients would be necessary. Thus, the principal question to be addressed is whether such high gradients are reasonable. Without question, such gradients are commonplace in areas undergoing active extension such as the Basin and Range province, where heat flow averages more than 80 mW/m². However, in intracratonic settings, such as that of the Michigan Basin, a case cannot be made easily for such elevated heat flow. One must look instead to special situations of thermal conductivity as the cause of high gradients.

We assume that the strata eroded from the Michigan Basin were Permo-Carboniferous in age and fluvial-deltaic in character, based on the presence of thick Carboniferous fluvial-deltaic deposits in the adjacent Illinois and Appalachian sedimentary basins. The characteristics of the Carbonifeous sections in these neighboring basins suggest that the eroded overburden in the Michigan Basin may have contained significant thicknesses of carbonaceous mudstones, dark shales, dispersed coaly material, and coal seams. Several investigators (Clark, 1966; Beck, 1976; Kayal and Christoffel, 1982; Ballard and others, 1987; Blackwell and Steele, 1988) have shown that such carboniferous material has a very low thermal conductivity. In a typical fluvio-deltaic sequence comprising sandstones, mudstones, and coal seams, the thermal conductivity can range over a full order of magnitude,

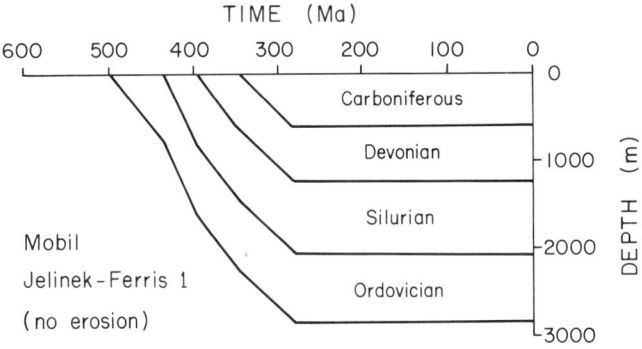

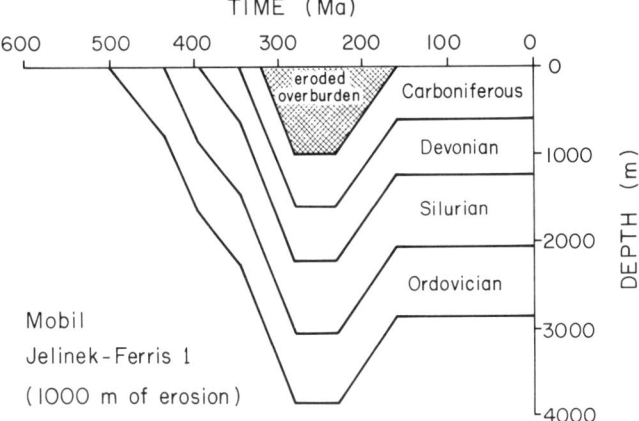

Figure 2. Burial-history curves for the Mobil Jelinek Ferris well, shown with and without reconstructed former overburden.

from 0.5 W/m/K for coal to 1.0 for carbonaceous mudstones to 5.0 for some clean quartzose sandstones (Ballard and others, 1987). Stated another way, for a given heat flow the temperature increase across 10 m of coal is the same as across 20 m of carbonaceous mudstone or 100 m of sandstone, thus underscoring the important role of carbonaceous material in creating a "thermal blanket." If our characterization of the lithology of the former Permo-Carboniferous overburden in the Michigan Basin is correct, the thermal blanket effect would have caused temperatures in underlying strata to rise, without any changes in basement heat flow or geothermal gradients in the underlying formation. Following erosion of the low conductivity strata in the Mesozoic, the thermal blanket effect would have dissipated, as the gradient in the remaining less-carbonaceous Carboniferous strata of the basin has a value of only 26°C/km.

Field observations of high gradients have been made by one of the authors in several of the Gondwana basins of southern Africa, where despite a heat flow of only 40 to 45 mW/m⁻² (Ballard and others, 1987) gradients of 60°C/km are observed over 200 to 300 m sections of the Ecca coal measures. Locally within the actual coal seams gradients as great as 200°C/km exist. Similarly high values were observed by Beck (1976) in the coal measures of Queensland. High gradients, ranging from 60° to 80°C/km$^{-1}$ have also been suggested for coal-bearing strata in the Ruhr Valley of Germany during Carboniferous time (Teichmuller and Teichmuller, 1986), although these workers attribute the high gradients to a higher heat flow.

The range of measured heat flow in the Michigan Basin is between 33 and 58 mW/m$^2$ (Combs and Simmons, 1973; Judge and Beck, 1973), but most of the observed values are between 42 and 54 mW/m$^2$. Taking a heat flow of 50 mW/m as roughly representative of the basin, lithologies with thermal conductivities in the range of 1.25 to 0.8 W/m/K would yield gradients of 40° to 60°C/km. Such conductivities would be common in the Permo-Carboniferous overburden we postulate for the Michigan Basin. Work done in coal-bearing Carboniferous strata of the Appalachian Basin by Hower and Davis (1981) suggests that gradients there ranged from 30° to 45°C/km during coal metamorphism. It should be pointed out, however, that these gradients, inferred from measurements of vitrinite anisotropy, require a minimum of 3.5 km of post-Pennsylvanian erosion in order to account for the observed maturity of coals in western Pennsylvania (Hower and Davis, 1981). If gradients in these coal-bearing units were actually similar to those observed in other carbonaceous sequences and proposed for the eroded overburden of the Michigan Basin, a lesser amount of overburden would need to have been present and later removed. Analogous arguments could also be made for the Illinois Basin, for which an overburden of 1.5 to 2.0 km has been estimated by Damberger (1971) from coal rank.

In summary, we believe a very reasonable case can be made for geothermal gradients in excess of 40°C/km in the Permo-Carboniferous overburden formerly present above preserved Paleozoic strata of the present day. Such gradients extending through 1,000 m of overburden would provide the necessary thermal blanket to raise organic maturities to their present levels.

## SUMMARY AND CONCLUSIONS

Much of the Paleozoic sedimentary fill of the Michigan Basin displays elevated levels of organic maturity that cannot be explained by present-day burial depths and geothermal gradient. Attributing the anomalous maturity to higher heat flow associated with the initiation of subsidence in the Ordovician fails to account for the elevated maturity of Devonian and younger strata, and a higher post-Devonian heat flow leads to excessive maturity of the Ordovician strata. Previous investigations have concluded that a significant thickness of Permo-Carboniferous sediments has been removed from this region by Mesozoic erosion. If this overburden had temperatures of 40° to 60°C at its base for a period of 160 m.y., it would provide a thermal blanket that would yield the observed maturities of the preserved strata beneath. A 2-km overburden with a geothermal gradient the same as exists presently in the preserved Carboniferous strata would yield the necessary temperatures. Alternatively, 1 km of overburden with a temperature gradient of 40° to 60°C/km would also yield the requisite conditions. We favor the lesser thickness of overburden for stratigraphic reasons and because it requires a less dramatic uplift and erosional event for its removal. The temperature gradient through the now-eroded overburden would have to have been greater than presently exists in the preserved basin strata, but could have been easily achieved without enhanced heat flow if the overburden comprised a fluvio-deltaic sequence with abundant carbonaceous mudstones, dispersed coaly detritus, and occasional coal seams, all characterized by low thermal conductivity.

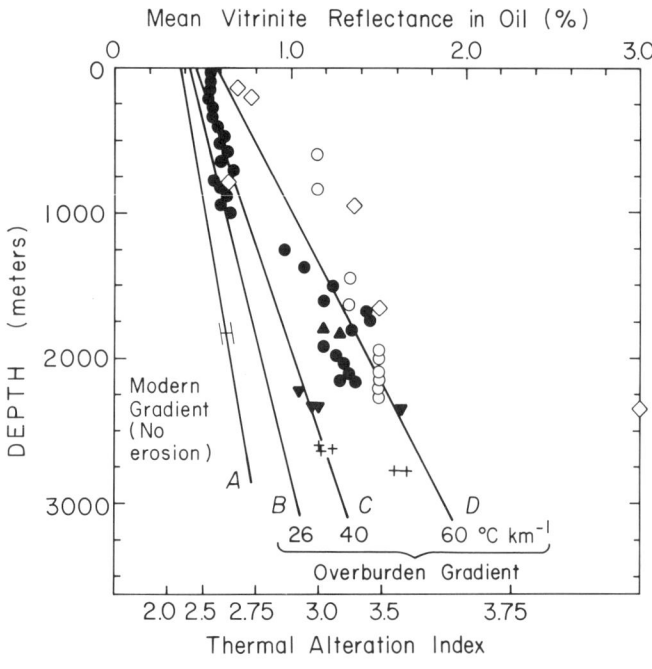

Figure 3. Comparison of observed maturities and calculated maturities from Lopatin models described in text. Observed maturities are as shown in Figure 1. Calculated maturities are shown as lines representing the average model predictions for all six wells. The variations caused by stratigraphic differences between wells fall within the range shown along model line A. Although changes in slope for different stratigraphic units are observed when maturity predictions are plotted as TTI values, they are too small to depict on the vitrinite reflectance scale used here. For all thermal models, present-day gradients are applied to existing units throughout their entire burial history. Line A shows maturities predicted by models that assume no former overburden above existing strata. Line B shows maturities predicted by models that assume 1,000 m of former overburden through which existed a gradient equivalent to the present-day gradient of Carboniferous strata (26°C/km). Lines C and D show maturities predicted by models that assume 1,000 m of former overburden through which existed gradients of 40° and 60°C/km, respectively.

## APPENDIX: ANNOTATED ORGANIC MATURITY DATA BASE

**I. Sleep, Nunn, and Chou (1980)**

The earliest reported thermal maturity data for the Michigan Basin are coal moisture contents (CMC) measured by Andrews and Huddle (1948). N. H. Sleep of Stanford Univeristy and his co-authors discuss these results and also present a single vitrinite reflectance index (VRI) measured by J. Castano of Shell Oil on a Saginaw coal sample from Grand Ledge, Michigan.

| FORMATION | WELL NAME/ PERMIT | DEPTH (M) | MATURITY | |
|---|---|---|---|---|
| Saginaw | Surface sample Bay City, Michigan | | 14.5–10.4% | CMC |
| | Surface sample, Grand Ledge, Michigan | | 0.54% | VRI |

**II. Moyer (1982)**

In an unpublished M.S. thesis study done at Michigan State University, R. B. Moyer analyzes more than 100 samples of cores and cuttings from throughout the Michigan Basin, measuring sporopollen coloration index (SCI) for strata ranging in age from Pennsylvanian to Precambrian. Sporopollen coloration index (SCI) values as assigned by Moyer appear to be equivalent to doubled TAI values.

| FORMATION | WELL NAME/ PERMIT | DEPTH (M) | MATURITY | |
|---|---|---|---|---|
| Saginaw | McClure-Sparks/ 29739 | 122–143 | 5.4 | SCI |
| | Surface sample Grand Ledge, Michigan | | 0.54% | VRI |
| Michigan | MCG-NH 162A/30884 | 291 | 5.0 | SCI |
| | MCG-Six Lakes/31497 | 387 | 5.0 | SCI |
| | MCG-Six Lakes/31497 | 383 | 5.6 | SCI |
| | MCG-Six Lakes/31530 | 376 | 5.8 | SCI |
| | MCG-Herring/26447 | 352 | 5.8 | SCI |
| | Sun-Yake/24239 | 457 | 4.8 | SCI |
| | MGSC-831/25995 | 432 | 5.0 | SCI |
| | MGSC-831/25462 | 409 | 5.2 | SCI |
| | Basin-Badour/15387 | 75–79 | 4.6 | SCI |
| | Cities-Methner/10639 | 299–308 | 5.2 | SCI |
| | Taggart-McCormick/ 11582 | 363–369 | 4.1 | SCI |
| | Ide-Cooley/3815 | 279–287 | 4.8 | SCI |
| | Taggart-Wooden/7770 | 420–423 | 4.7 | SCI |
| | Gordon-DeKraker/10595 | 427–433 | 5.0 | SCI |
| | Union-Newland/8775 | 465–469 | 5.1 | SCI |
| | MGSC 972/26002 | 443–470 | 5.0 | SCI |
| | Sohio-Hoffman/11394 | 430 | 5.7 | SCI |
| | McClure-Sparks/29739 | 204–213 | 5.5 | SCI |
| Antrim | Pure-Nix/9582 | 750 | 6.0 | SCI |
| | Riddell-Trautner/11597 | 694–706 | 5.1 | SCI |
| | Cities-Methner/10639 | 901–905 | 5.6 | SCI |
| | Ohio-Mio/11995 | 570–579 | 5.8 | SCI |
| | Sun-Secord/13632 | 676–686 | 5.6 | SCI |
| | Drake-Gladstone/8482 | 759–768 | 6.4 | SCI |
| | Hook-McClear/16690 | 359–365 | 4.8 | SCI |
| | Bridger-Basin/8270 | 466–475 | 4.3 | SCI |
| | Sinclair-McIntyre/10936 | 536–546 | 5.4 | SCI |
| | Columbia-Satterlee/3298 | 820–833 | 6.8 | SCI |
| | Hilliard-Fox/10661 | 846–856 | 4.5 | SCI |
| | Cities/16734 | 597–602 | 6.3 | SCI |
| | Sun-Rice/9897 | 885–894 | 7.0 | SCI |
| | Chamness/10792 | 228–237 | 4.1 | SCI |
| | McClure-Hewitt/30974 | 768–792 | 4.9 | SCI |
| | McClure-Sparks/29739 | 768–792 | 5.3 | SCI |
| Bell | Sun-Bradley/13816 | 838 | 6.8 | SCI |
| | MCGC-Shuttleworth/ 26799 | 988 | 6.6 | SCI |
| | Pure-Mulder/12376 | 1184 | 7.5 | SCI |
| | Benedum-Corlew/27390 | 1220 | 6.9 | SCI |
| | MCGC-Reiman/27191 | 1130 | 7.4 | SCI |
| | Bolger-Adams/10572 | 883 | 5.7 | SCI |
| | Gulf-Lassiter/10511 | 831–835 | 6.3 | SCI |
| | Strickler-Clear/17892 | 825–836 | 6.8 | SCI |
| | Hilliard-Fox/10661 | 1099–1105 | 6.8 | SCI |
| | Sun-Rice/9897 | 1181–1193 | 7.2 | SCI |
| | Degenther-Ellens/13275 | 793–794 | 6.9 | SCI |
| | Smith-Schulz/14418 | 926–933 | 7.0 | SCI |
| | Ohio-Mio/11995 | 905–910 | 6.4 | SCI |
| | Tamblyn-Jordan/10913 | 776–796 | 6.5 | SCI |
| | Sohio-Cummings/10977 | 1116–1125 | 7.8 | SCI |
| | Maguire-State/10015 | 1181–1185 | 7.6 | SCI |
| | McClure-Hewitt/30974 | 521–540 | 6.2 | SCI |
| | McClure-Sparks/29379 | 943–956 | 6.6 | SCI |
| Salina C | McClure-Hewitt/30974 | 1271–1289 | 6.5 | SCI |
| | McClure-Sparks/29739 | 1655–1676 | 7.0 | SCI |
| | Amoco-PCA/30475 | 1530–1545 | 6.6 | SCI |
| | Marathon-Cupp/31335 | 463–482 | 6.8 | SCI |
| | Marathon-Rzepka/31253 | 509–521 | 7.3 | SCI |
| | Marathon-Russel/31295 | 522–567 | 7.2 | SCI |
| | Amoco-Carpenter/31584 | 820–835 | 7.0 | SCI |
| | Amoco-Garland/28546 | 1893–1917 | 7.0 | SCI |
| | Amoco-DePeel/28999 | 985–997 | 7.3 | SCI |
| | Amoco-Charlton/29099 | 1366–1378 | 7.2 | SCI |
| | SME-Amoco-USA/29474 | 1707–1734 | 6.4 | SCI |
| | Amoco-Timm/30137 | 942–954 | 6.7 | SCI |
| | Amoco-Hansen/30814 | 911–927 | 7.3 | SCI |
| | Amoco-Sorgett/31067 | 802–817 | 7.3 | SCI |
| | Amoco-Mancelona/31293 | 1792–1811 | 7.2 | SCI |
| Utica | Sun-Bradley/13816 | 1818 | 7.5 | SCI |
| | CPC-204/25536 | 1257 | 7.2 | SCI |
| | Gulf-Bateson/5441 | 2858–2862 | 7.8 | SCI |
| | Violette-Warren/6364 | 394–401 | 7.1 | SCI |
| | Chamness-Rodden/ 10792 | 1238–1251 | 7.5 | SCI |
| | McClure-Hewitt/30974 | 1919–1943 | 7.6 | SCI |
| | McClure-Sparks/29739 | 2316–2347 | 7.8 | SCI |
| | Marathon-Cupp/31335 | 793–823 | 7.2 | SCI |
| | Marathon-Rzepke/31253 | 946 | 7.0 | SCI |
| | Marathon-Russell/31295 | 982 | 7.2 | SCI |
| | Amoco-Rau/31338 | 2978–2999 | 7.6 | SCI |
| | Amoco-Timm/30137 | 1417–1433 | 7.5 | SCI |
| | Hunt-Winterfield/33680 | 3013 | 7.0 | SCI |
| | Shell-Taratutta/29372 | 1445 | 9.0 | SCI |
| Eau Claire | Allied-Disposal/80604 | 1167 | 7.6 | SCI |
| | CPC-BD-152/80663 | 1394 | 7.8 | SCI |
| | Sun-Brazos-Foster/ 25099 | 3684 | 8.2 | SCI |
| | Marathon-Cupp/31335 | 1320–1359 | 8.9 | SCI |
| Precambrian | Allied-Disposal/80604 | 1232 | 9.0 | SCI |
| | McClure-Sparks/29739 | 4090 | 10.0 | SCI |

**APPENDIX: ANNOTATED ORGANIC MATURITY DATA BASE** (continued)

### III. Repetski and Harris (1981)

Preliminary estimates of conodont coloration were made by J. Repetski and A. G. Harris of the U.S. Geological Survey on the Brazos State Foster well from the Michigan Basin (USGS, unpublished data). Harris provided the following conodont coloration indices (CAI) by personal communication (1982):

| Formation | Well Name/Permit | Depth (m) | Maturity | |
|---|---|---|---|---|
| Early Ordov. | Sun-Brazos-Foster/25099 | 3546 | 2.5 | CAI |
| | Sun-Brazoc-Foster/25099 | 3692 | 2.5 | CAI |
| | Sun-Brazos-Foster/25099 | 3842 | 3.0 | CAI |
| | Sun-Brazos-Foster/25099 | 3960 | 3.0 | CAI |

### IV. Gardner and Bray (1984)

W. C. Gardner and E. E. Bray of Mobil Research and Development Corporation included a graphical profile of thermal alteration indices (TAI) for Mississippian to Ordovician strata in a recent study of Niagaran oil geochemistry. The following values are taken from their Figure 4:

| Formation | Well Name/Permit | Depth (m) | Maturity | |
|---|---|---|---|---|
| Coldwater | Mobil-Jelinek/27907 | 300 | 3.0 | TAI |
| Traverse | Mobil-Jelinek/27907 | 640 | 3.0 | TAI |
| Salina G | Mobil-Jelinek/27907 | 1060 | 3.1 | TAI |
| Niagara | Mobil-Jelinek/27907 | 1580 | 3.4 | TAI |
| Utica | Mobil-Jelinek/27907 | 1760 | 3.5 | TAI |
| Trenton | Mobil-Jelinek/27907 | 1930 | 4.0 | TAI |
| Glenwood | Mobil-Jelinek/27907 | 2000 | 4.0 | TAI |

### V. Hogarth and Sibley (1985)

A recent study of conodont coloration was carried out on the Trenton Formation by C. G. Hogarth and D. F. Sibley at Michigan State University. The data below are from Hogarth's MSU thesis:

| Formation | Well Name/Permit | Depth (m) | Maturity | |
|---|---|---|---|---|
| Trenton | Hunt-Winterfield/33680 | 3027 | 2.5 | CAI |
| | Hunt-Winterfield/33680 | 3028 | 2.5 | CAI |
| | Omni-Hirde/34268 | 1300 | 1.5 | CAI |
| | Omni-Hirde/34268 | 1302 | 1.5 | CAI |
| | Shell-Maidens/34277 | 1865 | 2.0 | CAI |
| | Sun-Bradley/13816 | 1830 | 2.0 | CAI |
| | Hunt-BigCreek/34070 | 2891 | 2.0 | CAI |
| | Hunt-BigCreek/34070 | 2894 | 2.0 | CAI |
| | Hunt-BigCreek/34070 | 2899 | 2.0 | CAI |
| | Total-Harmon/33129 | 1495 | 1.5 | CAI |
| | Total-Harmon/33129 | 1508 | 1.5 | CAI |
| | Total-Harmon/33129 | 1527 | 1.5 | CAI |
| | Humble-Riley/23039 | 1259 | 1.5 | CAI |
| | Humble-Riley/23039 | 1269 | 1.5 | CAI |
| | Humble-Riley/23039 | 1276 | 1.5 | CAI |
| | Mobil-Jelinek/27907 | 1869 | 2.0 | CAI |
| | Mobil-Jelinek/27907 | 1877 | 2.0 | CAI |
| | Mobil-Jelinek/27907 | 1881 | 2.0 | CAI |
| | Torosian-Nerreter/18940 | 1314 | 1.5 | CAI |
| | Torosian-Nerreter/18940 | 1328 | 1.5 | CAI |
| | Torosian-Nerreter/18940 | 1341 | 1.5 | CAI |
| | Torosian-Nerreter/18940 | 1354 | 1.5 | CAI |
| | Humble-Kryst/22213 | 1209 | 1.5 | CAI |
| | Humble-Kryst/22213 | 1217 | 1.5 | CAI |
| Trenton | Anderson-Whitaker/28407 | 928 | 1.5 | CAI |
| | Anderson-Whitaker/28407 | 949 | 1.5 | CAI |
| | Anderson-Whitaker/28407 | 939 | 1.5 | CAI |
| | Anderson-Whitaker/28407 | 971 | 1.5 | CAI |
| | Mammoth-Wooden/22168 | 1177–1181 | 1.5 | CAI |
| | Shell-Bidwell/37704 | 877 | 1.0 | CAI |
| | Texaco-Konkol/2654 | 1384 | 1.5 | CAI |
| | Texaco-Konkol/2654 | 1387 | 1.5 | CAI |
| | Texaco-Konkol/2654 | 1402 | 1.5 | CAI |
| | Shell-Taratuta/29372 | 1447 | 1.5 | CAI |
| | Shell Blue Lake/34673 | 2666 | 2.0 | CAI |
| | Shell-Allis/34957 | 1692 | 1.5 | CAI |
| | Shell-Allis/34957 | 1697 | 1.5 | CAI |
| | Shell-Weber/34132 | 2381 | 2.0 | CAI |
| | Shell-Weber/34132 | 2359 | 2.0 | CAI |
| | Amoco-Timm/30137 | 1524 | 1.5 | CAI |
| | Amoco-Timm/30137 | 1529 | 1.5 | CAI |
| | Amoco-Timm/30137 | 1540 | 1.5 | CAI |
| | Humble-Hoppinthal/25357 | 1761 | 1.5 | CAI |
| | Humble-Hoppinthal/25357 | 1765 | 1.5 | CAI |
| | Humble-Hoppinthal/25357 | 1777 | 1.5 | CAI |
| | Shell-Prevost/37779 | 2862 | 2.0 | CAI |
| | Shell-Prevost/37779 | 2867 | 2.0 | CAI |

### VI. Marzi and Rullkotter (personal communication, 1987)

Geochemical studies of the Michigan basin have been initiated by R. Marzi and J. Rullkotter of the Institute of Petroleum and Organic Geochemistry in Julich, West Germany. They have supplied the following vitrinite reflectance index (VRI) measurements by personal communication (1987):

| Formation | Well Name/Permit | Depth (m) | Maturity | |
|---|---|---|---|---|
| Saginaw | Surface sample | – | 0.46% | VRI |
| Coldwater | (not provided) | 366 | 1.11% | VRI |
| Bell | (not provided) | 1174 | 1.20% | VRI |
| Antrim | (not provided) | 750 | 1.25% | VRI |

## APPENDIX: ANNOTATED ORGANIC MATURITY DATA BASE (continued)

**VII. Mobil Data Base:**

Mobil Exploration and Producing U.S., Inc., has been very active in the Michigan Basin, and Mobil scientists have collected an extensive data base of organic geochemical parameters. These data have been provided for release by C. E. Griffith and W. G. Zempolich. The data consist mostly of thermal alteration index (TAI) and vitrinite reflectance index (VRI) analyses, although the reflectance measurements reported for Silurian and Ordovician strata were measured on chitinous and macrinitic organic matter instead of woody material. The values presented here have not been edited or corrected in any way. Zempolich (personal communication, 1988) notes that for vitrinite samples from the upper part of the Jelinek-Ferris well, Mobil interprets the low-gray vitrinite fraction (data marked loVRI, percent abundance shown in parentheses) as the indigenous vitrinite population. The high-gray vitrinite fraction (hiVRI), although commonly a much larger fraction of the population, is interpreted as oxidized and recycled vitrinite.

| Formation | Well Name/Permit | Depth (m) | Maturity | | |
|---|---|---|---|---|---|
| Undif. Miss | Mobil-Jelinek/27907 | 31 | (2%) | 0.55% | loVRI |
| | Mobil-Jelinek/27907 | 31 | (98%) | 1.14% | hiVRI |
| | Mobil-Jelinek/27907 | 91 | (8%) | 0.55% | loVRI |
| | Mobil-Jelinek/27907 | 91 | (92%) | 1.15% | hiVRI |
| | Mobil-Jelinek/27907 | 150 | (14%) | 0.54% | loVRI |
| | Mobil-Jelinek/27907 | 150 | (86%) | 1.05% | hiVRI |
| | Mobil-Jelinek/27907 | 213 | (5%) | 0.53% | loVRI |
| | Mobil-Jelinek/27907 | 213 | (95%) | 1.20% | hiVRI |
| | Mobil-Jelinek/27907 | 274 | (5%) | 0.56% | loVRI |
| | Mobil-Jelinek/27907 | 274 | (95%) | 1.12% | hiVRI |
| | Mobil-Jelinek/27907 | 335 | (66%) | 0.55% | loVRI |
| | Mobil-Jelinek/27907 | 335 | (34%) | 1.17% | hiVRI |
| | Mobil-Jelinek/27907 | 402 | (1%) | 0.59% | loVRI |
| | Mobil-Jelinek/27907 | 402 | (99%) | 1.23% | hiVRI |
| Berea | Mobil-JHelinek/27907 | 477 | (3%) | 0.63% | loVRI |
| | Mobil-Jelinek/27907 | 477 | (97%) | 1.26% | hiVRI |
| Antrim | Surface sample, Kettle Point, Ontario | | | 3.25 | TAI |
| | Mobil-Bucher/30615 | 539 | | 0.36% | VRI |
| | Mobil-Bucher/30615 | 581 | | 0.40% | VRI |
| | Mobil-Bucher/30615 | 608 | | 0.39% | VRI |
| | Mobil-Jelinek/27907 | 518 | (17%) | 0.60% | loVRI |
| | Mobil-Jelinek/27907 | 518 | (83%) | 1.15% | hiVRI |
| | Mobil-Jelinek/27907 | 579 | (34%) | 0.64% | loVRI |
| | Mobil-Jelinek/27907 | 579 | (66%) | 1.17% | hiVRI |
| Traverse | Gulf-Alderman/13791 | 959 | | 3.0 | TAI |
| | Mobil-Jelinek/27907 | 640 | (16%) | 0.61% | loVRI |
| | Mobil-Jelinek/27907 | 640 | (84%) | 1.29% | hiVRI |
| | Mobil-Jelinek/27907 | 701 | (16%) | 0.69% | loVRI |
| | Mobil-Jelinek/27907 | 701 | (84%) | 1.26% | hiVRI |
| Bell | Amoco-Sheldon/29157 | 594–610 | | 3.0 | TAI |
| | Bolger-Adama/10572 | 882 | | 3.0 | TAI |
| | Gulf-Alderman/13791 | 1174 | | 3.25 | TAI |
| Dundee | Bolger-Adams/10572 | 893 | | 3.0 | TAI |
| | Gulf-Schweitzer/9272 | 846 | | 3.0 | TAI |
| | Gulf-Schweitzer/9272 | 870 | | 3.0 | TAI |
| | Sun-Bradley/13816 | 857 | | 3.0 | TAI |
| | Sun-Bradley/13816 | 876 | | 4.0 | TAI |
| | Carter-Lauber/17549 | 674 | | 3.75 | TAI |
| | Surface sample, Goderich, Ontario | | | 3.25 | TAI |
| Detroit River | Mobil-Jelinek/27907 | 762 | (83%) | 0.56% | loVRI |
| | Mobil-Jelinek/27907 | 762 | (17%) | 1.55% | hiVRI |
| | Mobil-Jelinek/27907 | 823 | (1%) | 0.61% | loVRI |
| | Mobil-Jelinek/27907 | 823 | (99%) | 1.37% | hiVRI |
| | Mobil-Jelinek/27907 | 884 | (7%) | 0.64% | loVRI |
| Detroit River | Mobil-Jelinek/27907 | 884 | (93%) | 1.43% | hiVRI |
| Amherstburg | Amoco-Sheldon/29157 | 838–844 | | 3.0 | TAI |
| | Surface sample, Formosa Quarry, Ontario | | | 3.25 | TAI |
| Bois Blanc | Surface sample, Cargill, Ontario | | | 3.25 | TAI |
| | Mobil-Jelinek/27907 | 945 | (1%) | 0.60% | loVRI |
| | Mobil-Jelinek/27907 | 945 | (99%) | 1.05% | hiVRI |
| | Mobil-Jelinek/27907 | 1006 | (1%) | 0.66% | loVRI |
| | Mobil-Jelinek/27907 | 1006 | (99%) | 1.30% | hiVRI |
| Salina C(?) | Mobil-Jelinek/27907 | 1250 | | 0.97% | VRI |
| Salina A-2 | Amoco-Sheldon/29157 | 1457–1463 | | 3.25 | TAI |
| | Humble-Mulder/23096 | 806 | | 3.5 | TAI |
| | Humble-Mulder/23096 | 810 | | 3.5 | TAI |
| | Mobil-Jelinek/27907 | 1372 | | 1.08% | VRI |
| Salina A-1 | Amoco-Sheldon/29157 | 1628–1634 | | 3.25 | TAI |
| | Panam-Trapp/27920 | 1030 | | 3.5 | TAI |
| | Mobil-Jelinek/27907 | 1493 | | 1.23% | VRI |
| Niagara | Lawton-Rose/23711 | 2500 | | 3.25 | TAI |
| | Panam-Simpson/27932 | 2185 | | 3.5 | TAI |
| | Panam-Simpson/27932 | 2213 | | 3.5 | TAI |
| | Panam-Trapp/27920 | 1063 | | 3.5 | TAI |
| | Panam-Trapp/27920 | 1074 | | 3.5 | TAI |
| | Panam-Trapp/27920 | 1080 | | 3.5 | TAI |
| | Panam-Trapp/27920 | 1093 | | 3.5 | TAI |
| | Panhandle-Umlauf/23394 | 655 | | 3.25 | TAI |
| | Panhandle-Umlauf/23394 | 674 | | 3.25 | TAI |
| | Panhandle-Umlauf/23394 | 701 | | 3.25 | TAI |
| | Panam-VanAken/27643 | 1432 | | 3.25 | TAI |
| Clinton | Mobil-Jelinek/27907 | 1615 | | 1.19% | VRI |
| Richmond | Mobil-Jelinek/27907 | 1676 | | 1.44% | VRI |
| | Mobil-Jelinek/27907 | 1737 | | 1.45% | VRI |
| Utica | Amoco-Sheldon/29157 | 1950–1957 | | 3.5 | TAI |
| | Amoco-Sheldon/29157 | 1966–1972 | | 3.5 | TAI |
| | Mobil-Kelly/29117 | 1524 | | 3.25 | TAI |
| | Mobil-Jelinek/27907 | 1798 | | 1.35% | VRI |
| Trenton | Amoco-Sheldon/29157 | 1978–1984 | | 3.5 | TAI |
| | Amoco-Sheldon/29157 | 2033–2048 | | 3.5 | TAI |
| | Amoco-Sheldon/29157 | 2097–2106 | | 3.5 | TAI |
| | Sun-Buss-Halb/19231 | 1149 | | 3.5 | TAI |
| | Sun-Buss-Halb/19231 | 1159 | | 3.5 | TAI |
| | Sun-Buss-Halb/19231 | 1289 | | 3.5 | TAI |
| | Sun-Buss-Halb/19231 | 1327 | | 3.5 | TAI |
| | Sun-Buss-Halb/19231 | 1334 | | 3.5 | TAI |
| | Sun-Buss-Halb/19231 | 1341 | | 3.5 | TAI |
| | CPC-Evans/26100 | 1166 | | 3.5 | TAI |
| | CPC-Evans/26100 | 1196 | | 3.5 | TAI |
| | CPC-Evans/26100 | 1216 | | 3.5 | TAI |
| | Carter-Lauber/17549 | 1508 | | 3.5 | TAI |
| | Carter-Lauber/17549 | 1551 | | 3.5 | TAI |
| | Mobil-Kelly/29117 | 1588 | | 3.5 | TAI |
| | Mobil-Kelly/29117 | 1624 | | 3.5 | TAI |
| | Mobil-Jelinek/27907 | 1920 | | 1.19% | VRI |
| | Mobil-Jelinek/27907 | 1981 | | 1.26% | VRI |
| | Mobil-Jelinek/27907 | 2042 | | 1.30% | VRI |
| | Mobil-Jelinek/27907 | 2103 | | 1.33% | VRI |
| Black River | Amoco-Sheldon/29157 | 2149–2164 | | 3.5 | TAI |
| | Amoco-Sheldon/29157 | 2176–2192 | | 3.5 | TAI |
| | Amoco-Sheldon/29157 | 2210–2225 | | 3.5 | TAI |
| | Amoco-Sheldon/29157 | 2262–2271 | | 3.5 | TAI |
| | Carter-Lauber/17549 | 1573 | | 3.5 | TAI |
| | Mobil-Kelly/29117 | 1658 | | 3.75 | TAI |
| | Mobil-Kelly/29117 | 1713 | | 3.75 | TAI |
| Glenwood | Mobil-Kelly/29117 | 1753 | | 3.75 | TAI |

## APPENDIX: ANNOTATED ORGANIC MATURITY DATA BASE (continued)

| Formation | Well Name/Permit | Depth (m) | Maturity | |
|---|---|---|---|---|
| Prairie du Chien | Mobil-Kelly/29117 | 1774 | 3.75 | TAI |
|  | Mobil-Kelly/29117 | 1826 | 3.75 | TAI |
|  | Mobil-Kelly/29117 | 1838 | 3.75 | TAI |
| Trempeleau | Mobil-Kelly/29117 | 1883 | 4.0 | TAI |
|  | Mobil-Kelly/29117 | 1908 | 4.0 | TAI |
|  | Mobil-Jelinek/27907 | 2146 | 1.28% | VRI |
|  | Mobil-Jelinek/27907 | 2149 | 1.37% | VRI |

### VIII. Shell Data Base (1988)

J. R. Straccia of Shell Western Exploration and Production, Inc., has authored a recent company report entitled "Thermal history of the Michigan Basin." With Straccia's assistance, the vitrinite reflectance data from this report have been released by Shell managers. For pre-Devonian units, these values are assumed to represent measurements on chitonous and macrinitic material. For more information, contact Straccia.

| Formation | Well Name/Permit | Depth (m) | Maturity | |
|---|---|---|---|---|
| Antrim | Amoco-Union/31364 | 495 | 0.53% | VRI |
|  | Amoco-Union/31364 | 507 | 0.53% | VRI |
|  | Amoco-Union/31364 | 524 | 0.51% | VRI |
|  | Wolverine-Club/33405 | 396 | 0.52% | VRI |
|  | Wolverine-Club/33405 | 399 | 0.53% | VRI |
|  | Wolverine-Club/33405 | 409 | 0.53% | VRI |
|  | Wolverine-Club/33405 | 413 | 0.47% | VRI |
|  | Welch-Chester/33875 | 437 | 0.45% | VRI |
|  | Welch-Chester/33875 | 453 | 0.52% | VRI |
|  | Welch-Chester/33875 | 462 | 0.51% | VRI |
|  | Welch-Chester/33875 | 465 | 0.59% | VRI |
| Niagara | Miller-Con-Power/27530 | 1583 | 1.16% | VRI |
|  | NMEC-Kalkaska/28024 | 1986 | 1.08% | VRI |
|  | NMEC-Kalkaska/28024 | 2061 | 1.09% | VRI |
|  | Panam-Mancelona/28194 | 2002 | 1.08% | VRI |
|  | Panam-Mancelona/28194 | 2037 | 1.12% | VRI |
| Clinton | Shell-Mayfield/37403 | 2091 | 1.00% | VRI |
|  | Shell-Mayfield/37403 | 2100 | 1.17% | VRI |
|  | Sun-Hoening/18701 | 938 | 0.98% | VRI |
|  | Sun-Hoening/18701 | 942 | 0.92% | VRI |
|  | Sun-Hoening/18701 | 950 | 0.45% | VRI |
| Cataract | Shell-Blue Lake/34613 | 2591 | 1.16% | VRI |
|  | Shell-Blue Lake/34613 | 2612 | 1.23% | VRI |
|  | Shell-Blue Lake/34613 | 2623 | 1.17% | VRI |
| Utica | Shell-Blair/34319 | 2224 | 1.05% | VRI |
|  | Shell-Visser/35704 | 2046 | 1.10% | VRI |
|  | Shell-Visser/35704 | 2048 | 1.15% | VRI |
|  | Shell-Visser/35704 | 2049 | 1.08% | VRI |
|  | Shell-Visser/35704 | 2053 | 1.09% | VRI |
|  | Shell-Visser/35704 | 2061 | 1.15% | VRI |
|  | Shell-Maidens/34277 | 1797 | 1.19% | VRI |
|  | Shell-Maidens/34277 | 1836 | 1.28% | VRI |
|  | Shell-Blair/34292 | 2224 | 1.40% | VRI |
| Glenwood | Shell-Foster/37765 | 3172 | 1.95% | VRI |
|  | Shell-Web-Sharon/34132 | 2454 | 1.63% | VRI |
|  | Shell-Web-Sharon/34132 | 2455 | 1.57% | VRI |
|  | Shell-Blair/34292 | 2333 | 1.16% | VRI |
|  | Shell-Blair/34292 | 2333 | 1.12% | VRI |
|  | Shell-Blair/34292 | 2339 | 1.64% | VRI |
|  | Shell-Blue Lake/34613 | 2770 | 1.58% | VRI |
|  | Shell-Blue Lake/34613 | 2771 | 1.65% | VRI |
|  | Jem-Doorn/34376 | 3233 | 1.50% | VRI |
| Prairie du Chien | Jem-Liberty/35099 | 2708 | 1.48% | VRI |
|  | Jem-Doorn/34376 | 3546 | 1.51% | VRI |
|  | Jem-McCormick/34536 | 2964 | 1.63% | VRI |
|  | Sun-Bradley/13816 | 2021 | 1.51% | VRI |
| Franconian | Jem-Doorn/34376 | 4342 | 1.94% | VRI |

### IX. Exlog Data Base

A 22-well study entitled "Regional petroleum geochemistry of the Michigan Basin" by Exlog/Brown and Ruth is currently available for purchase and may be publicly released within the next few years. For more information, contact A. R. Daly.

## ACKNOWLEDGMENTS

We thank C. E. Griffith and W. G. Zempolich of Mobil Exploration and Producing U.S. Inc., J. R. Straccia of Shell Western Exploration and Production Inc., A. G. Harris of the United States Geological Survey, N. H. Sleep of Stanford University, and J. Rullkotter of the Institute of Oil and Organic Geochemistry (Julich) for their invaluable assistance in gathering organic maturity data for the Michigan Basin. We also thank W. G. Zempolich, J. A. Nunn, S. Jacobson, J. R. Straccia, and J. Leonard for critical comments and assistance with organic modeling. This research was supported in part by a grant to the senior author from the Petroleum Research Foundation of the American Chemical Society.

# REFERENCES CITED

Andrews, D. A., and Huddle, J. W., 1948, Analyses of Michigan, North Dakota, South Dakota, and Texas coals: U.S. Bureau of Mines Technical Paper 700, 106 p.

Ballard, S., Pollack, H. N., and Skinner, N. J., 1987, Terrestrial heat flow in Botswana and Namibia: Journal of Geophysical Research, v. 92, p. 6291–6300.

Beaumont, C., Quinlan, G., and Hamilton, J., 1988, Orogeny and stratigraphy; Numerical models of the Paleozoic in the eastern interior of North America: Tectonics, v. 7, p. 389–416.

Beck. A. E., 1976, The use of thermal resistivity logs in stratigraphic correlation: Geophysics, v. 41, p. 300–309.

Blackwell, D., and Steele, J., 1988, Thermal conductivity of sedimentary rocks; Measurement and significance, in Naeser, N., and McCulloch, eds., Thermal history of sedimentary basins: New York, Springer-Verlag (in press).

Cercone, K. R., 1984, Thermal history of Michigan Basin: American Association of Petroleum Geologists Bulletin, v. 68, p. 130–136.

Clark, S. P., Jr., 1966, Thermal conductivity, in Clark, S. P., Jr., ed., Handbook of physical constants, revised edition: Geological Society of America Memoir 97, p. 461–482.

Combs, J., and Simmons, G., 1973, Terrestrial heat flow determinations in the north central United States: Journal of Geophysical Research, v. 78, p. 441–461.

Crowley, K. D., Ahern, J. L., and Naeser, C. W., 1984, Major epeirogenic events in the Williston Basin region; Evidence from fission-track dating: Geological Society of America Abstracts with Programs, v. 16, p. 480.

Crowley, K. D., Kuhlman, S. L., and Naeser, C. W., 1986a, Apatite fission-track analysis of basement from the Canadian shield: Two episodes of Phanerozoic orogeny: Geological Society of America Abstracts with Programs, v. 18, p. 577.

Crowley, K. D., Naeser, C. W., and Babel, C. A., 1986b, Tectonic significance of Precambrian apatite fission-track ages from the midcontinent United States: Earth and Planetary Science Letters, v. 79, p. 329–336.

Damberger, H. H., 1971, Coalification patterns of the Illinois Basin: Econonmic Geology, v. 66, p. 488–494.

Dorr, J. A., Jr., and Eschman, D. F., 1971, Geology of Michigan: Ann Arbor, University of Michigan Press, 95 p.

Droste, J. B., and Shaver, R. H., 1983, Atlas of early and middle Paleozoic paleogeography of the southern Great Lakes area: Indiana Geological Survey Special Report 323, 32 p.

Dykstra, J., 1987, Compaction correction for burial-history curves; Application to Lopatin's method for source rock maturation determination: Geobyte, v. 2, p. 16–23.

Friedman, G. M., 1987, Vertical movements of the crust; Case histories from the northern Appalachian basin: Geology, v. 15, p. 1130–1133.

Gardner, W. C., and Bray, E. E., 1984, Oils and source rocks of Niagaran reefs (Silurian) in the Michigan Basin, in Palacas, J. G., ed., Petroleum geochemistry and source rock potential of carbonate rocks: American Association of Petroleum Geologists Studies in Geology, no. 18, p. 33–44.

Hogarth, C. G., and Sibley, D. F., 1985, Thermal history of the Michigan Basin; Evidence from conodont color alteration indices, in Cercone, K. R., and Budai, J. M., eds., Ordovician and Silurian rocks of the Michigan Basin and its margins: Michigan Basin Geological Society Special Paper 4, p. 45–57.

Hower, J. C., and Davis, A., 1981, Application of vitrinite reflectance anisotropy in the evaluation of coal metamorphism: Geological Society of America Bulletin, v. 92, p. 350–366.

Illich, H. A., and Grizzle, P. L., 1983, Comment on 'Comparison of Michigan Basin crude oils': Geochimica et Cosmochimica Acta, v. 47, p. 1151–1155.

—— , 1985, Thermal subsidence and generation of hydrocarbons in Michigan Basin; Discussion: American Association of Petroleum Geologists Bulletin, v. 69, p. 1401–1403.

Judge, A. S., and Beck, A. E., 1973, Analysis of heat flow data; Several holes in a sedimentary basin: Canadian Journal of Earth Sciences, v. 10, p. 1494–1507.

Kayal, J. R., and Christoffel, D. A., 1982, Relationship between electrical and thermal resistivities for differing grades of coal: Geophysics, v. 47, p. 121–129.

Lopatin, N. V., 1971, Temperature and geologic time as factors in coalification: Akademiya Nauk SSR Isvestiya Seriya Geologicheskikh, no. 3, p. 95–106.

Mackenzie, A. S., and McKenzie, D., 1983, Isomerization and aromatization of hydrocarbons in sedimentary basins formed by extension: Geological Magazine, v. 120, p. 417–528.

Moyer, R. B., 1982, Thermal maturity and organic content of selected Paleozoic formations; Michigan Basin [M.S. thesis]: East Lansing, Michigan State University, 62 p.

Nunn, J. A., 1986, Subsidence and thermal history of Michigan Basin, in Burruss, J., ed., Thermal modeling in sedimentary basins: Paris, Editions Technip, p. 417–436.

Nunn, J. A., Sleep, N. H., and Moore, W. E., 1984, Thermal subsidence and generation of hydrocarbons in Michigan Basin: American Association of Petroleum Geologists Bulletin, v. 68, p. 296–315.

Pruitt, J. D., 1983, Comment on 'Comparison of Michigan Basin crude oils': Geochimica et Cosmochimica Acta, v. 47, p. 1157–1159.

Quigley, T. M., and Mackenzie, A. S., 1988, The temperature of oil and gas formation in the subsurface: Nature, v. 333, p. 549–552.

Rullkotter, J., Meyers, P. A., and Schaefer, R. G., and Dunham, K. W., 1986, Oil generation in the Michigan Basin; A biological marker and carbon isotope approach: Organic Geochemistry, v. 10, p. 359–375.

Rullkotter, J., Marzi, R., and Meyers, P. A., 1990, Biological markers in Paleozoic sedimentary tocks and crude oils from the Michigan Basin; Reassessment of sources and thermal history of organic matter: Contribution to IGCP Project 157 Conference Volume (in press).

Sleep, N. H., Nunn, J. A., and Chou, L., 1980, Platform basins: Annual Review of Earth and Planetary Science, v. 8, p. 17–34.

Speece, M. A., Bowen, T. D., Folcik, J. L., and Pollack, H. N., 1985, Analysis of temperatures in sedimentary basins; The Michigan Basin: Geophysics, v. 50, p. 1318–1334.

Teichmuller, R., and Teichmuller, M., 1986, Relations between coalification and paleogeothermics in Variscan and Alpidic foredeeps of western Europe, in Buntebath, G., and Stegena, L., eds., Paleogeothermics: New York, Springer-Verlag, p. 53–78.

Tissot, B. P., Ford, J. F., and Espitalie, J., 1980, Principle factors controlling the timing of petroleum generation, in Facts and principles of world petroleum occurrence: Canadian Society of Petroleum Geologists Memoir 6, p. 143–152.

Tissot, B. P., Pelet, R., and Ungerer, Ph., 1987, Thermal history of sedimentary basins, maturation indices, and kinetics of oil and gas generation: American Association of Petroleum Geologists Bulletin, v. 71, p. 1445–1466.

Vogler, E. A., Meyers, P. A., and Moore, W. A., 1981, Comparison of Michigan Basin crude oils: Geochimica et Cosmochimica Acta, v. 45, p. 2287–2293.

Waples, D. W., 1980, Time and temperature in petroleum formation; Application of Lopatin's method to petroleum exploration: American Association of Petroleum Geologists Bulletin, v. 64, p. 916–926.

Wood, D. A., 1988, Relationships between thermal maturity indices calculated using Arrhenius equation and Lopatin method; Implications for petroleum exploration: American Association of Petroleum Geologists Bulletin, v. 72, p. 115–134.

MANUSCRIPT ACCEPTED BY THE SOCIETY JUNE 1, 1990

Printed in U.S.A.

# Economic geology and history of metallic minerals in the Northern Peninsula of Michigan

**Robert C. Reed**
*Geological Survey Division, Michigan Department of Natural Resources, Box 30028, Lansing, Michigan 48909*

## ABSTRACT

A substantial section of Precambrian rock is exposed over an area of approximately 19,400 km$^2$ (7,500 mi$^2$) in the western part of the Northern Peninsula of Michigan. This province is a portion of the exposed southern terminus of the Canadian Precambrian Shield and contains a large variety of igneous, sedimentary, and metamorphic rocks. Significant amounts of iron and copper from Precambrian rocks of Michigan have provided important contributions to the growth of the state and national economy for nearly 150 years.

Archean rocks consist of volcanics, sediments, and younger felsic and mafic intrusives, some of batholitic dimensions. Volcanic and associated sedimentary rocks occur as greenstone belts included in the Ramsay Formation, Gogebic County; Dickinson Group, Dickinson and Iron Counties; and Marquette Greenstone Belt, Marquette County. Volcanic rocks consist of mafic to felsic lava flows and pyroclastics and sediments derived from volcanic rocks. Volcanic flows include amygdaloidal and ellipsoidal varieties. Pyroclastics consist of agglomerate, conglomerate, breccia, and tuff. Sediments are described as graywacke, argillite, siltstone, conglomerate, quartzite, iron formation, and chert. Granite and granitic gneiss, principally tonalite and granodiorite, intrude the periphery and interiors of the greenstone belts. Mafic intrusives, including peridotite, are subordinate. Shearing is prominent in some areas, and metamorphic grade ranges from lower-greenschist to upper-amphibolite facies. Minor amounts of gold and silver have been produced from the Marquette Greenstone Belt.

Early Proterozoic strata are subdivided into four groups, in ascending order: the Chocolay, Menominee, Baraga, and Paint River Groups. Copper mineralization occurs in the Kona Dolomite of the Chocolay Group. The Menominee Group contains three major iron formations of equivalent age, the Negaunee Iron Formation of the Marquette Iron Range; the Vulcan Iron Formation of the Menominee Iron and Felch Mountain Districts and the Ironwood Iron Formation of the Gogebic Iron Range. In the Baraga Group, the Goodrich Quartzite contains concentrations of monazite, and the Michigamme Formation has vast amounts of graphitic carbon. The Paint River Group includes the highly productive Riverton Iron Formation.

Iron was discovered in 1844 on the Marquette Iron Range, and an early pig-iron industry flourished. The east-west–trending Marquette syncline, containing the Negaunee Iron Formation, is more than 65 km (40 mi) long. The Negaunee has a maximum thickness of 1,060 m (3,500 ft), and iron-formation resources have been estimated at 205 billion long tons. There are four iron formations on this range, three of which have been productive. However, 97 percent of the 588 million tons mined came from the Negaunee. The east-west–trending Menominee Iron-bearing District, in southern Dickinson

---

Reed, R. C., 1991, Economic geology and history of metallic minerals in the Northern Peninsula of Michigan, *in* Catacosinos, P. A., and Daniels, P. A., Jr., eds., Early sedimentary evolution of the Michigan Basin: Geological Society of America Special Paper 256.

County, consists of a north and south range segmented by longitudinal faulting. The Vulcan Iron Formation is exposed over a strike length of 28 km (16 mi) and has a maximum thickness of 180 m (600 ft). Production amounted to nearly 82 million long tons. In the Felch Mountain District of central Dickinson County, only eroded remnants of the Vulcan Iron Formation remain. Production of 36 million tons was principally from the Groveland low-grade iron mine. The Gogebic Iron Range, in Gogebic County, is an essentially east-west-trending, northward-dipping sequence of sediments containing the Ironwood Iron Formation. In Michigan, the Ironwood has a strike length of about 40 km (25 mi) and a maximum thickness of about 490 m (1,600 ft). Iron ore production totals 255 million long tons. The Negaunee, Vulcan, and Ironwood iron formations are considered to be stratigraphically equivalent.

The Iron River–Crystal Falls District in Iron County is primarily composed of the Paint River Group containing the Riverton Iron Formation. The Paint River Group is outlined in a triangular-shaped basin approximately 260 km$^2$ (100 mi$^2$) in area. The Riverton has a maximum thickness of 240 m (800 ft) and has been intensely and complexly folded. A high phosphorous and manganese content characterizes the Riverton and its naturally derived iron ores. Production amounted to 207 million long tons.

Middle Proterozoic rocks in Michigan consist of a very thick sequence of volcanics and sediments. For the most part, strata dip uninterrupted toward Lake Superior at varying degrees. Native copper was the exclusive mineral produced from the Portage Lake Volcanics in Michigan's Keweenaw Peninsula. Stratabound native copper mineralization forms ore bodies in amygdaloidal and brecciated tops of lava flows, and in interflow conglomerates. Minor amounts were produced from transverse fissures. Production of refined copper through 1976 amounted to 4,769,465 metric tons (5,257,438 short tons). Sulfide copper (chalcocite) with some native metal is mined from the Nonesuch Formation several thousand feet about the Portage Lake Volcanics in the Porcupine Mountain area. Copper mineralization is confined to siltstone and shale of the basal portion of the Nonesuch. Small amounts of disseminated native copper are produced from the uppermost sandstone of the underlying Copper Harbor Conglomerate. Through 1987, 1,364,800 metric tons (1,504,433 short tons) of refined copper has been produced.

## INTRODUCTION

On January 26, 1837, Michigan achieved statehood, and on that date a bill was introduced to create a State Geological Survey. This bill was signed into law on February 23, 1837. Douglass Houghton, a physician, was appointed as Michigan's first state geologist and managed the first Geological Survey from 1837 to 1842. The most powerful motivation of the first geological survey was the urgent need for locating larger, cheaper supplies of salt. However, it was not until 1860 that salt production was recorded. In the mid 1840s, discovery, followed by development and production, of iron and copper precipitated a mining boom in the Northern Peninsula of Michigan (Fig. 1), which has been likened to the California gold rush.

Despite current rapid transitions in the nonfuel minerals industry, Michigan remains a substantial producer. The value of production of metallic and nonmetallic minerals for the past 25 years is given in Table 1. Nationally, Michigan ranked fourth in the value of mineral production in 1988. The state ranked first in the sales value of calcium chloride, magnesium compounds, and peat; second in bromine, sand and gravel, industrial sand and iron ore, and third in gypsum. Currently, 15 nonfuel minerals are produced from a variety of geologic formations developed over the past 2,700 m.y. This diverse geologic province may contain additional valuable, undiscovered mineral resources.

Metallic minerals in Michigan are produced from rocks of Precambrian age exposed in the western part of the Northern Peninsula (Fig. 1). The vast, early Proterozoic, sedimentary banded-iron formations were discovered in 1844, and middle Proterozoic native (elemental) copper deposits in 1845. Initially slowed by inadequate transportation to markets, production gradually increased, and Michigan became a primary producer of the raw materials necessary in the formative years of a rapidly expanding industrial nation.

Early surveys and reports of the Michigan Geological Survey and the U.S. Geological Survey were instrumental in the rapid expansion of iron and copper discovery. In the 1940s, a cooperative program between the Michigan Geological Survey and the U.S. Geological Survey was initiated. The geology of the iron ranges, the copper range, and adjacent areas was mapped in detail. Aeromagnetic surveys eventually covered all of the Northern Peninsula. However, the metallic mining industry began to

Figure 1. Location of exposed rocks of Precambrian age in the Northern Peninsula of Michigan (shaded).

TABLE 1. VALUE OF NONFUEL MINERALS PRODUCTION 1963–1987

| Year | Metallic | Nonmetallic | Total |
|------|----------|-------------|-------|
|      | (in millions of dollars) |  |  |
| 1987 | 552 | 790 | 1,342 |
| 1986 | 451 | 802 | 1,256 |
| 1985 | 591 | 748 | 1,387 |
| 1984 | 620 | 733 | 1,353 |
| 1983 | 508 | 653 | 1,161 |
| 1982 | 349 | 629 | 978 |
| 1981 | 767 | 732 | 1,499 |
| 1980 | 736 | 832 | 1,568 |
| 1979 | 677 | 771 | 1,448 |
| 1978 | 610 | 753 | 1,363 |
| 1977 | 414 | 641 | 1,055 |
| 1976 | 487 | 584 | 1,071 |
| 1975 | 425 | 493 | 918 |
| 1974 | 369 | 498 | 867 |
| 1973 | 460 | 277 | 737 |
| 1972 | 407 | 233 | 640 |
| 1971 | 380 | 226 | 606 |
| 1970 | 383 | 249 | 632 |
| 1969 | 355 | 249 | 604 |
| 1968 | 347 | 214 | 561 |
| 1967 | 334 | 208 | 542 |
| 1966 | 324 | 211 | 535 |
| 1965 | 304 | 204 | 508 |
| 1964 | 300 | 189 | 489 |
| 1963 | 272 | 155 | 427 |

change materially. In 1952, 37 iron mines, 12 native copper mines, and 3 native copper reclamations were in production. Currently, there are 2 active iron mines, 1 copper mine, and 1 gold mine. Yet iron ore production is now about the same as it was in 1952, and copper production is considerably larger (Appendices A and B). Facilities to conduct concentration (beneficiation) research on the vast iron resources began in the 1940s. In the 1950s, the construction of open-pit low-grade iron-ore facilities commenced. These were designed to produce a higher grade, more uniform iron-ore pellet. Concentrates and pellets gradually displaced natural ores. The last natural-ore iron mine closed in 1979. All native copper mines were closed by strikes in 1968.

## ARCHEAN GEOLOGY

Rocks of Archean age, formerly called lower Precambrian or Precambrian W, exposed in the western part of the Northern Peninsula, consist of two major types: an older succession of volcanics and sediments, and younger granites and gneisses. The volcanics and sediments (Fig. 2), about 2,700 m.y. old, have been intruded by large volumes of felsic rock (Fig. 3) and smaller dikes, sills, plutons, etc., of mafic to felsic composition. The four major areas of Archean layered rocks are indicated on Figure 2 and comprise some 972 km$^2$ (375 mi$^2$).

Archean volcanic rocks have been intruded by felsic rocks such as tonalite and granodiorite. These volcanic rocks have been assimilated or engulfed to varying degrees, causing widespread distribution of gneisses that contain inliers of preexisting rocks. Major areas of granitic rock and gneiss are illustrated on Figure 3. They comprise a total of 3,475 km$^2$ (1,340 mi$^2$).

The possibility of Archean gneiss as a host rock for economic mineral deposits has generally not been accepted, and thus, such rocks have received little attention from the exploration community. A concept that high-grade metamorphism remobilizes and redistributes previous mineral concentrations or that the gneiss has been so highly eroded as to remove preexisting ore deposits, may have reduced exploration interests. Also, the "1 mile from contact" concept of Emmons (1937), that mineral deposits are not formed in granitic rocks more than 1 mi from contact with the host rock, may have been an effective deterrent. The potential of Archean gneiss for containing economic ore deposits has recently been championed by Sims and others (1987). They cite several examples of major ore deposits hosted by Archean gneiss. Certainly, over the past few decades, Archean gneiss has generally been avoided with regard to its economic mineral potential. The Northern Complex of Michigan (Fig. 3, location 2) is an outstanding example of disinterest in batholithic interiors. In many areas surrounding the gneiss complex, the sediments and volcanics have been extensively studied, while the granite gneiss has barely been investigated. Archean gneisses do contain remnants of rocks normally considered favorable for prospecting, and so the stigma attached to high-grade metamorphic areas, labeled as unfavorable hosts, is gradually dissolving.

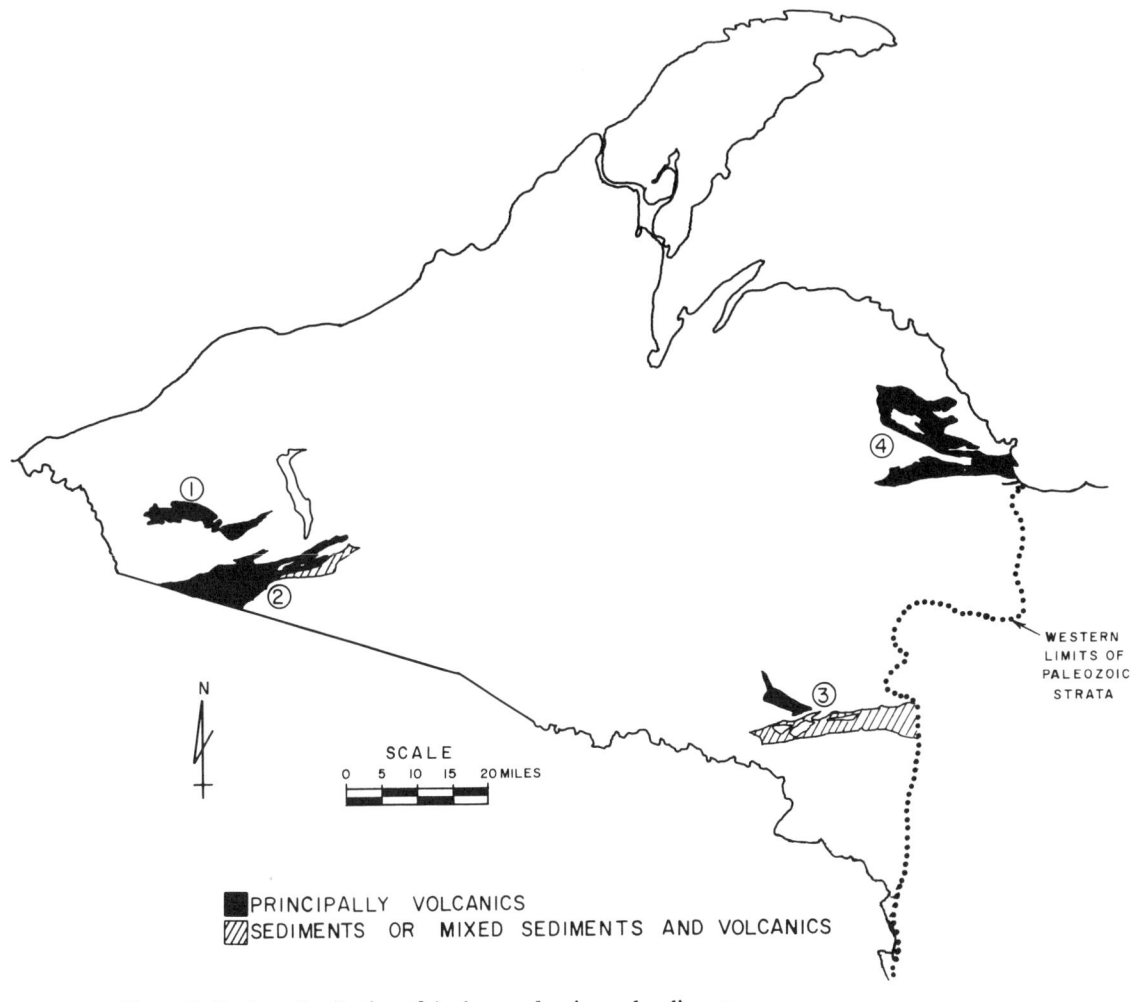

Figure 2. Surface distribution of Archean volcanics and sediments.

| LOCATION | NAME | (m²) | (km²) |
|---|---|---|---|
| 1. Gogebic County | Ramsay Formation | 40 | 104 |
| 2. Gogebic County | Cup Lake–Banner Lake Area | 80 | 207 |
| 3. Dickinson and Iron Counties | Dickinson Group | 130 | 337 |
| 4. Marquette County | Marquette Greenstone Belt | <u>125</u> | <u>324</u> |
| | TOTALS | 375 | 972 |

## Archean volcanic belts

*Ramsay Formation.* The Ramsay Formation (Fig. 2, location 1) is a typical Archean greenstone belt. It is more than 32 km (20 mi) long and at least 6,100 m (20,000 ft) thick. The Ramsay (Fig. 4) is located immediately south of the Gogebic Iron Range.

Schmidt (1976) mapped the rocks of the Ironwood area and described the Ramsay Formation, a sequence of metamorphosed mafic flows, volcanic breccia, tuff, tuffaceous clastics, argillite, siltstone, and minor amounts of quartzite. This sequence has been intruded from the south by the Puritan Quartz Monzonite, producing a fringe of highly modified country rock. The contact metamorphic zone is named the Whiskers Creek Gneiss. Areas of extensive injection of the Puritan Quartz Monzonite into older layered rocks are called the Van Buskirk Gneiss.

The Ramsay, at least a few thousand feet thick, dips about 55 degrees to the south. Schmidt (1976) states that no sulfide deposits have been found as yet in the Ramsay Formation; however, this volcanic sequence does fit a model indicating the possible presence of volcanogenic sulfides. Since little of the greenstone is exposed in the study area, future exploration may yet discover economic mineral deposits.

In the adjacent Wakefield area, Prinz and Hubbard (1975) report that the basal Archean volcanic rocks are mafic to intermediate pyroclastics consisting of tuff, volcanic conglomerate, and breccia with some graywacke. To the south, above the pyroclastics, are 3,000 m (9,850 ft) of ellipsoidal and amygdaloidal andesitic to dacitic lava flows, which include some beds of pyroclastics as much as 45 m (150 ft) thick. Above the flows are felsic schists (rhyolite or rhyodacite) with some pyroclastics estimated

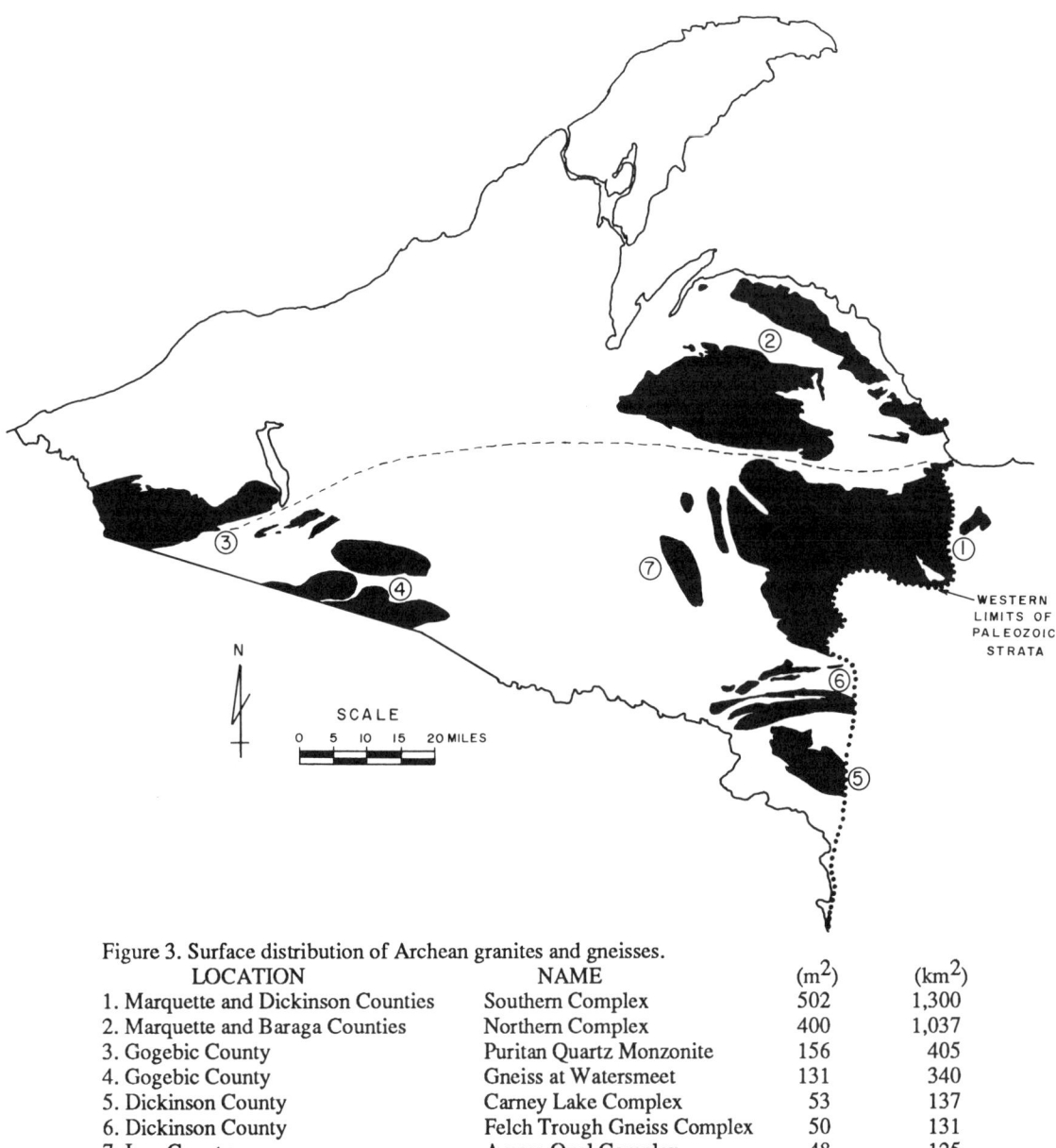

Figure 3. Surface distribution of Archean granites and gneisses.

| LOCATION | NAME | (m²) | (km²) |
|---|---|---|---|
| 1. Marquette and Dickinson Counties | Southern Complex | 502 | 1,300 |
| 2. Marquette and Baraga Counties | Northern Complex | 400 | 1,037 |
| 3. Gogebic County | Puritan Quartz Monzonite | 156 | 405 |
| 4. Gogebic County | Gneiss at Watersmeet | 131 | 340 |
| 5. Dickinson County | Carney Lake Complex | 53 | 137 |
| 6. Dickinson County | Felch Trough Gneiss Complex | 50 | 131 |
| 7. Iron County | Amasa Oval Complex | 48 | 125 |
| | TOTALS | 1,340 | 3,475 |

to be 2,500 m (8,200 ft) to 3,000 m (9,850 ft) thick. Most of the felsic rocks are quartz-feldspar-biotite schists, some strongly lineated and laminated. Pyrite is locally abundant. In the Wakefield area, metamorphic grade increases from greenschist in the northeast to amphbolite to the southwest, where the Ramsay Formation has been invaded from the south by the Puritan Quartz Monzonite.

The area east of Wakefield was mapped by Trent (1973). The Ramsey Formation is described as amygdaloidal basalt, ellipsoidal basalt, and flow breccia with a thickness of 1,066 m (3,500 ft). Metamorphic grade varies from low to high greenschist facies toward the contact zone with the Puritan Quartz Monzonite.

*Cup Lake–Banner Lake area.* East of Lake Gogebic, Fritts (1969) mapped all layered rocks, sediments and volcanics, in this area (Fig. 2, location 2) as middle Precambrian (now early Proterozoic) in age. The structure of these formations was interpreted as a northeast-trending, southeast-dipping monocline, with younger formations overlapping toward the east. At least part of the evidence for a middle Precambrian age was radiometric dating of samples from the gneiss near Watersmeet of about 1.8 Ga. A reevaluation of the area east of Lake Gogebic by Prinz (1981) and Sims and others (1984) resulted in revision of the geologic age of part of this area. Rock formations previously described by Fritts (1969) and provisionally named the Strata near Cup Lake and Strata near Banner Lake, were reclassified to Archean age.

|  | Maximum Thickness | |
|---|---:|---:|
|  | (m) | (ft) |
| **EARLY PROTEROZOIC** | | |
| Baraga Group | | |
|    Michigamme Formation | 2,900 | 9,500 |
| Menominee Group | | |
|    Ironwood Iron Formation | 490 | 1,600 |
|      Anvil Member | | |
|      Pence Member | | |
|      Norrie Member | | |
|      Yale Member | | |
|      Plymouth Member | | |
|    Emperor Volcanic Complex | | |
|    Palms Formation | 335 | 1,100 |
| **ARCHEAN** | | |
| Puritan Quartz Monzonite | | |
| Ramsay Formation | 6,100 | 20,000 |

Figure 4. Stratigraphic column of Gogebic Iron Range area.

The Strata near Cup Lake was described as mainly mafic lavas and tuff interlayered with graywacke about 2,440 m (8,000 ft) thick. The overlying Strata near Banner Lake is predominantly biotitic and garnetiferous graywacke approximately 3,500 m (11,500 ft) thick. Sims and others (1984) described the Strata near Cup Lake as interlayered gneiss and amphibolite interpreted as originally felsic to intermediate pyroclastic deposits and mafic flows. The Strata near Banner Lake, largely graywacke, includes some tuff and minor iron formation. The gneiss and amphibolite have been intruded by small bodies of Puritan Quartz Monzonite, thought to be 2,650 m.y. old.

*Dickinson Group.* The Dickinson Group (Fig. 2, location 3; Fig. 5) in central Dickinson County and adjacent Iron County to the west have been mapped as Archean in age and described by James and others (1961). When projected to the east under Paleozoic rock cover (based on interpretation of aeromagnetic data), the group covers about 130 mi$^2$. The Dickinson Group, principally a series of sedimentary and volcanic rocks some 3,650 m (12,000 ft) thick, contains three formations. In ascending order, these are: (1) East Branch Arkose, (2) Solberg Schist, and (3) Six Mile Lake Amphibolite.

*East Branch Arkose.* The oldest of the Dickinson Group formations was deposited unconformably upon an older granite gneiss. Quartz is the major mineral present in the Arkose, followed by microcline and white mica. The Arkose contains distinctive beds of conglomerate, largely quartzite but with some gneiss and sericitic slate clasts. In the upper part of the formation to the south, the East Branch Arkose is interbedded with mafic tuffs and basalt flows.

*Solberg Schist.* The lower (northern) part of this unit is chiefly hornblende and biotite schist with muscovite locally abundant. The upper portion consists of quartz-mica schist and/or micaceous quartzite. The Solberg Schist contains the Skunk Creek Member, an iron formation about 30 m (100 ft) thick. The Solberg Schist is thought to be of sedimentary and volcanic origin.

*Six Mile Lake Amphibolite.* Originally mafic lava, this formation is composed chiefly of hornblende and some plagioclase. It contains dikes, pods, and irregular bodies of younger pegmatite. Pegmatite content increases to the south, culminating in a banded gray gneiss. The banded gray gneiss is a reaction zone created as granite intruded the Six-Mile Lake Amphibolite from the south.

**Marquette Greenstone Belt.** The southern portion of the Marquette Greenstone Belt (Fig. 2, location 4; Fig. 6) was mapped by Gair and Thaden (1968), Puffett (1974), Clark and others (1975), and Cannon and Klasner (1977). The geology of the belt was summarized by Morgan and DeCristoforo (1980). The geology of the northern part of the belt is under investigation by personnel of the Department of Geology and Geological Engineering, Michigan Technological University. Recent information on this area is provided by Owens and Bornhorst (1985), Boben and others (1986), Johnson and others (1987), Baxter and others (1987), and MacLellan and Bornhorst (1989).

The Marquette Greenstone Belt is composed principally of volcanic and plutonic rocks of varied compositions. There is some volcanic ejecta that appears to have been water deposited or water worked, and there are some minor amounts of chemical sediments present. The original broad classification of the rocks as the Kitchi and Mona Schists has continued with more recent mapping.

|  | Maximum Thickness | |
|---|---:|---:|
|  | (m) | (ft) |
| **EARLY PROTEROZOIC** | | |
| Baraga Group | | |
|    Badwater Greenstone | 760 | 2,500 |
|    Michigamme Formation | 1,525 | 5,000 |
|    Hemlock Formation | 915 | 3,000 |
| Menominee Group | | |
|    Vulcan Iron Formation | 120 | 400 |
|    Felch Formation | 60 | 200 |
| Chocolay Group | | |
|    Randville Dolomite | 610 | 2,000 |
|    Sturgeon Quartzite | 565 | 1,850 |
|    Fern Creek Formation | 20 | 70 |
| **ARCHEAN** | | |
| Gneissic Granite | | |
| Hardwood Gneiss | | |
| Dickinson Group | 3,660 | 12,000 |
|    Six-mile Lake Amphibolite | | |
|    Solberg Schist | | |
|    East Branch Arkose | | |
| Granite Gneiss | | |

Figure 5. Stratigraphic column of central Dickinson County.

*Kitchi Schist.* There is general agreement that the Kitchi Schist in the southwest corner of the Marquette Greenstone Belt (Figs. 6 and 7) is the oldest rock unit. The basal Kitchi, an amphibolite, was formed by regional metamorphism of basalt, diabase, and gabbro. The most important characteristic of the Kitchi is pyroclastic material—fragmental volcanic rocks of latite to rhyodacite composition, ranging from coarse agglomerate to fine-grained tuff. Some larger fragments are rounded, but many are angular, resembling both conglomerate and breccia. Basalt and felsic volcanics are also present. The Kitchi is about 1,370 m (4,500 ft) thick and separated from the overlying Mona Schist by a shear zone (Puffett, 1974; Clark and others, 1975; Morgan and DeCristoforo, 1980).

*Mona Schist.* The Mona Schist (Figs. 6 and 7) has been subdivided into an Undifferentiated Greenstone, Lower Member, Nealy Creek Member, Sheared Rhyolite Tuff Member, and Lighthouse Point Member. The Mona is reported to be 6,400 m (21,000 ft) thick in the Marquette Quadrangle (Gair and Thaden, 1968) and 7,315 m (24,000 ft) thick in the Negaunee Quadrangle (Puffett, 1974).

The Undifferentiated Greenstone (Fig. 6) occurs at the base of the Mona Schist in the Negaunee Quadrangle (Puffett, 1974). It is a zone of sheared rock, largely basalt, containing masses of felsic rock and some sericitic slate. This unit is thought to be composed of the Lower Member of the Mona Schist, the Kitchi Schist, and younger rocks. It is about 450 m (1,500 ft) thick.

The Lower Member of the Mona Schist (Fig. 6) is as much as 3,600 m (11,800 ft) thick in the Marquette Quadrangle (Gair and Thaden, 1968) and 2,740 m (9,000 ft) thick in the Negaunee Quadrangle (Puffett, 1974). Most of the Lower Member consists of massive basalt flows interlayered with chloritic or actinolitic slate or schist, and felsite. Large pillow structures in the flows are common, indicating subaqueous extrusion. The rocks dip north 75 degrees to the vertical. Metamorphic grade is of the lower greenschist facies.

Most of the upper part of the Lighthouse Point Member is layered amphibole schist. The distinctive layering has prompted speculation of a sedimentary origin, but the lenticular form of the layering suggests greatly stretched and flattened pillows. Gair and Thaden (1968) interpret the upper Lighthouse Point Member as metamorphosed mafic tuff. Morgan and DeCristoforo (1980) report that, in addition to layered amphibolites, the unit includes felsic tuff and interbedded chemical sediments, chert, and iron formation. The amphibolite is believed by them to be metamorphosed gabbro. The member is estimated to be as much as 3,535 m (11,600 ft) thick in the Marquette Quadrangle. The lower part of the Lighthouse Point Member is described by Gair and Thaden (1968) as a chloritic and actinolitic quartz-sericite schist and slate, water-laid tuff, and/or mixtures of volcanic ash and chemically precipitated silica. In the Negaunee Quadrangle to the west, Puffett (1974) divided the Lower Lighthouse Point Member into the Nealy Creek Member and the Sheared Rhyolite Tuff Member. This classification was adopted by Clark and others (1975) in the Negaunee SW Quadrangle and by Morgan and DeCristoforo (1980).

The Nealy Creek Member, located in the Negaunee Quadrangle, is a quartz-feldspar chlorite schist interpreted by Puffett (1974) to be an air-fall rhyodacite tuff. Morgan and DeCristoforo (1980), however, consider its origin to be shale and graywacke or vitric crystal tuff. It is about 915 m (3,000 ft) thick.

The Sheared Rhyolite Tuff Member occupies a major shear zone between the Lower Member and the Lighthouse Point Member in the Negaunee Quadrangle. It is a quartz-rich volcanic tuff, near rhyolite in composition, and contains as much as 50 percent quartz and feldspar phenocrysts. Maximum reported thickness is 915 m (3,000 ft); Puffett (1974).

Reference has been made previously to a cooperative mapping program between the Michigan Geological Survey and Michigan Technological University, in the northern part of the Marquette Greenstone Belt (Fig. 6). The objective is to evaluate the geology and mineral potential of the Marquette Greenstone Belt north of the Dead River Basin. To 1987, about 25 mi$^2$ had been completed in four areas, resulting in open-file reports and geologic maps by Owens and Bornhorst (1985), Johnson and others (1987), and Baxter and others (1987). Additionally, some old mineral prospects were studied (Boben and others, 1986).

The oldest rock formations present are a sequence of layered rocks, mostly pillow basalts, but also pyroclastics, mud flows, iron formation, and laminated schist. Parts of the layered rocks are altered, and some greatly so. These have been intruded by gabbro and basalt, which in areas associated with faults or shears, are highly altered. The mafic rocks were in turn intruded by multiple stocks of granodiorite and rhyolite, assumed to have a common parentage. During and/or after felsic intrusion, all rocks were highly deformed, faulted, sheared, and foliated, offering conduits for hydrothermal gases and fluids that altered surrounding rocks, introducing other elements and compounds. Locally, rocks have been chloritized, epidotized, sericitized, carbonatized, and silicified. The metamorphic grade ranges from lower greenschist to upper amphibolite facies.

*Deer Lake Peridotite.* Northwest of Deer Lake (Figs. 6 and 7), the Kitchi Schist is bisected by a serpentinized peridotite called the Deer Lake Peridotite. In addition to serpentization, the Deer Lake is locally asbestos bearing and sometimes altered to talc-carbonate. The Deer Lake Peridotite is further described in the discussion of the Ropes Gold Mine.

*Compeau Creek Gneiss.* This is largely foliated tonalite and granodiorite with chloritic, biotitic, and hornblende varieties intruding the Mona Schist. The Compeau Creek contains scattered inclusions, interpreted to be Mona Schist, as much as 610 m (2,000 ft) long (Gair and Thaden, 1968). Puffett (1974) described extensive zones of silicification in the Compeau Creek of the Negaunee Quadrangle.

*Dead River Pluton.* South of the Dead River Storage Basin (Fig. 6), the Dead River Pluton, about 8.8 km (5.5 mi) long and 7.8 km$^2$ (3 mi$^2$) in area, intrudes the Lower Member and Neely

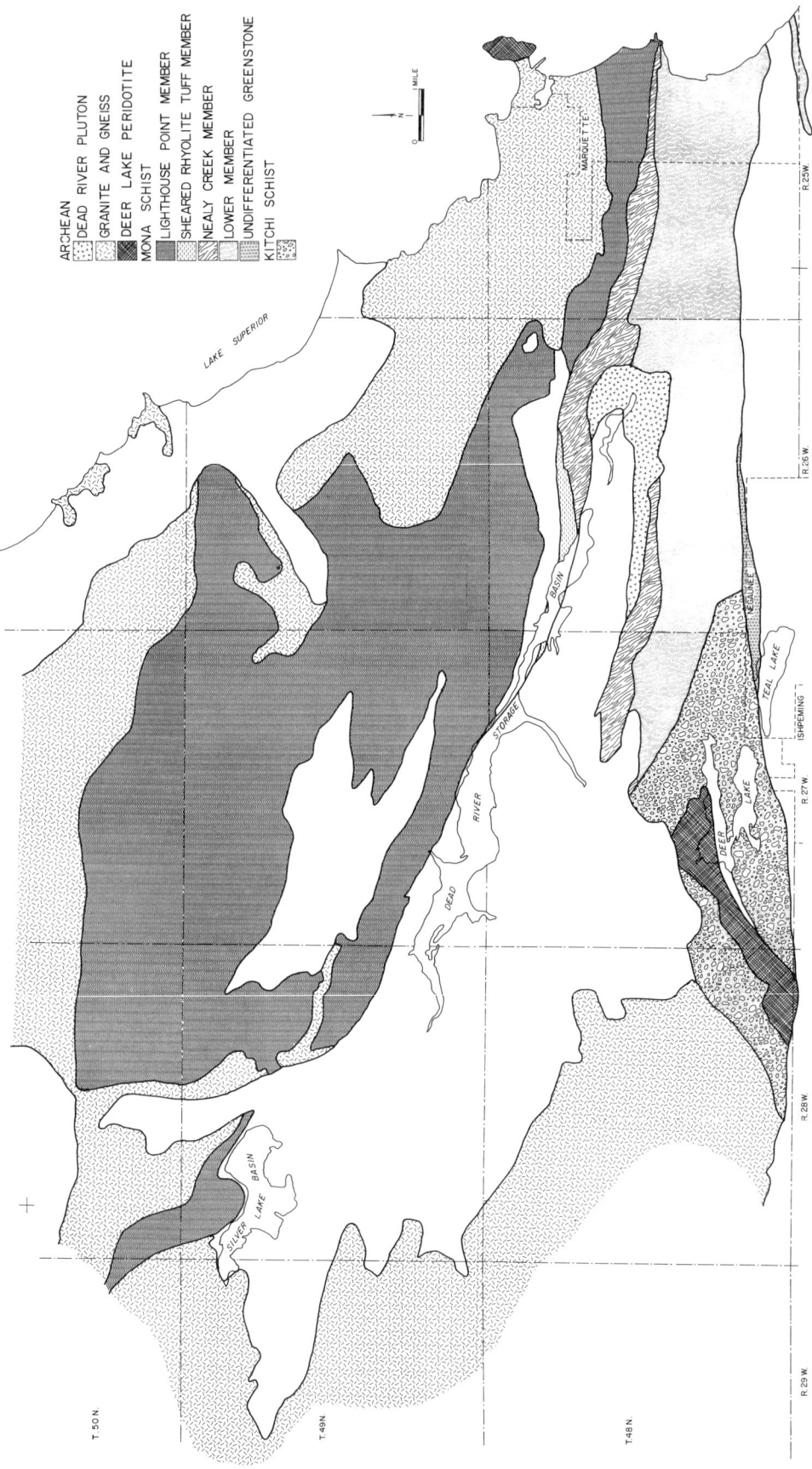

Figure 6. Geologic map of Marquette Greenstone Belt.

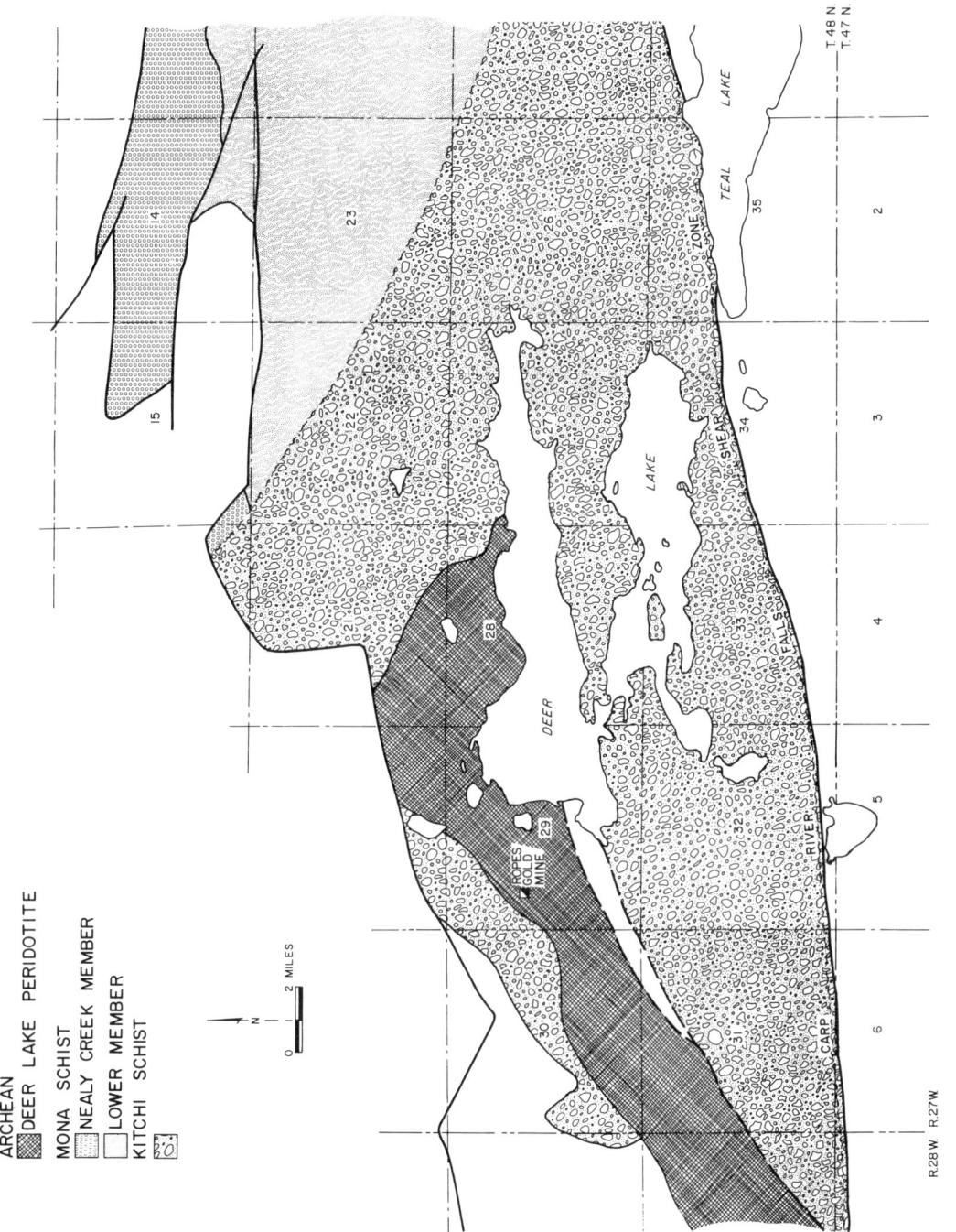

Figure 7. Geologic map of Ropes gold mine area, Marquette Greenstone Belt.

Creek Member of the Mona Schist and, in part, is covered by the early Proterozoic Michigamme Formation. This pluton is composed of three distinct rock types: (1) hornblende diorite, (2) porphyritic syenite, and (3) granodiorite porphyry. The pluton is further described by Puffett (1974) and Morgan and DeCristoforo (1980).

## Ropes Gold Mine

*History.* Gold was discovered in Marquette County in 1880 in Section 29, T48N, R27W, about 3.5 mi northwest of the city of Ishpeming (Fig. 7). In late 1880, Julius Ropes, a chemist, postmaster, and prospector, was exploring in the vicinity of a serpentinized peridotite for Verde Antique marble, when he located a small quartz vein. When tested for gold and silver, it was not considered commercial. In May 1881, Ropes discovered another quartz outcrop from which specimens assayed gold with a value of $21.00 per ton. The Ropes Gold mine was organized in 1881 and began production in 1883. Although the mine produced gold and silver until 1897 (Table 2), it was not an economic success. Annual reports indicate costs greater than sales value for each year of operations; production continued through assessments of stockholders.

In 1899, Corrigan McKinney and Company purchased the Ropes property and processed about 30,000 tons of waste tailings recovering 2,640 ounces of gold, the equivalent of 0.088 ounces per ton. In 1934, the property was leased by Calumet and Hecla Consolidated Copper Company. Calumet and Hecla dewatered the Ropes mine, extended drifts and cut new crosscuts on the 15th (850-foot) level, and conducted a program of sampling and exploration drilling from the 9th, 13th, and 15th levels. On the surface, a series of exploration holes was drilled, and an extensive program of trenching was conducted on the Ropes and adjacent properties. The Ropes was later purchased by Louis Koeppel, co-owner of the Arcadian Copper Mine, a tourist mine near Hancock, for possible use as a tourist gold mine. In 1975, Callahan Mining Corporation, Phoenix, Arizona, purchased the property, renewed exploration, and developed the mine for production. The mill of the Humboldt Iron Mine, located 16 mi from the Ropes, was acquired, revamped, and readied for production, which began late in 1985. Gold and silver production data through 1987 is provided in Table 3.

*Geology.* At the Ropes mine, gold mineralization occurs in a narrow, elongated, highly sheared and altered rock sliver. These Archean-age volcanic flows and associated fragmentals are sandwiched between, and enveloped by, serpentinized peridotite. The mineralized host and surrounding rocks are a portion of a relatively large area of bedrock composed of Archean volcanics, sediments, and intrusives exposed over an area of about 324 km$^2$ (125 mi$^2$). This Archean province is currently referred to as the Marquette Greenstone Belt (Fig. 6). Volcanic rocks of the belt date from about 2,700 Ma. Early accounts of the geologic features of the mineralized zone and environs were quite general—principally descriptive. Fortunately, the comprehensive and detailed geological and exploratory evaluations of Calumet and Hecla were carefully recorded and presented by Broderick (1945). A summary of geologic features of the Ropes mine area that are deemed significant follows.

1. The orebody of the Ropes mine is contained in a narrow, nearly vertical, tabular body of Archean Keewatin schist separating two masses of intrusive peridotite. The schist is more than 610 m (2,000 ft) long and as much as 46 m (150 ft) wide. The schist is highly sheared and altered, and has been referred to as the "mineral dike."

2. A shear zone several hundred feet wide, transgressing the volcanic schist, is a major structural feature whose lateral extent is several thousand feet and whose limits have yet to be established. The shear zone was the conduit for fluids that altered and mineralized the volcanic schist.

**TABLE 2. GOLD AND SILVER PRODUCTION OF ROPES GOLD MINING COMPANY 1883–1897**

| Year | Tons Milled | Ounces Gold | Ounces per Ton | Ounces Silver | Ounces per Ton |
|---|---|---|---|---|---|
| 1883 | 100 | 40 | 0.40 | 155 | 1.55 |
| 1884 | 60 | 34 | 0.57 | 177 | 3.00 |
| 1885 | 5,413 | 1,422 | 0.26 | 3,454 | 0.64 |
| 1886 | 6,959 | 1,859 | 0.27 | 4,700 | 0.65 |
| 1887 | 10,216 | 1,560 | 0.15 | 2,645 | 0.26 |
| 1888 | 16,855 | 2,528 | 0.15 | 5,670 | 0.34 |
| 1889 | 31,365 | 4,349 | 0.14 | 10,270 | 0.33 |
| 1890 | 31,578 | 3,435 | 0.11 | 8,110 | 0.26 |
| 1891 | 21,355 | 3,079 | 0.14 | 8,720 | 0.41 |
| 1892 | 21,794 | 2,640 | 0.12 | 6,130 | 0.28 |
| 1893 | 15,080 | 1,960 | 0.13 | 4,560 | 0.30 |
| 1894 | 21,185 | 2,050 | 0.10 | 4,770 | 0.22 |
| 1895 | 16,800 | 1,682 | 0.10 | 2,113 | 0.13 |
| 1896 | 16,686 | 1,740 | 0.11 | 4,000 | 0.25 |
| 1897 | 4,300 | 430 | 0.10 | 1,000 | 0.23 |
| Total | 219,746 | 28,808 |  | 66,474 |  |
| Average |  |  | 0.131 |  | 0.303 |

**TABLE 3. GOLD AND SILVER PRODUCTION OF CALLAHAN MINING CORPORATION 1985–1987**

| Year | Tons Milled | Ounces Gold | Ounces per Ton | Ounces Silver | Ounces per Ton |
|---|---|---|---|---|---|
| 1985 | 154,190 | 4,945 | 0.032 | 3,401 | 0.022 |
| 1986 | 608,660 | 46,255 | 0.076 | 62,624 | 0.103 |
| 1987 | 669,240 | 42,646 | 0.064 | 44,483 | 0.066 |
| Total | 1,432,090 | 93,846 |  | 110,508 |  |
| Average |  |  | 0.066 |  | 0.077 |

3. The volcanic schist has been subjected to extreme alteration, principally silicification, sericitization, and pyritization with subordinate chloritization. The adjacent peridotite has been serpentinized and, approaching the contact with the volcanic schist, steatized (formation of talc) and carbonatized (dolomite).

4. Silicification of the volcanic schist formed eight major quartz lenses with associated smaller, unsheared lenticular quartz: tetrahedrite masses arranged en echelon with maximum dimensions of 61 m (200 ft) long, 76 m (250 ft) deep, and 11 m (38 ft) wide. Accessory minerals associated with the veins consist primarily of pyrite and tetrahedrite with minor chalcopyrite, galena, sphalerite, native gold, and silver. Tourmaline, molybdenite, and scheelite are rare.

5. Surrounding the quartz-tetrahedrite veins is an envelope of finely disseminated pyrite. Gold values within the pyrite increase with the intensity of silicification, sericitization, and pyritization. Exploration drilling from the 9th, 13th, and 15th levels indicates that the gold-bearing pyrite halo widens and plunges eastward with depth. Bulk samples from crosscuts surrounding existing exploration drill holes assay 60 percent greater gold content than from the drill holes. This is a confirmation of the perils of obtaining a representative sample where the mineral content is erratically dispersed throughout the host medium.

6. Estimates of the unmined pyritized fringe zone indicate more than 1 million tons of mineralized rock averaging 0.13 ounces per ton gold and 0.7 ounces per ton silver are present above the 850-foot level.

Renewed exploration and development at the Ropes mine and resumption of mining in late 1985 introduced the only "gold only" operation east of the Black Hills. Additionally, several studies were initiated to further evaluate the geology and mineralization of the area. Selected conclusions of some of the investigations are summarized below:

Brozdowski and others (1986) envisioned the country rock as fine-grained, felsic to intermediate, volcaniclastic rocks and thick flows. The volcaniclastics consist of tuff and coarse tuff breccia. The host rock has been intruded by a swarm of sill-like masses of the Deer Lake serpentinite, which have conformable contacts and altered margins. Part of the serpentinite may possibly be flows.

The orebody is in a strata-bound and strata-form alteration zone, reflecting original stratigraphic and structural relations of volcanic rocks enclosed by intrusive sills. The mineralized zone is contained in a quartz–white mica–chlorite phyllite, containing 5 to 8 percent pyrite, successively enclosed in a carbonate-quartz-chlorite phyllite and a carbonate-talc phyllite bounded by sill-like masses of the Deer Lake serpentinite. Relic crystal and lithic fragments in the quartz–white mica–chlorite phyllite are interpreted as evidence that these rocks are not intensely deformed. About 300 m (985 ft) west of the Ropes orebody, the lateral equivalent of the quartz–white mica–chlorite phyllite, is a layered quartz-magnetite iron formation and a cherty, microcrystalline, ferroan dolomite quartz rock interpreted as chemical sediments.

Bornhorst and others (1986) described the parent rock, the Kitchi Schist, as a foliated, intermediate to felsic, pyroclastic volcanic unit that includes agglomerates, tuffs, breccias, and local basaltic to dacitic lavas. The rocks are metamorphosed at least to the lower or middle greenschist facies. The northeast-trending Deer Lake Peridotite is oblique to the regional east-west trend of the Kitchi Schist (Figs. 6 and 7). Contacts are discordant, and where exposed, the Kitchi consists of a soft zone of actinolite-chlorite schist with numerous quartz veins. At the contact, the serpentinite contains abundant carbonate and talc. The oblique relation of the Deer Lake Peridotite within the Kitchi Schist, and the altered external contacts, lead to the conclusion that the peridotite was an intrusive, structurally emplaced as a relatively cold body in a solid or semisolid state. The mineralized zone is referred to as the Ore Host Rock, a nearly vertical, tabular body of highly altered schistose rock whose relation to the Kitchi Schist has not been established.

The Ore Host Rock can be divided into four mappable units: (1) Sericitic, (2) Sericitic-Chloritic, (3) Chloritic, and (4) Chlorite-Carbonate units. The Ore Host Rock is interpreted to represent mafic to intermediate volcanic rocks tectonically emplaced as a relatively rigid body within the Deer Lake Peridotite along a shear zone. The volcanic rocks were intensely deformed and sheared. Extensive fluid introduction resulted in complete alteration of primary minerals forming secondary minerals, principally sericite and chlorite. The adjacent serpentinite was extensively altered to talc-carbonate for a distance as great as 120 m (400 ft).

## EARLY PROTEROZOIC GEOLOGY

Early Proterozoic sediment and volcanic rock in Michigan formed some 2,100 to 1,900 m.y. ago. They were formerly called Huronian, middle Precambrian, Animikie, Precambrian X, and Marquette Range Supergroup. Sedimentation and volcanism, including minor intrusions, culminated with the Penokean orogeny about 1,850 Ma. The Penokean orogeny caused large-scale regional metamorphism (James, 1955). Early Proterozoic rock contains four groups. They are the: (1) Chocolay, (2) Menominee, (3) Baraga, and (4) Paint River Groups.

The Chocolay Group (Fig. 8) contains beds of quartzite, dolomite, and slate. Widespread copper sulfides, chalcocite, bornite, and chalcopyrite occur in the Kona Dolomite portion of the group (Reed, 1965).

The Menominee Group (Fig. 9) consists of a basal quartzite and/or slate succeeded by iron formation. The iron formation, a chemical sediment, attains a maximum thickness of 1,065 m (3,500 ft) at the eastern end of the Marquette Iron Range. Menominee Group iron formations provided more than 80 percent of the iron ore produced in Michigan, which has historically ranked second in the nation in iron-ore production. The amount of iron formation present is roughly estimated to be between 300 and 400 billion long tons. Veins of copper sulfides, bornite, and

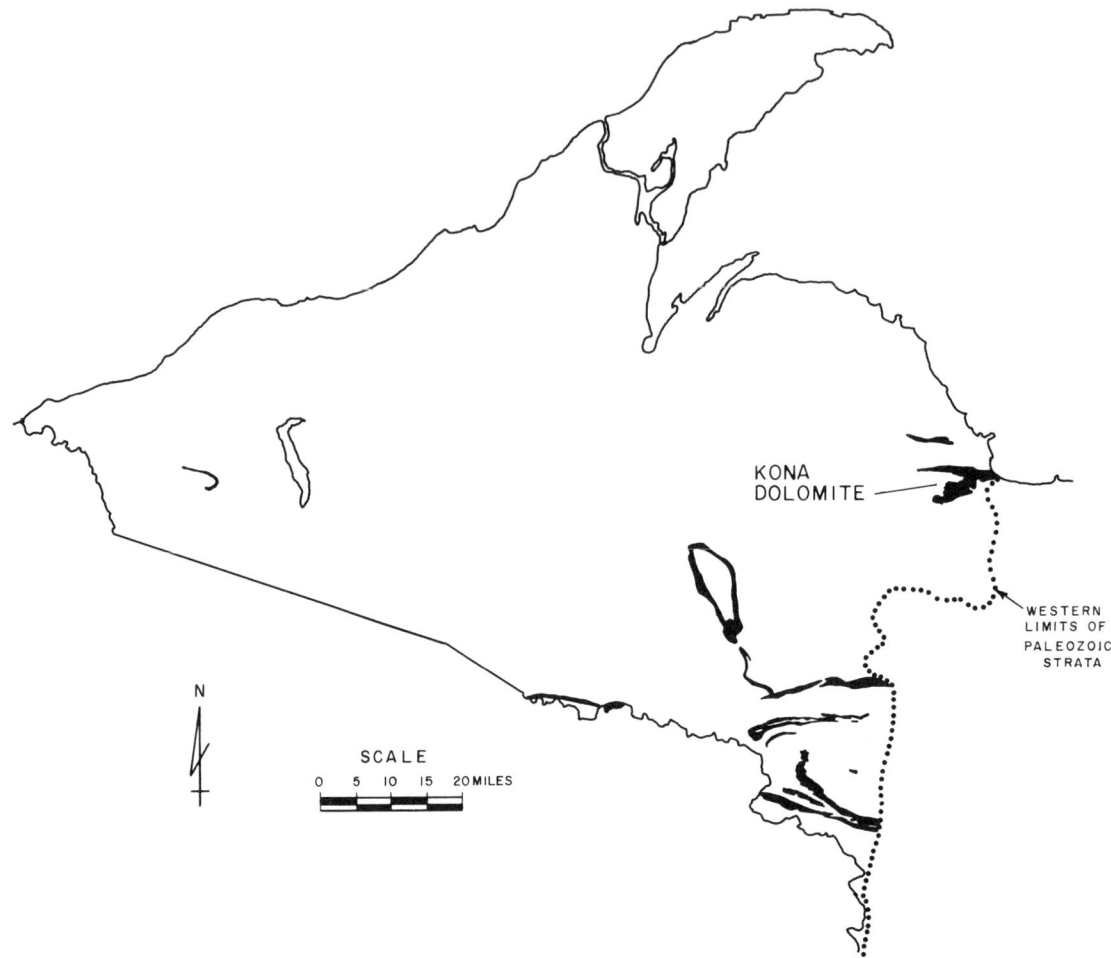

Figure 8. Surface distribution of early Proterozoic Chocolay Group, shaded black.

chalcopyrite up to 0.6 m (2 ft) thick crosscut the Negaunee Iron Formation at the Cliffs Shaft iron mine in the City of Ishpeming (Reed, 1965). This mineralization is believed by the writer to be related to copper mineralization in the Kona Dolomite of the Chocolay Group. On the eastern Gogebic Iron Range, Trent (1973) descibes volcanic extrusives, the Emperor Volcanic Complex, consisting of lava flows and fragmental rocks, flow breccias and tuffs, interbedded with the iron formation. Trent's geologic map indicates the maximum surface width of the complex to be about 2,440 m (8.000 ft).

The Baraga Group (Figs. 10 and 11) denotes changing geologic conditions from principally shelf sediments, coarse clastic and chemical, to eugeosynclinal slate-graywacke-argillite rocks and extrusive volcanics. The basal member is the Goodrich Quartzite, and in the Palmer area of Marquette County (Fig. 10), the group contains abnormal amounts of monazite. Vickers (1956) reports monazite contents of local Goodrich-derived glacial erratics as much as 110 pounds per ton. Ten samples of Goodrich, analyzed for a Michigan Geological Survey unpublished manuscript, indicate large amounts of Cerium (950 to 5,050 ppm), Lanthanum (490 to 2,825 ppm), Thorium (450 to 2,300 ppm), and Neodymium (250 to 1,500 ppm). Uranium contents ranged from 18 to 62 ppm and gold from 10 to 40 ppb. Although not metallic, the Baraga Group Michigamme Formation contains high concentrations of graphitic carbon. The graphitic slates have been under investigation by the Michigan Geological Survey and Michigan Technological University. Preliminary estimates indicate in excess of three billion tons of rock containing 20 to 33 percent carbon (unpublished report).

The Paint River Group (Fig. 12) is contained in an isolated triangular-shaped basin that covers about 260 km$^2$ (100 mi$^2$). This sedimentary package includes the Riverton Iron Formation, from which about 18 percent of the iron ore produced in Michigan was mined.

### Iron resources

***Marquette Iron Range. History.*** The early history of the Marquette Iron Range is the early history of iron in Michigan and the Lake Superior region. The discovery, development, production, and transportation of Precambrian iron ores revolutionized the industrialization of Michigan and vitalized the economies of

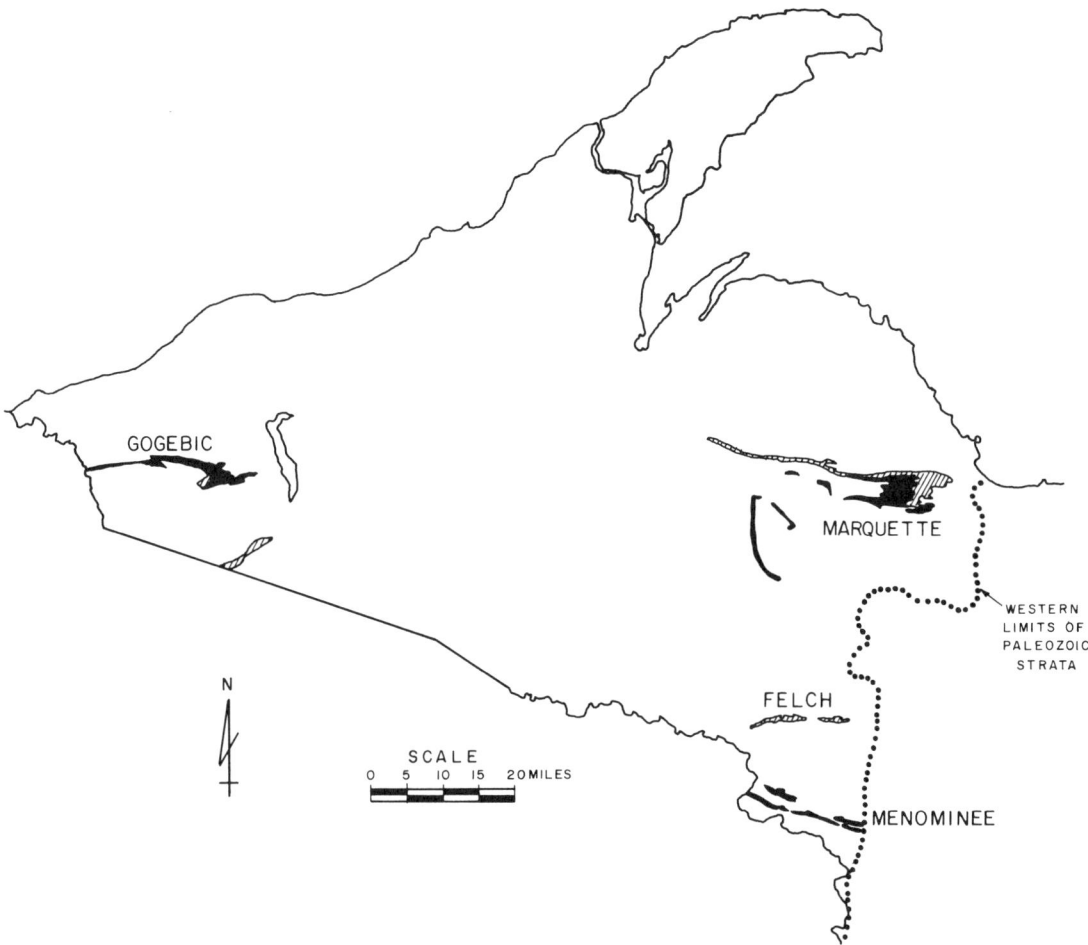

Figure 9. Surface distribution of early Proterozoic Menominee Group indicating location of iron ranges and districts.

the northeastern United States. These iron ores provided enormous volumes of a superior quality, low-cost, raw material resource.

Douglass Houghton, appointed Michigan's first state geologist in 1837, began his geological and mineral investigations of the Northern Peninsula in 1831. In his first annual report to the legislature (Houghton, 1841), he stated, "Although hematite ore is abundantly disseminated through all the rocks of the metamorphic group, it does not appear in sufficient quantity at any one point that has been examined" (Brooks, 1873). By 1841, Houghton had traversed the south shore of Lake Superior five times in a small boat or canoe. However, the iron formations discovered later are at least 7 mi from the shore of Lake Superior.

In 1843, severe financial problems faced the state of Michigan. This prompted the legislature to withhold the annual appropriation for the Geological Survey. In order to continue geological surveys of the state, Houghton conceived of combining geological surveys with ongoing federal linear surveys. Linear surveys consisted of dividing land areas into geographic townships and sections. Houghton succeeded in obtaining an allowance per mile of linear surveys to cover the additional expenses of geological surveys. He personally contracted with the federal government to complete both linear and geological surveys of the Northern Peninsula. Linear surveys were previously started by William A. Burt, deputy surveyor. In the spring of 1844, Houghton commenced operations of his contract, with Burt in charge.

In his official diary of the year 1844, Burt stated:

East boundary of Township 47 North, Range 27 West. This line is very extraordinary, on account of the great variations of the needle, and the circumstances attending the survey of it. Commenced in the morning, the 19th of September; weather clear; the variation high and fluctuating, on the first mile, section one. On sections 12 and 13, variations of all kinds, from south 87 degrees east, to north 87 degrees west. In some places the north end of the needle would dip to the bottom of the box, and would not settle anywhere. In other places it would have variations 40, 50, and 60 degrees east, then west variation alternating in the distance of a few chains. Camped on a small stream in section 13. [Brooks, 1873]

Jacob Houghton, brother of Douglass and barometer operator for the survey party, related the story of the discovery of iron:

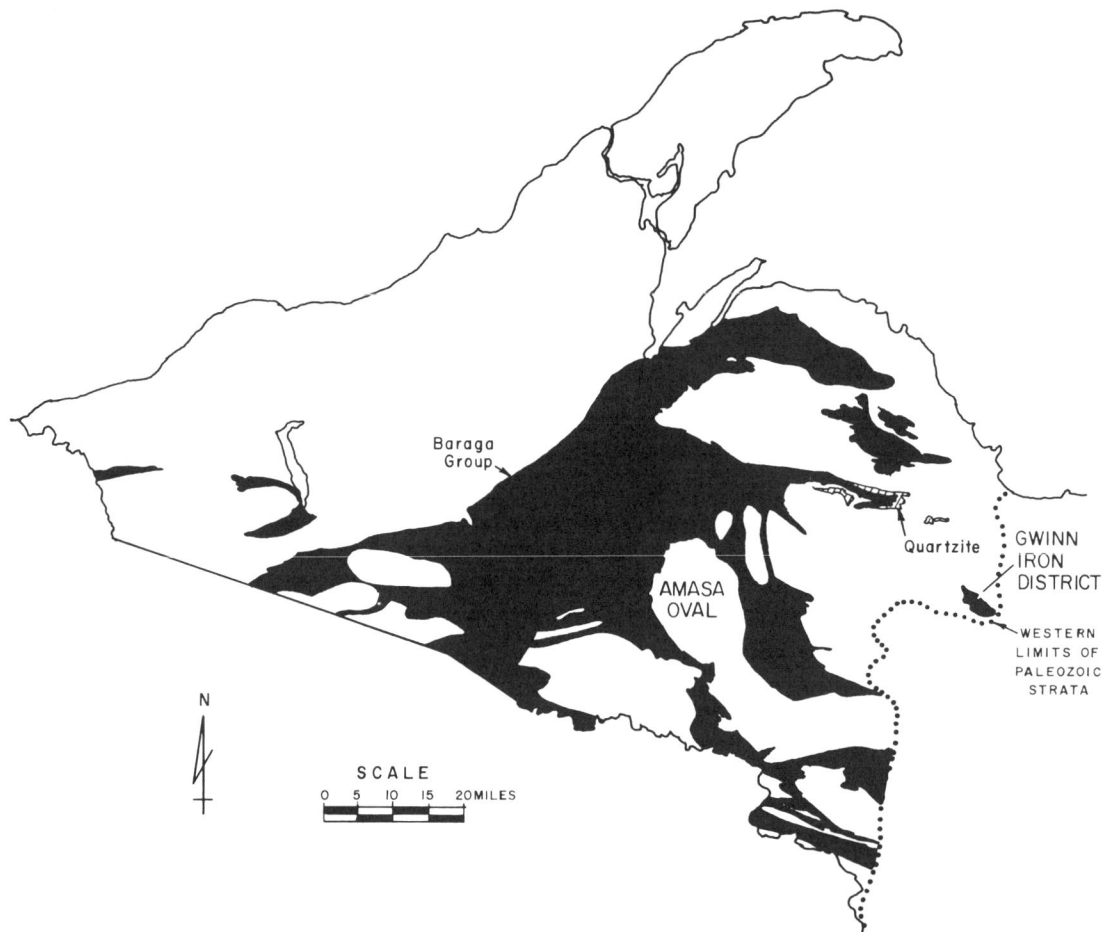

Figure 10. Surface distribution of early Proterozoic Baraga Group sediments.

On the evening of the 15th of September, we reached the lake and established the north-east corner of Town 47 North, Range 25 West, between the Chocolate and Carp Rivers. We thence ran west the township line, between Towns 47 and 48, and camped at the town corner on the east side of Teal Lake, on the 18th of September. On the morning of the 19th we started, running the line south, between Ranges 26 and 27. So soon as we reached the hill to the south of the Lake, the compass-man began to notice the fluctuation in the variation of the magnetic needles. We were, of course, using the Solar Compass, of which Mr. Burt was the inventor, and I shall never forget the excitement of the old gentleman when viewing the changes of the variation—the needle not actually traversing alike in any two places. He kept changing his position to take observations, all the time saying, "How would they survey this country without my compass? What could be done here without my compass?" It was the full and complete realization of what he had foreseen when struggling through the first stages of his invention. At length the compass-man called for all to "come and see a variation that will beat them all." As we looked at the instrument, to our astonishment the north end of the needle was traversing a few degrees to the south of west. Mr. Burt called out, "Boys, look around and see what you find!" We all left the line, some going to the east, and some to the west, and all of us returning with specimens of iron ore, mostly gathered from outcrops. This was along the first mile from Teal Lake. We carried out all the specimens we could conveniently. [Brooks, 1783]

The survey contractor, Houghton, was not with the party at that time. He was not aware of this important discovery of vast quantities of iron ore until the return of Burt's survey party to Detroit in late 1844 (Brooks, 1873).

The first iron-mining venture, the Jackson mine, operated by the Jackson Mining Company from Jackson, Michigan, formed in 1845. Their property included all of Section 1, T47N, R27W, located in what is now the City of Negaunee, later to be the location of the vast Mather B underground mine. In early 1848, the Jackson Mining Company completed a forge on the Carp River, 5 mi east of Negaunee, to produce wrought iron. This would provide an outlet for some of the ore from the Jackson mine.

Other early mining companies formed on the Marquette Iron Range included the Cleveland Iron Mining Company in 1847, the Marquette Iron Company in 1849, and the Lake Superior Iron Company in 1853. The Marquette Iron Company constructed a forge at Marquette in 1849–50. Two additional forges were constructed on the Dead River north and west of Marquette in 1855.

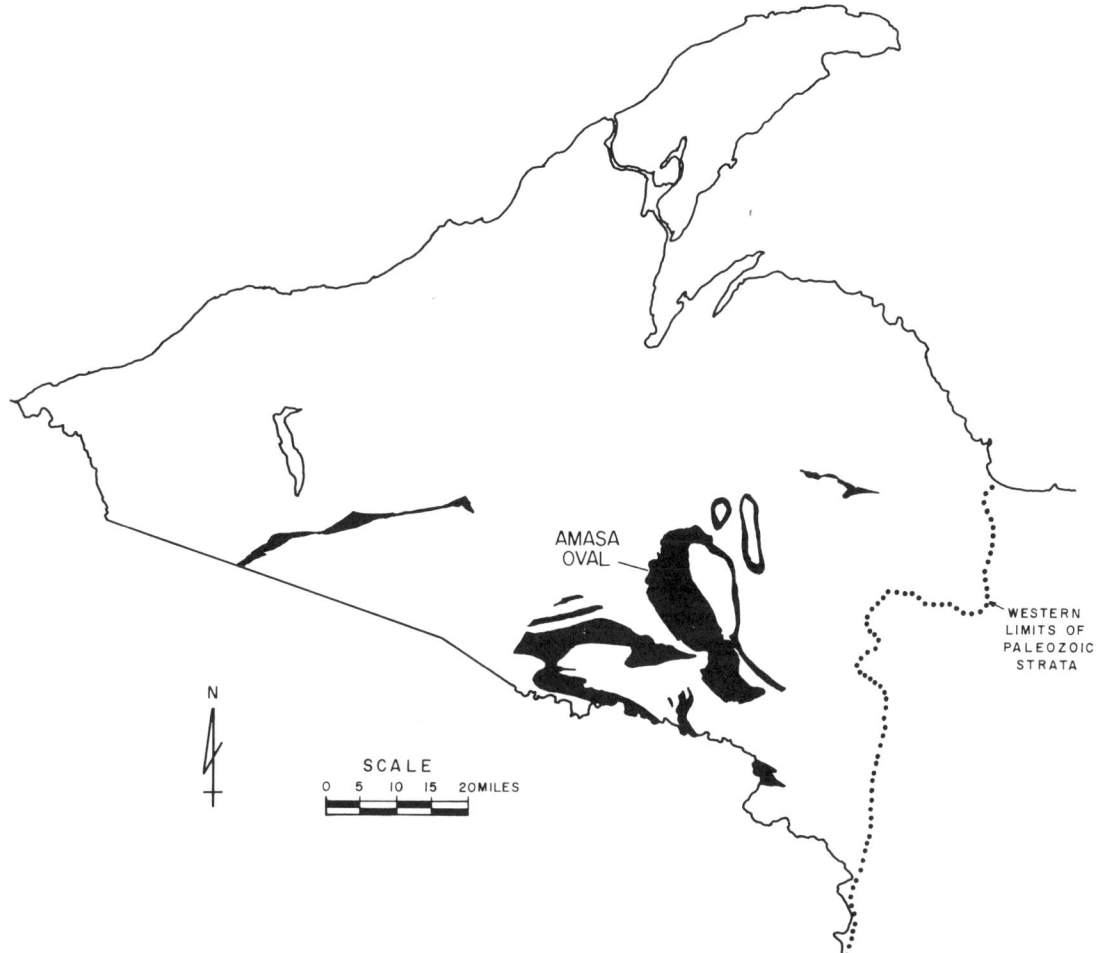

Figure 11. Surface distribution of early Proterozoic Baraga Group volcanics.

A unique iron-mining venture was conducted by the Eureka Iron Company, which was organized October 29, 1853. The Eureka Iron Company was to mine ore and manufacture pig iron from Lake Superior ores. Initially, a furnace was to be constructed in Marquette County, but the location was later changed to Wyandotte. Land for the mine was purchased near Marquette. When a few hundred tons of ore had been mined, however, it became evident that ore did not exist in quantity, and the land was resold to the original owner (Brooks, 1873).

The so-called Eureka iron mine was located along the contact of the Lower Member and the Lighthouse Point Member of the Mona Schist of Archean age. It was located in the north-central part of Section 21, T48N, R25W. This is approximately 7.5 mi northeast of the nearest contact with the Negaunee Iron Formation. Several hundred tons of earthy goethitic iron ore was mined from a 5-ft-wide shear zone. The iron ore, originally iron-bearing carbonate, resulted from oxidation of carbonate and chlorite, concentration of iron, and removal of silica, magnesium, calcium, and alumina (Gair and Thaden, 1968).

Mining companies formed rapidly to produce iron ore, but production was constrained by markets and particularly by transportation to those markets. In 1855, the Jackson Iron Company and Lake Superior Iron Company constructed iron-ore docks at Marquette and a plank road leading from the mines. The first railroad, the Iron Mountain, from the mines to Marquette, was completed in 1857.

At Sault Ste. Marie, the St. Mary's River connected Lake Superior with Lake Huron. Rapids of the river, about 1 mi long with a fall of 17 ft, severely constricted transportation. The importance of overcoming this obstacle was recognized early. Congressional concern was reported by the Detroit Free Press in 1840:

*Washington, April 21st, 1840.* This day in the senate, the bill granting to the State of Michigan 100,000 acres of land to aid her in the construction of a canal around the falls of Ste. Marie, came up again on third reading. Mr. Norvel and others advocated the bill. Mr. Clay, of Kentucky, took occasion to speak of the work as one beyond the range of the remotest settlements in the United States, or in the moon. Senator Norvel advocated the bill mainly on the ground that the completion of the canal would stimulate the fisheries of Lake Superior, estimated to be worth one

Figure 12. Surface distribution of early Proterozoic Paint River Group.

million dollars per annum. The honorable Senator added, "In the country bordering on the southern shore of Lake Superior copper ore and other minerals are believed to exist in abundance." [Swineford, 1882]

Congress granted 750,000 acres of land to aid in construction of the Ste. Marie's Falls Ship Canal in 1852, and in 1853, the Michigan legislature approved construction. The canal was completed in 1855 and opened in June 1856.

The Peninsula Railroad from Negaunee to Escanaba, later to become the Peninsula Division of the Chicago and Northwestern Railroad (C&NW), was completed in 1864. Prior to 1870, the C&NW reached Ft. Howard, Wisconsin, and in 1872 extended to Escanaba where ore docks were established to provide shipping on Lake Michigan. Also in 1872, the Iron Mountain Railroad, now the Marquette and Ontonagon Railroad, was extended westward to L'Anse where ore docks were constructed (Boyum 1975, 1977; Brooks, 1873; Swineford, 1882; Wright, 1879).

It was previously noted that four charcoal iron forges were constructed by 1855. The larger pig-iron blast furnaces required hundreds of tons of castings and machinery, which before the opening of the canal and locks, was too costly to portage the rapids. With the opening of the canal, pig-iron blast furnaces, mostly fueled by charcoal, proliferated, as illustrated by Tables 4 and 5 (LaFayette 1977).

*Geology.* The Marquette Range (Figs. 9, 13, and 14) is a relatively narrow, essentially east-west–striking, westward-plunging syncline more than 65 km (40 mi) long and 3 to 10 km (2 to 6 mi) wide. Structural deformities extend the range south and southeast of Lake Michigamme. The principal iron formation, the Negaunee, is shown on Figure 13. The Negaunee Iron Formation ranges in thickness from 135 to 1,060 m (450 to 3,500 ft) or more and is commonly about 300 m (1,000 ft). Dominant primary iron minerals are siderite, and locally hematite and magnetite. Most magnetite is primary or diagenetic. Iron silicates, minnesotaite, and stilpnomelane are locally abundant in low-grade metamorphic zones (Gair, 1973). Structurally controlled areas of the Negaunee Iron Formation have been subject to oxidation and leaching, resulting in naturally concentrated iron ores generally referred to as direct-shipping ores or soft ores. This was the product mined almost exclusively until 1952, the exception being experimental concentrates and siliceous iron ore. Naturally

## TABLE 4. LIST OF UPPER PENINSULA FORGES AND BLAST FURNACES OPENING PRIOR TO 1900

|  | Start Up | Last Known Operation |
|---|---|---|
| **Forges** | | |
| Carp River Forge | Feb. 10, 1848 | June 1857, last reference |
| Marquette Iron Co. | About July 6, 1850 | Burned, Dec. 14, 1853 |
| Forest Iron Co. | July 1855 | May 1862, last reference |
| Collins Forge | August 1855 | Converted into experimental blast furnace Jan. 1858 |
| **Blast Furnaces** | | |
| Collins Experimental Furnace | Jan. 21, 1858 | Abandoned Jan. 25, 1858 |
| Pioneer No. 1 stack | April 26, 1858 | Major breakdown in June 1892, never started again |
| Collins Blast Furnace | Dec. 13, 1858 | Exhausted fuel supply in May 1873 |
| Northern Furnace | Summer 1860 | Made iron in 1910 for a short while |
| Pioneer No. 2 stack | May 1859 | Blown out May 18, 1893, completely worn out |
| Bancroft Furnace | May 1861 | River overflowed May 1876, destroyed flumes shutdown |
| Morgan Furnace | Nov. 27, 1863 | Blown out December 1876 |
| Greenwood Furnace | June 1865 | Bad roads, exhausted ore and fuel, shut down April 9, 1875 |
| Michigan Furnace | Feb. 10, 1867 | Blown out for repairs Jan. 12, 1875, never restarted |
| Schoolcraft Furnace (Munising) | Jun 28, 1868 | Blew out Nov. 1877 |
| Jackson No. 1 stack (Fayette) | Dec. 25, 1867 | Company closed down Dec. 1890 |
| Champion Furnace | Dec. 4, 1867 | Burned April 9, 1874 |
| Deer Lake No. 1 stack | About Sept. 1, 1868 | Shut down late fall of 1891 |
| Jackson No. 2 stack (Fayette) | May 2, 1870 | Company closed down Dec. 1890 |
| Bay Furnace No. 1 stack | March 5, 1870 | Burned May 31, 1877 |
| Marquette and Pacific | July 13, 1871 | Fall of 1881 or early in 1882 |
| Peat Furnace (Excelsior) | March 1872 | Shut down Nov. 5, 1897, hurting Ishpeming's water supply |
| Grace Furnace | Dec. 10, 1872 | Out of blast March 24, 1874, never used again |
| Bay Furnace No. 2 stack | Dec. 15, 1872 | Burned May 31, 1877 |
| Escanaba Furnace (Cascade) | Early in 1873 | Blown out late in 1874, dismantled 1879 |
| Menominee Furnace | July 8, 1873 | Possibly in late 1883 |
| Carp River Furnace | April 26, 1874 | Shut down 1907 |
| Deer Lake No. 2 stack | Jan. 1874 | Shut down late fall of 1891 |
| Cliffs Furnace | March 14, 1874 | Abandoned about 1877 |
| Martel Furnace | Aug. 14, 1881 | Blown out 1903, fire destroyed stock house |
| Vulcan Furnace | May 21, 1883 | Renovated 1902, possibly 1945 |
| Iron River Furnace (Gogebic) | Feb. 2, 1886 | Exhausted ore and fuel May 1888 |
| Weston Furnace | April 30, 1891 | Closed down 1922 |
| Gladstone Furnace | April 16, 1896 | Closed down 1922 |

## TABLE 5. PIG IRON PRODUCTION FROM UPPER PENINSULA FORGES AND BLAST FURNACES

|  | (short tons) |
|---|---|
| Bancroft | 55,608 |
| Bay Furnace (two stacks) | 50,706 |
| Carp River | 83,500 |
| Champion | 31,048 |
| Cliffs | 8,209 |
| Collins | 41,997 |
| Deer Lake (two stacks) | 93,579 |
| Escanaba | 8,650 |
| Excelsior | 68,634 |
| Fayette (two stacks) | 229,288 |
| Gogebic | 3,700 |
| Grace | 11,346 |
| Greenwood | 40,202 |
| Manistique | 150,904 |
| Martel | 58,349 |
| Marquette and Pacific | 41,857 |
| Menominee | 59,553 |
| Michigan | 41,531 |
| Morgan | 57,573 |
| Munising (Schoolcraft) | 28,312 |
| Northern (15,059 to 1881, est. for 1891 and 1910 operation 15,000) | 30,059 |
| Pioneer (two stacks at Negaunee and the Pioneer at Gladstone through 1902) | 637,299 |
| Vulcan | 73,829 |
| Total | 1,905,733 |

concentrated ores generally occurred in two structural environments, at the base of the iron formation above the footwall Siamo Slate, and ores related to intrusive sills and dikes. Boyum (1988) reports that production of footwall-related ores amounted to 186,607,540 tons and intrusive-related ores at 54,092,030 tons, principally from underground mining operations.

Immediately below the Negaunee Iron Formation, the Siamo Slate contains a ferruginous magnetic layer—the Goose Lake Member (Fig. 13). The Goose Lake, an iron formation, is 15 to 30 m (50 to 100 ft) thick and is composed principally of siderite, chert, magnetite, chlorite, and stilpnomelane (Gair, 1975). One mine, located west of the City of Ishpeming, produced from the Goose Lake Member (Table 6).

Above the Negaunee Iron Formation, within the Michigamme Formation, is a banded grunertic cherty iron formation, the Bijiki (Fig. 13). The Bijiki is 30 to 50 m (100 to 165 ft) thick, and near the surface has been oxidized to hematite and goethite, locally enriched to iron ore (Cannon and Klasner, 1977). The writer favors a correlation of the iron formation of the Taylor mine area in Baraga County and the Gwinn Iron District southeast of the Marquette Range (Fig. 10) with the Bijiki Iron Formation. Therefore, Marquette Range production tabulations

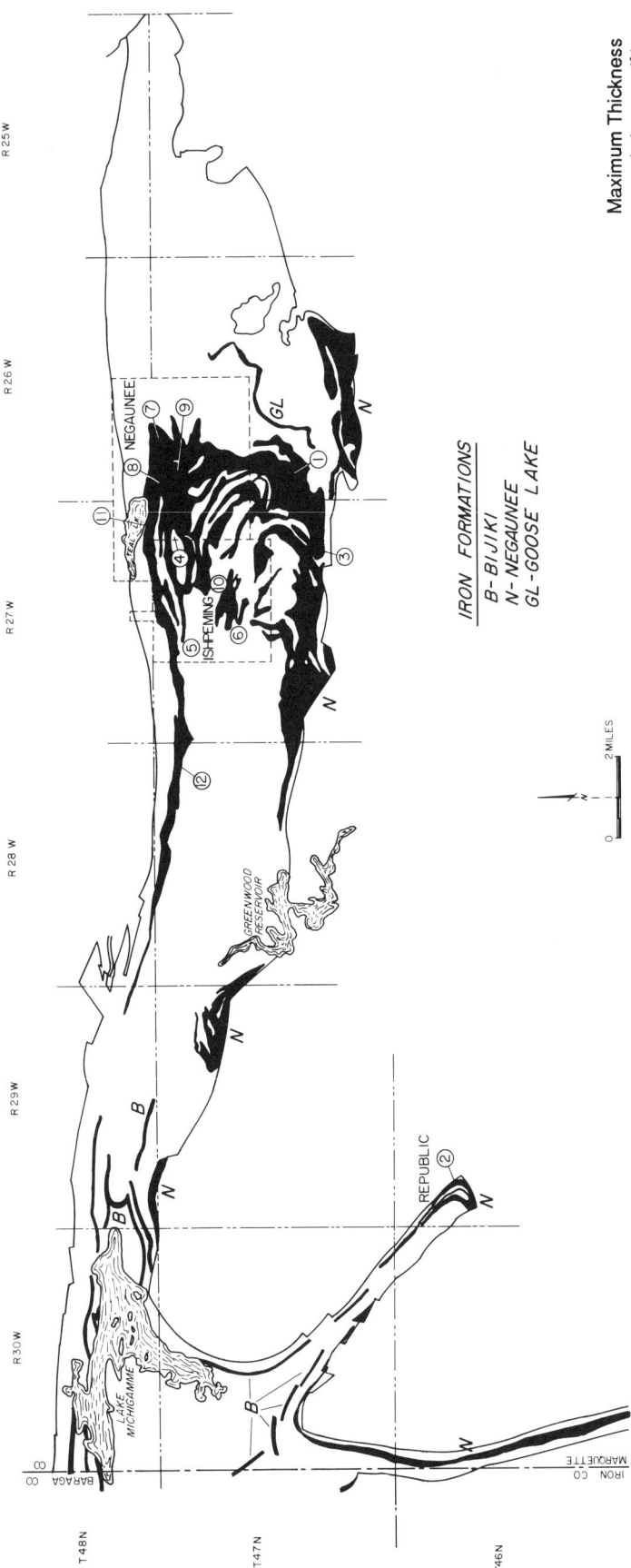

Figure 13. Marquette Iron Range indicating location or iron formations and major producing mines. Major producing mines are listed on Table 7.

| | Maximum Thickness | |
|---|---|---|
| | (m) | (ft) |
| **EARLY PROTEROZOIC** | | |
| Baraga Group | | |
| Michigamme Formation | | |
| Upper Slate Member | ND | ND |
| Bijiki Iron Formation Member | 150 | 500 |
| Lower Slate Member | ND | ND |
| Clarksburg Volcanics Member | 150 | 500 |
| Greenwood Iron Formation Member | 610 | 2,000 |
| Lower Argillite Member | 365 | 1,200 |
| Goodrich Quartzite | 460 | 1,500 |
| Menominee Group | | |
| Negaunee Iron Formation | 1,065 | 3,500 |
| Siamo Slate | 915 | 3,000 |
| Ajibik Quartzite | 200 | 650 |
| Chocolay Group | | |
| Wewe Slate | 275 | 900 |
| Kona Dolomite | 365 | 1,200 |
| Mesnard Quartzite | 150 | 500 |
| Enchantment Lake Formation | 150 | 500 |

Figure 14. Stratigraphic column of Marquette Iron Range area. ND = no data.

(Table 6) include the Taylor mine and Gwinn Iron District as Bijiki production.

With respect to Negaunee Iron Formation resources of the Marquette Range, Cannon and others (1978) concluded:

> The Negaunee iron-formation in the Marquette district of Michigan contains about 205 billion long tons of material averaging about 32 percent iron. About 49 billion long tons is within 1,000 feet (304.8 m) of the surface.
>
> Considering only material that might be minable from open pits and that has suitable mineralogy and grain size to be potentially processible by current concentrating processes or by modifications of them, a maximum of about 11 billion short tons of iron is potentially recoverable from the district. However, many geologic, economic, technologic, political, and environmental restraints are likely to prevent utilization of much of it.

*Production.* Appendix A lists annual iron-ore production of principal Michigan iron ranges through 1987. As indicated, production from the Marquette Range totals 587,641,341 long tons or 50.3 percent of the total state production. Table 6 subdivides production by source iron formation and producing districts. Table 7 indicates those mines whose individual total production exceeded 10 million tons and that are located on Figure 13. The 12 mines listed on Table 7, all of which are produced from the Negaunee Iron Formation, represent the major portion of total production. They produced 77.6 percent of the total from the Negaunee Iron Formation and 75.4 percent from the Marquette Range. Table 8 lists mines whose products were concentrated or pelletized from low-grade iron formation. The amount from the Negaunee Iron Formation represents 44 percent of total range production.

Another iron product from the Negaunee Iron Formation was siliceous iron ore. Siliceous ore averaged 36 to 43 percent iron and 30 to 43 percent silica (Gair, 1975). The principal use was to control viscosity in the blast furnace and in the production of ferrosilicon. Sixteen siliceous ore properties of the 76 mining operations utilizing the Negaunee Iron Formation produced 4.7 percent of the total production (Table 9). Still another product, generally referred to as "hard ore" was produced from the Negaunee Iron Formation. Hard ore or "lump ore" normally contained 58 to 62.5 percent iron (Gair, 1974) at the Cliffs Shaft mine (Table 7; Fig. 13), the largest producer. Total production of hard ore is indicated on Table 9. All occurrences of hard ore were in the upper portion of the Negaunee Iron Formation immediately below the Goodrich Quartzite. This ore was desirable as a direct feed into the open-hearth furnace.

***Menominee Iron–Felch Mountain District.*** The Menominee Iron-bearing District is located in extreme southern Dickinson County, and the Felch Mountain District in central Dickinson County (Figs. 9 and 15). They are included as a unit because the principal iron formation, the Vulcan, apparently occurs in both districts.

*History.* The Felch Mountain District was the first to be discovered. As reported by James and others (1961), the first published record of discovery (in a 1849 Senate document) contains geological notes by Bela Hubbard (Michigan Geological Survey) based on surveys of William A. Burt in 1846. The notes described a bed of iron near what is now the abandoned Groveland iron mine. Iron of the Menominee Iron-bearing District was

**TABLE 6. IRON-ORE PRODUCTION BY FORMATION, MARQUETTE IRON RANGE**

| Formation | Number of Mines | Production (Long Tons) | Percent |
|---|---|---|---|
| Bijiki Iron Formation of Michigamme Formation | 21 | 17,085,180 | 2.9 |
| Negaunee Iron Formation | 76 | 570,658,136 | 97.1 |
| Goose Lake Member of Siamo Slate | 1 | 23,395 | |
| Total | | 587,766,711 | |
| Negaunee Iron Formation | 76 | 570,658,136 | |
| Principal Range | 58 | 480,680,075 | 84.2 |
| Palmer District | 12 | 18,287,071 | 3.2 |
| Republic District | 6 | 71,690,990 | 12.6 |
| Bijiki Iron Formation | 21 | 17,085,180 | |
| Principal Range | 13 | 4,266,952 | 25.0 |
| Gwinn District | 7 | 12,785,258 | 74.8 |
| Taylor mine | 1 | 32,970 | 0.2 |

**TABLE 7. INDIVIDUAL MINE PRODUCTION EXCEEDING TEN MILLION LONG TONS, MARQUETTE IRON RANGE***

| Mine | Production (million long tons) | Type of Operation | | Years of Shipments |
|---|---|---|---|---|
| 1. Empire | 107.1 | C | OP | 1964–1987 |
| 2. Republic | 63.0 | C | OP | 1956–1981 |
| 3. Tilden | 60.4 | C | OP | 1974–1987 |
| 4. Mather-Pioneer | 54.2 | DS-C | U | 1888–1905, 1943–1980 |
| 5. Cliffs Shaft | 29.0 | DS | U | 1868–1972 |
| 6. Lake Superior Group | 25.1 | DS | U | 1858–1937 |
| 7. Negaunee | 22.7 | DS | U | 1887–1949 |
| 8. Maas | 21.3 | DS | U | 1907–1967 |
| 9. Athens-Bunker Hill | 16.5 | DS | U | 1918–1966 |
| 10. Cleveland Lake | 16.3 | DS | U | 1854–1927 |
| 11. Cambria-Jackson | 16.2 | DS | U | 1874–1859 |
| 12. Morris | 11.0 | DS | U | 1912–1961 |
| Total tons | 442.8 | | | |

*Located on Figure 13.
C = Concentrate/Pellet Product; DS = Direct Shipping Ore; OP = Open-Pit Operation; U = Underground Operation.

**TABLE 8. TOTAL MINE PRODUCTION OF CONCENTRATED AND/OR PELLETIZED IRON PRODUCTS, MARQUETTE IRON RANGE**

| Mine | Production (long tons) | Years of Shipments |
|---|---|---|
| **Bijiki Iron Formation Member** | | |
| Ohio | 745,620 | 1952–1962 |
| **Negaunee Iron Formation** | | |
| Negaunee Construction Works | 12,708 | 1882–1886 |
| Edison | 893 | 1889 |
| Humboldt | 9,433,305 | 1954–1972 |
| Republic | 62,959,898 | 1956–1981 |
| Empire | 107,053,267 | 1964–1987 |
| Mather | 19,226,357 | 1965–1980 |
| Tilden | 60,418,789 | 1974–1987 |
| Total | 259,105,217 | |

**TABLE 9. IRON-ORE PRODUCTION SUMMARY, MARQUETTE IRON RANGE**

| | Long Tons | Long Tons | Percent |
|---|---|---|---|
| Direct shipping ores* | | | |
| Footwall soft ores | 186,607,540 | | 31.7 |
| Intrusive soft ores | 54,092,030 | | 9.2 |
| Hard (lump) ores | 60,350,944 | | 10.3 |
| | | 301,050,514 | |
| Concentrates and Pellets | | 259,105,217 | 44.1 |
| Siliceous | | 27,557,975 | 4.7 |
| Total | | 587,713,706 | |

*Boyum, 1988

discovered in 1848 by J. W. Foster and S. W. Hill (Bayley and others, 1966). However, iron production did not commence until 1877 in the Menominee Iron-bearing District and 1882 from the Felch Mountain District.

*Geology.* The Menominee Iron-bearing District (Figs. 15 and 16) is composed of two linear, WNW-trending iron ranges. The Vulcan Iron Formation of the north range is exposed for about 16 km (10 mi), and of the south range for about 29 km (18 mi), from the Menominee River, through Iron Mountain and Norway, east to Waucedah. The Vulcan extends east another 56 km (35 mi) to Escanaba, beneath younger Paleozoic rocks, as indicated by magnetic surveys and a few exploration drill holes.

The Vulcan Iron Formation is part of two trough-like basins where the older Randville Dolomite to the north has been thrust upward along major faults (Fig. 15). The younger Michigamme Formation overlies the Vulcan to the south. Therefore, the Vulcan Iron Formation, unless vertical or overturned, dips to the south an unknown distance.

The Vulcan Iron Formation ranges in thickness from 90 to 180 m (300 to 600 ft) and averages about 135 m (450 ft) (Gair, 1973). It is composed of four members: two cherty members, the Traders and Curry; and two slaty members, the Briar and Loretto (Bayley and others, 1966). The Traders and Curry Members are chemical sediments with layers of iron alternating with layers of chert or jasper. In the Briar and Loretto Members, clastic material is dominant over chemical. The iron minerals of the cherty members are principally hematite with lesser amounts of magnetite. The slaty members contain principally magnetite. Oxidation has, in part, converted magnetite to hematite (martite). Grades and thicknesses of each member are listed in Table 10.

In the Felch Mountain District (Fig. 9), central Dickinson County, the Groveland iron mine has been most productive. Bedrock in the district has been greatly eroded so only remnants of iron formation are exposed at the surface (Fig. 15). At the Groveland mine, the original interpretation of the structure was that of a syncline. After considerable mapping and development drilling, the structure was found to be monoclinal, dipping to the north. The Vulcan Iron Formation members of the Menominee Iron-bearing district cannot be differentiated in the Felch Mountain District. Here, the iron formation is described as an upper gray-banded unit and a lower oolitic unit. The upper gray-banded iron formation consists of chert and iron oxides, mostly magnetite with minor specular hematite. In the lower member, the chert is typically oolitic. Specular hematite is the principal iron mineral, and there is some magnetite present. The Felch Mountain District has been subjected to moderate to high-grade metamorphism, and grunerite is fairly abundant (James and others, 1961).

*Production.* Iron-ore production by district and product is given in Table 11. Table 12 lists those mines whose production exceeded 10 million tons and are located on Figure 15. The four mines on Table 12 produced 81 percent of the total.

***Gogebic Iron Range.*** *History.* As early as 1847, U.S. Geological Survey geologists G. O. Barnes and J. D. Whitney traversed parts of Gogebic County, in particular the east and west sides of Range 46 West, some 5 to 11 mi east of the Montreal River. In these traverses, they recorded Archean crystalline rocks and Keweenawan volcanic rocks, but no sediments were exposed. The following year, 1848, another U.S. Geological Survey geologist, A. Randall, observed exposures of lean magnetic iron near Upson, Wisconsin, while surveying the fourth principal meridian.

Although surveys and explorations continued on the Wisconsin side of the range, it was not until 1871 that T. B. Brooks and R. Pumpelly surveyed the Michigan side from the Montreal River to Lake Gogebic (Brooks and Pumpelly, 1872; Brooks, 1873). Iron ore, i.e., naturally concentrated direct-shipping ore, was first discovered at the Colby mine, located in what is now the city of Bessemer and subsequently included in the Peterson Group (Table 13; Fig. 17). A trapper and hunter, Richard Langford, found the ore-bearing outcrop beneath an overturned tree

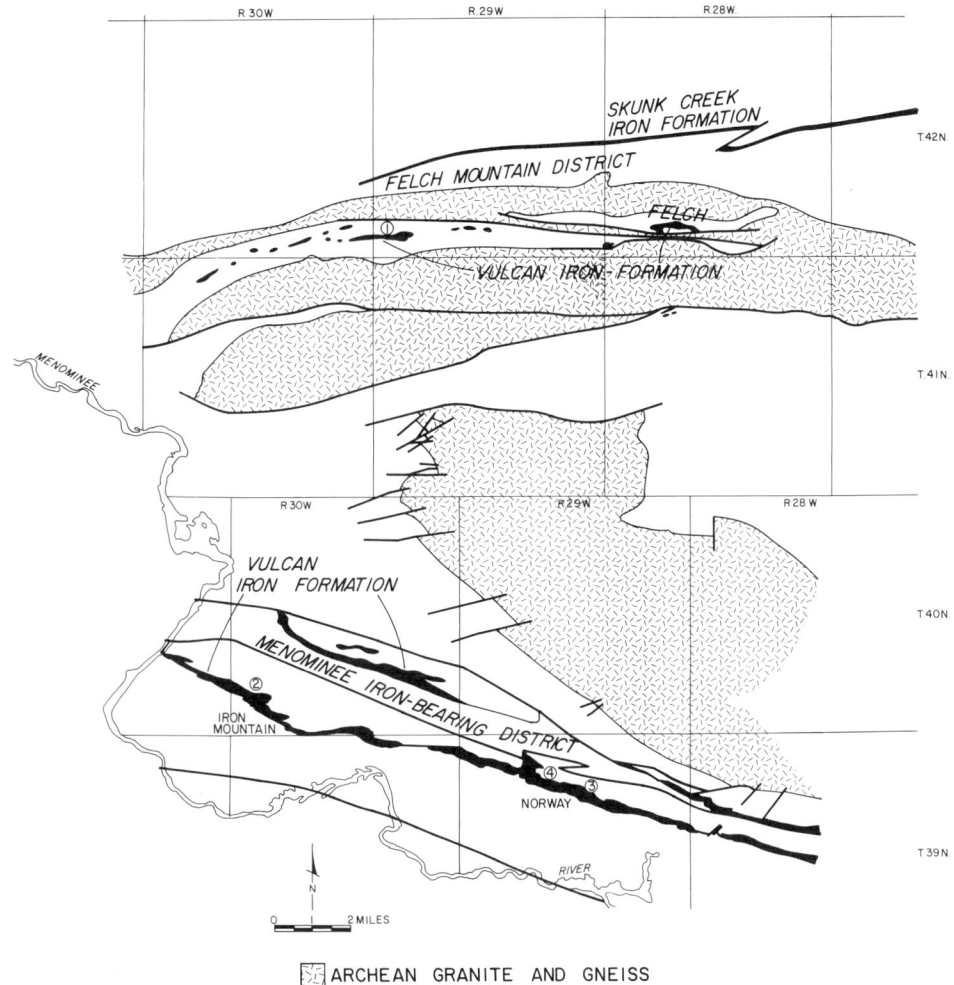

Figure 15. Menominee Iron-bearing District and Felch Mountain District indicating location of iron formations and major producing mines. Major producing mines are listed on Table 12.

### TABLE 10. REPORTED IRON CONTENT AND THICKNESS OF INDIVIDUAL MEMBERS OF THE VULCAN IRON FORMATION, MENOMINEE IRON-BEARING DISTRICT*

| Member | Percent Iron | | Thickness | |
|---|---|---|---|---|
| | Average | Range | Average | Range |
| Loretto | | 16–30 | Not reported | |
| Curry | 31 | | 130 | 60–200 |
| Brier | 18 | 7–34 | 100 | (North Range) |
| | | | 300 | (South Range) |
| Traders | | | 135 | 80–200 |

*Bayley and others, 1966.

| | | Maximum Thickness | |
|---|---|---|---|
| | | (m) | (ft) |
| EARLY PROTEROZOIC | | | |
| Baraga Group | | | |
|   Badwater Greenstone | | 2,440 | 8,000 |
|   Michigamme Formation | | 1,525 | 5,000 |
| Menominee Group | | | |
|   Vulcan Iron Formation | | 185 | 600 |
|     Loretto Slate Member | | | |
|     Curry Iron-bearing Member | | | |
|     Briar Slate Member | | | |
|     Traders Iron-bearing Member | | | |
|   Felch Formation | | 150 | 500 |
| Chocolay Group | | | |
|   Randville Dolomite | | 610 | 2,000 |
|   Sturgeon Quartzite | | 610 | 2,000 |
|   Fern Creek Formation | | 30 | 100 |

Figure 16. Stratigraphic column of Menominee Iron-bearing District.

TABLE 11. IRON-ORE PRODUCTION BY DISTRICT AND PRODUCT, MENOMINEE IRON-BEARING DISTRICT AND FELCH MOUNTAIN DISTRICT

| District and Product | Number of Mines | Production (long tons) | Percent Production |
|---|---|---|---|
| Menominee Iron-Bearing District | | | |
| Direct Shipping | 12 | 74,443,273 | 63.1 |
| Siliceous | 11 | 7,286,220 | 6.2 |
| Total District | | 81,729,493 | 69.3 |
| Felch Mountain District | | | |
| Direct Shipping | 3 | 162,946 | 0.1 |
| Siliceous | 2 | 311,840 | 0.3 |
| Concentrates and Pellets | 1 | 35,682,199 | 30.3 |
| Total District | | 36,156,985 | 30.7 |
| Total Range | | 117,886,478 | |

TABLE 12. INDIVIDUAL MINE PRODUCTION EXCEEDING TEN MILLION LONG TONS, MENOMINEE IRON-BEARING DISTRICT AND FELCH MOUNTAIN DISTRICT*

| Mine | Production Million Long Tons | Type of Operation | | Years of Shipments |
|---|---|---|---|---|
| 1. Groveland | 35.7 | C | OP | 1891–1913, 1952–1959, 1974–1982 |
| 2. Chapin | 27.5 | DS | U | 1880–1934 |
| 3. Penn | 21.6 | DS | U | 1877–1945 |
| 4. Aragon | 11.2 | DS | U | 1889–1931 |
| Total tons | 96.0 | | | |

*Located on Figure 15.
C = Concentrate/Pellet Production; DS - Direct-Shipping Ore; OP = Open-Pit Operation; U = Underground Operation.

and displayed specimens for several years before mining began in 1884 (Zapffe, 1938).

*Geology.* The Gogebic Iron Range (Figs. 9 and 17) is nearly 130 km (80 mi) long, of which about 85 km (53 mi) is in Wisconsin (Aldrich, 1929). In Michigan, the Ironwood Iron Formation trends eastward for another 40 km (25 mi) or more. As indicated on Figures 9 and 17, much of the range is a rather straight, simple monocline, dipping northward. In the westernmost part of Michigan, iron-formation thickness averages about 260 m (850 ft) with a rather uniform dip of 60 degrees north. This consistency is interrupted in the vicinity of Wakefield by numerous faults. The largest fault, the Sunday Lake, has displaced the footwall about 2,286 m (7,500 ft), and structural complexity led to expansion of the surface area of the iron formation (Lake, 1917). In the Wakefield area, the iron formation thickens to 450 to 500 m (1,475 to 1,640 ft) (Prinz and Hubbard, 1975). Normal structural conditions resume east of Wakefield, dipping about 60°N. In Range 44 West (Fig. 17) the iron formation bends into a complexly faulted, northeast-trending drag fold, and dips flatten to about 35°N (Prinz, 1967).

Numerous dikes, approximately normal to bedding, occur throughout the productive portion of the range. Most dike-bedding intersections pitch east at low angles, although some are in the opposite direction. Faults are a prominent feature of the iron formation. Transverse faults displace both the formation and faults parallel to dikes (normal to the formation). Bedding faults that parallel the bedding exist near the base of the Plymouth Member, within or adjacent to the Yale Member and within the Pence Member (Hotchkiss, 1919). Faults, dikes, and iron formation lithology were the instrumental features responsible for channeling circulating waters to form naturally concentrated direct-shipping iron ores.

As indicated (Fig. 4), the Ironwood Iron Formation overlies the Palms Formation, a thin-bedded quartzite locally termed "quartz slate." The Ironwood in turn is overlain by the Tyler Formation, a very thick argillaceous unit containing ferruginous material at or near the base. The former Tyler and Copps Formations of the Gogebic Range are now considered equivalents of the Michigamme Formation (Fig. 4).

The Ironwood Iron Formation contains five members (Fig. 4). Three members, the Plymouth, Norrie, and Anvil, are generally referred to as cherty members. The Yale and Pence Members are called slaty members, although they are not necessarily argillaceous. Cherty members consist of chert that is irregularly (wavy) bedded, ferruginous, and granular (oolitic). Beds may be as much as several inches thick and contain magnetite and iron silicates. The wavy-bedded, granular chert alternates with thin, evenly bedded, ferruginous slaty material containing carbonate, magnetite, iron silicates, and thin chert layers. The slaty members consist of evenly bedded, finely laminated, ferruginous slate, along with carbonate, chert, magnetite, and iron silicates. Minnesotaite and stilpnomelane are the major iron silicates. More detailed descriptions may be gleaned from Irving and Van Hise (1892), Hotchkiss (1919), Aldrich (1929), Huber (1959a, b), and Schmidt (1980).

Huber (1959a) reported the range of percent carbonate composition from nine carbonate and chert-carbonate iron formation samples as: $FeCO_3$, 58.9 to 84.8 percent; $MgCO_3$, 12.0 to 19.7 percent; $CaCO_3$, 1.8 to 17.7 percent; and $MnCO_3$, 1.4 to 7.4 percent. In Michigan, the Ironwood Iron Formation has been extensively oxidized from the Montreal River east to the Sunday Lake mine at Wakefield (Fig. 17). This oxidation is known to exceed depths of 1,220 m (4,000 ft).

Downward-percolating meteoric waters removed most minerals other than iron, which was naturally concentrated into direct-shipping iron ores. Such waters were responsible for oxida-

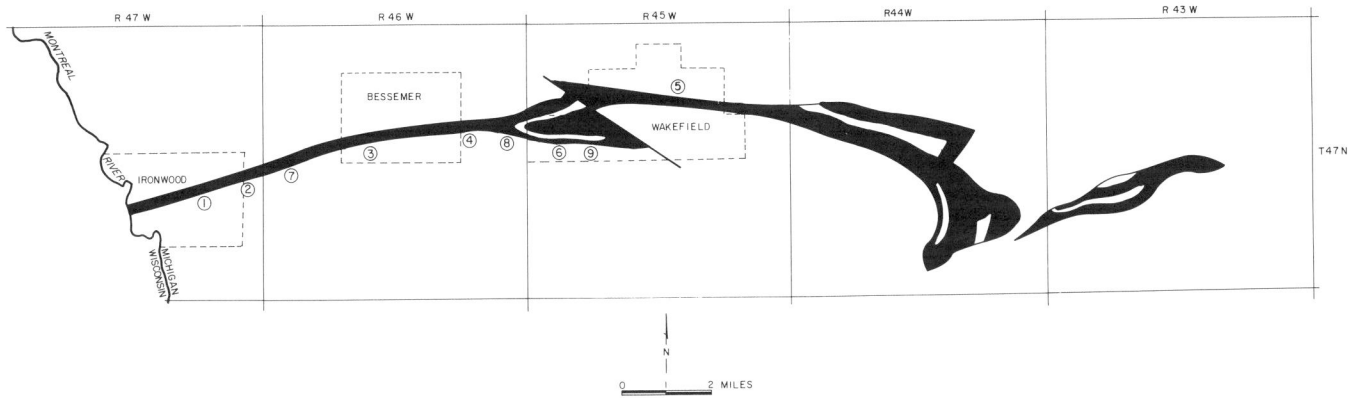

Figure 17. Gogebic Iron Range indicating location of iron formation and major producing mines. Major producing mines are located on Table 13.

tion of principal primary iron minerals; the carbonate, magnetite and silicates to a residuum, essentially hematite, and dissolved and removed silica and carbonate. Ore-forming structures developed at intersections of pitching dikes and iron formation with a steeply dipping, less permeable footwall, whether it be the base of the Plymouth Member, the top of a slaty member, other less permeable horizons, or bedding fault. Such structures formed pitching troughs within which iron ore was naturally concentrated and subsequently mined. Although not numerous, the loci of eastward- and westward-pitching dikes were particularly suited for amassing large orebodies. The physical condition of iron formation beds was also a factor in ore localization. Those beds that contained a greater amount of primary iron and were more porous resulted in concentration of greater amounts of iron (Hotchkiss, 1919; Royce, 1938; Schmidt, 1980).

*Production.* Production data from individual iron formation members are not available for Michigan. However, Hotchkiss (1919) provides estimates for production rankings from adjacent Wisconsin mines as follows:

1. Lower part of Plymouth Member.
2. Upper part of Plymouth Member.
3. In the horizon of the "great bedding fault" within the Yale Member or basal Norrie Member.
4. Top of Norrie Member and base of Pence Member.
5. Within the Anvil Member.

Appendix A lists annual production from the Gogebic Iron Range. Table 13 indicates those mines whose total production exceeded 10 million tons and are located on Figure 17. The nine largest mines produced 95 percent of total production of which 87 percent was from underground mining operations.

*Iron River–Crystal Falls Iron District.* The Paint River Group (Figs. 12 and 18) outlines the core of the Iron River–Crystal Falls District. It contains the Riverton Iron Formation (Fig. 18), the youngest banded iron formation of Precambrian age in Michigan. This district also includes production from the Mansfield Iron-bearing Slate Member of the Hemlock Formation east of Crystal Falls and the Amasa Formation, sandwiched between the Hemlock Formation and the overlying Michigamme Formation around the Amasa Oval (Fig. 10). The Mansfield and Amasa are part of the Baraga Group (Fig. 19).

*History.* The discovery of iron in this district is credited to a U.S. land surveyor, Harry Mellen, who in 1851 discovered an outcrop of Riverton Iron Formation on the west side of Stambaugh Hill at what is now the city of Stambaugh (Allen, 1910). However, it was not until 1881, 30 years later, that production began. In 1882, the C&NW completed railroads into the Iron River and Crystal Falls areas and shipments began from six mines in the Crystal Falls area and two mines from the Iron River area (James and others, 1968).

TABLE 13. INDIVIDUAL MINE PRODUCTION EXCEEDING TEN MILLION LONG TONS, GOGEBIC IRON RANGE*

| Mine | Production Million Long Tons | Type of Operation | Years of Shipments |
|---|---|---|---|
| 1. Penokee Group | 64.6 | DS U | 1885–1962 |
| 2. Newport-Bonnie | 36.7 | DS U | 1886–1961 |
| 3. Peterson Group | 36.1 | DS U | 1884–1967 |
| 4. Anvil-Palms-Keweenaw | 22.0 | DS U | 1887–1957 |
| 5. Sunday Lake | 21.2 | DS U | 1885–1961 |
| 6. Plymouth | 18.0 | DS OP | 1895–1952 |
| 7. Geneva-Davis | 16.3 | DS U | 1887–1905, 1913–1961 |
| 8. Eureka-Asteroid | 14.5 | DS U | 1890–1896, 1906–1950 |
| 9. Wakefield | 13.9 | DS OP | 1913–1944, 1956–1958 |
| Total tons | 243.3 | | |

*Located on Figure 17.
DS = Direct-Shippng Ore; OP = Open-Pit Operation; U = Underground Operation.

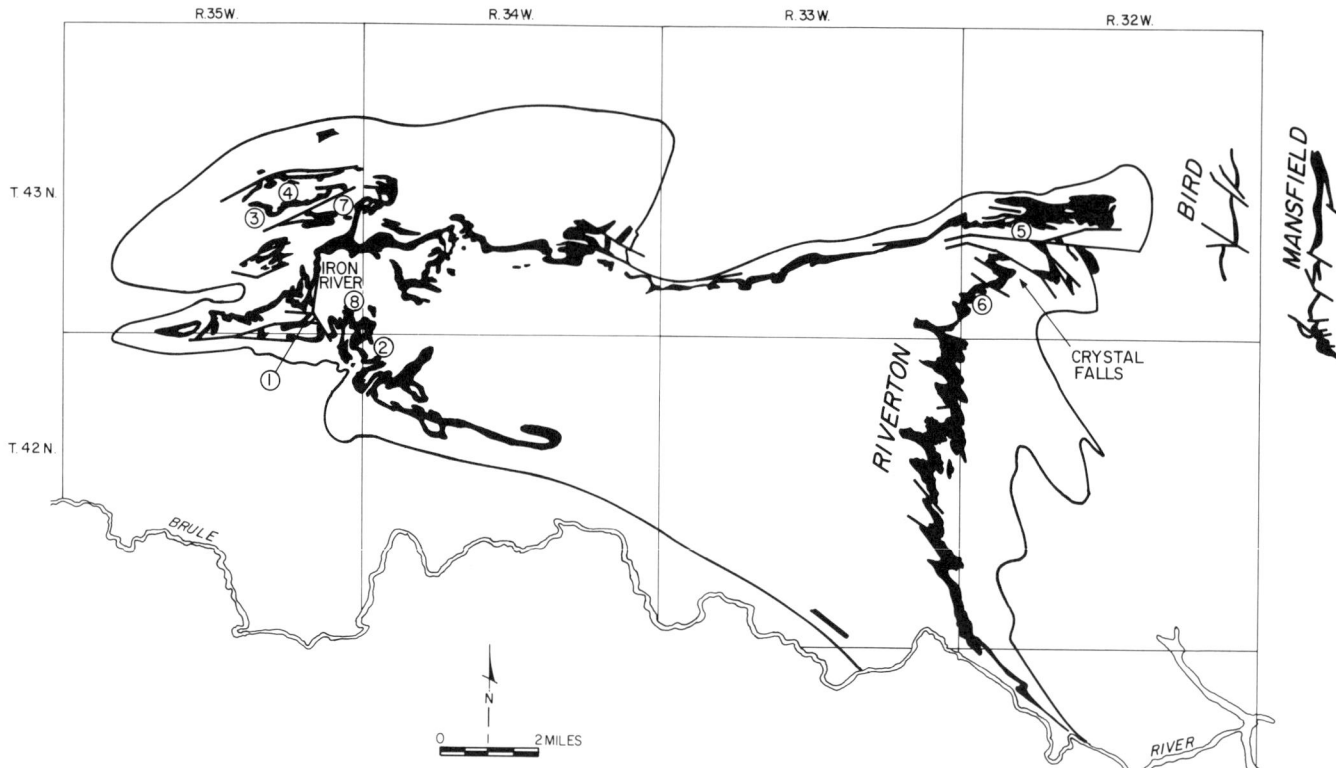

Figure 18. Iron River–Crystal Falls Iron District indicating location of iron formations and major producing mines. Major producing mines are located on Table 16.

|  | Maximum Thickness | |
|---|---:|---:|
|  | (m) | (ft) |
| **EARLY PROTEROZOIC** | | |
| Paint River Group | | |
|   Fortune Lakes Slates | 1,220 | 4,000 |
|   Stambaugh Formation | 30 | 100 |
|   Hiawatha Graywacke | 150 | 500 |
|   Riverton Iron Formation | 245 | 800 |
|   Dunn Creek Slate | 460 | 1,500 |
|     Wauseca Pyritic Member | 15 | 50 |
| Baraga Group | | |
|   Badwater Greenstone | 4,570 | 15,000 |
|   Michigamme Formation | 1,830 | 6,000 |
|   Amasa Formation | 550 | 1,800 |
|   Hemlock Formation | 1,830 | 6,000 |
|     Bird Iron Formation Member | 60 | 200 |
|     Mansfield Iron-bearing Slate Member | 150 | 500 |
|   Goodrich Quartzite | 150 | 500 |

Figure 19. Stratigraphic column of Iron River–Crystal Falls District.

*Geology.* The sequence of layered rock formations within the broad Iron River–Crystal Falls District is given in Figure 19. Within the Baraga Group (Fig. 10), the Mansfield Iron-bearing Slate Member of the Hemlock Formation (Bayley, 1959) is found. It consists of a lower, thick-bedded, pyritic, gray slate and an upper chert-carbonate iron formation. The member has a maximum thickness of about 150 m (500 ft) and averages 60 m (200 ft). The upper chert carbonate ranges in thickness from 10 to 45 m (30 to 150 ft). Along its eastern edge, the Mansfield is at or near the contact of the West Kiernan Sill, which intrudes the Hemlock Formation. At or near the contact, the iron minerals are grunerite and magnetite; farther west, iron is represented by stilpnomelane and magnetite. The iron formation is variously oxidized at or near the surface for the 4 mi of its length (Fig. 18). Production is indicated on Table 14.

The Bird Iron-bearing Member of the Hemlock Formation (Figs. 18 and 19) occurs 365 to 425 m (1,200 to 1,400 ft) below the top of the Hemlock. The member has been described as chert interbedded with magnetite and hard blue hematite and is in part ferruginous slate and quartzite. Oolitic granules of specular hematite occur in the chert (James and others, 1968). The Bird has a maximum thickness of about 60 m (200 ft) and is about 4.8 km (3 mi) long. There is no recorded production from the Bird Iron-bearing Member.

Overlying the Hemlock Formation is the Amasa Formation, some 550 m (1,800 ft) thick. The Amasa, located east of Crystal

Falls, trends northwest and wraps around the west side of the Amasa Oval (Figs. 10 and 11). It is believed to be correlative with the Fence River Formation on the east side of the Amasa Oval. The Amasa contains a lower slate overlain by iron formation. In some places, there are two iron formations separated by amygdaloidal greenstone and slate. Production from the Amasa is indicated on Table 14.

Sedimentary rocks of the Paint River Group occupy a triangular-shaped basin (Fig. 12) that has been intensely deformed and squeezed into tight complex folds (Fig. 18). Faulting is also prevalent; many have displacements of thousands of feet (James and others, 1968). Although highly deformed structurally, the group is only slightly metamorphosed to the biotite grade. James (1951) reported that the entire sequence of the Paint River Group has an average iron content of 20 percent, believed to represent an "era of iron-rich sedimentation."

The Riverton Iron Formation (Fig. 18) extends about 31 km (19 mi) east of Iron River to Crystal Falls, 26 km (16 mi) south to Florence, Wisconsin, and 40 km (25 mi) northwest to Iron River. The iron-formation basin is roughly estimated to cover an area of about 260 km$^2$ (100 mi$^2$). The Riverton is estimated to be about 45 to 90 m (150 to 300 ft) thick in the Iron River area, increasing to 150 to 240 m (500 to 800 ft) in the Crystal Falls area (James and others, 1968). Intense folding, however, has modified the apparent thickness and, in places, post-Riverton erosion has reduced thickness to less than 3 m (10 ft).

The primary or unoxidized iron formation of the district consists of alternating layers of chert and siderite. Iron content ranges from 10 to 40 percent and averages about 25 percent. The estimated mineralogical composition is given in Table 15. In unoxidized iron formation, manganese content was considered to be about 7 percent of the iron content, indicating a significantly high manganese content of the original chemical sediment.

Oxidation of the primary iron formation extends to depths greater than 760 m (2,500 ft). Extreme oxidation and removal of silica formed naturally concentrated, direct-shipping orebodies in the keel and on the flanks of synclinal structures. The goethite-hematite ore is a hard, competent rock that would not readily cave and is ideally suited for the sub-level stoping method of mining.

One detrimental feature of the orebodies was that they had a relatively high phosphorus content, which resulted in a selling-price penalty. Phosphorus imparts brittleness in iron and steel products. Phosphorus content of ore shipped normally ranged between 0.1 and 0.3 percent. Some intersections of ore contained several percent phosphorus and were not mined.

Below the Riverton Iron Formation is the Dunn Creek Slate, the upper 6 to 15 m (20 to 50 ft) of which is a highly sheared, graphitic, pyritic slate, the Wauseca Pyritic Member (Fig. 19). The Wauseca contains about 8 to 15 percent carbon and 30 to 45 percent fine-grained pyrite (James, 1951; James and others, 1968). Dutton (1971) reports carbon contents of 15.9 and 29.4 percent from the Wauseca Pyritic Member in the Florence area of Wisconsin. The basal part of the Wauseca is a breccia, the so-called "speckled gray," a district-wide marker bed. The pyritic slate oxidizes when exposed, the source of "fires" in the mines, portions of which had to be sealed. The oxidized pyrite also resulted in acid mine waters, prevalent throughout the district.

The Stambaugh Formation (Fig. 19) is a magnetic slate about 30 m (100 ft) thick. It is principally composed of magnetite, siderite, chlorite, and chert containing about 20 percent iron. The Stambaugh was preserved as local basins in synclinal structures and, as the only magnetic formation of the Paint River Group, was important in delineating structures of the area through ground magnetometer surveys.

*Production.* Appendix A lists mine shipments from the Menominee Range, which includes the Iron River–Crystal Falls District. Table 14 provides total production from formations of the district. Table 16 indicates major producing mines of the district, which are located on Figure 18. The only concentrated

TABLE 14. IRON-ORE PRODUCTION BY FORMATION, IRON RIVER–CRYSTAL FALLS DISTRICT

| Formation | Number of Mines | Production (long tons) | Percent Production |
|---|---|---|---|
| Riverton Iron Formation | | | |
| Iron River Area | 26 | 148,198,575 | 71.6 |
| Crystal Falls Area | 21 | 50,026,588 | 24.2 |
| Total | | 198,225,163 | 95.8 |
| Amasa Formation | 9 | 7,238,780 | 3.5 |
| Mansfield Iron-Bearing Slate Member of Hemlock Formation | 1 | 1,462,504 | 0.17 |
| Total | 57 | 206,926,447 | |

TABLE 15. AVERAGE MINERALOGICAL COMPOSITION, RIVERTON IRON FORMATION, IRON RIVER–CRYSTAL FALLS DISTRICT*

| Carbonate | Percent | Weight Percent |
|---|---|---|
| FeCO$_3$ | 81.0 | |
| MgCO$_3$ | 9.5 | |
| MnCO$_3$ | 4.9 | |
| CaCO$_3$ | 4.6 | 62.4 |
| Chert | | 32.0 |
| Graphite | | 1.8 |
| Miscellaneous– phosphate, chlorite, iron oxide | | 3.8 |

*James, 1951.

TABLE 16. INDIVIDUAL MINE PRODUCTION EXCEEDING TEN MILLION LONG TONS, IRON RIVER–CRYSTAL FALLS DISTRICT*

| Mine | Production Million Long Tons | Type of Operation | | Years of Shipments |
|---|---|---|---|---|
| 1. Hiawatha | 22.2 | DS | U | 1893–1967 |
| 2. Buck Group | 21.7 | DS | U | 1901–1962 |
| 3. Homer-Cardiff-Minckler | 17.5 | DS | U | 1915–1971 |
| 4. Wauseca-Aronson | 15.4 | DS | U | 1926–1929, 1940–1972 |
| 5. Bristol-Youngstown | 14.8 | DS | U | 1890–1934, 1950–1969 |
| 6. Sherwood | 13.7 | DS | U | 1931–1979 |
| 7. Tobin-Columbia-Monongahela | 13.5 | DS | U | 1882–1963 |
| 8. Cannon | 12.0 | DS | U | 1910–1963 |
| Total tons | 130.8 | | | |

*Located on Figure 18.
DS = Direct-Shipping Ore; U = Underground Operation.

(beneficiated) iron product from the Iron River–Crystal Falls District was from the Book mine located south of Crystal Falls. Between 1954 and 1958, 428,233 tons of concentrate were produced.

## MIDDLE PROTEROZOIC GEOLOGY

### General geology

The Michigan copper district is located on the southern rim of the Lake Superior basin (Fig. 20). Michigan's sulfide (chalcocite) and native copper deposits are enclosed in Keweenawan rocks within a portion of what is now known to be a large crustal rift through the north-central United States. Originally named the Mid-continent Gravity High (King and Zietz, 1971), the rift is now referred to as the Mid-continent Rift System (MRS). Based on gravity and magnetic data, it has been inferred that mafic rocks of the Keweenawan rift zone extend from eastern Lake Superior southeast through the Michigan Basin and that this structure may have influenced the development of the basin (Hinze and Merritt, 1969; Fowler and Kuenzi, 1978; Catacosinos, 1981; Dickas, 1986).

Michigan copper was produced from two formations: native copper was mined from the Portage Lake Volcanics (Figs. 21 and 22) and sulfide copper, principally chalcocite with native copper, from and immediately underlying the Nonesuch Formation (Figs. 21, 23, and 24). Native copper deposits from the Keweenaw Peninsula are unique among the world's copper-producing districts.

The Portage Lake Volcanics (PLV) is a thick series of lava flows. The flows include basalts, olivine basalts, and basaltic andesites. They contain interflow sediments consisting of conglomerates, sandstones, and siltstones of intermediate to felsic composition. Maximum thickness is as much as 5,600 m (18,400 ft) (Broderick and others, 1946). There are at least 200 individual flows. The thickest, the Greenstone, is at least 490 m (1,600 ft) and is conservatively estimated to contain 200 mi$^3$ of lava. It has been estimated that 37 percent of the total volume of lava is in individual flows more than 30 m (100 ft) thick, and 12 percent in flows more than 60 m (200 ft) thick (White, 1960). The uppermost 5 to 20 percent of most flows contains 5 to 50 percent visicles, now filled with secondary minerals. In 21 percent of the flows, the crust was brecciated. There are 21 numbered sedimentary beds between certain flows. Some lava flows and sediments can be traced for distances as great as 145 km (90 mi) (White, 1968). Radiometric dating of the Portage Lake indicates volcanic activity between 1,098 and 1,086 Ma (Bornhorst and others, 1988).

The Portage Lake Volcanics is overlain by the Oak Bluff Formation (formerly unnamed formation; Figs. 21, 22, and 23). The Oak Bluff differs from the Portage Lake Volcanics in that the Oak Bluff flows are more felsic in composition. The Oak Bluff consists of basalts, andesites, felsites, and quartz-feldspar porphyries. They are less ophitic and more porphyritic and brecciated. Some flow tops are scoriaceous or fragmental. The Oak Bluff contains interflow conglomerates, sandstones, and siltstones similar to those within the underlying Portage Lake Volcanics and the overlying Copper Harbor Conglomerate. Within the central area south of White Pine (Figs. 22 and 23), the Oak Bluff is about 2,440 m (8,000 ft) thick. The thickness diminishes east and west, terminating to the northeast in the Mass-Greenland area and west near the Wisconsin border (Johnson and White, 1969; Whitlow, 1974; Hubbard, 1975a).

The succeeding formations, the Copper Harbor Conglomerate, Nonesuch Formation, and Freda Formation (Figs. 21 and 23), are predominantly sediments. Daniels (1982) emphasizes the color and oxidation state of these formations. The Nonesuch is dark and unoxidized, while the surrounding Copper Harbor and Freda are reddish brown and oxidized. Additionally, the Nonesuch Formation exhibits greater textural maturity, contains sulfides and hydrocarbons, and is locally enriched in chlorite.

In the Keweenaw Peninsula, the Copper Harbor Conglomerate is described as a pebble-to-boulder conglomerate with some beds of sandstone. It is composed predominantly of rhyolite or rhyolite porphyry with some fragments of amygdaloidal or massive basalt (Cornwall, 1955). The maximum reported thickness is about 1,830 m (6,000 ft) (Broderick and others, 1946). In the Porcupine Mountain area (Fig. 25), the Copper Harbor Conglomerate is mostly sandstone and siltstone with subordinate conglomerate. In the northern part, thickness is about 1,525 m (5,000 ft), which thins southward to about 155 m (500 ft) (Hubbard, 1975b). South of White Pine (Fig. 23), conglomerate is predominant near the base, and the amount of sandstone increases toward the top. Thickness ranges from 1,830 m (6,000 ft) in the west to 107 m (350 ft) in the east.

Thinning of the Copper Harbor occurs above thickened

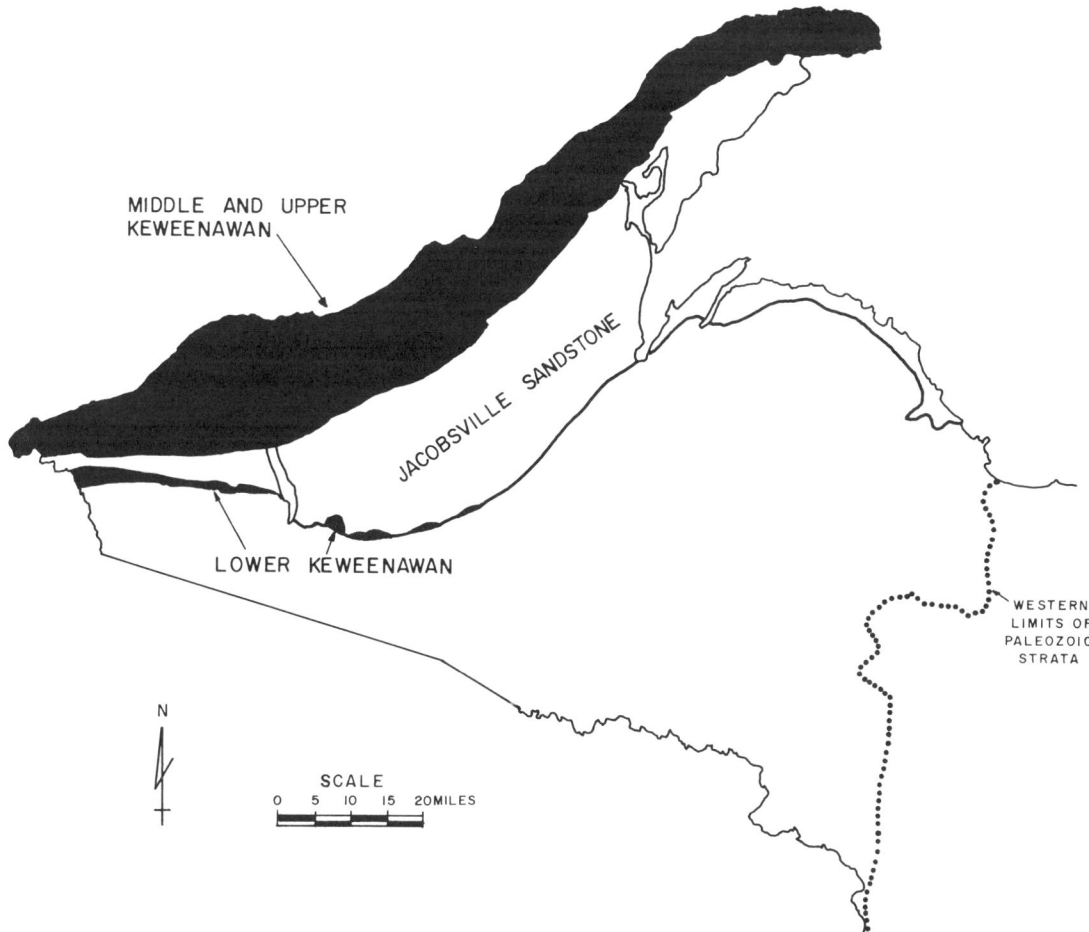

Figure 20. Surface distribution of middle Proterozoic volcanics and sediments.

portions of the Oak Bluff Formation (Hubbard, 1975a; Johnson and White, 1969; White and Wright, 1960). Daniels (1982) compiled data indicating that, for the most part, conglomerate clasts are from mafic volcanics. In the sandstone, rock fragments predominate; the remainder is largely quartz and opaque minerals. In most areas reported, the amount of mafic rock fragments is greater than silicic plus intermediate. The Copper Harbor Conglomerate contains scattered lava flows. In the northern Keweenaw Peninsula, andesite and dacite flows are up to 610 m (2,000 ft) thick (Cornwall, 1955). In the Porcupine Mountains area (Fig. 25), Hubbard (1975b) reports six andesitic and basalt flows interbedded with sediments. Whitlow (1974) reports mafic, rhyolite, and felsite flows in the Copper Harbor Conglomerate in the Mass-Greenland area that are believed to correlate and intertongue with the Oak Bluff Formation.

The Nonesuch Formation (Fig. 24) extends from southeast of Superior, Wisconsin (White, 1978, Fig. 5), northeast to north of Calumet, Michigan, a distance of about 240 km (150 mi). The Nonesuch ranges in thickness from 75 m (250 ft) in the southwest to 215 m (700 ft) to the northeast (White, 1971). It averages about 180 m (600 ft) thick (Daniels, 1982). The Nonesuch Formation is basically a gray-black, thinly laminated siltstone and thin-bedded sandstone with some shale. It is largely composed of fragments of pre-Keweenawan rocks with lesser amounts of quartz, volcanic rocks, feldspar, and opaques. Cementing materials include calcite, silica, laumontite, and hematite. To the northeast, toward Lake Superior, there is an increasing percentage of sand in the basal 45 m (150 ft) of the formation (Hubbard, 1975a; Daniels, 1982).

|   | Maximum Thickness (m) | (ft) | Percent |
|---|---|---|---|
| 1. Jacobsville Sandstone | 3,000 | 9,850 | 12 |
| 2. Freda Formation | 3,660 | 12,000 | 15 |
| 3. Nonesuch Formation | 215 | 700 | 1 |
| 4. Copper Harbor Conglomerate | 1,825 | 6,000 | 7 |
| 5. Oak Bluff Formation | 2,440 | 8,000 | 10 |
| 6. Portage Lake Volcanics | 5,605 | 18,400 | 23 |
| 7. Powder Mill Group | 7,440 | 24,400 | 31 |
| 8. Bessemer Quartzite | 90 | 300 | 1 |
| TOTALS | 24,275 | 79,650 |  |
| Sediments (no's. 1, 2, 3, 4, and 8) | 8,790 | 28,850 | 36 |
| Volcanics (no's. 5, 6, and 7) | 15,485 | 50,800 | 64 |

Figure 21. Stratigraphic column of middle Proterozoic sediments and volcanics.

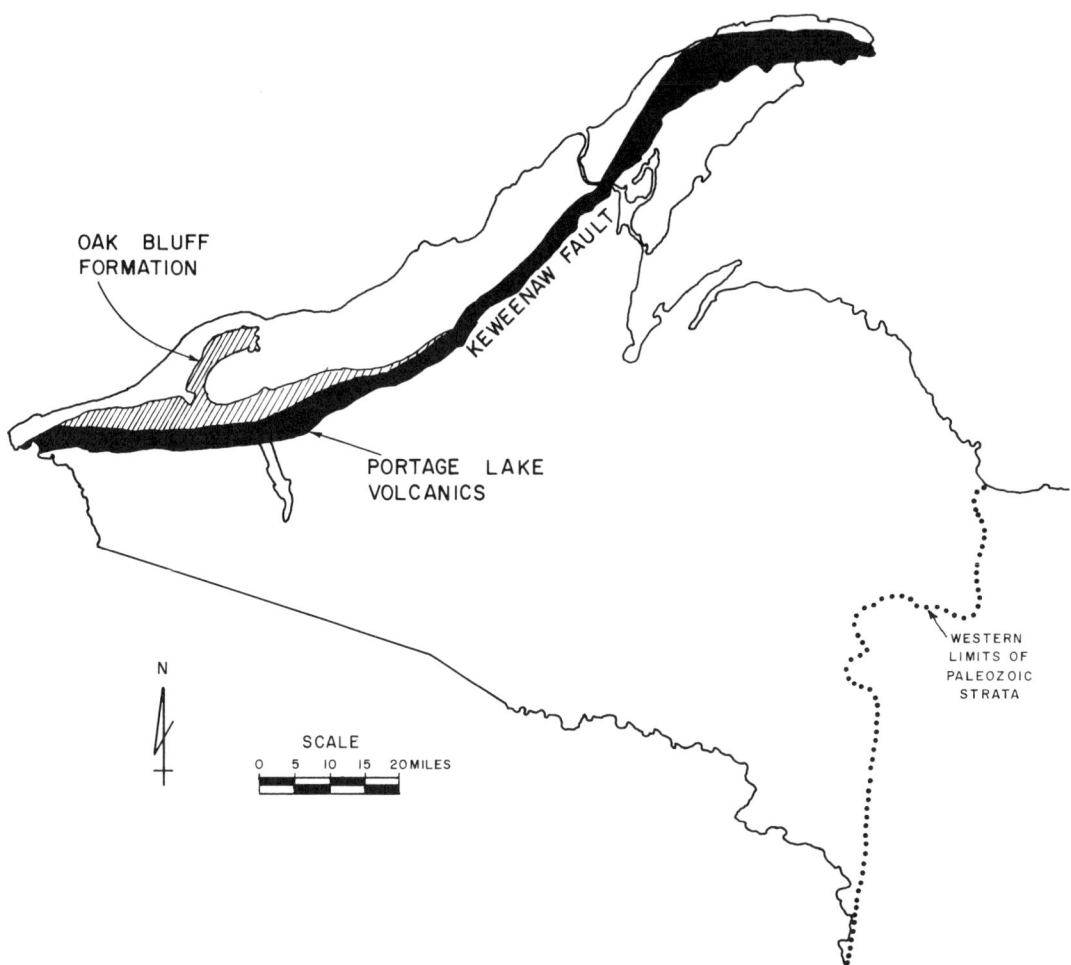

Figure 22. Surface distribution of middle Proterozoic Portage Lake Volcanics and Oak Bluff Formation.

As reported by Hubbard (1975a), the lower 915 m (3,000 ft) of the Freda Formation (Fig. 23) southwest of White Pine is 90 percent sandstone, with siltstone increasing upward from 5 percent to 25 percent. The middle 1,830 m (6,500 ft) is mostly siltstone with some sandstone, and the upper 760 m (2,500 ft) is sandstone. The Freda averages about 53 percent rock fragments, mostly felsic volcanic rock; 35 percent quartz; and 12 percent feldspar. Cementing materials are quartz, calcite, laumontite, hematite, and chlorite. The formation contains minor amounts of shale and conglomerate (Daniels, 1982).

Keweenawan rock structure is, for the most part, quite simple (Fig. 20). The rocks are inclined, at varying dips, toward Lake Superior. In general, dips flatten toward the lake. Large-amplitude anticlines and synclines pitch downdip on the larger southern synclinal limb. Perhaps the major structural feature is the Keweenaw fault (Fig. 22), which forms the south and southeast boundary of the Portage Lake Volcanics. Faults branch from the Keweenaw fault, two of which are indicated on Figure 26. Additionally, numerous transverse faults cross the formations, as indicated on Figure 23. Transverse faults were host to fissure copper in the Keweenaw Peninsula. A distinctive, prominent structural bulge interrupts the generally homoclinal layered rock sequence in Ontonagon County as the Porcupine Mountains (Figs. 22, 23, and 25). Complimentary synclines as a result of this flexure are located on Figure 23. Another folded structure, not shown on the accompanying maps, is located at Mass, south of the termination of the Oak Bluff Formation (Fig. 23). The structure is about 6 km (4 mi) long and called the Mass anticline and Lake mine syncline (Butler and Burbank, 1929, plate 13).

*Native copper. History.* Elemental copper played a major role in Michigan's early industrial history. This native metal, associated with minor amounts of native silver, was highly prized, even revered, by Indian tribes who eagerly displayed specimens to early explorers. Masses of "float copper" were common in extensive deposits of glacial drift of the Keweenaw Peninsula, some of which were transported as far south as Illinois, Indiana, and Ohio. One mass weighing 63 kg (140 lb) was found near Green Bay, Wisconsin. The largest piece of float copper, weighing 16 metric tons (18 short tons), was located near Hancock, Michigan (Drier and Du Temple, 1961). Prior to 1844, the existence of native copper from bedrock in this area was unknown to the new explorers.

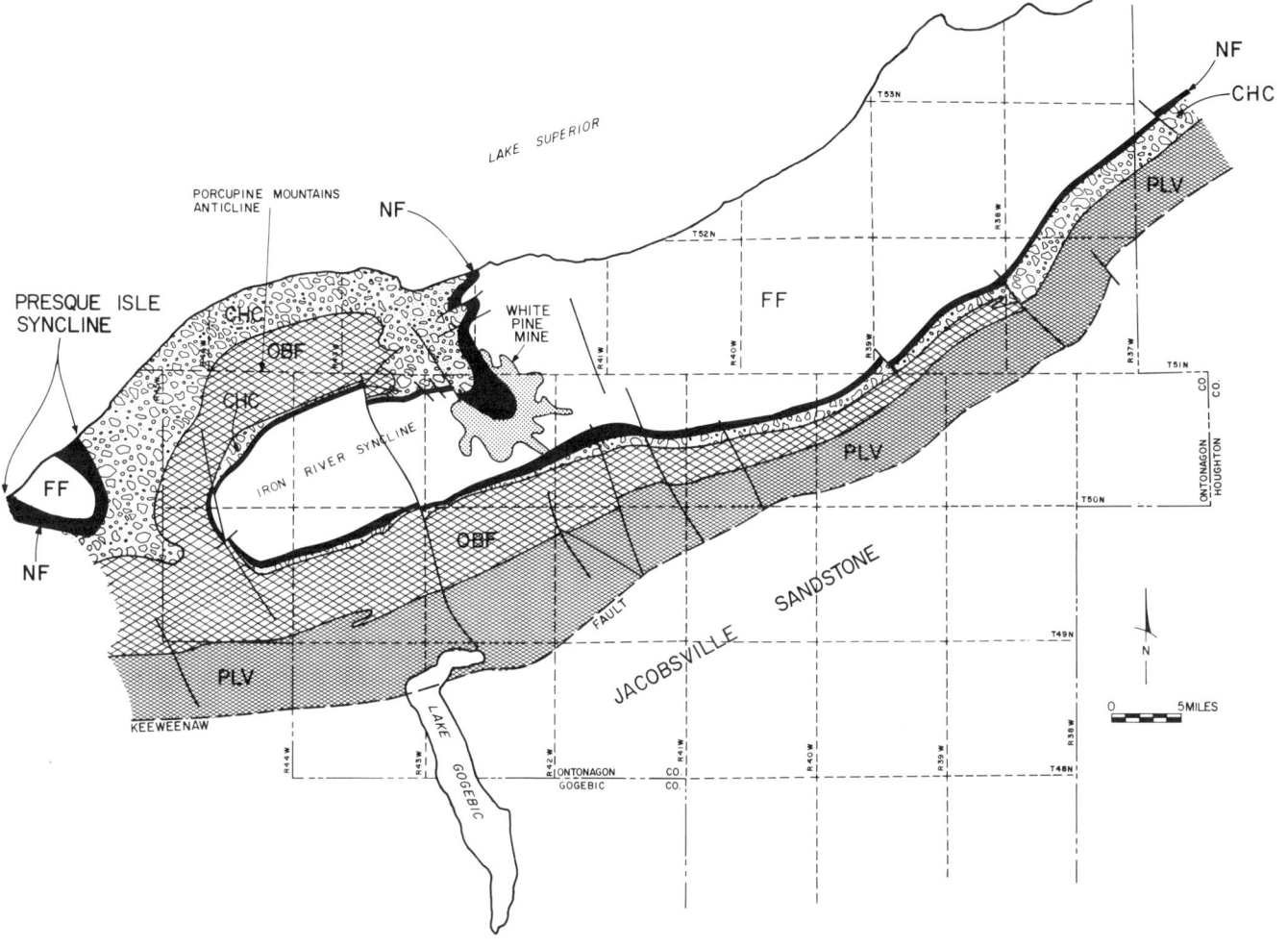

Figure 23. Geologic map of the White Pine area.

| | |
|---|---|
| Freda Formation | FF |
| Nonesuch Formation | NF |
| Copper Harbor conglomerate | CHC |
| Oak Bluff Formation | OBF |
| Portage Lake Volcanics | PLV |

Native copper in rock formations of the Lake Superior district was known to an early race of people—a vanished prehistoric race. The only remaining evidence of these early miners are the pits they excavated in rock in their search for native copper and their debris remaining in the pits. It has been estimated that more than 5,000 such pits occur around Lake Superior, principally on the Keweenaw Peninsula and Isle Royale (Drier and Du Temple, 1961). The pits were 6 to 9 m (20 to 30 ft) in diameter and 6 to 18 m (20 to 60 ft) deep (Griffin, 1961). Rock containing native copper was broken by fire and water and with crude stone hammers. Ancient pits contained numerous grooved and ungrooved mauls from local glacial boulders, weighing as much as 18 kg (40 lb). Charcoal, copper implements, and remains of wooden shovels and pans were found in the pits (Drier, 1964).

The antiquity of ancient mining of native copper was revealed by the Michigan College of Mining and Technology (now Michigan Technological University) through 1953 and 1954 expeditions investigating prehistoric copper mining on Isle Royale. Radiocarbon dating of wood and charcoal recovered from the pits ranged from 3,000 ± 350 B.P. to 3,800 ± 500 B.P.

The toil and perseverance of these ancient people was truly remarkable. Masses of copper remaining in pits on Isle Royale weighing 2,595, 1,939, and 1,505 kg (5,720, 4,175, and 3,317 lb) have been reported (Griffin, 1961). In several pit operations on the Keweenaw Peninsula, large masses of native copper were found partially exposed. At the Central mine, a mineralized fissure not far from Eagle Harbor, a 42-metric-ton (46 short ton) mass was found. Farther south, at the Minesota mine in Ontonagon County, examiniation of a pit revealed that a mass weighing 5,256 kg (11,588 lb) had been raised a few feet along the slope of the vein and supported by logs. Every major native copper mine opened in the Lake Superior area, except for the

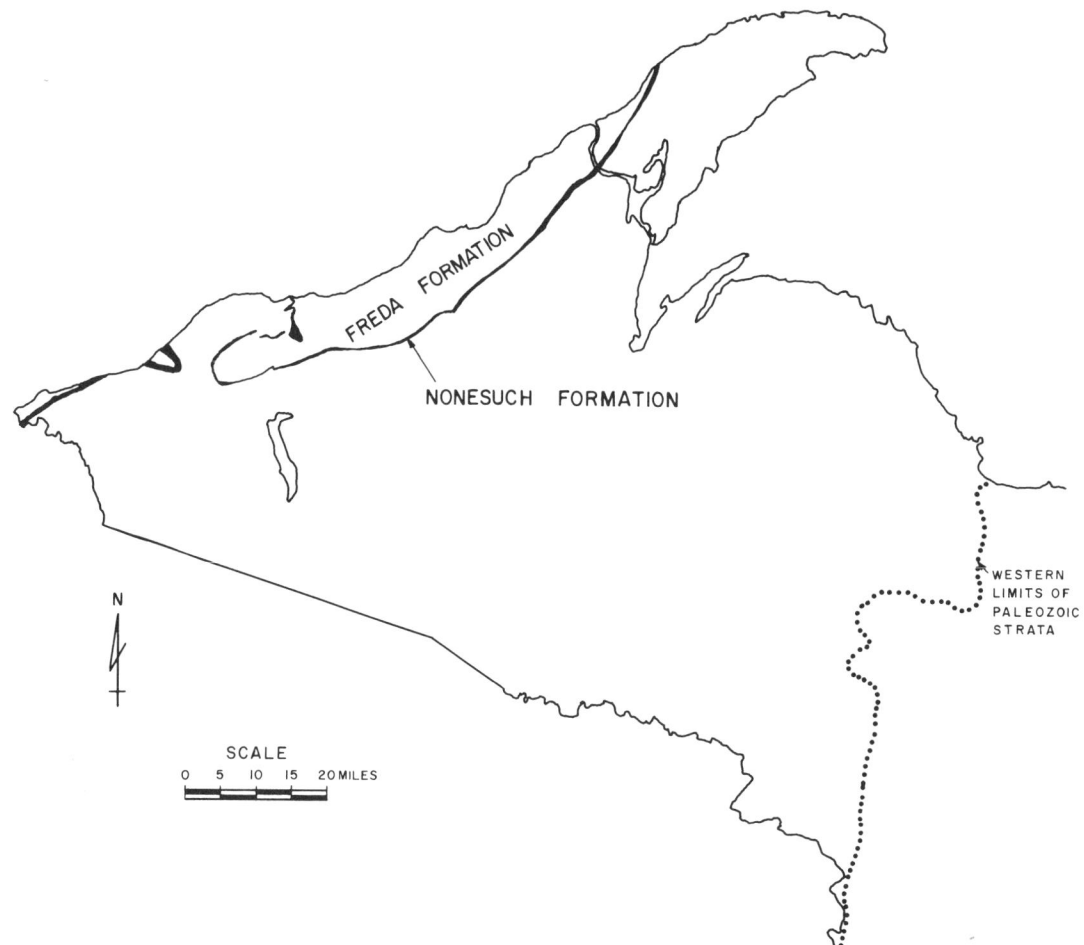

Figure 24. Surface distribution of Middle Proterozoic Nonesuch and Freda Formations.

Kingston, is known to have been worked in prehistoric time (Drier and Du Temple, 1961).

The first published account of the existence of native copper on the shore of Lake Superior was in Paris, France, in 1636. In 1766, Indian guides led Alexander Henry, an English explorer, to a large mass of copper along the Ontonagon River (Fisher, 1929). This specimen later became known as the famed "Ontonagon boulder," which resides at the Smithsonian Institution, Washington, D.C.

Douglass Houghton, Michigan's first state geologist, began the first systematic and scientific exploration for copper in the Keweenaw Peninsula in 1831. His first report (Houghton, 1841) indicated the existence of copper in bedrock. Until this time, masses and boulders from glacial deposits were the only known source of native copper.

The first exploration for and mining of copper was in 1844 from a fissure at what is now Fort Wilkins State Park near Copper Harbor. This ore was a mixture of copper oxides, cuprite, and tenorite, with some native copper (Cornwall, 1955). About 27 metric tons (30 short tons) of copper from the oxides is reported to have been produced. Approximately 1.3 km (0.8 mi) south of the copper oxide vein, manganese ore was mined between 1881 and 1883. About 1,088 metric tons (1,200 short tons) of ore, averaging 53.73 percent manganese and 1.36 percent copper, was shipped (Cornwall, 1955).

Michigan's native copper deposits occur as: (1) stratiform beds or lodes within portions of some amygdaloidal and brecciated tops of lava flows or in interflow conglomerates; (2) cross or strike fissures within the volcanics. Within these horizons and structures, native copper has filled intergranular space and fracture openings and has replaced host rock and earlier-formed secondary minerals (White, 1968).

Four types of amygdaloidal flow tops have been recognized (Butler and Burbank, 1929):

1. Cellular or banded amygdaloid. Flow tops are relatively smooth, highly vesicular, elongated in layers parallel to the surface but not interconnected. They have been likened to Swiss cheese, with high porosity and low permeability. Therefore, while they are the most common flow top, they contain little copper.

2. Coalescing cellular amygdaloid. A variation of the cellular type where vesicles are much larger and consist of long, parallel strings of gas bubbles. These form a series of almost unbroken

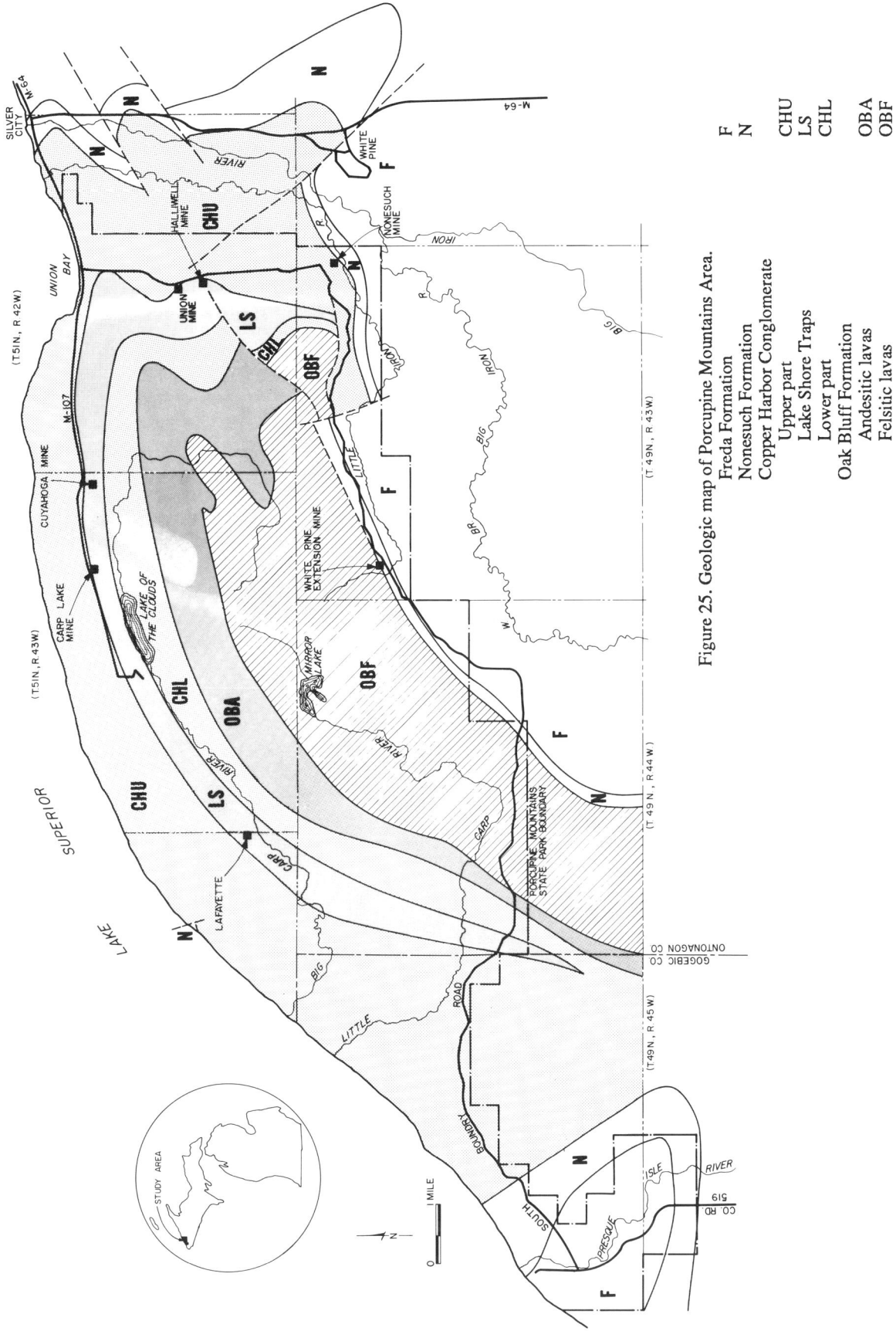

Figure 25. Geologic map of Porcupine Mountains Area.

F    Freda Formation
N    Nonesuch Formation
CHU Copper Harbor Conglomerate
     Upper part
LS    Lake Shore Traps
CHL   Lower part
OBA Oak Bluff Formation Andesitic lavas
OBF Felsitic lavas

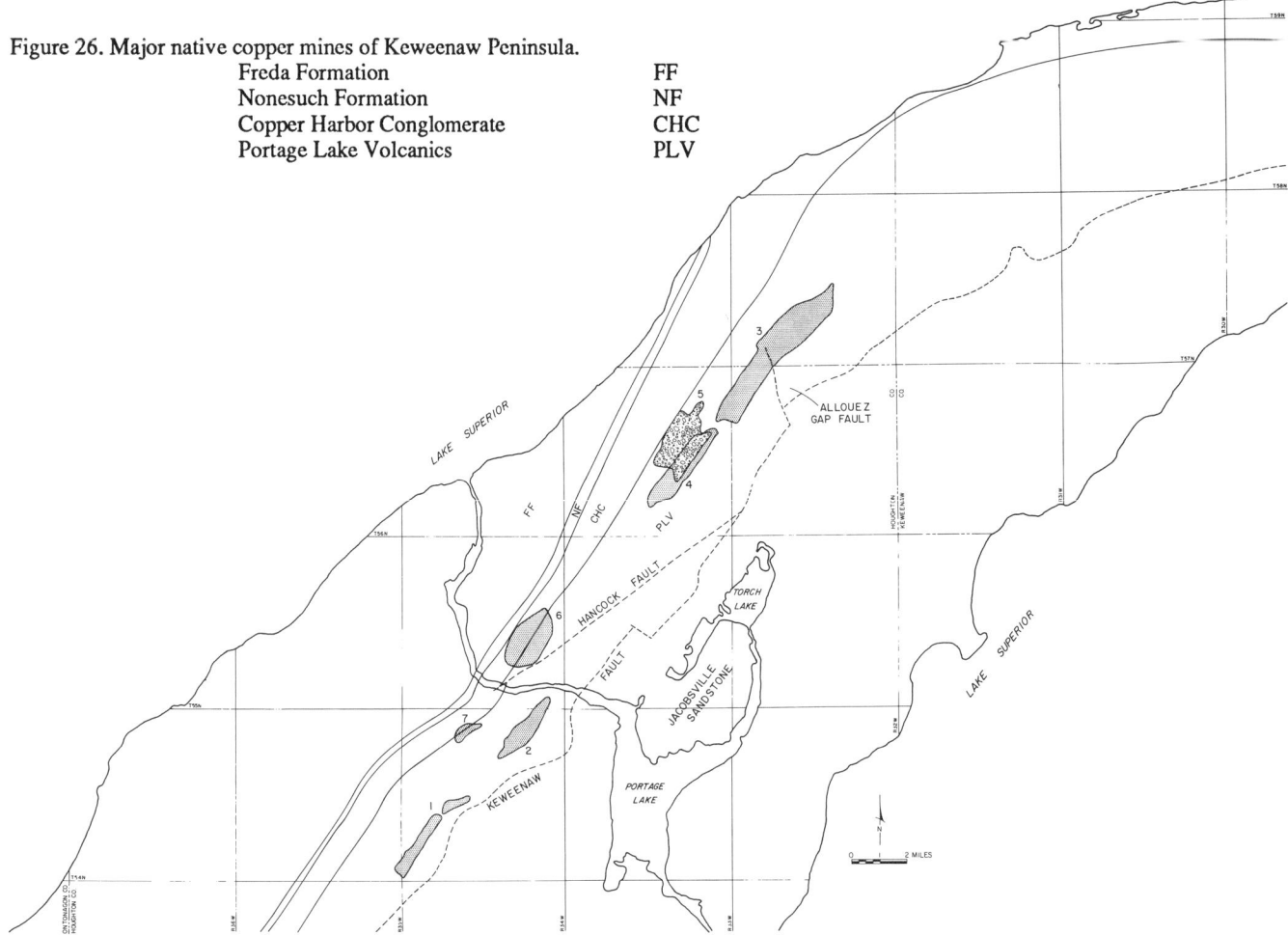

Figure 26. Major native copper mines of Keweenaw Peninsula.
Freda Formation FF
Nonesuch Formation NF
Copper Harbor Conglomerate CHC
Portage Lake Volcanics PLV

openings along the plane of the flow top. The Pewabic lode (Fig. 27) at the Quincy mine (Fig. 26), consisting of a series of thin flows, is representative of coalescing cellular amygdaloids.

3. Brecciated fragmental amygdaloid. Consists of irregular angular to rounded rubble of broken amygdaloid, commonly 15 to 20 cm (6 to 8 in) or smaller, jumbled together. Spaces between fragments contain finer material of similar composition. They are lava flow tops, brecciated during solidification, are highly permeable, and form the most common ore host rock. Copper mineralization of major mines, which produced from the Baltic, Isle Royale, Kearsarge, and Osceola lodes and from parts of the Pewabic lode (Fig. 27), is from the fragmental-type amygdaloid.

4. Scoriaceous amygdaloid. Consists of amygdaloid fragments in a matrix of basic sand grading upward into a fine-grained basic sediment. Fragments are often rounded, and this type is sometimes referred to as an amygdaloid conglomerate. It is believed to be a product of fragmentation, erosion, and sedimentation. The Ashbed amygdaloid (Fig. 27), most productive at the Atlantic mine (Fig. 26), is of this type. The Ashbed flow is toward the andesitic end of basaltic composition and contains clusters of feldspar phenocrysts.

Interflow conglomerates, consisting of consolidated boulders, pebbles, gravel, and sand composed chiefly of felsite and quartz porphyry materials, occur as interlayered sedimentary rock units between individual lava flows. Conglomerates are lenticular, pinching out or represented by thin shaley or sandy sediments, or swell to thicknesses greater than 30 m (100 ft). Conglomerate materials are dark red in color due to inclusions of small primary crystals of hematite (Butler and Burbank, 1929). Interflow conglomerates make up 3 to 8 percent of the Portage Lake Volcanics (White, 1971).

*Production.* The first native copper production (1845) from the Keweenaw Peninsula was from the Cliff fissure, located about 3.2 km (2 mi) southeast of Eagle River. The Phoenix mine, about 4.8 km (3 mi) southwest of Eagle River, began production from a fissure in 1846. Another major fissure mine, the Central, about 6.4 km (4 mi) southwest of Eagle Harbor, began in 1855. These three early fissure mines recorded production of 48 million kg (107 million lb) and paid dividends totaling $4.7 million. The fissures were essentially vertical and at right angles to the trend of the volcanics and sediments (Butler and Burbank, 1929).

The Minesota mine in Ontonagon County, previously referred to in the discussion of ancient mining, began production in 1848. The largest complete mass of native copper ever found was discovered in 1856 in a fissure parallel to the strike of the formations. It had maximum dimensions of 14 × 5.6 × 2.6 m (46 × 18.5 × 8.5 ft) and was reported to weight 453 to 530 metric tons (500 to 585 short tons) (Fisher, 1924).

The major producing mines located on Figure 26 produced from what was called the Portage Lake District, north and south of the Portage Lake Ship Canal. Production was mainly from amygdaloidal tops of lava flows and interflow conglomerates, commonly referred to as lodes. Production began on the Isle Royale amygdaloid (Isle Royale mine) in 1853. Next was the Pewabic amygdaloid (Quincy mine) in 1856, the Ashbed lode (Atlantic mine) and Calumet and Hecla Conglomerate (Calumet and Hecla mine) in 1866, the Kearsarge amygdaloid in 1882, and the Baltic lode in 1898 (Butler and Burbank, 1929). The stratigraphic horizon and total copper production from the principal lodes are illustrated on Figure 27.

Annual production of native copper from Michigan mines is listed on Appendix B. Of total production through 1987, 94.4 percent was from seven lodes indicated on Figure 27. Amygdaloidal horizons contributed 55.8 percent, conglomerates 42.2 percent, and fissures 2.0 percent. Production from the Calumet and Hecla Conglomerate amounted to 40.1 percent.

*Nonesuch Formation. History.* Jamison (1950) reports that prospectors were in the Porcupine Mountain area as early as 1844, and by 1845 there were two mining operations. Several early mines are located on Figure 25. These operations were located at or near the contact of the upper part of the Lake Shore Traps and the adjoining Copper Harbor Conglomerate, which in this area is a sandstone. Copper was in the form of the native metal. The only recorded production from Lake Shore Trap–related mines was 13.6 metric tons (15 short tons) from the Carp Lake mine (Fig. 25), from 1860 through 1920 (Jamison, 1950). Jamison (1950) also lists 27 mining corporations of the 1870s, located between Silver City and White Pine (Fig. 25). Spectacular discoveries of silver were reported, but no production. At least 53 mining location permits were granted in the Porcupine Mountain area (Jamison, 1950).

Copper-bearing rock of the Nonesuch Formation was discovered at the Nonesuch mine (Fig. 25), located about 5 km (3 mi) west of White Pine, in 1865. The Nonesuch Mining Company was organized in 1867 and produced from 1868 through 1885 (Appendix B) from native copper-bearing sandstone (Butler and Burbank, 1929).

The White Pine mine (Fig. 23) was first opened in 1881 from a shallow shaft. The White Pine Copper Company was organized in 1909 and operated as a subsidiary of the Calumet and Hecla Mining Company. In 1929, Copper Range Company purchased the White Pine mine. Exploration began in 1937 with metallurgical testing of White Pine rock conducted by Michigan Technological University, American Cyanamid Company, and Copper Range (Leone and others, 1971). Chalcocite was recognized in early drilling and mining; however, the chalcocite was

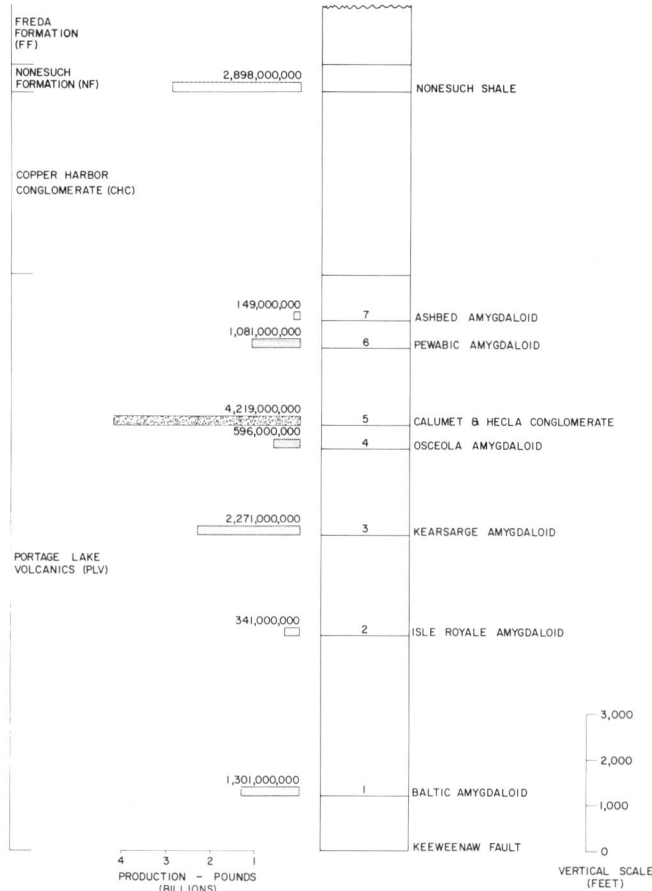

Figure 27. Stratigraphic column indicating location of major producing lodes and lode production.

too fine grained for recovery with then-existing metallurgical methods. Only the native copper was recovered in early operations. By 1946, metallurgical research indicated that 85 percent of the copper could be recovered from the Nonesuch (Leone and others, 1971).

In 1914, exploration on another Nonesuch Formation mining venture began at the White Pine Extension mine, located 13 to 15 km (8 to 9 mi) west of White Pine (Fig. 25). No production was recorded (Jamison, 1950; Butler and Burbank, 1929).

*Geology.* Copper mineralization is essentially restricted to the uppermost fraction of the Copper Harbor Conglomerate and lowermost Nonesuch Formation (Fig. 28). Mineralization is widespread. White and Wright (1966) state that the dark shale laminae of the lowermost inch or two (Nonesuch Formation) contains copper-bearing sulfides over many tens and possibly hundreds of square miles. They also point out that the top of the cupriferous zone varies between 0.3 and 15 m (1 and 50 ft) in thickness above the base. This zone varies in position and rock type on a regional scale. The upper limit of copper mineralization (fringe zone) contains a variety of sulfide minerals, including chalcocite, bornite, chalcopyrite, pyrite, and greenockite. So far as is known, however, only the basal 6 m (20 ft) of the ore horizon in the White Pine area and the basal Nonesuch in the Presque Isle basin (Fig. 23) can be considered for mining. The average copper distribution in the mineralized portion of the Nonesuch Formation in the White Pine area is indicated on Figure 28.

The Lower Sandstone (Fig. 28) is the uppermost portion of the Copper Harbor Conglomerate immediately below the Nonesuch Formation. This green to greenish gray sandstone is composed of detrital quartz, acidic and basic rock fragments, and feldspar. The upper 0.3 to 6 m (1 to 20 ft) of the sandstone contains interstitial chlorite and, below the chloritic zone, the sandstone is hematitic. There is a close spatial relation between native copper and carbonaceous material. In some areas, the top 1 to 2 m (3 to 7 ft) contains native copper that rims or cuts carbonaceous grains. Although the content of carbonaceous material is greater, native copper (and native silver) increases as the amount of carbonaceous material increases. Native copper and carbonaceous material also occur in chloritic lenses within the lower hematitic facies (Hamilton, 1965).

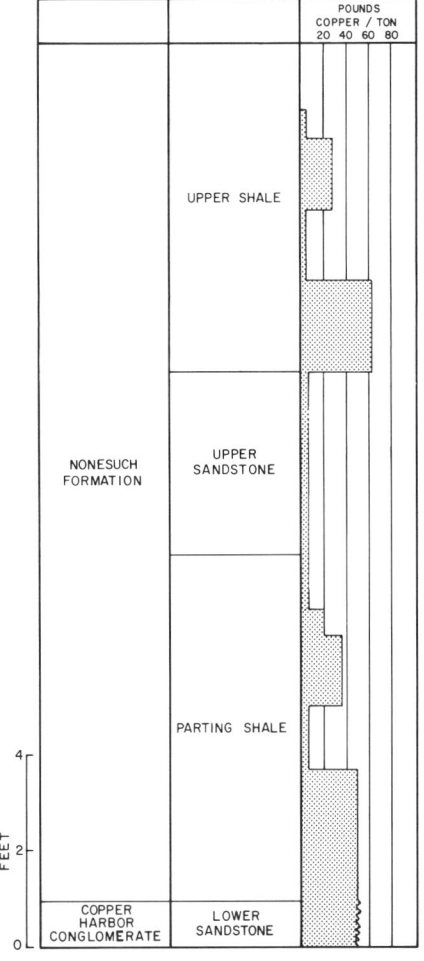

Figure 28. Stratigraphic column of ore-bearing horizon on Nonesuch Formation. Generalized from data contributed by the Copper Range Company.

Figure 28 and the following description are generalized from data contributed by the Copper Range Company. Lithology of the Parting Shale and Upper Shale is an almost mirror-like repeated sequence, strikingly similar in both units. With reference to Figure 28, the basal portions of the Parting and Upper Shales with appreciable copper content are fine-grained gray sandstones with black shale partings followed by thinly laminated, gray siltstones and black shales. Above these, the copper content diminishes in the massive red to gray or red-brown to gray siltstones. Copper content then increases in the overlying massive, dark gray siltstones. The succeeding copper-poor sediments are widely laminated, gray siltstones and black shales. The uppermost Parting Shale consist of a widely laminated, gray siltstone and red shale.

The Parting and Upper Shales are mostly siltstones composed in large part by quartz, chlorite, plagioclase, and muscovite (sericite) (Ensign and others, 1968). The darker-colored carbonaceous beds consistently contain the most copper (Leone and others, 1971).

In addition to the Lower Sandstones, native copper occurs abundantly in black siltstones and shales of the Parting Shale. The Parting Shale averages 20 to 30 percent native copper, and the Upper Shale 1 to 5 percent (Leone and others, 1971). The average grain size of native copper is greater than the average size of chalcocite; hence, the earlier mining of native copper from the sandstone. Native copper in sandstone averages 1.3 mm, and in shale, 0.2 mm. Chalcocite generally averages 2 to 20 $\mu$m (White and Wright, 1966). It has been noted that where lower sandstone copper is abundant, the copper content of the cupriferous zone is notably higher than normal (Ensign and others, 1968).

*Production.* Early mine production from the Porcupine Mountain area at the Nonesuch and White Pine mines (Figs. 23 and 25) is indicated on Appendix B. At that time, native copper from the sandstones was produced because of the larger average grain size of native copper, measured in millimeters as compared to the average chalcocite size, measured in microns. Until metallurgical research resolved the problem of the recovery of ultrafine chalcocite from its host rock in the mid-1940s, sulfide copper in the Nonesuch Formation was only a recognized resource, not an economically produceable product.

The Nonesuch Formation of the Presque Isle basin (Fig. 23) was explored in the 1950s. Copper-bearing resources of 104 million tons averaging 1.27 percent copper (25.4 lb per ton) were reported (Bodwell, 1972). An exploratory shaft was sunk in the basin, and two experimental stopes were opened. Despite exhaustive efforts, including high-density roof bolting, the hanging wall (roof) of the mining area could not be maintained, and efforts were abandoned.

At White Pine, experimental production from the Schacht shaft, between 1946 and 1951 (Appendix B) amounted to 272, 633 kg (601,054 lb). Development of the ore body began, and a new mining facility was constructed with production beginning in 1955. Through 1987, 1,364,800,000 kg (3,008,867,267 lb), 1.36 million metric tons (1.5 million short tons), of refined copper had been produced. The ore body at White Pine is layered and strata bound with varying dips. It is mined by the room and pillar method. Orientation and design of rooms to be extracted are dictated by interpretation of rock stress, faults, and fractures. Reported reserves are about 184 million extractable tonnes with grades of 1.1 percent (22 lb) copper and 6.77 g (0.22 troy ounces) silver per ton (Seasor and Brown, 1986).

## SUMMARY

Michigan has been a premier producer of iron ore and copper. It currently ranks second in the nation in the production of iron ore and is the fifth largest producer of copper. The Precambrian rocks in which these metals formed represent part of the southern exposed terminus of the Canadian Precambrian Shield from which a much larger variety of metals is produced in Canada. The Michigan portion of this province contains most classes of igneous, sedimentary, and metamorphic rock, some of which have been dated to 3.56 ga. This area of about 19,400 km$^2$ (7,500 mi$^2$) includes approximately 4,400 km$^2$ (1,700 mi$^2$) of rocks of Archean age, 7,000 km$^2$ (2,700 mi$^2$) of rocks of early Proterozoic age, and 8,000 km$^2$ (3,100 mi$^2$) of rocks of middle Proterozoic age. Compilation of the maximum thickness of layered rocks, volcanics, and sediments indicates a total of 56,800 m (186,500 ft) or 56 km (35 mi) present.

Most past exploration efforts have been directed toward the search for the metals first discovered in the area: the early Proterozoic iron ore and middle Proterozoic copper. As a result of intensive diamond drilling and geophysical surveys, the locations of iron formations are well known, and resources exceed 400 billion tons. The factor limiting their production is availability of markets.

Although extensively drilled in many areas, native copper resources are more enigmatic and difficult to evaluate because of the erratic distribution of the metal in the host rock. Results of exploration drilling are not reliable as indicators of the actual grade of copper in a mineralized area. This makes resource estimation difficult. This difficulty is compounded by a lack of effective exploration concepts.

Some of the exploration problems are the domination of interests in exploring for iron and copper, the paucity of bedrock exposures, and the mantling of most bedrock by glacial deposits. Regardless of the problems, it is proposed that a variety of economic metallic mineral deposits remain to be discovered in Precambrian rocks in Michigan.

## ACKNOWLEDGMENTS

The writer acknowledges critical review of the draft manuscript by James Trow of Michigan State University and Allan Johnson of Michigan Technological University. Suggestions by Paul Catacosinos, Delta College, significantly improved organization and content of this chapter.

Illustrations, maps, and charts were modified from publications of the U.S. Geological Survey developed from a cooperative program between the USGS and the Michigan Geological Survey. Statistical data presented were compiled from the files of the Michigan Geological Survey. Figure 28 and the accompanying stratigraphic description was provided by Robert Seasor, Chief Geologist, Copper Range Company. Maps and graphs were finalized by Mick Jones, Michigan Geological Survey.

## APPENDIX A. IRON-ORE PRODUCTION IN MICHIGAN THROUGH 1987 (LONG TONS)

| Year | Marquette Range | Menominee Range | Gogebic Range | Total Tons | Year | Marquette Range | Menominee Range | Gogebic Range | Total Tons |
|---|---|---|---|---|---|---|---|---|---|
| Unknown | 73,553 | | | 73,553 | 1925 | 4,197,846 | 5,199,031 | 5,961,215 | 15,358,092 |
| 1854 | 3,000 | | | 3,000 | 1926 | 4,435,029 | 5,946,377 | 6,428,754 | 16,810,160 |
| 1855 | 1,449 | | | 1,449 | 1927 | 4,147,777 | 5,213,256 | 5,254,037 | 14,615,070 |
| 1856 | 6,790 | | | 6,790 | 1928 | 4,298,717 | 4,841,637 | 5,151,465 | 14,291,819 |
| 1857 | 25,646 | | | 25,646 | 1929 | 5,409,712 | 5,566,305 | 5,912,431 | 16,889,448 |
| 1858 | 22,876 | | | 22,876 | 1930 | 3,633,968 | 3,546,544 | 3,976,968 | 11,157,480 |
| 1859 | 68,832 | | | 68,832 | 1931 | 1,809,445 | 1,461,443 | 2,286,362 | 5,577,250 |
| 1860 | 114,401 | | | 114,401 | 1932 | 357,255 | 307,721 | 313,388 | 978,364 |
| 1861 | 49,909 | | | 49,909 | 1933 | 2,807,325 | 1,510,985 | 1,757,587 | 6,075,897 |
| 1862 | 124,169 | | | 124,169 | 1934 | 2,473,847 | 1,335,027 | 1,690,897 | 5,499,771 |
| 1863 | 203,055 | | | 203,055 | 1935 | 3,265,537 | 1,634,022 | 2,341,985 | 7,241,544 |
| 1864 | 247,059 | | | 247,059 | 1936 | 4,627,889 | 2,163,679 | 3,710,468 | 10,502,036 |
| 1865 | 198,758 | | | 198,758 | 1937 | 5,747,812 | 2,647,042 | 4,243,391 | 12,638,245 |
| 1866 | 296,713 | | | 296,713 | 1938 | 1,476,257 | 978,419 | 1,652,904 | 4,107,580 |
| 1867 | 465,504 | | | 465,504 | 1939 | 4,907,623 | 2,157,122 | 4,175,150 | 11,239,895 |
| 1868 | 506,505 | | | 506,505 | 1940 | 5,920,463 | 3,101,751 | 4,749,065 | 13,771,279 |
| 1869 | 649,097 | | | 649,097 | 1941 | 6,254,391 | 4,127,964 | 4,818,966 | 15,201,321 |
| 1870 | 856,245 | | | 856,245 | 1942 | 6,540,731 | 4,927,609 | 4,691,940 | 16,160,280 |
| 1871 | 818,966 | | | 818,966 | 1943 | 5,601,418 | 4,880,679 | 4,094,722 | 14,576,819 |
| 1872 | 949,073 | | | 949,073 | 1944 | 4,790,177 | 4,876,210 | 4,067,879 | 13,734,266 |
| 1873 | 1,174,972 | | | 1,174,972 | 1945 | 4,585,436 | 4,240,546 | 3,007,935 | 11,833,917 |
| 1874 | 935,604 | | | 935,604 | 1946 | 3,270,344 | 2,590,499 | 2,619,349 | 8.480,192 |
| 1875 | 898,974 | | | 898,974 | 1947 | 5,543,126 | 3,667,547 | 3,709,856 | 12,920,529 |
| 1876 | 995,224 | | | 995,224 | 1948 | 4,898,044 | 4,085,777 | 3,914,634 | 12,898,455 |
| 1877 | 1,013,144 | 10,405 | | 1,023,549 | 1949 | 4,253,381 | 3,587,067 | 3,156,466 | 10,996,914 |
| 1878 | 1,039,368 | 82,824 | | 1,122,192 | 1950 | 4,055,002 | 4,144,431 | 3,827,323 | 12,926,756 |
| 1879 | 1,135,396 | 247,135 | | 1,382,531 | 1951 | 5,647,423 | 4,707,931 | 3,318,519 | 13,673,873 |
| 1880 | 1,384,010 | 537,164 | | 1,921,174 | 1952 | 4,516,509 | 4,277,880 | 3,003,861 | 11,798,250 |
| 1881 | 1,579,834 | 541,076 | | 2,120,910 | 1953 | 5,571,502 | 4,620,902 | 3,188,352 | 13,380,756 |
| 1882 | 1,829,394 | 894,802 | | 2,724,196 | 1954 | 3,675,429 | 3,669,710 | 2,377,743 | 9,722,882 |
| 1883 | 1,305,425 | 1,013,688 | | 2,319,113 | 1955 | 6,639,966 | 4,325,625 | 3,182,532 | 14,148,123 |
| 1884 | 1,558,034 | 861,660 | 1,022 | 2,420,716 | 1956 | 5,689,013 | 3,889,213 | 2,958,076 | 12,536,302 |
| 1885 | 1,430,442 | 650,003 | 114,122 | 2,194,547 | 1957 | 5,992,752 | 4,296,567 | 2,837,407 | 13,126,726 |
| 1886 | 1,627,380 | 832,749 | 656,041 | 3,116,170 | 1958 | 3,722,139 | 3,095,239 | 1,393,528 | 8,210,906 |
| 1887 | 1,851,417 | 1,060,035 | 1,053,305 | 3,964,757 | 1959 | 3,529,949 | 2,477,980 | 1,250,786 | 7,258,715 |
| 1888 | 1,923,755 | 986,698 | 1,249,415 | 4,159,846 | 1960 | 4,944,715 | 4,121,165 | 1,889,986 | 10,955,866 |
| 1889 | 2,642,814 | 1,486,009 | 1,555,989 | 5,684,812 | 1961 | 4,182,973 | 3,885,902 | 1,361,855 | 9,430,730 |
| 1890 | 3,000,805 | 1,944,435 | 2,226,012 | 7,221,252 | 1962 | 4,500,447 | 3,462,371 | 1,480,383 | 9,443,201 |
| 1891 | 2,512,242 | 1,640,764 | 1,559,249 | 5,712,255 | 1963 | 5,850,347 | 4,304,194 | 812,630 | 10,967,171 |
| 1892 | 2,665,169 | 1,964,140 | 2,553,035 | 7,182,344 | 1964 | 7,944,840 | 4,624,274 | 1,403,137 | 13,972,251 |
| 1893 | 1,837,140 | 1,305,272 | 1,228,138 | 4,370,550 | 1965 | 8,925,165 | 4,360,694 | 772,569 | 14,058,428 |
| 1894 | 2,060,260 | 960,302 | 1,668,729 | 4,689,291 | 1966 | 9,659,989 | 4,327,914 | 364,407 | 14,352,310 |
| 1895 | 2,093,791 | 1,692,098 | 2,126,090 | 5,911,979 | 1967 | 10,164,895 | 3,630,145 | 238,851 | 14,033,891 |
| 1896 | 2,606,790 | 1,431,624 | 1,434,006 | 5,472,420 | 1968 | 9,366,744 | 3,448,687 | | 12,815,431 |
| 1897 | 2,712,947 | 1,801,136 | 1,865,130 | 6,379,213 | 1969 | 11,162,671 | 3,389,741 | | 14,552,412 |
| 1898 | 3,119,461 | 2,177,915 | 2,072,356 | 7,369,732 | 1970 | 10,335,545 | 2,924,875 | | 13,260,420 |
| 1899 | 3,738,192 | 3,109,522 | 2,444,362 | 9,292,076 | 1971 | 9,330,521 | 2,834,443 | | 12,164,964 |
| 1900 | 3,479,242 | 3,172,123 | 2,444,169 | 9,095,534 | 1972 | 9,250,747 | 2,577,402 | | 11,828,149 |
| 1901 | 3,246,611 | 3,525,914 | 2,419,144 | 9,191,669 | 1973 | 9,940,525 | 2,373,631 | | 12,314,156 |
| 1902 | 3,865,350 | 4,369,004 | 3,022,438 | 11,256,792 | 1974 | 9,008,916 | 2,521,880 | | 11,530,796 |
| 1903 | 3,040,092 | 3,648,639 | 2,465,263 | 9,153,994 | 1975 | 12,142,758 | 2,235,469 | | 14,378,227 |
| 1904 | 2,851,745 | 2,919,779 | 2,042,398 | 7,813,922 | 1976 | 13,361,958 | 2,325,952 | | 15,687,910 |
| 1905 | 4,235,651 | 4,253,508 | 3,215,352 | 11,704,511 | 1977 | 9,683,403 | 2,332,305 | | 12,015,708 |
| 1906 | 4,057,226 | 4,934,114 | 3,113,980 | 12,105,320 | 1978 | 15,933,903 | 2,180,922 | | 18,114,825 |
| 1907 | 4,388,073 | 4,785,773 | 3,093,083 | 12,266,929 | 1979 | 14,992,543 | 2,357,978 | | 17,350,521 |
| 1908 | 2,413,575 | 2,538,802 | 2,348,737 | 7,301,114 | 1980 | 13,867,535 | 1,865,996 | | 15,733,531 |
| 1909 | 4,252,622 | 4,593,407 | 3,402,567 | 12,248,596 | 1981 | 15,182,900 | 182,009 | | 15,364,909 |
| 1910 | 4,392,726 | 3,909,461 | 3,663,438 | 11,965,625 | 1982 | 6,927,261 | 99,898 | | 7,027,250 |
| 1911 | 2,835,902 | 3,815,908 | 2,258,666 | 8,910,476 | 1983 | 10,147,941 | | | 10,147,941 |
| 1912 | 4,202,723 | 4,561,833 | 4,103,532 | 12,868,088 | 1984 | 13,052,985 | | | 13,052,985 |
| 1913 | 3,967,918 | 4,839,140 | 3,873,517 | 12,680,575 | 1985 | 12,413,310 | | | 13,413,310 |
| 1914 | 2,491,857 | 3,133,672 | 3,164,420 | 8,789,949 | 1986 | 10,879,016 | | | 10,879,016 |
| 1915 | 4,106,202 | 4,822,157 | 4,613,190 | 13,541,549 | 1987 | 11,872,330 | | | 11,872,330 |
| 1916 | 5,409,582 | 6,179,920 | 7,340,582 | 18,930,084 | | | | | |
| 1917 | 4,874,150 | 5,928,141 | 7,052,579 | 17,854,870 | Total | 587,641,341 | 324,812,923 | 255,224,103 | 1,167,678,369 |
| 1918 | 4,354,297 | 6,108,886 | 7,150,636 | 17,613,819 | | | | | |
| 1919 | 2,922,245 | 4,314,536 | 5,572,484 | 12,879,265 | | | | | |
| 1920 | 4,608,323 | 6,428,149 | 7,956,459 | 18,992,931 | | | | | |
| 1921 | 1,116,560 | 1,584,466 | 2,269,827 | 4,970,853 | | | | | |
| 1922 | 2,818,374 | 4,079,444 | 5,542,571 | 12,440,389 | | | | | |
| 1923 | 3,891,801 | 4,830,222 | 5,557,508 | 14,279,531 | | | | | |
| 1924 | 3,174,835 | 3,836,826 | 4,329,803 | 11,341,464 | | | | | |

## APPENDIX B. COPPER PRODUCTION IN MICHIGAN THROUGH 1987 (POUNDS)

| Year | Native Copper | Sulfide Copper | Total | Year | Native Copper | Sulfide Copper | Total |
|---|---|---|---|---|---|---|---|
| 1845 | 26,880 | | 26,880 | 1918 | 231,096,150 | 3,273,680 | 234,369,830 |
| 1846 | 58,240 | | 58,240 | 1919 | 177,594,135 | 1,979,268 | 179,573,403 |
| 1847 | 477,120 | | 477,120 | 1920 | 153,483,952 | 1,850,787 | 155,334,739 |
| 1848 | 1,032,640 | | 1,032,640 | 1921 | 100,918,001 | 30,311 | 100,948,312 |
| 1849 | 1,505,280 | | 1,505,280 | 1922 | 122,545,126 | | 122,545,126 |
| 1850 | 1,281,280 | | 1,281,280 | 1923 | 137,691,306 | | 137,691,306 |
| 1851 | 1,744,960 | | 1,744,960 | 1924 | 145,333,227 | | 145,333,227 |
| 1852 | 1,774,080 | | 1,774,080 | 1925 | 154,799,247 | | 154,799,247 |
| 1853 | 2,905,280 | | 2,905,280 | 1926 | 175,441,565 | | 175,441,565 |
| 1854 | 4,074,560 | | 4,074,560 | 1927 | 177,537,775 | | 177,537,775 |
| 1855 | 5,808,320 | | 5,808,320 | 1928 | 178,442,704 | | 178,442,704 |
| 1856 | 8,211,840 | | 8,211,840 | 1929 | 186,393,974 | | 186,393,974 |
| 1857 | 9,531,200 | | 9,531,200 | 1930 | 169,297,775 | | 169,297,775 |
| 1858 | 9,157,120 | | 9,157,120 | 1931 | 118,495,005 | | 118,495,055 |
| 1859 | 8,926,400 | | 8,926,400 | 1932 | 54,396,108 | | 54,396,108 |
| 1860 | 12,069,120 | | 12,069,120 | 1933 | 46,853,130 | | 46,853,130 |
| 1861 | 15,037,120 | | 15,037,120 | 1934 | 48,215,859 | | 48,215,859 |
| 1862 | 13,585,600 | | 13,585,600 | 1935 | 64,107,889 | | 64,107,889 |
| 1863 | 12,985,280 | | 12,985,280 | 1936 | 95,968,019 | | 95,968,019 |
| 1864 | 12,490,240 | | 12,490,240 | 1937 | 94,649,805 | | 94,649,805 |
| 1865 | 14,385,400 | | 14,385,400 | 1938 | 94,075,588 | | 94,075,588 |
| 1866 | 13,749,120 | | 13,749,120 | 1939 | 87,969,153 | | 87,969,153 |
| 1867 | 17,525,760 | | 17,525,760 | 1940 | 90,402,594 | | 90,402,594 |
| 1868 | 20,935,040 | 3,502 | 20,935,040 | 1941 | 92,864,678 | | 92,864,678 |
| 1869 | 26,624,640 | 3,261 | 26,624,640 | 1942 | 91,365,660 | | 91,365,660 |
| 1870 | 24,622,080 | | 24,622,080 | 1943 | 92,736,686 | | 92,736,686 |
| 1871 | 26,750,080 | 257 | 26,750,337 | 1944 | 84,843,078 | | 84,843,078 |
| 1872 | 24,552,640 | | 24,552,640 | 1945 | 60,801,716 | | 60,801,716 |
| 1873 | 30,089,920 | | 30,089,920 | 1946 | 43,203,563 | 2,741 | 43,206,304 |
| 1874 | 34,332,480 | 27,450 | 34,359,930 | 1947 | 46,581,000 | 473,473 | 47,054,473 |
| 1875 | 36,039,360 | 49,667 | 36,089,027 | 1948 | 54,309,661 | 124,381 | 54,434,042 |
| 1876 | 38,270,400 | | 38,270,400 | 1949 | 38,498,167 | | 38,498,167 |
| 1877 | 39,025,280 | | 39,025,280 | 1950 | 51,788,017 | | 51,788,017 |
| 1878 | 39,690,560 | | 39,690,560 | 1951 | 49,297,405 | 459 | 49,297,864 |
| 1879 | 42,848,960 | 31,973 | 42,880,933 | 1952 | 43,939,304 | | 43,939,304 |
| 1880 | 49,736,960 | 55,584 | 49,792,544 | 1953 | 47,999,690 | | 47,999,690 |
| 1881 | 54,573,120 | 119,061 | 54,692,181 | 1954 | 44,796,824 | | 44,796,824 |
| 1882 | 56,982,765 | 46,450 | 57,029,215 | 1955 | 36,265,073 | 58,943,499 | 95,208,572 |
| 1883 | 59,702,404 | | 59,702,404 | 1956 | 47,649,651 | 75,599,096 | 123,348,747 |
| 1884 | 69,353,202 | 23,867 | 69,377,069 | 1957 | 45,118,291 | 68,562,019 | 113,680,310 |
| 1885 | 72,148,172 | 28,484 | 72,176,656 | 1958 | 40,402,115 | 81,656,908 | 122,059,023 |
| 1886 | 79,890,798 | | 79,890,798 | 1959 | 43,434,301 | 69,647,232 | 113,081,533 |
| 1887 | 75,471,890 | | 75,471,890 | 1960 | 40,314,264 | 75,120,243 | 115,434,507 |
| 1888 | 86,472,034 | | 86,472,034 | 1961 | 38,517,943 | 103,756,178 | 142,274,121 |
| 1889 | 88,175,675 | | 88,175,675 | 1962 | 37,147,227 | 106,613,451 | 143,760,678 |
| 1890 | 101,410,277 | | 101,410,277 | 1963 | 35,880,333 | 113,900,191 | 149,780,524 |
| 1891 | 114,222,709 | | 114,222,709 | 1964 | 29,326,043 | 108,458,751 | 137,784,794 |
| 1892 | 123,198,460 | | 123,198,460 | 1965 | 22,174,047 | 123,743,392 | 145,917,439 |
| 1893 | 112,605,078 | | 112,605,078 | 1966 | 25,181,717 | 121,181,717 | 146,744,040 |
| 1894 | 114,308,870 | | 114,308,870 | 1967 | 24,458,983 | 92,731,257 | 117,190,240 |
| 1895 | 129,330,749 | | 129,330,749 | 1968 | 11,288,306 | 137,092,516 | 148,380,822 |
| 1896 | 143,524,069 | | 143,524,069 | 1969 | | 149,952,283 | 149,952,283 |
| 1897 | 145,282,059 | | 145,282,059 | 1970 | | 131,373,570 | 131,373,570 |
| 1898 | 158,491,703 | | 158,491,703 | 1971 | | 112,155,147 | 112,155,147 |
| 1899 | 147,400,338 | | 147,400,338 | 1972 | | 131,937,378 | 131,937,378 |
| 1900 | 145,461,498 | | 145,461,498 | 1973 | | 151,678,939 | 151,678,939 |
| 1901 | 156,289,481 | | 156,289,481 | 1974 | | 131,037,602 | 131,037,602 |
| 1902 | 170,699,228 | | 170,699,228 | 1975 | 1,455,651 | 140,819,850 | 142,275,501 |
| 1903 | 192,400,577 | | 192,400,577 | 1976 | 3,866,218 | 83,044,081 | 86,910,299 |
| 1904 | 208,309,130 | | 208,309,130 | 1977 | | 85,960,551 | 85,960,551 |
| 1905 | 230,287,992 | | 230,287,992 | 1978 | | 78,301,136 | 78,301,136 |
| 1906 | 229,695,730 | | 229,695,730 | 1979 | | 87,469,886 | 87,469,886 |
| 1907 | 219,131,503 | | 219,131,503 | 1980 | | 71,540,437 | 71,540,437 |
| 1908 | 222,289,584 | | 222,289,584 | 1981 | | 86,723,895 | 86,723,895 |
| 1909 | 227,005,923 | | 227,005,923 | 1982 | | 45,202,832 | 45,202,832 |
| 1910 | 221,462,984 | | 221,462,984 | 1983 | | | |
| 1911 | 218,185,236 | | 218,185,236 | 1984 | | 10,299,016 | 10,299,016 |
| 1912 | 231,112,228 | | 231,112,228 | 1985 | | | |
| 1913 | 155,715,286 | | 155,715,286 | 1986 | | 62,625,432 | 62,625,432 |
| 1914 | 158,009,748 | | 158,009,748 | 1987 | | 106,516,621 | 106,516,621 |
| 1915 | 238,956,411 | 2,824,145 | 238,956,411 | | | | |
| 1916 | 269,794,531 | 4,207,449 | 274,001,980 | Totals | 10,514,876,219 | 3,028,101,046 | 13,542,977,265 |
| 1917 | 268,508,098 | 4,067,529 | 272,575,627 | | | | |

# REFERENCES CITED

Aldrich, H. R., 1929, The geology of the Gogebic iron range of Wisconsin: Wisconsin Geological and Natural History Survey Bulletin 71, 279 p.

Allen, R. C., 1910, The Iron River iron-bearing district of Michigan: Michigan Geological and Biological Survey Publication 3, Geological Series 2, 151 p.

Baxter, D. A., Bornhorst, T. J., and Van Alstine, J. L., 1987, Geology, structure, and associated precious metal mineralization of Archean rocks in the vicinity of Clark Creek, Marquette County, Michigan: Michigan Geological Survey Division Open-File Report OFR-87-8, 62 p.

Bayley, R. W., 1959, Geology of the Lake Mary Quadrangle, Iron County, Michigan: U.S. Geological Survey Bulletin 1077, 112 p.

Bayley, R. W., Dutton, C. E. and Lamey, C. A., 1966, Geology of the Menominee Iron-bearing District, Dickinson County, Michigan, and Florence and Marinette Counties, Wisconsin: U.S. Geological Survey Professional Paper 513, 96 p.

Boben, C. L., Bornhorst, T. J., and Van Alstine, J. L., 1986, Detailed geological study of three precious metal prospects in Marquette County and one in Gogebic County, Michigan: Michigan Geological Survey Division Open-File Report OFR-86-1, 40 p.

Bodwell, W. A., 1972, Geological compilation and nonferrous metal potential, Precambrian secton, northern Michigan [M.S. theses]: Houghton, Michigan Technological University, 109 p.

Bornhorst, T. J., Shepeck, A. W., and Rossell, D. M., 1986, The Ropes gold mine, Marquette County, Michigan, U.S.A.; An Archean hosted lode gold deposit, in Macdonald, A. J., ed., Proceedings of Gold '86 Symposium, Toronto, Ontario, Canada: Toronto, GOLD '86, p. 213–227.

Bornhorst, T. J., Paces, J. B., Grant, N. K., Obradovich, J. D., and Huber, N. K., 1988, Age of copper mineralization, Keweenaw Peninsula, Michigan: Economic Geology, v. 83, p. 619–625.

Boyum, B. H., 1975, The Marquette Mineral District of Michigan: Ishpeming, Michigan, The Cleveland Cliffs Iron Company, 59 p.

——, 1977, The Saga of iron mining in Michigan's Upper Peninsula: Marquette, Michigan, J. M. Longyear Research Library, 48 p.

——, 1988, The origin and extent of the hard and soft iron ores of the Marquette Range, Michigan; Tectonic control of ore deposits and the vertical and horizontal extent of ore systems: Rolla, University of Missouri, p. 301–311.

Broderick, T. M., 1945, Geology of the Ropes gold mine, Marquette County, Michigan: Economic Geology, v. 40, p. 115–128.

Broderick, T. M., Hohl, C. D., and Eidemiller, H. N., 1946, Recent contributions to the geology of the Michigan copper district: Economic Geology, v. 41, no. 7, p. 675–725.

Brooks, T. B., 1873, Iron-bearing rocks (economic): Michigan Geological Survey, Upper Peninsula, v. 1, pt. 1, 319 p.

Brooks, T. B., and Pumpelly, R., 1872, On the age of the copper-bearing rocks of Lake Superior: American Journal Science, 3rd series, v. 3, p. 428–432.

Brozdowski, R. A., Gleason, R. J., and Scott, G. W., 1986, The Ropes mine; A pyritic gold deposit in Archean volcaniclastic rock, Ishpeming, Michigan, U.S.A., in Proceedings of Gold '86 Symposium, Toronto, Ontario, Canada: Toronto, GOLD '86, p. 228–242.

Butler, B. S., and Burbank, W. S., 1929, The copper deposits of Michigan: U.S. Geological Survey Professional Paper 144, 237 p.

Cannon, W. F., and Klasner, J. S., 1977, Bedrock geologic map of the southern part of the Diorite and Champion 7½-Minute Quadrangles, Marquette County, Michigan: U.S. Geological Survey Map I-1058, scale 1:24,000.

Cannon, W. F., Powers, S. L., and Wright, N. A., 1978, Computer-aided estimates of concentrating-grade iron resources in the Negaunee Iron Formation, Marquette District, Michigan: U.S. Geological Survey Professional Paper 1045, 21 p.

Catacosinos, P. A., 1981, Origin and stratigraphic assessment of pre-Mt. Simon clastics (Precambrian) of Michigan Basin: American Association of Petroleum Geologists Bulletin, v. 69, p. 1617–1620.

Clark, L. D., Cannon, W. F., and Klasner, J. S., 1975, Bedrock geologic map of the Negaunee SW Quadrangle, Marquette County, Michigan: U.S. Geological Survey Map GQ-1226, scale 1:24,000.

Cornwall, H. R., 1955, Bedrock geology of the Fort Wilkins Quadrangle, Michigan: U.S. Geological Survey Map GQ-74, scale 1:24,000.

Daniels, P. A., 1982, Upper Precambrian sedimentary rocks; Oronto Group, Michigan–Wisconsin, in Wold, R. J., and Hinze, W. J., eds., Geology and tectonics of the Lake Superior Basin: Geological Society of America Memoir 156, p. 107–133.

Dickas, A. B., 1986, Comparative Precambrian stratigraphy and structure along the Mid-Continent rift: American Association of Petroleum Geologists Bulletin, v. 70, p. 225–238.

Drier, R. W., 1964, Prehistoric copper mining in Michigan: Skillings Mining Review, v. 53, no. 34, p. 1, 4, 5, and 21.

Drier, R. W., and DuTemple, O. J., 1961, Prehistoric copper mining in the Lake Superior region; A collection of reference articles: Privately published by authors, 209 p.

Dutton, C. E., 1971, Geology of the Florence area, Wisconsin and Michigan: U.S. Geological Survey Professional Paper 633, 53 p.

Emmons, W. H., 1937, Gold deposits of the world: New York, McGraw-Hill, 562 p.

Ensign, C. O., Jr., and 8 others, 1968, Copper deposits in the Nonesuch Shale, White Pine, Michigan, in Ridge, J. D., ed., Ore deposits of the United States, 1933–1967; Graton-Sales Volume: American Institute Mining, Metallurgical, and Petroleum Engineers, p. 460–488.

Fisher, J., 1924, Historical sketch of the Lake Superior copper district; The Keweenawan: Houghton, Michigan College of Mines, p. 217–288.

——, 1929, Historical sketch of the Lake Superior copper district: Proceedings of the Lake Superior Mining Institute, v. 27, p. 54–67.

Fowler, J. H., and Kuenzi, W. D., 1978, Keweenawan turbidites in Michigan (deep borehole redbeds); A foundered basin sequence developed during evolution of a protoceanic rift system: Journal of Geophysical Research, v. 83, p. 5833–5843.

Fritts, C. E., 1969, Bedrock geologic map of the Marenisco–Watersmeet area, Gogebic and Ontonagon Counties, Michigan: U.S. Geological Survey Miscellaneous Geologic Investigations Map I-576, scale 1:48,000.

Gair, J. E., 1973, Iron deposits of Michigan (United States of America); Genesis of Precambrian iron and manganese deposits: Proceedings Kiev Symposium UNESCO, p. 365–375.

——, 1974, Hard iron ore at the Cliffs shaft mine, Ishpeming, Michigan: U.S. Geological Survey Open-File Report 74–227, 5 p.

——, 1975, Bedrock geology and ore deposits of the Palmer Quadrangle, Marquette County, Michigan: U.S. Geological Survey Professional Paper 769, 159 p.

Gair, J. E., and Thaden, R. E., 1968, Geology of the Marquette and Sands Quadrangles, Marquette County, Michigan: U.S. Geological Survey Professional Paper 397, 77 p.

Griffin, J. B., ed., 1961, Lake Superior copper and the Indians; Miscellaneous studies of Great Lakes prehistory: Ann Arbor, University of Michigan Museum of Anthropology, no. 17, 133 p.

Hamilton, S. K., 1965, Copper mineralization in the upper part of the Copper Harbor Conglomerate at White Pine, Michigan [M.S. thesis]: Madison, University of Wisconsin, 56 p.

Hinze, W. J., and Merritt, D. W., 1969, Basement rocks of the Southern Peninsula of Michigan, in Stonehouse, H. B., ed., Studies of the Precambrian of the Michigan Basin: Michigan Basin Geological Society Annual Field Excursion Guidebook, p. 28–59.

Hotchkiss, W. O., 1919, Geology of the Gogebic Range and its relation to recent mining developments: Engineering and Mining Journal, v. 108, p. 443–452, 501–507, 537–541, and 577–582.

Houghton, D., 1841, Fourth annual report of the state geologist: Michigan House of Representatives, Document 2, p. 3–89.

Hubbard, H. A., 1975a, Keweenawan geology of the North Ironwood, Ironwood and Little Girls Point Quadrangles, Gogebic County, Michigan: U.S. Geological Survey Open-File Report 75-152, 14 p.

——— , 1975b, Geology of Porcupine Mountains in Carp River and White Pine Quadrangles, Michigan: U.S. Geological Survey Journal of Research, v. 3, no. 5, p. 519–528.

Huber, N. K., 1959a, Some aspects of the origin of the Ironwood Iron Formation of Michigan and Wisconsin: Economic Geology v. 54, p. 82–118.

——— , 1959b, The environmental control of sedimentary iron minerals and its relation to the origin of the Ironwood Iron Formation: U.S. Geological Survey Open-File Report, 95 p.

Irving, R. D., and Van Hise, C. R., 1892, The Penokee iron-bearing series of Michigan and Wisconsin: U.S. Geological Survey Monograph 19, 534 p.

James, H. L., 1951, Iron Formation and associated rocks in the Iron River District, Michigan: Geological Society of America Bulletin, v. 62, p. 251–266.

——— , 1955, Zones of regional metamorphism in the Precambrian of northern Michigan: Geological Society of America Bulletin, v. 66, p. 1455–1488.

James, H. L., Clark, L. D., Lamey, C. A., and Pettijohn, F. J., 1961, Geology of central Dickinson County, Michigan: U.S. Geological Survey Professional Paper 310, 176 p.

James, H. L., Dutton, C. E., Pettijohn, F. J., and Wier, K. L., 1968, Geology and ore deposits of the Iron River–Crystal Falls district, Iron County, Michigan: U.S. Geological Survey Professional Paper 570, 134 p.

Jamison, J. K., 1950, The mining adventures of this Ontonagon Country: Ontonagon, Michigan, Ontonagon Herald Company, 90 p.

Johnson, R. C., Bornhorst, T. J., and VanAlstine, J. L., 1987, Geologic setting of precious metal mineralization in the Silver Creek to Island Lake area, Marquette County, Michigan: Michigan Geological Survey Division Open-File Report OFR-87-4, 134 p.

Johnson, R. F., and White, W. S., 1969, Preliminary report of the bedrock geology and copper deposits of the Matchwood Quadrangle, Ontonagon County, Michigan: U.S. Geological Survey Open-File Report, 29 p.

King, E. R., and Zietz, I., 1971, Aeromagnetic study of the Mid-continent gravity high of central United States: Geological Society of America Bulletin, v. 82, p. 2187–2208.

LaFayette, K. D., 1977, Flaming brands; Fifty years of iron making in the Upper Peninsula of Michigan, 1848–1898: Marquette, Northern Michigan University Press, 52 p.

Lake, M. C., 1917, Geology of the Wakefield area of the eastern Gogebic, in The iron ores of Lake Superior, 3rd ed.: Cleveland, Ohio, The Penton Press, p. 85–96.

Leone, R. J., Seasor, R. W., and Rohrbacher, T. J., 1971, Geology of the White Pine area; Guidebook for Field Conference, Michigan Copper District: Society of Economic Geologists, p. 45–64.

MacLellan, M. L., and Bornhorst, T. J., 1989, Bedrock geology of the Reany Lake area, Marquette County, Michigan: Michigan Geological Survey Division Open-File Report OFR-89-2, 111 p.

Morgan, P. F., and DeCristoforo, D. T., 1980, Geological evolution of the Ishpeming greenstone belt, Michigan, U.S.A.: Precambrian Research, v. 11, p. 23–41.

Owens, E. O., and Bornhorst, T. J., 1985, Geology and precious metal mineralization of the Fire Center and Holyoke mines area, Marquette County, Michigan: Michigan Geological Survey Division Open-File Report OFR-85-2, 105 p.

Prinz, W. C., 1967, Pre-Quaternary geologic and magnetic map and sections of part of the eastern Gogebic Iron Range, Michigan: U.S. Geological Survey Map I-497, scale 1 inch equals 800 feet.

——— , 1981, Geologic map of the Gogebic Range–Watersmeet area, Gogebic and Ontonagon Counties, Michigan: U.S. Geological Survey Miscellaneous Investigations Series Map I-1365, scale 1:125,000.

Prinz, W. C., and Hubbard, H. A., 1975, Preliminary geologic map of the Wakefield Quadrangle, Gogebic County, Michigan: U.S. Geological Survey Open-File Report 75-119, 10 p., map scale 1:24,000.

Puffett, W. P., 1974, Geology of the Negaunee Quadrangle, Marquette County, Michigan: U.S. Geological Survey Professional Paper 788, 53 p.

Reed, R. C., 1965, Copper mineralization in Animikie sediments of the Marquette Range, Marquette County, Michigan [M.S. thesis]: East Lansing, Michigan State University, 55 p.

Royce, S., 1938, Geology of the iron ranges: Cleveland, Ohio, Lake Superior Iron Ores, p. 27–45.

Schmidt, R. C., 1976, Geology of Precambrian W (Lower Precambrian) rocks in western Gogebic County, Michigan: U.S. Geological Survey Bulletin 1407, 40 p.

——— , 1980, The Marquette Range Supergroup in the Goebic Iron District, Michigan and Wisconsin: U.S. Geological Survey Bulletin 1460, 96 p.

Seasor, R. W., and Brown, A. C., 1986, White Pine stratiform copper deposit, in Proterozoic sediment hosted stratiform copper deposits of upper Michigan and Belt Supergroup of Idaho and Montana, Field Trip no. 1 Guidebook: Ottawa, Ontario, Carlton University, p. 37–39.

Sims, P. K., Peterman, Z. E., Prinz, W. C., and Benedict, F. C., 1984, Geology, geochemistry, and age of Archean and early Proterozoic rocks in the Marenisco–Watersmeet area, northern Michigan: U.S. Geological Survey Professional Paper 1292-A, 41 p.

Sims, P. K., Kisvarsonyi, E. B., and Morey, G. B., 1987, Geology and metallogeny of Archean and Proterozoic basement terranes in the northern midcontinent U.S.A.; An overview: U.S. Geological Survey Bulletin 1815, 51 p.

Swineford, A. P., 1882, Annual review of the iron mining and other industries of the Upper Peninsula for the year ending Dec. 31, 1881: Marquette, Michigan, Mining Journal, 193 p.

Trent, V. A., 1973, Geologic map of the Marenisco and Wakefield NE Quadrangles, Gogebic County, Michigan: U.S. Geological Survey Open-File Map G-29112, scale 1:20,000.

Vickers, R. C., 1956, Geology and monazite content of the Goodrich Quartzite, Palmer area, Marquette County, Michigan: U.S. Geological Survey Bulletin 1030-F, p. 171–185.

White, W. S., 1960, The Keweenawan lavas of Lake Superior; An example of flood basalts: American Journal of Science, v. 258A, Bradley Volume, p. 367–374.

——— , 1968, The native copper deposits of northern Michigan in Ore deposits of the United States, 1933–1967; Graton-Sales Volume: American Institute of Mining, Metallurgical, and Petroleum Engineers, v. 2, p. 303–325.

——— , 1971, Geologic setting of the Michigan Copper District, in Guidebook for Field Conference, Michigan Copper District: Society of Economic Geologists, p. 3–17.

——— , 1978, A theoretical basis for exploration for native copper in northern Wisconsin: U.S. Geological Survey Circular 769, 19 p.

White, W. S., and Wright, J. C., 1960, Lithofacies of the Copper Harbor Conglomerate, Northern Michigan: U.S. Geological Survey Professional Paper 400B, p. B5–B8.

——— , 1966, Sulphide mineral zoning in the basal Nonesuch Shale, northern Michigan: Economic Geology, v. 61, p. 1171–1190.

Whitlow, J. W., 1974, Geologic map of the Greenland and Rockland Quadrangles, Ontonagon County, Michigan: U.S. Geological Survey Map MF-596, scale 1:62,500.

Wright, C. E., 1879, First annual report of the Commissioner of mineral statistics of the State of Michigan for 1877–1878 and previous years: Marquette, Michigan, Mining Journal Stream Printing House, 229 p.

Zapffe, C., 1938, Discovery and early development of the iron ranges: Cleveland, Ohio, Lake Superior Iron Ores, p. 13–26.

MANUSCRIPT ACCEPTED BY THE SOCIETY JUNE 1, 1990

Printed in U.S.A.

Geological Society of America
Special Paper 256
1991

# Stratigraphy of Middle Proterozoic to Middle Ordovician formations of the Michigan Basin

**Paul A. Catacosinos**
*Department of Geology, Delta College, University Center, Michigan 48710*
**Paul A. Daniels, Jr.**
*Petro-Hunt Corporation, 1601 Elm Street, Dallas, Texas 75201*

## ABSTRACT

Continental rifting in the area now known as the Michigan Basin occurred some 1.1 b.y. ago (Van Schmus and Hinze, 1985), along with similar tectonism in other portions of the mid-continental United States. Although little is known of the subsequent 500 m.y., it appears that a change from a continental to a marine depositional regime took place during the Late Cambrian (Dresbachian) when northerly transgressing epeiric seas advanced into a slowly developing ancestral Michigan Basin.

The record of those seas is documented by Late Cambrian to Middle Ordovician formations. These are, in ascending order, the Mt. Simon, Eau Claire, Galesville, Franconia, Trempealeau–Prairie du Chien (T-PDC), St. Peter, and Glenwood. On the margins of the basin, the basal Mt. Simon Sandstone rests disconformably on older Precambrian basement. There, also, T-PDC rocks (Late Cambrian–Early Ordovician age) were eroded, producing a major interregional unconformity (the post-Sauk unconformity) on which the St. Peter Sandstone lies and which marks the top of the Sauk sequence. In the central Michigan Basin, however, deposition of the Sauk sequence was continuous and the post-Sauk unconformity was not developed.

Again, on the margins of the basin, a younger (Middle Ordovician) post–St. Peter unconformity was developed between the St. Peter and Glenwood Formations, but again is not present in the central basin where essentially continuous deposition of the entire section took place.

The configuration of the present-day Michigan Basin was established during the Early Ordovician. Since that initial configuration, however, significant structural elements have been added during subsequent Paleozoic time.

## INTRODUCTION

Although details of the middle Precambrian–earliest Paleozoic history of the Michigan Basin remain to be clarified, recent petroleum exploration in the central Michigan Basin has provided considerable new data on this portion of the stratigraphic section. In 1981, natural gas in commercial quantities was discovered in the central basin in sandstones immediately below the Middle Ordovician Glenwood Formation. The subsequent exploration and drilling activity has provided much new information on these lowermost sedimentary rocks. Older stratigraphic models have been superceded by this data, and newer models (e.g., Brady and Dehaas, 1987; Wheeler, 1987) have been proposed. The interpretation presented here, while dealing primarily with the section in the southern (lower) peninsula of Michigan, also relates to surrounding areas.

Long considered to be the classic cratonic basin, the Michigan Basin was thought to occupy an area of some 315,980 km$^2$ (122,000 mi$^2$; Cohee and Landes, 1958; Ells, 1969). Our boundaries (Fig. 1), however, suggest an area of about 207,200

Catacosinos, P. A., and Daniels, P. A., Jr., 1991, Stratigraphy of middle Proterozoic to Middle Ordovician formations in the Michigan Basin, *in* Catacosinos, P. A., and Daniels, P. A., Jr., eds., Early sedimentary evolution of the Michigan Basin: Geological Society of America Special Paper 256.

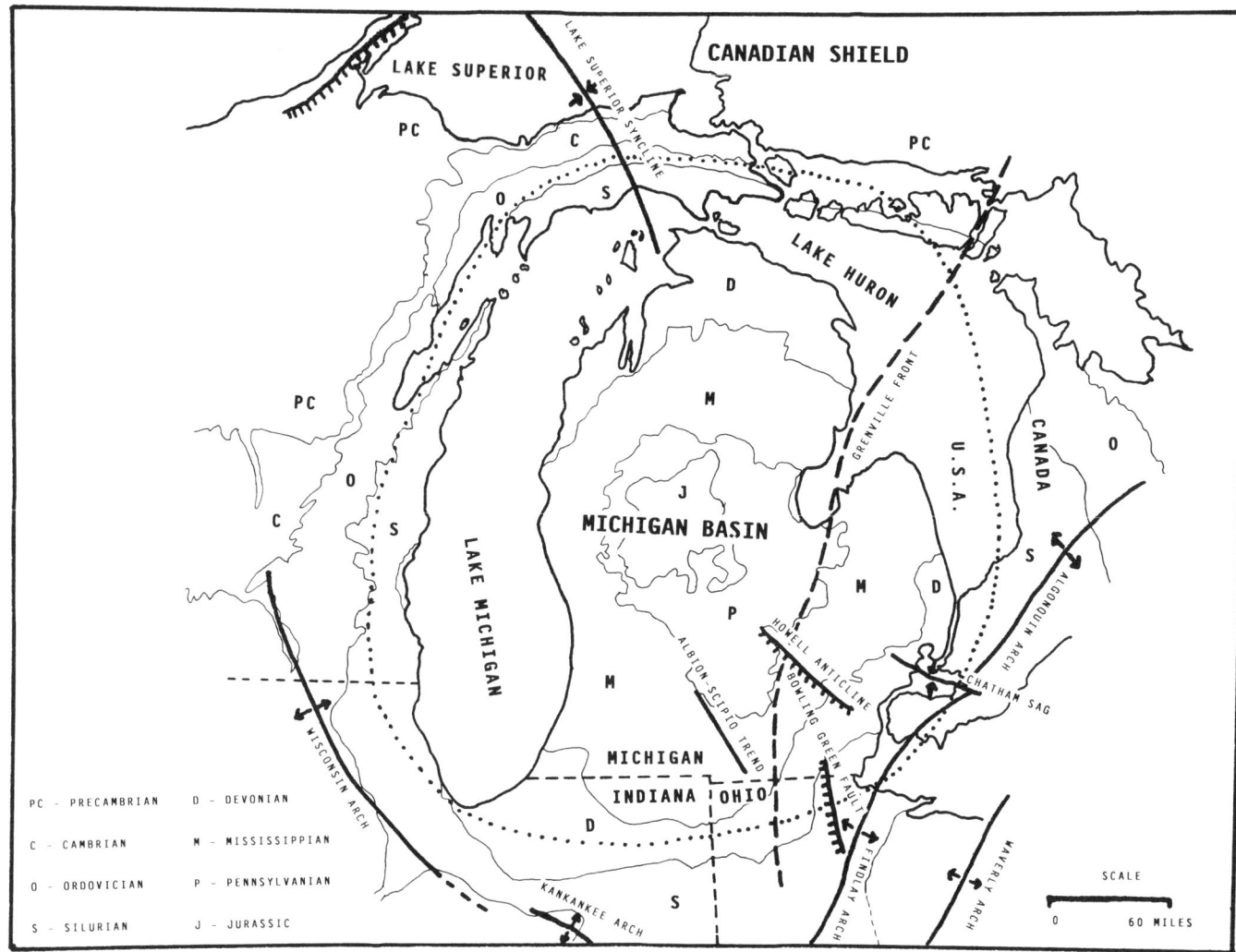

Figure 1. Index map, Michigan Basin and environs. Dotted line shows approximate basin limits. After Catacosinos and others (1991).

km² (80,000 mi²). (Please note that in calculating feet to meters a factor of 0.3 was used.) The index map (Fig. 1) shows the relation of the basin to other nearby regional structures. A brief history of the evolution of ideas concerning the stratigraphy and structure of the basin is presented below.

### Previous stratigraphic interpretations

Although study of the Precambrian and lowermost Paleozoic strata of the Michigan Basin may be said to have commenced with the pioneering work of Houghton (1841), the modern era of investigation clearly began with the work of Hamblin (1958) in the northern (upper) peninsula outcrop area. Subsequently, as more information accumulated from hydrocarbon exploration, it became clear that the lowermost Paleozoic rocks of the southern peninsula were more varied in lithology than those of the northern peninsula.

Ells (1967) illustrated the problems of Cambrian and Ordovician correlation facing stratigraphers up to that time. In 1969, Fisher constructed broad regional isopach maps of the entire Cambrian and Ordovician section as then defined and understood. Also in 1969, stratigraphic cross sections were prepared by the Michigan Basin Stratigraphic Committee. Catacosinos (1973) provided a detailed synthesis of the basic lithologic framework within the basin; while in 1978, Lilienthal prepared a major compilation of basin stratigraphy, complete with numerous cross sections. Additional correlations of the Cambrian and Ordovician strata were provided by Bricker and others (1983). It was a paper by Fisher and Barratt (1985), however, reporting age data provided by Repetski and Harris (1981), that permitted the refinement of the various stratigraphic models currently being developed.

### Structural interpretation

Studies of the structural history have evolved in a similar manner, beginning essentially with the work of Newcombe

(1933), and further chronicled in the cogent work of Ells (1969). A Precambrian rifting model was first presented by Hinze and Merritt (1969) and then Hinze and others (1975) for the Mid-Michigan gravity anomaly. These works also related the Mid-Michigan anomaly to the Lake Superior Basin and the Midcontinent gravity high. In 1978, a number of papers were published together in the *Journal of Geophysical Research* that analyzed the information obtained from the deepest hole drilled in the Michigan Basin—the McClure Oil, Sparks and others 1-8, Gratiot County, Michigan. Among these was the paper by Fowler and Kuenzi (1978) that incorporated the rift model into the Precambrian structural history of the basin. The rift model concept was further supported by the COCORP (Consortium for Continental Reflection Profiling) study of Brown and others (1982).

More recently, Dickas (1986) showed that the Precambrian rift zone of Michigan was part of the larger Midcontinent rift system of the central United States. This interpretation, that the Mid-Michigan gravity anomaly marks a zone that was initially a rift, and that this rift is also related to the large Mid-Continent rift system of North America, is generally accepted at this time. Other important new data on the Midcontinent Rift System (MRS) can be found in Cannon and others (1989) and Chandler and others (1989).

## STRATIGRAPHY

### Precambrian sedimentary rocks

A description of the crystalline rocks and basement provinces that underlie the Michigan Basin (Fig. 2) is in Hinze and others (1975). Keweenawan sedimentary rocks of middle Proterozoic age crop out along the northern and northwestern edges of the basin in Wisconsin and the northern peninsula of Michigan (Fig. 1).

In the northern peninsula of Michigan, Keweenawan sedimentary rocks include the Oronto Group (Daniels, 1982, 1986; Daniels and Elmore, 1988), the Jacobsville Sandstone (Kalliokoski, 1982, 1988), and their equivalents. The Oronto Group consists, in ascending order, of the Copper Harbor Conglomerate, Nonesuch Formation, and Freda Sandstone. The Jacobsville is considered to unconformably overlie the Oronto Group (Cannon and others, 1989).

All of these upper Keweenawan clastic units are heavily oxidized, with the exception of the Nonesuch Formation. To provide the reader with an idea regarding the types of Precambrian sedimentary rock that may be present in the Michigan Basin, a brief review of the Oronto Group and the Jacobsville follows. The Precambrian terminology used in this chapter is given in Figure 3. Note also that within the southern peninsula of Michigan, the Keweenawan sedimentary rocks include pre–Mt. Simon clastics.

***Northern peninsula of Michigan. Copper Harbor Conglomerate.*** The Copper Harbor Conglomerate is a basinward-thickening, ferruginous wedge of volcanogenic clastics (and subordinate volcanics), which become finer-grained both distally basinward and vertically upsection. Maximum known thickness of this unit exceeds 1,300 m (4,333 ft). The Copper Harbor represents a prograding alluvial fan complex that may well have been deposited under monsoonal conditions (Daniels, 1982; Elmore, 1983, 1984; and others). Portions of this complex fit fan-delta models, while other parts exhibit braid-delta components (Daniels and Elmore, 1988).

*Nonesuch Formation.* Overlying and interfingering with the Copper Harbor Conglomerate is the Nonesuch Formation, an unoxidized sequence of gray-black siltstone, fine-grained sandstone, and shale. Petroliferous and metalliferous, the Nonesuch is perceived to have been deposited in a series of anoxic, rift-flanking, lacustrine basins (Daniels, 1982; Daniels and Elmore, 1988; Elmore and others, 1988; and others). In part coeval with the Copper Harbor Conglomerate, the Nonesuch is also gradational with the overlying Freda Sandstone.

*Freda Sandstone.* Lithologically, the Freda Sandstone is a lithic, red-brown, cyclic sequence of sandstones, mudstones, and shales (conglomerates are rare) that conformably overlies the Nonesuch Formation. Environmentally, the Freda represents various braided stream deposits that initially interfingered with, and ultimately prograded over, the Nonesuch sediments (Daniels, 1982; Daniels and Elmore, 1988; and others).

*Jacobsville Sandstone.* The Jacobsville Sandstone is the stratigraphically highest of the Keweenawan (middle Proterozoic) sedimentary rocks that crop out near the margins of the Michigan Basin. Recent work reported by Cannon and others (1989) provides additional information supporting the interpretation of an unconformable contact between the Jacobsville and the underlying Oronto Group.

Generally a ferruginous sandstone, commonly containing secondary reduction zones, the Jacobsville comprises various lithologic facies representing fluvial to lacustrine environments (Hamblin, 1958). The Jacobsville is the lateral equivalent of the Bayfield Group of Wisconsin. More recent information is available in the works of Ojakangas and Morey (1982) and Kalliokoski (1982, 1988).

Overall, the Oronto Group and the overlying Jacobsville sediments, stratigraphically upward, exhibit increases in both textural and compositional maturity, with the Jacobsville containing the highest maturity quotients.

***Southern peninsula of Michigan. Pre–Mt. Simon Clastics.*** Within the southern peninsula of Michigan, sedimentary rocks of undoubted Precambrian age are present (Fig. 2) in the McClure Oil Company, Sparks and others 1-8 well, Gratiot County, which penetrated more than 1,587 m (5,290 ft) of pre–Mt. Simon clastics (Fowler and Kuenzi, 1978). An age of 1,000 Ma for these rocks has been determined by Van der Voo and Watts (1978); they are most likely time equivalents (Fig. 3) of the Freda Sandstone (Catacosinos, 1981a; Fisher and others, 1988). The origin of these red beds is problematic. Fowler and Kuenzi (1978) considered them to be marine turbidites; however,

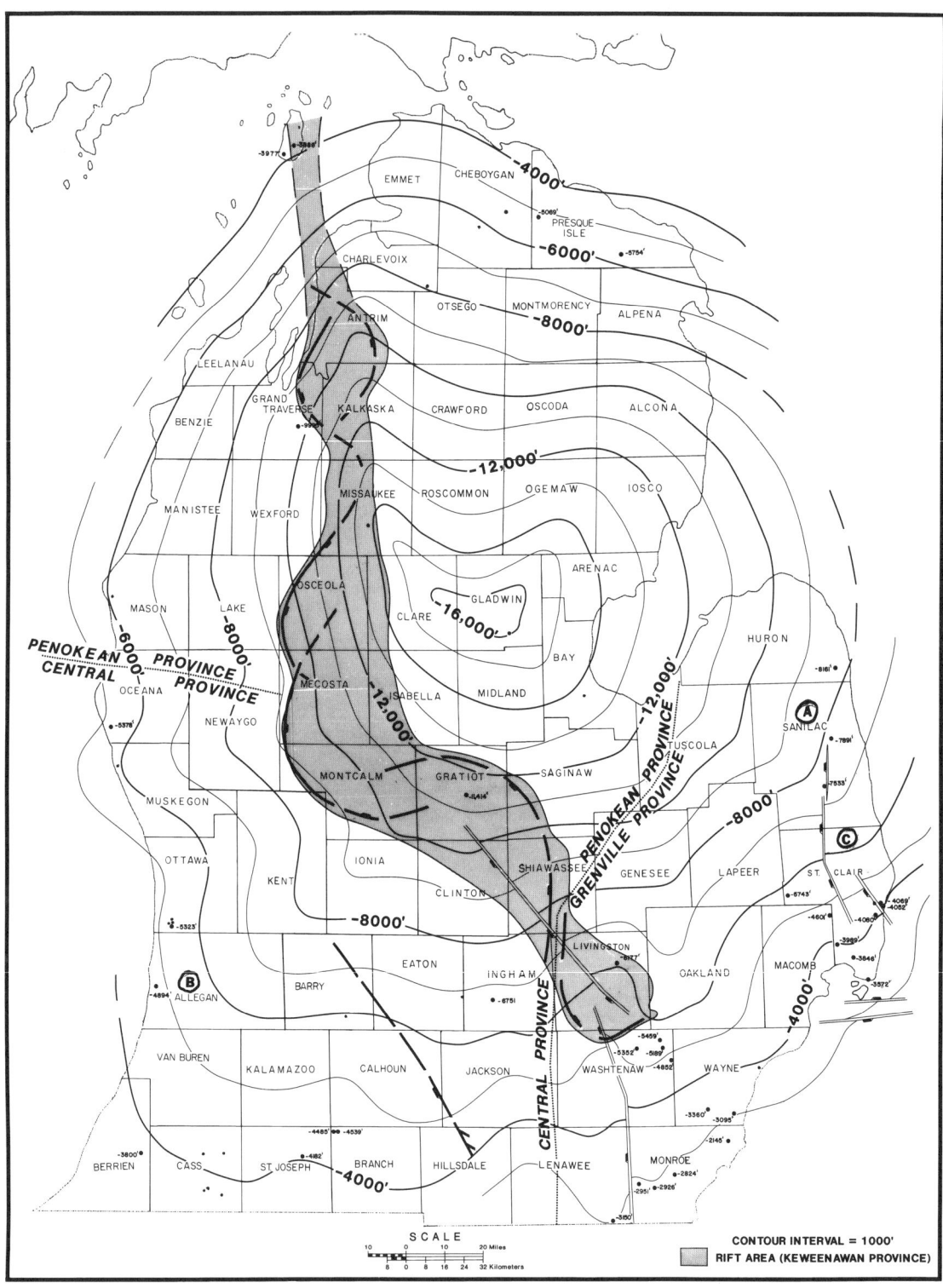

Figure 2. Precambrian structure contour map. Shaded areas shows location of buried Precambrian rift zone. A, B, and C refer, respectively, to the approximate locations of the Sanilac, Allegan, and St. Clair structural platforms. The Beaver Island wells are located in the northern portion of the rift shown; the Sparks well is in the rift portion of Gratiot County. Other Precambrian tests are scattered around the basin, as shown. See text for details. Modified from Hinze and Merritt (1969); Kellogg (1971); and Reynolds (1984).

| Northern Peninsula Michigan | Charlevoix Co., MI State-Beaver Island #1 | Gratiot Co., MI Sparks et al #1 |
|---|---|---|
| **Upper Cambrian** Munising Fm. Mt. Simon Ss | (Rift Sequence) Mt. Simon Ss | Mt. Simon Ss |
| **Precambrian** Jacobsville Sandstone | Pre-Mt. Simon (Jacobsville Equivalents) | None Identified |
| Oronto Group Freda Sandstone Nonesuch Fm. Copper Harbor Cong. | Basement | Pre-Mt. Simon "Red Beds" (Freda Sandstone Equivalents) |

Figure 3. Middle Proterozoic correlations of the pre–Mt. Simon clastics, southern peninsula, with equivalent rocks of the northern peninsula. See text for descriptions and discussion.

Catacosinos (1981a) has argued for continental deposition in a lacustrine environment, whereas Ojakangos and Morey (1982) suggest a fluvial origin for the rocks.

The age and origin of pre–Mt. Simon strata in the McClure Oil, Beaver Island #1 well, Charlevoix County (Fig. 2), is less certain. Based on analyses of the local geology (Catacosinos, 1973; Van der Voo and Watts, 1980) and an age date for granites present in the nearby McClure Oil, Beaver Island #2 well (Lidiak and others, 1966), these strata are thought to be the time equivalent of the middle Proterozoic Jacobsville Sandstone (Fig. 3) by the authors.

Fisher and others (1988) consider the pre–Mt. Simon section in the Beaver Island #1 well to be granite wash; however, Catacosinos (1973) has described them as sandstone, siltstone, and shale more than 240 m (813 ft) thick. Pale reddish purple in color, the lithology of the unit is varied; however, fine to coarse, rounded to subrounded quartz and feldspar clasts predominate. Catacosinos (1973, 1981a) suggested a marine origin for this section because initial examination indicated the possible presence of glauconite. Reexamination of the samples has not confirmed the presence of glauconite, so the section may be of continental (perhaps fluvial) origin rather than marine. A continental origin would more closely follow the sedimentary environments known for similar rocks in the region. As the evidence is insufficient at this time to make a firm judgement, the authors provisionally assign both a fluvial origin and a Precambrian age to the Beaver Island pre–Mt. Simon section. Figure 3 provides the correlations we presently prefer.

Several other wells have been drilled to the Precambrian, particularly in the southern portion of the rift zone, but to date only the two discussed above have definitively confirmed the presence of pre–Mt. Simon sediments within the Michigan Basin proper.

*Paleozoic sedimentary rocks*

In 1981, the Dart, Edwards 7-36 well in Falmouth Field, Missaukee County, Michigan (*not* Clare County as reported by Catacosinos, 1987), discovered commercial quantities of natural gas in Middle Ordovician sandstones below the Glenwood Formation. The subsequent exploration has greatly extended knowledge of the deeper formation that make up the central Michigan Basin. Unfortunately, it has also led to many improper changes to the existing lower Paleozoic lithostratigraphic nomenclature of the basin. Consequently, we have chosen to use only stratigraphic terminology widely accepted and used throughout the basin. While new terminology will certainly be required as basin studies continue, it should be developed in accordance with the rules provided by the North American Commission on Stratigraphic Nomenclature (1983).

The stratigraphic nomenclature used in this chapter is given in Figure 4. In ascending order the units are: Mt. Simon Sandstone; Eau Claire Formation; Galesville Sandstone; Franconia Formation; Trempealeau–Prairie du Chien (T-PDC); St. Peter Sandstone; and Glenwood Formation. These formations represent a predominantly transgressive sequence of sediments of shallow-marine origin, although periods of minor regression are also documented in the section.

*Mt. Simon Sandstone.* The Mt. Simon Sandstone unconformably overlies the Precambrian basement complex of the Michigan Basin. Because the Mt. Simon of the Michigan Basin is of probable Late Cambrian age (Catacosinos, 1973), the post–middle Proterozoic through Middle Cambrian hiatus represents about 500 m.y.

As defined here, the Mt. Simon consists of two sandstones. This is in contrast to the tripartite subdivision present in the Wisconsin outcrop areas (Driese and others, 1981). In the subsurface, the lower portion of the unit generally contains medium-to coarse-grained, subrounded to rounded, light-colored quartz grains that are predominantly silica cemented (Catacosinos, 1973). Toward the base it is somewhat pinkish or reddish in color, reflecting increases in both feldspar content and oxidation state (Lilienthal, 1978). Subordinate amounts of green shale, glauconite, anhydrite, and dolomite (as cement) have also been noted (Catacosinos, 1973). In contrast, the upper unit is a finer-grained sandstone that becomes glauconitic along the basin margins (the nearshore environment). Because of this high glauconite content and the gradational nature of the overlying Mt.

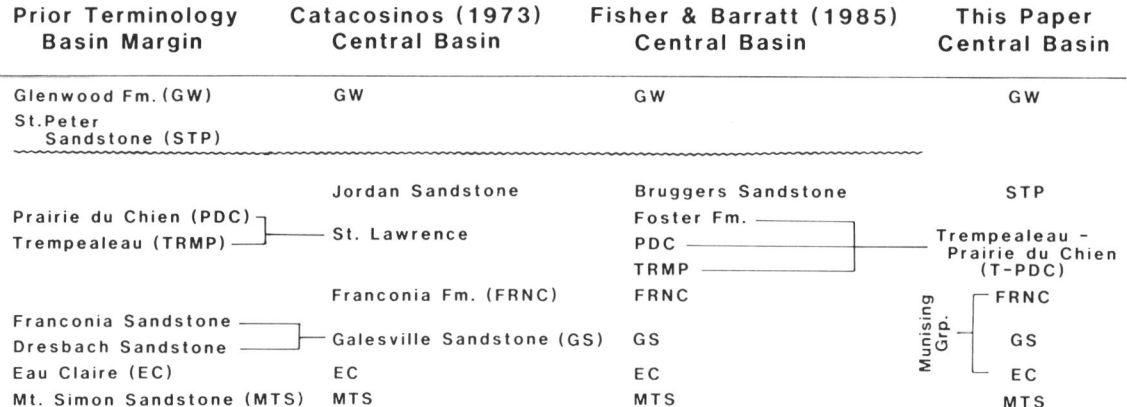

Figure 4. Lower Paleozoic stratigraphic nomenclature, Michigan Basin. The unconformity shown is the post-Sauk unconformity. See text for discussion.

Simon–Eau Claire contact, Catacosinos (1973, 1974, 1981b) included and mapped this sandy upper unit as the lower Eau Claire Formation. Others studying this unit elsewhere (e.g., Janssens, 1973, in Ohio) have followed this usage also.

Toward the central portion of the basin, however, this upper unit contains progressively less glauconite, which results in a lower gamma-ray response on logs. This in turn has permitted some workers to select an increasingly higher lithologic pick for the top of the Mt. Simon. In order to maintain consistency, the entire sandstone sequence has been referred to the Mt. Simon, regardless of glauconite content. This follows the usage of others (e.g., Calvert, 1963; Michigan Basin Stratigraphic Committee, 1969).

The Mt. Simon isopach map (Fig. 5) displays a marked thickening more than 420 m [1,400 ft]) in the west-central portion of the basin. This indicates the presence of an embryonic Michigan Basin. The thickest section penetrated so far is in the Amoco, Schiller 1-10, Oceana County, Michigan (some 416 m [1,388 ft] into Precambrian basement). The actual depocenter for the formation appears to be farther eastward into the basin (Fisher and Barratt, 1985). Two smaller lows are also present: one located southwest of Beaver Island, Charlevoix County (the northwestern quarter of the basin), and one in the southeastern portion of the basin in Lenewee County (Fig. 5). The Mt. Simon is considered to be missing in portions of southeastern Michigan by Catacosinos (1981b, 1987) because of erosion and/or non-deposition on preexisting Precambrian basement highs. These positive areas are located in two general areas (Fig. 5): in Livingston County where no Mt. Simon is present in the Mobil Oil, Messmore #1; and to the northeast in the "Thumb" area of Michigan, where the Mt. Simon is absent in three wells drilled into Precambrian basement in eastern Huron and Sanilac Counties. The absence of the Mt. Simon in these wells may be related to their proximity to the Sanilac platform ("A" on Fig. 2).

The Mt. Simon is also missing locally in the northern portion of the basin as, for example, in the Pan Am, Drasey #1 well, Presque Isle County, where the T-PDC directly overlies a (presumed) Precambrian quartzite (Catacosinos, 1973). Along the northern rim of the Michigan Basin, the Mt. Simon reportedly extends into the eastern portion of the northern peninsula, perhaps as far as the village of Grand Marais, Alger County, near the southern shore of Lake Superior (James Seglund, written communication, 1987). Mt. Simon sediments were not identified in the study area of Hamblin (1958).

The Mt. Simon is interpreted to be the basal unit of a marine transgressive succession initiated in probable Late Cambrian (Dresbachian) time. Outcrop studies of the unit in Wisconsin (Driese and others, 1981) show it to contain both thin, braided-fluvial and shallow-marine facies. The overall interpretation by Driese and others (1981) is that the depositional character of the Mt. Simon was truly transgressive marine only initially, becoming increasingly progradational in character as deposition continued. Within the Michigan Basin, the Mt. Simon was probably deposited in a wide variety of shallow-marine environments.

Few cores of the Mt. Simon are available for study in Michigan. One such core (Consumers Power, 1-7, St. Clair County) shows the Mt. Simon to be a fine-grained sandtone that is extremely well cemented and containing intercalated green shale partings suggestive of a low-energy (shallow embayment?) environment (Fig. 5). Elsewhere in the basin, the formation exhibits greater grain size, suggesting deposition under higher-energy conditions. Mt. Simon core analyses reported by Briggs (1968) show porosities ranging from 4 to 20 percent and permeabilities of as much as 32 millidarcies. The formation appears to have greater porosity along the western margins of the basin.

*Eau Claire Formation.* This unit consists of dark gray, locally glauconitic shales intercalated with thinly bedded siltstones and sandy dolomites. The distribution of the Eau Claire in the Michigan Basin is illustrated in Figure 6. Throughout much of the southern peninsula, this formation maintains a thickness of 30 to 60 m (100 to 200 ft). This consistent thinness suggests that stable tectonic conditions prevailed across much of the basin during its time of deposition (Dresbachian). Because no cores of the Eau Claire Formation are known to be available for public examina-

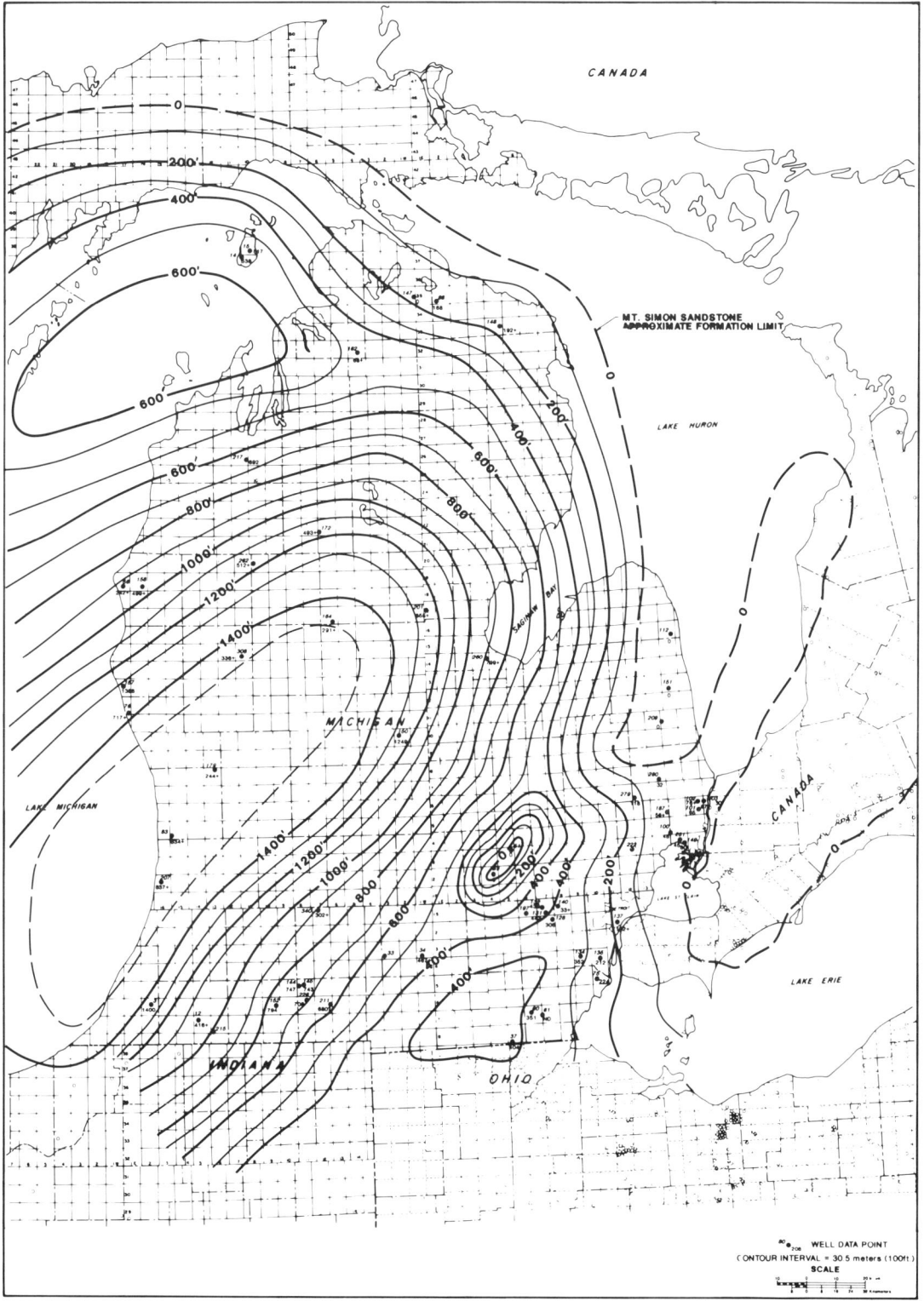

Figure 5. Isopach map, Mt. Simon Sandstone. The major depositional center is flanked by a subsidiary satellite to the northwest. The three Precambrian tests shown in the "thumb" area of Michigan locate a Precambrian high near the present-day Sanilac platform. A smaller basin and structural high are both shown in southeastern Michigan. The Mt. Simon isopach geometry strongly suggests the existence of the ancestral Michigan Basin during Dresbachian time.

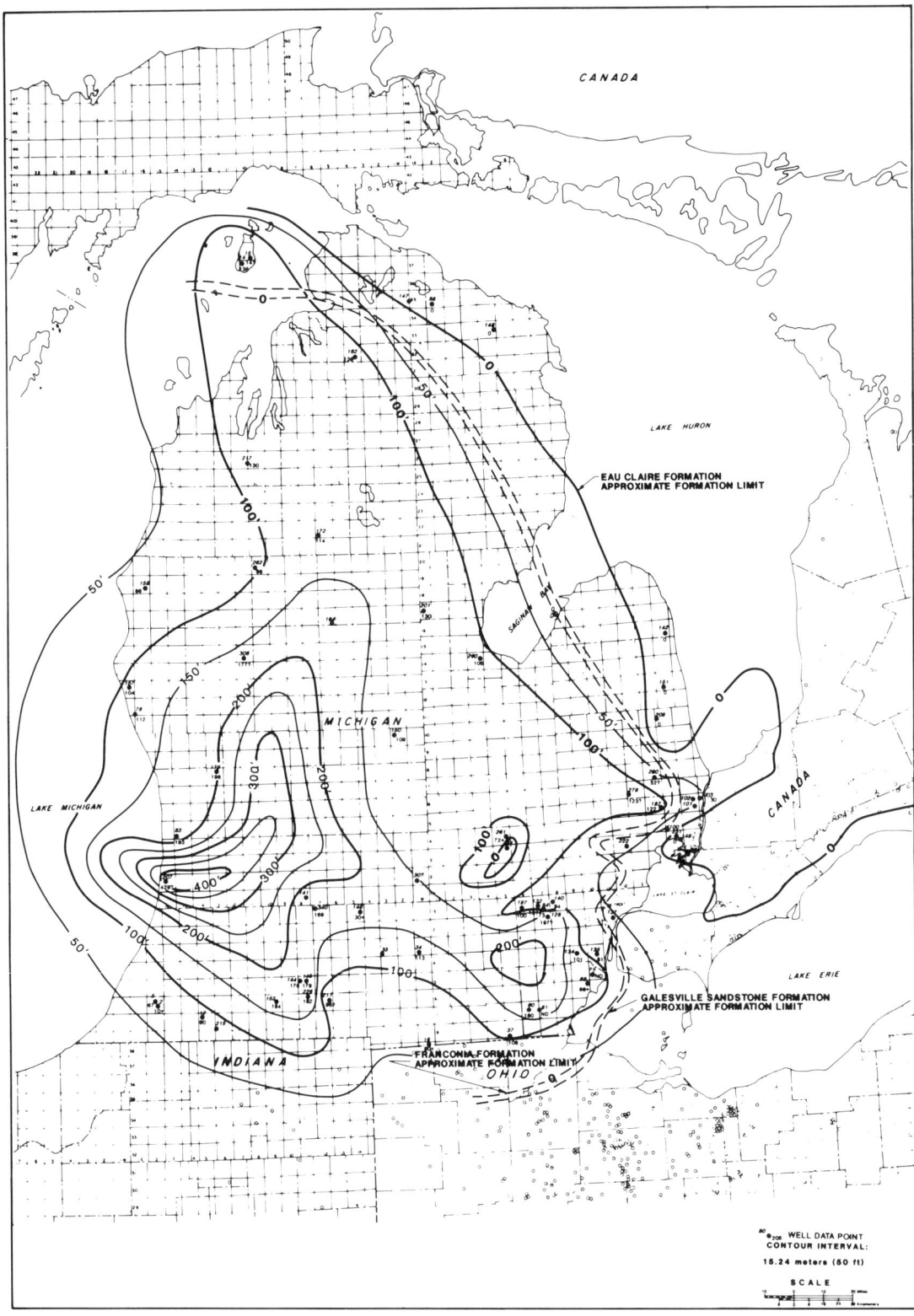

Figure 6. Isopach map, Eau Claire Formation. The depocenter is located in the southwestern portion of the southern peninsula. The dashed double line indicates the approximate edge of the overlying Galesville Sandstone. The "thin" in the southeastern quadrant of the mapped area indicates a basement high.

tion, it is difficult to characterize its environment of deposition in the central basin areas as other than shallow marine. In outcrop along the western margin of the basin, in Wisconsin, much of Eau Claire deposition appears to have occurred in tidal-flat environments and related shallow-marine settings (Driese and others, 1981). They further suggested that, taken together, the Mt. Simon and Eau Claire Formations constitute "a single progradational shoaling- and fining-upward sequence, truncated by an unconformity" (Driese and others, 1981).

As it is difficult to distinguish between the Eau Claire and the Franconia Formation where the intervening Galesville Sandstone is absent, the limits of the Galesville are also shown on Figure 6. Thus, where the Galesville is absence (indicated as the area between the double-dashed line and the zero isopach line on Fig. 6), the Eau Claire and Franconia Formations are mapped together. In the central Michigan Basin, the upper contact of the Eau Claire Formation appears to be conformable with either the Galesville or the Franconia Formation. Toward the margins of the basin, however, this contact is expected to become unconformable (as reported by Dreise and others, 1981). On wireline logs, the top of the Eau Claire is picked where there is a marked increase in gamma-ray response in comparison to the lower response of the overlying Galesville Sandstone (Michigan Basin Stratigraphic Committee, 1969).

*Galesville Sandstone.* The Galesville is lithologically identifiable as a light-colored, medium-grained, nonglauconitic sandstone, which is generally between 30 to 60 m (100 to 200 ft) in thickness (Fig. 7), and dimensionally resembles a sheet sandstone. In the Michigan Basin, the Galesville is an extension of the Ironton and Galesville Formations studied by Emrich (1966) in Illinois and Wisconsin (Catacosinos, 1973). In the central portion of the Michigan Basin, the Galesville Sandstone is gradational with both the underlying Eau Claire and overlying Franconia Formations. Toward the margins of the basin, however, the contacts may become unconformable.

The Galesville appears to mark a marine regressive interlude in the basin that took place toward the close of Dresbachian time. Slight thinning of the formation in the west-central portion of the Michigan Basin (Fig. 7) may denote broad tectonic (isostatic?) adjustments of the underlying Precambrian rift zone. For reference purposes, the edge of the underlying Eau Claire Formation is also shown on Figure 7.

Outcrop studies in Illinois and Wisconsin indicate that Galesville deposition occurred mainly in shallow-marine shelfal environments, as evidenced by macrofossils (e.g., *Elvina sp.*) and sedimentary features such as cross-stratification, oolites, and pebble conglomerates (Emrich, 1966). The Galesville tends to decrease upward in both textural and compositional maturity, though exhibiting a stable suite of heavy minerals throughout the section. The dominance of generally well-rounded quartz sand grains, low feldspar percentage, and a stable heavy mineral suite indicate a multicyclic history for the framework grains, their provenance being older sedimentary units (Emrich, 1966).

*Franconia Formation.* This formation consists primarily of gray to tan, sandy dolomites interbedded with a variable section of gray, fine-grained, glauconitic sandstones and shales. The dolomite content increases toward the eastern portion of the Michigan Basin. The maximum known thickness of the Franconia Formation (Fig. 8) is more than 60 m (200 ft). The two slightly basinal areas shown on Figure 8 may be the result of reactivated tectonic (isostatic?) activity along the underlying Precambrian rift zone during Franconian time.

Of shallow-marine origin, the formation represents a transgressive phase of the sea. The unit grades upward into the overlying carbonates of the T-PDC. As in the two previous figures, the underlying Galesville edge is indicated on Figure 8 for reference and orientation purposes.

In the southern peninsula of Michigan, the Eau Claire, Galesville, and Franconia Formations have been collectively referred to as the Munising Group because of their apparent genetic relation (Catacosinos, 1973, 1974). The age and stratigraphic relations of the Munising Group of the central Michigan Basin to the Munising Formation of the northern peninsula of Michigan are shown in Figure 9.

*Trempealeau–Prairie du Chien (T-PDC).* In this chapter, we combine the Trempealeau Formation (Upper Cambrian) with the overlying Lower Ordovician Prairie du Chien Group, which is perhaps as young as early Whiterockian (Harrison and Barnes, 1988), in the Michigan Basin. In our view, the carbonates of the Trempealeau and the overlying Prairie du Chien form a continuous depositional succession that is time transgressive. Differentiation of the Trempealeau from the Prairie du Chien in the Michigan Basin can be done (e.g., Brady and DeHaas, 1988a, b; Barnes and others, 1988; and others), but a definitive study of the unit has yet to be formally published in the literature as required by the Stratigraphic Code (North American Commission on Stratigraphic Nomenclature, 1983). Formal subdivision may also be possible for some of the various facies present within the Prairie du Chien of the Michigan Basin but, until this is done and properly reported in the literature, we prefer to combine these units and use the acronym "T-PDC" for clarity in communication.

The T-PDC of this chapter is the equivalent of the St. Lawrence of Catacosinos (1973), the Trempealeau and Foster Formations of Fisher and Barratt (1985), and the (informal) Trempealeau and Umlor formations of Brady and DeHaas (1988a, b). It also includes all or parts of units informally referred to as the "Brazos" shale and Munising formation (southern peninsula) in earlier literature and oil field terminology.

The unit consists of dolomite at the base and, in the central basin, grades upward into dolomitic siltstone and variously colored, but mostly black, shale (informally referred to as the "Brazos shale") at the top. The T-PDC is conformably overlain in an interfingering relation by the St. Peter Sandstone in the central basin, but on the edges of the basin it is separated from the St. Peter by the major regional unconformity that marks the top of

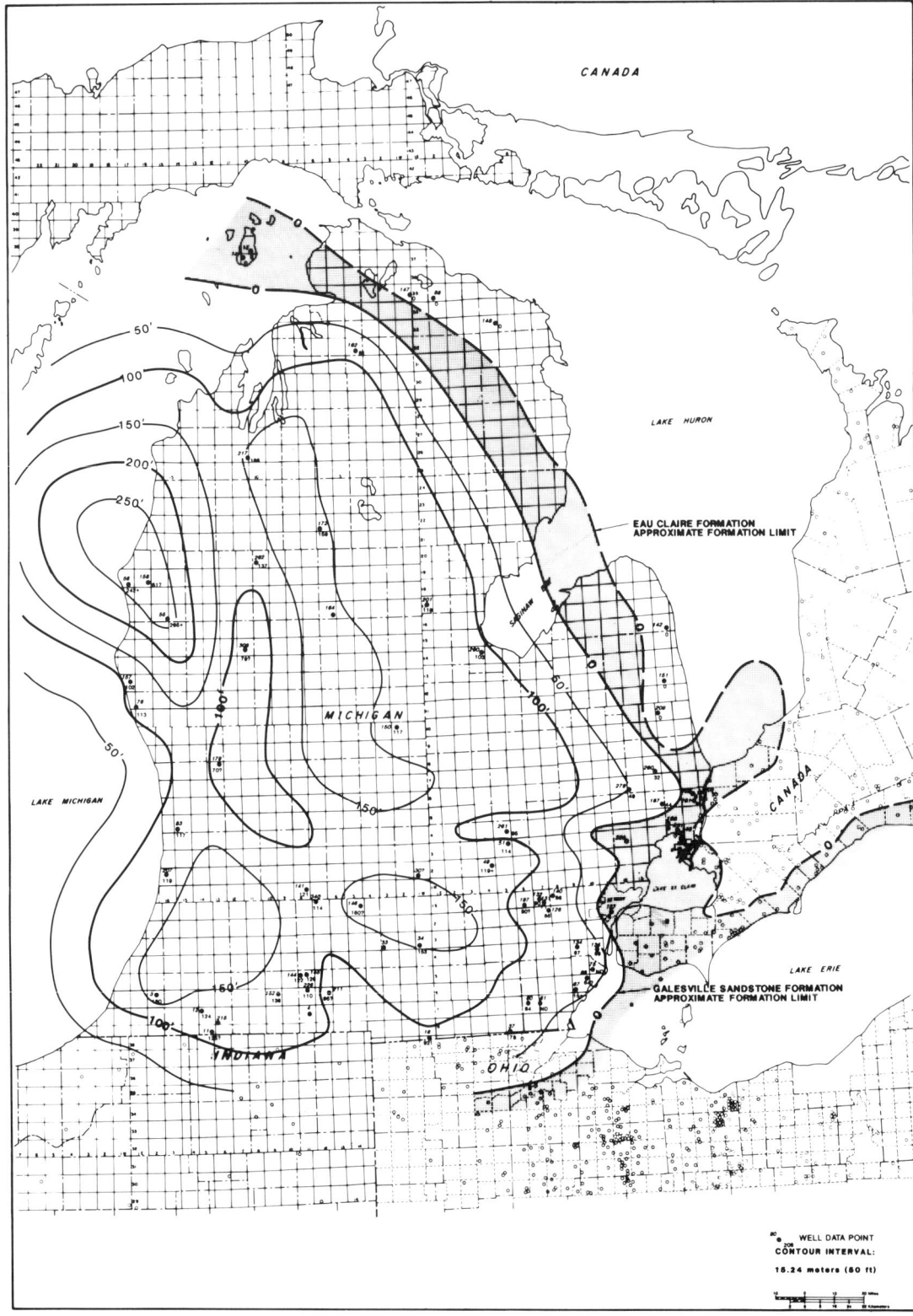

Figure 7. Isopach map, Galesville Sandstone. The configuration of this regressive marine sandstone suggests possible adjustment of the underlying Precambrian rift zone. The area where it is difficult to distinguish between the Eau Claire and Franconia Formations because of the absence of the intervening Galesville is shaded.

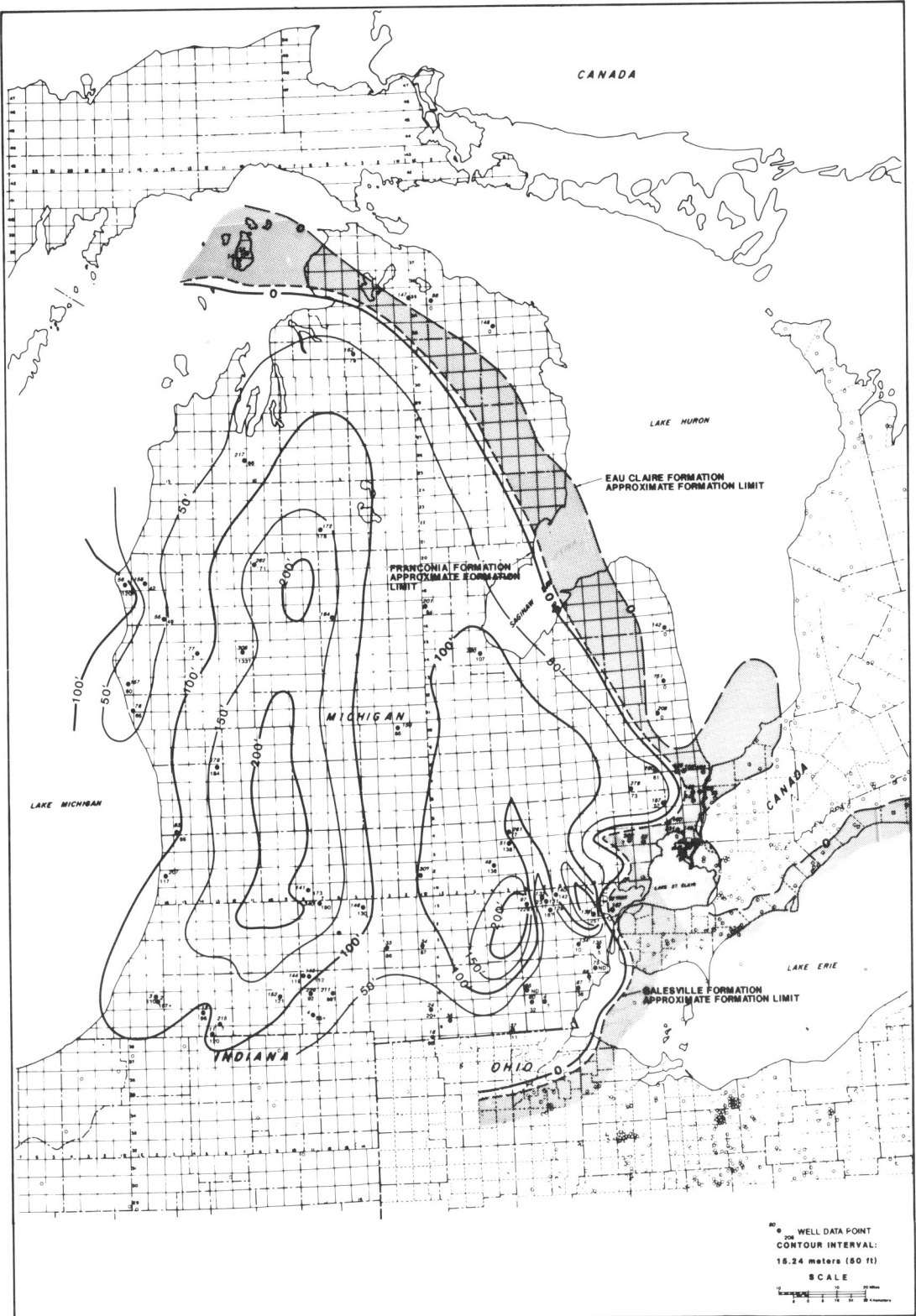

Figure 8. Isopach map, Franconia Formation. Shaded area is as described on Figure 7. Dashed line indicates the approximate position of the edge of the underlying Galesville Sandstone.

| Northern Peninsula | | Central Basin | |
|---|---|---|---|
| Formation | Age | Formation | Age |
| Glenwood | Chazy to Mohawkian | Glenwood | Chazy to Mohawkian |
| | | St. Peter Sandstone | Whiterockian to Chazy |
| Au Train (PDC Equivalent) | Canadian | Trempealeau – Prairie du Chien | Trempealeaun to Canadian |
| Munising Fm. | Dresbachian to Franconian | Mun. Grp. { Franconia Fm. / Galesville Sandstone / Eau Claire Fm. / Mt. Simon Sandstone | Dresbachian to Franconian |
| Mt. Simon Sandstone | | | |
| Jacobsville Sandstone | Middle Proterozoic | Pre-Mt. Simon Sediments or Crystalline Basement | Middle Proterozoic |

Figure 9. Age and correlation of the middle Proterozoic to Middle Ordovician section, northern peninsula, to that of the central Michigan Basin. In the northern peninsula, Whiterockian to Chazy beds were removed by the combined post-Sauk (middle) unconformity and post–St. Peter (upper) erosional surface (see text). The formational status of the Au Train Formation (Canadian) of Hamblin (1958) is unclear at this time.

the Sauk sequence (Sloss, 1963). Our correlation of the unit is shown in Figure 9.

The T-PDC has a known thickness of 726 m (2,420 ft) in the Hunt Energy, Martin 1-15 well in Gladwin County, and may exceed 750 m (2,500 ft) in thickness. It is apparent from its isopach geometry and thickness (Fig. 10) that T-PDC deposition marks the inception of the modern Michigan Basin. That deposition may have occurred as early as the Trempealeaun (Late Cambrian), but more likely is Canadian (Early Ordovician).

As indicated by well logs, the T-PDC contains complex facies, particularly in the central basin where recent drilling has found commercial quantities of hydrocarbons (primarily gas) in various parts of the sequence. The T-PDC is time transgressive where, at its base, it is Trempealeaun (Late Cambrian), whereas most of the overlying strata in the central basin (i.e., the Foster Formation of Fisher and Barratt, 1985) were reported by them to be Canadian-Whiterockian (Early and Middle Ordovician) based on conodont analysis by Repetski and Harris (1981). Also based on conodont biostratigraphy, Shaw (1988) reports the Early/Middle Ordovician boundary to be either within, or just below, the Brazos shale. In the northern peninsula of Michigan, the equivalent Au Train Formation (Hamblin, 1958) has been assigned a Canadian age (Early Ordovician) by Guldenzopf (1967) on the basis of his conodont studies (Fig. 9).

*St. Peter Sandstone.* This formation was originally thought to be completely missing in the Michigan Basin by some workers (e.g., Catacosinos, 1973) or at least of limited thickness and extent by others (e.g., Bricker and others, 1983). Recently, however, Harrison (1987) and Barnes and others (1988) have presented convincing arguments as to the identification and correlation of this unit. The terminology developed for this sandstone in recent years in the Michigan Basin is quite complex (Fig. 4). Although an overview is provided by Figure 4, the following examples may serve to make the point. This sandstone has been called the Bruggers Formation by Fisher and Barratt (1985); and also previously identified as a facies of the Prairie du Chien (or St. Lawrence) and called the Jordan Sandstone by Catacosinos (1973). It is also commonly called the Prairie du Chien Sand, Massive Sand, or Knox Sandstone in Michigan Basin oil field usage. Other terms applied include New Richmond Sandstone and the St. Peter Sandstone. We believe that the initial results of the on-going work by Harrison (1987) and Barnes and others (1988) provide adequate arguments to preclude the usage of all names other than St. Peter Sandstone, and consequently we have used that terminology in this chapter and recommend abandoning the others.

In the Michigan Basin, the St. Peter attains a thickness exceeding 330 m (1,100 ft) in the central basin area (Fig. 11). The lower portion of this environmentally complex, "massive" sandstone typically contains dolomite interbeds. Within the basin, the St. Peter is a transgressive marine succession that was deposited in peritidal to storm-dominated outer-shelf environments (Lundgren and Barnes, 1989). Harrison (1987) delineated the following depositional environments: (1) a high-energy beach and shoreface, and (2) an uppermost bioturbated, lower energy tidal flat or shallow subtidal zone. More recent work indicates the existence of at least four depositional environments, including: coastal to shallow shelf; paralic; storm-dominated, lower shoreface to upper offshore; and the marine shelf (Barnes and others, 1988).

The age of the St. Peter of the central Michigan Basin is Middle Ordovician (Whiterockian-Chazyan; Repetski and Harris, 1981). Shaw (1988) notes that conodont biostratigraphy documents this thick sandstone unit to be coeval, in part, with the Everton Dolomite and the St. Peter Sandstone of the Illinois Basin and should not be designated Prairie du Chien.

The St. Peter rests conformably upon, and interfingers with, the T-PDC in the central basin, but this is not the case regionally (Fig. 12). The St. Peter was deposited atop the major regional post-Sauk (also known as Knox) unconformity both along the basin margins and elsewhere in the mid-continental United States. In the central Michigan Basin, the top of the formation grades upward into the lower Glenwood Formation. There is no post-Sauk (i.e., major) unconformity present there as had been postulated by Catacosinos (1973).

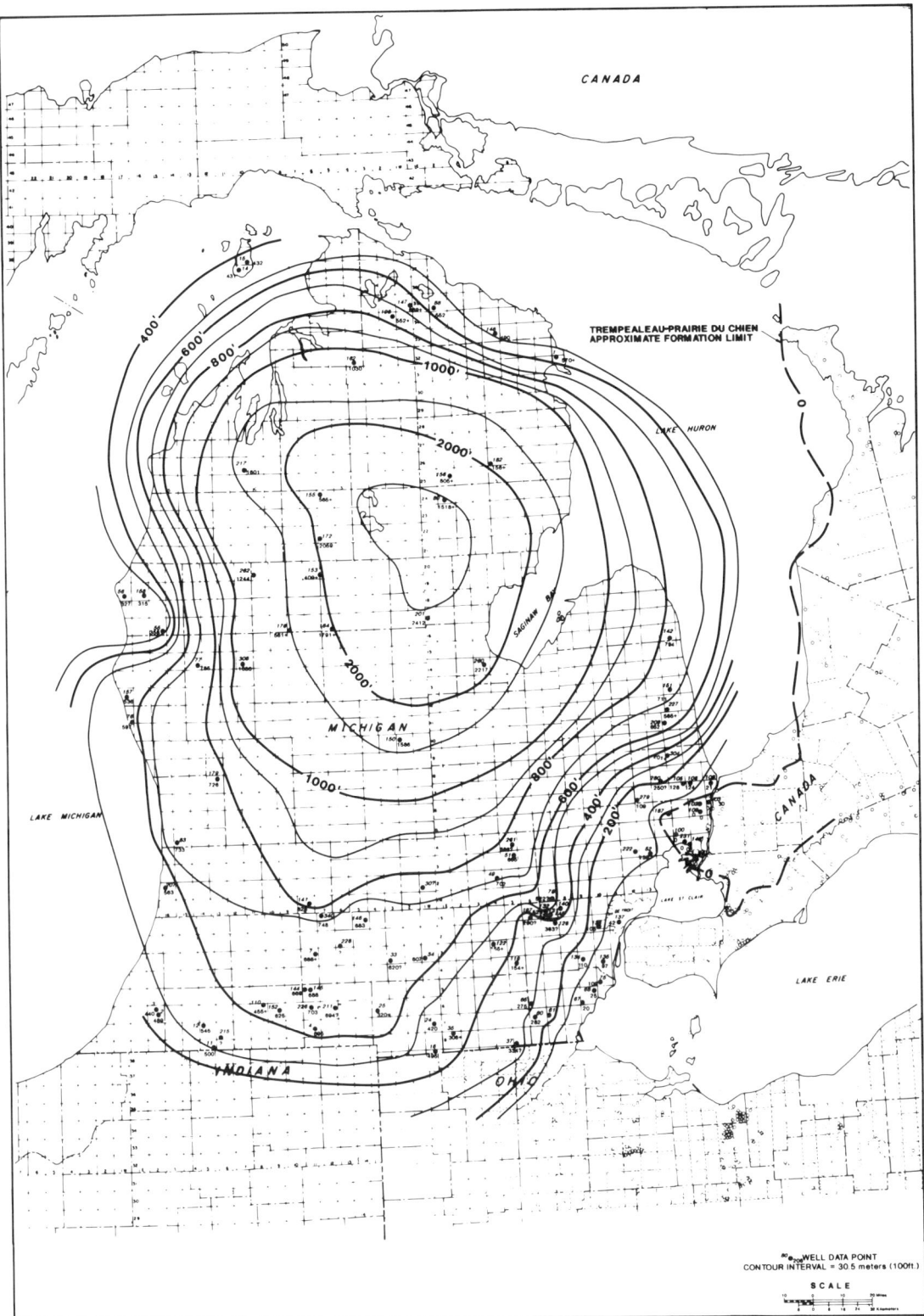

Figure 10. Combined isopach map of Trempealeau–Prairie du Chien (T-PDC) rocks. Basin depocenter has shifted away from the west-central to the north-central portion of the southern peninsula. These strata record the initiation of the modern structural Michigan Basin, probably in Canadian (Early Ordovician) time.

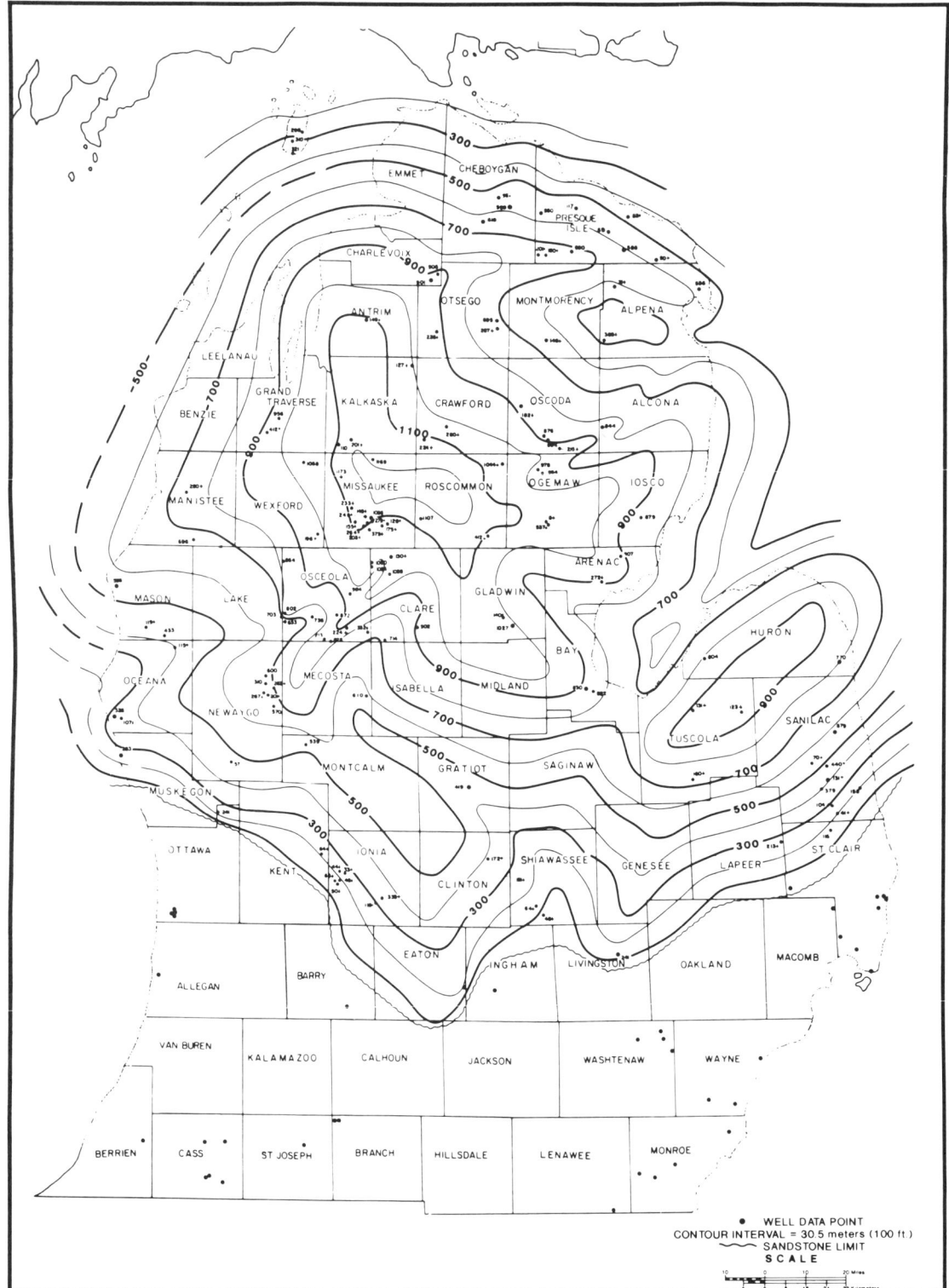

Figure 11. Isopach map, St. Peter Sandstone. Basin depocenter is in the north-central southern peninsula. This thick, complex, marine sandstone is the principal hydrocarbon target of the current "deep play" in Michigan. Thin erosional remnants of the St. Peter preserved in southwestern Michigan are not shown.

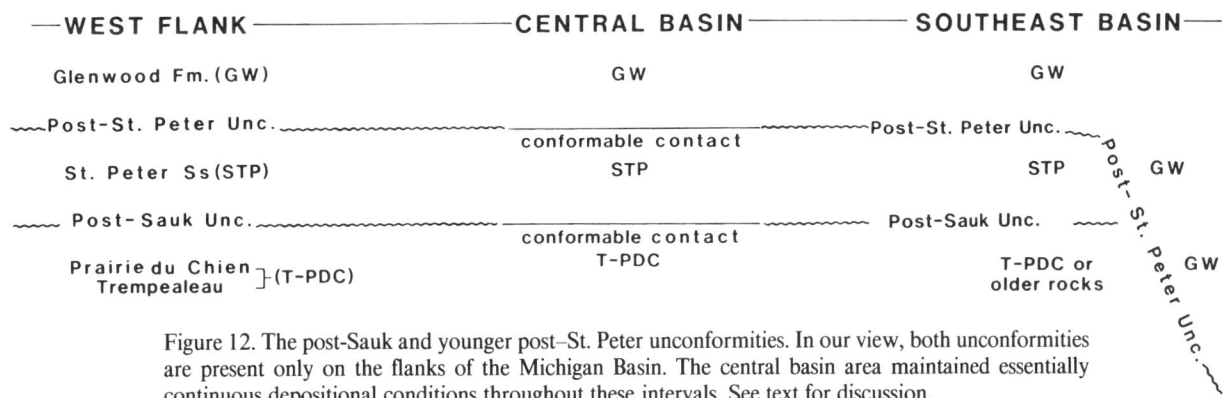

Figure 12. The post-Sauk and younger post-St. Peter unconformities. In our view, both unconformities are present only on the flanks of the Michigan Basin. The central basin area maintained essentially continuous depositional conditions throughout these intervals. See text for discussion.

*Glenwood Formation.* The Glenwood Formation of the Michigan Basin (Middle Ordovician: Chazy-Mowhawkian) is a varied succession of shale, gray argillaceous dolomite, and thin sandstone and limestone interbeds, which has been described by Harrison and Barnes (1988) as a condensed section of marine shelf deposits. In the central basin, the Glenwood Formation attains a thickness of 39 m (130 ft; Harrison and Barnes, 1988) and interfingers with the underlying St. Peter Sandstone. Toward the edges of the basin, however, the Glenwood unconformably overlaps progressively older Ordovician, Cambrian, and Precambrian rocks.

The lower portion of the Glenwood Formation has been informally referred to as the "zone of unconformity" (Bricker and others, 1983). Although rip-up clasts and other sedimentary structures that indicate a high-energy environment are present within the lower Glenwood at some localities, deposition of the Glenwood in the central Michigan Basin is interpreted by the writers to have been essentially continuous, and any hiatuses are considered to be minor. Only along the basin margins did a significant post–St. Peter erosional event occur.

## The upper boundary of the Sauk Sequence in the Michigan Basin

At this time there is no universal agreement on the placement of the major unconformity marking the top of the Sauk sequence (Sloss, 1963) within the Michigan Basin. In our view, this major regional unconformity (the post-Sauk Unconformity) is not present in the central portion of the Michigan Basin, but is present along the basin margin and elsewhere in the mid-continental United States. In the central Michigan Basin, the contact between the T-PDC and the overlying St. Peter Sandstone is marked by an intercalated zone of sandstone and dolomite just above a dark shale (improperly called the Brazos shale) at the top of the T-PDC. It is probable that deposition within this sandstone and dolomite zone represents a period of fluctuating sea level that accompanied a marked regression along the basin's margins. This resulting "central basin flanking" erosional surface marks the major (post-Sauk) unconformity and is the surface upon which the St. Peter Sandstone was deposited in all areas of the region, except in the central Michigan Basin.

It appears, however, that a younger unconformity of lesser magnitude does separate the St. Peter Sandstone from the overlying Glenwood Formation on the eastern, western, and northern margins of the Michigan Basin. Although this younger post–St. Peter unconformity does not extend into the central basin area, it may be coeval with the lower Glenwood zone that contains numerous rip-up clasts and may possibly correlate with the "pre-Glenwood unconformity" of Wheeler (1987). Along the basin margins, this post–St. Peter unconformity progressively truncates the lower Glenwood Formation and the St. Peter and merges with the underlying post-Sauk unconformity upon which the St. Peter Sandstone lies, along the margins of the Michigan Basin. Thus, where the lower Glenwood and St. Peter have been removed, the remaining upper portion of the Glenwood Formation rests unconformably upon either the T-PDC or lower formations. Figure 12 schematically presents our concept of the geometry of these two Ordovician unconformities within the Michigan Basin.

## GEOLOGIC HISTORY

As outlined by Fowler and Kuenzi (1978), during middle Proterozoic ("X") time, continental rifting began in the Michigan Basin. This rifting was associated with similar events taking place in the mid-continent region (Dickas, 1986). In Michigan, the rifting ceased approximately 1 Ga. Within the rift zone, pre–Mt. Simon sediments would have been deposited in a variety of continental settings; erosion during the next 500 m.y. apparently removed almost all traces of those sedimentary rocks.

By the Late Cambrian (Dresbachian), initial formation of an ancestral Michigan Basin had taken place, and the area was inundated by the waters of a northerly transgressing sea. Although probably initially deposited upon a post-Precambrian erosional surface under shallow-marine transgressive conditions present in the central basin area, the Upper Cambrian Mt. Simon Sandstone was also accumulating in a wide variety of other environments along the basin's margins. Various examples, such as braided fluvial deposits, occur in Wisconsin (Dreise and others, 1981).

Basinal subsidence continued at a slow pace during deposition of the Eau Claire Formation. Essentially stable tectonic conditions prevailed until near the close of Dresbachian time, when a marine regression took place. This is recorded by the progradation of the Galesville Sandstone into the ancestral Michigan Basin. The Galesville is similar to the Mt. Simon Sandstone in that deposition occurred under shallow-marine condition, with a wide variety of sub-environments present along the margins of the basin. A range of such possible depositional environments has been documented for cratonic sheet (or blanket) sandstones by Dott and others (1986). By Franconian time, shallow-marine transgressive conditions were reestablished, with subsidence still proceeding at a slow pace during deposition of the Franconia Formation.

During the Trempealeauan (Late Cambrian) and into Canadian time (Early Ordovician), shallow-water carbonate deposition prevailed as the Trempealeau–Prairie du Chien (T-PDC) was deposited. The thick accumulation of T-PDC sediments marks the beginning of the formation of the modern-day Michigan Basin, with probable initiation occurring during Canadian (Early Ordovician) time.

Toward the close of Canadian time and into Whiterockian time (Middle Ordovician), a relative lowering of base level occurred, as indicated by the increase in the amount of fine-grained clastics, including dark shales, in the central basin. This condition developed into a major lowering of sea level in the basin, exposing the margins of the basin to subsequent erosion of not only the T-PDC but also lower strata in some areas. This period of marine regression terminated Sauk sequence deposition and created the post-Sauk (also known as Knox) unconformity, upon which the sands of the St. Peter prograded. At the same time, within the central portion of the developing Michigan Basin, deposition continued without major interruption. Only minor fluctuations in water depth marked the interfingering of sand with dolomite or shale interbeds near the contact between the T-PDC and the St. Peter. Very slow basinal sinking continued; more than 330 m (1,100 ft) of St. Peter sediments accumulated throughout the Whiterockian and much of the Chazyean.

Near the close of the Chazyean, marine regression occurred, again exposing the margins of the basin. As this regressive phase continued, some St. Peter strata were progressively removed along the basin margins, producing the post–St. Peter unconformity. Deposition continued without major interruption in the central basin area, as indicated by the gradational contact between the St. Peter and the overlying Glenwood Formation. Thus, this Chazyean or post–St. Peter erosion surface that developed was a "basin-margin" event. Toward the basin margins, this post–St. Peter erosion surface merged with the older post-Sauk (or Knox) unconformity (Whiterockian), thereby removing the lower Glenwood, St. Peter, and, in some places, still lower rocks. The amount of such removal has not yet been determined due to complicating factors such as regional thinning, truncation, and facies changes.

By the close of the Chazyean and during early Mohawkian time, marine transgression of the basin again occurred, marked by widespread deposition of the uppermost member of the Glenwood Formation (i.e., the Glenwood shale zone). Continued subsidence and deepening of the basin is indicated by deposition of the overlying Trenton–Black River carbonates. During the deposition of these strata, a major shift in sedimentation patterns occurred, with migration of the Michigan Basin depocenter to the far southeastern portion of the basin. By the close of deposition of the Trenton Limestone during the Middle Ordovician (Mohawkian), the present geometry of the modern Michigan Basin had been essentially established, although it was to be further modified by subsequent Paleozoic structuring events.

## SUMMARY AND CONCLUSIONS

Whereas recent drilling for oil and gas in the Glenwood, St. Peter, and Trempealeau–Prairie du Chien has increased our knowledge of these strata many fold, there is still a lack of information on the Cambrian System, particularly in the central Michigan Basin area. Complex facies relations among Cambrian formations exist and are under study (e.g., Barnes and others, 1988). Such studies, however, remain to be documented and formally published in the literature. Figure 13, a fence diagram of the southern peninsula of Michigan, is intended to provide a

### TABLE 1. WELLS USED IN CONSTRUCTING FIGURE 13, FENCE DIAGRAM

| Permit No. | Well Name | Co. Location |
|---|---|---|
| 1. 26112 | Security-Thalman #1 | 6S–17W,10/Berrien |
| 2. BD* | Holland–Suco 32 | 5N–15W,30/Ottawa |
| 3. 34885 | Gulf–Umlor #1-3 | 8N–13W,3/Ottawa |
| 4. 33134 | Amoco–Shiller 1-10 | 13N–18W,10/Oceana |
| 5. 39984 | Miller Bros.–Vict. #2-26 | 19N–17W,26/Mason |
| 6. 34319 | Shell–St. Blair #2-24 | 26N–11W,24/Grand Traverse |
| 7. 34824 | Energy Acq.–NML&O #1-27 | 32N–4W,27/Charlevoix |
| 8. 27199 | Pan Am.–Draysey #1 | 35N–2E,29/Presque Isle |
| 9. 29372 | Shell–Taratuta #31-13 | 33N–5E,13/Presque Isle |
| 10. 25690 | PEPL–Ford Motor #1-5 | 31N–9E,5/Alpena |
| 11. 23435 | McClure–Beaver Isle. #1 | 38N–10W,27/Charlevoix |
| 12. 34376 | Jem–Doornbos #5-30 | 22N–6W,30/Missaukee |
| 13. 35090 | Hunt Energy–Martin #1-15 | 17N–1E,15/Gladwin |
| 14. 37779 | Shell–Prevost #1-11 | 14N–4E,11/Bay |
| 15. 29191 | Mobil–Volmering #1 | 15N–15E,26/Huron |
| 16. 39856 | Wolv.–Patr.&Norw. #2-28 | 15N–11W,28/Newaygo |
| 17. 29739 | McClure–Sparks and others #1-8 | 10N–2W,8/Gratiot |
| 18. 27986 | Mobil–Messmore #1 | 3N–5E,11/Livingston |
| 19. 33737 | Energy Acq.–Gri'son #1-24 | 5N–14E,24/Macomb |
| 20. 33999 | Mid-Amer–Woodruff #1-19 | 10N–15E,19/Sanilac |
| 21. 29966 | Consumers Power–Clark #1 | 5S–8W,8/Branch |
| 22. 22275 | Collin and Black–Dancer #1 | 3S–1W,29/Jackson |
| 23. 34223 | Hunt Energy–Worrell #1-28 | 1S–6E,28/Washtenaw |

*BD = brine disposal well

*Stratigaphy of Middle Proterozoic to Middle Ordovician formations* 69

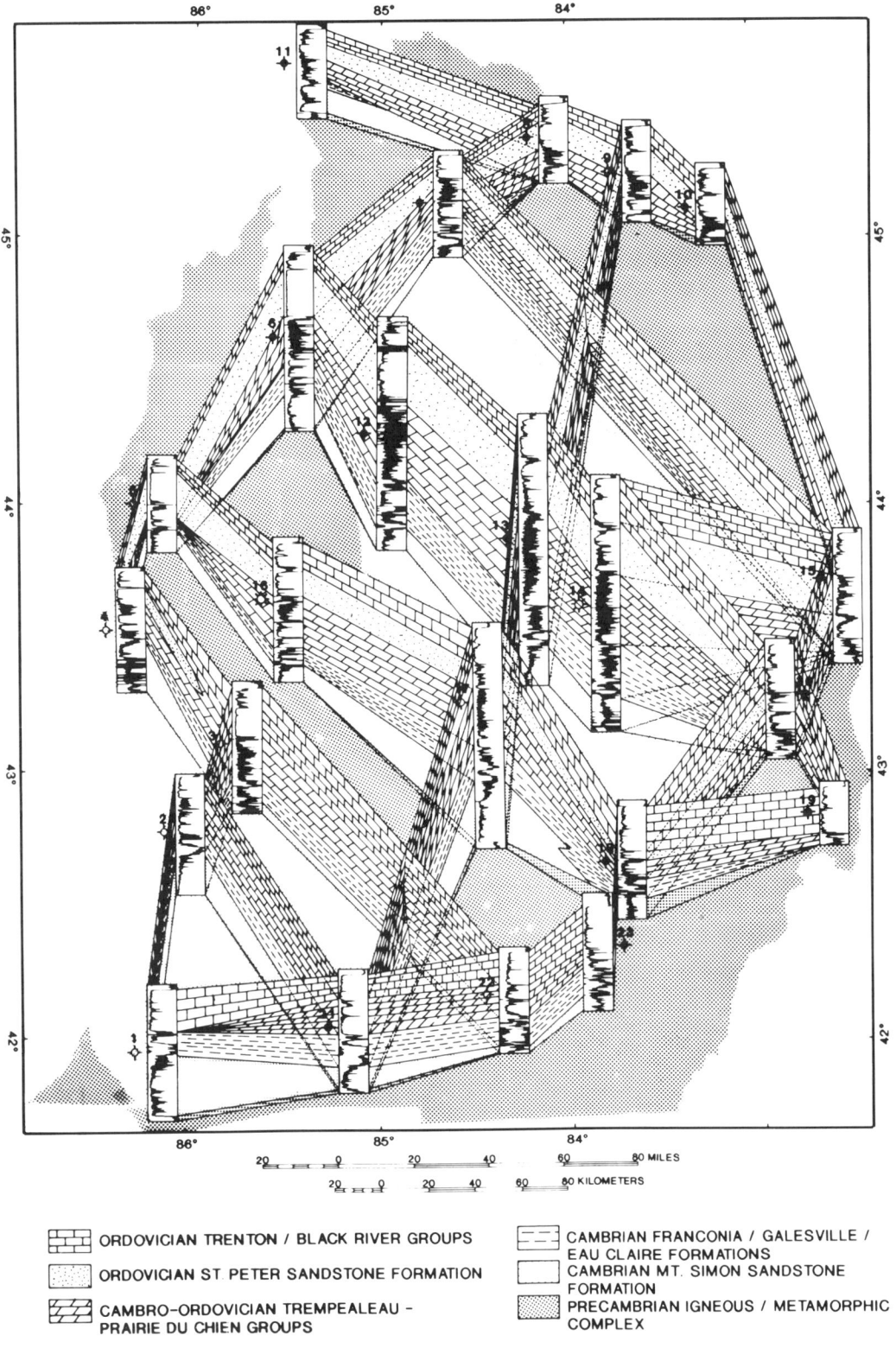

Figure 13. Fence diagram. The diagram illustrates the distribution of the major stratigraphic units discussed. The Glenwood Formation is too thin to be shown at the vertical scale used for construction of this figure (1 in equals 1,240 ft prior to its reduction). The 23 wells utilized are given in Table 1. Computer graphics are courtesy of Aangstrom Precision Corporation, Mt. Pleasant, Michigan.

guide or framework for future stratigraphic investigations. The wells utilized in its construction are listed in Table 1.

Although rifting occurred in the Michigan Basin area during the middle Proterozoic, the role it played, if any, in the subsequent structural development of the basin is as yet unclear, as subsurface information from this interval (i.e., about 1,000 to 500 Ma) is presently available from only two wells in the basin.

A change in depositional environment from continental to marine occurred during the Late Cambrian, caused by northerly transgressing epeiric seas. As yet, no cores from this series are available for study, but such data are essential for the biostratigraphic studies needed. If this core data is obtained by future drilling, detailed temporal correlations and facies analyses within the Sauk sequence will become possible and will provide a rich harvest of information.

## ACKNOWLEDGMENTS

The writers gratefully acknowledge the assistance of Delta College, Michigan, and Petro-Hunt Corporation, Dallas, Texas, in the preparation of this manuscript. We also thank C. W. "Neal" Barnes of Coastal Oil and Gas Corporation, Jackson, Mississippi, and James A. Seglund, consulting geologist, Metairie, Louisiana, for information and ideas on the structure and stratigraphy of the Michigan Basin. F. Bryan Davies and John Esch of Aangstrom Precision Corp., Mt. Pleasant, Michigan, kindly provided computer graphics and services. Tim Turmelle, Peninsular Oil and Gas Company, Grand Rapids, Michigan, and Rob Chapman, Petrostar Energy, Traverse City, Michigan, provided key well logs to us. We deeply appreciate our colleagues Dave Barnes and Bill Harrison and their students at the Core Research Laboratory of Western Michigan University, Kalamazoo, for their constructive criticisms and valuable insights, which added much to the quality of this chapter. The comments and constructive criticisms of reviewers Donald E. Owen, Lamar University, Beaumont, Texas, and David J. Poche, Amoco Production Company, Houston, Texas, greatly improved this manuscript. Sole responsibility, however, rests with the authors.

## REFERENCES CITED

Barnes, D. E., Harrison, W. B., Lundgren, C. E., and Wieczorek, L. M., 1988, The lower Paleozoic of the Michigan Basin, a core workshop: Kalamazoo, Western Michigan University Core Research Laboratory, 65 p.

Brady, R. V., and DeHaas, R. J., 1987, Updated isopach maps of Prairie du Chien, Trempealeau, and Munising of Michigan [abs]: American Association of Petroleum Geologists Bulletin, v. 71, p. 1102.

——, 1988a, The "deep" (pre-Glenwood) formations of the Michigan Basin: Part 4, The Umlor Formation: Michigan's Oil and Gas News, v. 94, no. 22, p. 35–43.

——, 1988b, The "deep" (pre-Glenwood) formations of the Michigan Basin; Part 5, The Trempealeau Formation: Michigan's Oil and Gas News, v. 94, no. 26, p. 32–40.

Bricker, D. M., Milstein, R. L., and Reszka, C. R., Jr., 1983, Selected studies of Cambro–Ordovician sediments within the Michigan Basin: Michigan Geological Survey Report of Investigation 26, 54 p.

Briggs, L. I., Jr., 1968, Geology of subsurface waste disposal in Michigan Basin, in Galley, J. E., ed., Subsurface disposal in geologic basins; A study of reservoir strata: American Association of Petroleum Geologists Memoir 10, p. 128–153.

Brown, L., Jensen, L., Oliver, J., Kaufman, S., and Steiner, D., 1982, Rift structure beneath the Michigan Basin from COCORP profiling: Geology, v. 10, p. 645–649.

Calvert, W. L., 1963, Sub-Trenton rocks of Ohio in cross sections from West Virginia and Pennsylvania to Michigan: Ohio Geological Survey Report of Investigation 49, 5 p.

Cannon, W. F., and 15 others, 1989, The North American Midcontinent rift beneath Lake Superior from GLIMPSE seismic reflection profiling: Tectonics, v. 8, no. 2, p. 305–332.

Catacosinos, P. A., 1973, Cambrian lithostratigraphy of Michigan Basin: American Association of Petroleum Geologists Bulletin, v. 57, p. 2404–2418.

——, 1974, Cambrian and Cambro–Ordovician stratigraphic framework of the Michigan Basin: Ontario Petroleum Institute, 13th Annual Conference, Technical Paper 11, 18 p.

——, 1981a, Origin and stratigraphic assessment of pre-Mt. Simon clastics (Precambrian) of Michigan Basin: American Association of Petroleum Geologists Bulletin, v. 65, p. 1617–1620.

——, 1981b, Various Cambrian maps, plate 11, in Hydrology for underground injection control in Michigan; Part 1, Hydrogeologic atlas of Michigan: Kalamazoo, Western Michigan University Department of Geology, 778 p.

——, 1987, East Michigan potential hydrocarbon horizons: Oil and Gas Journal, v. 85, no. 27, p. 58–59.

Catacosinos, P. A., Daniels, P. A., Jr., and Harrison, W. B., III, 1991, Structure, stratigraphy and petroleum geology of the Michigan Basin, in Leighton, M. W., Kolata, R. D., Oltz, D. F., and Eidel, J. J., eds., Interior cratonic basins: Tulsa, Oklahoma, American Association of Petroleum Geologists Memoir (World Petroleum Basins), Chapter 6 (in press).

Chandler, V. W., McSwiggen, P. L., Morey, G. B., Hinze, W. J., and Anderson, R. R., 1989, Interpretation of seismic reflection, gravity, and magnetic data across middle Proterozoic Mid-Continent rift system, northwestern Wisconsin, eastern Minnesota, and central Iwoa: American Association of Petroleum Geologists Bulletin, v. 73, p. 261–275.

Cohee, G. V., and Landes, K. K., 1958, Oil in the Michigan Basin, in Weeks, L. G., ed., Habitat of oil: Tulsa, Oklahoma, American Association of Petroleum Geologists, p. 473–493.

Daniels, P. A., Jr., 1982, Upper Precambrian sedimentary rocks; Oronto Group; Michigan–Wisconsin, in Wold, R. J., and Hinze, W. J., eds., Geology and tectonics of the Lake Superior Basin: Geological Society of America Memoir 156, p. 107–133.

——, 1986, Oronto Group; A petroliferous rift-fill sequence of the Mid-Continent rift system [abs]: American Association of Petroleum Geologists, v. 70, p. 579.

Daniels, P. A., Jr., and Elmore, R. D., 1988, Upper Keweenawan rift-fill sequence, Mid-Continent rift system, in Wollensak, M. S., ed., Upper Keweenawan rift fill sequence, Mid-Continent rift, Michigan: Michigan Basin Geological Society, Annual Field Excursion, p. 1–23.

Dickas, A. B., 1986, Comparative Precambrian stratigraphy and structure along the Mid-Continent rift: American Association of Petroleum Geologists Bulletin, v. 70, p. 225–238.

Dott, R. H., Jr., Byers, C. W., Fielder, G. W., Stenzel, S. R., and Winfree, K. E., 1986, Aeolian to marine transition in Cambro–Ordovician cratonic sheet sandstones of the northern Mississippi Valley, U.S.A.: Sedimentology, v. 3, p. 345–367.

Driese, S. G., Byers, C. W., and Dott, R. H., Jr., 1981, Tidal deposition in the basal Upper Cambrian Mt. Simon Formation in Wisconsin: Journal of Sedimentary Petrology, v. 51, p. 367–381.

Ells, G. D., 1967, Correlation of Cambro–Ordovician rocks in Michigan: Michigan Basin Geological Society Annual Field Excursion Guidebook, p. 42–57.

——, 1969, Architecture of the Michigan Basin: Michigan Basin Geological Society Annual Field Excursion Guidebook, p. 60–88.

Elmore, R. D., 1983, Precambrian non-marine stromatolites in alluvial fan deposits, the Copper Harbor Conglomerate, upper Michigan: Sedimentology, v. 30, p. 829–842.

——, 1984, The Copper Harbor Conglomerate; A Late Precambrian fining-upward alluvial fan sequence in northern Michigan: Geological Society of America Bulletin, v. 95, p. 610–617.

Elmore, R. D., Engel, M. H., and Daniels, P. A., Jr., 1988, The Nonesuch Formation; Lacustrine sedimentation in a Precambrian mid-continent rift basin, in Wollensack, M. S., ed., Upper Keweenawan rift fill sequence, Mid-Continent rift, Michigan: Michigan Basin Geological Society Annual Field Excursion, p. 103–125.

Emrich, G. H., 1966, Ironton and Galesville (Cambrian) Sandstones in Illinois and adjacent areas: Illinois State Geological Survey Circular 403, 55 p.

Fisher, J. H., 1969, Early Paleozoic history of the Michigan Basin: Michigan Basin Geological Society Annual Field Excursion Guidebook, p. 89–93.

Fisher, J. H. and Barratt, M. W., 1985, Exploration in Ordovician of central Michigan Basin: American Association of Petroleum Geologists Bulletin, v. 69, p. 2065–2076.

Fisher, J. H., Barratt, M. W., Shaver, J. B. Droste, and Shaver, R. H., 1988, Michigan Basin in Sloss, L. L., ed., Sedimentary cover–North America Craton; U.S.: Boulder, Colorado, Geological Society of America, The Geology of North America, v. D-2, p. 361–382.

Fowler, J. H., and Kuenzi, W. D., 1978, Keweenawan turbidites in Michigan (deep bore hole red beds); A foundered basin sequence developed during evolution of a protoceanic rift system: Journal of Geophysical Research, v. 83, p. 5833–5843.

Guldenzopf, E. C., 1967, Conodonts from the Prairie du Chien of Northern Peninsula of Michigan, preliminary report, in Ostrom, M. E., and Slaughter, A. E., eds., Correlation problems of the Cambrian and Ordovician outcrop areas, Northern Peninsula of Michigan: Michigan Basin Geological Society Annual Field Excursion Guidebook, p. 58–64.

Hamblin, W. K., 1958, The Cambrian sandstones of northern Michigan: Michigan Geological Survey Publication 51, 146 p.

Harrison, W. B., III, 1987, Michigan's "deep" St. Peter gas play continues to expand: World Oil, v. 204, no. 4, p. 56–61.

Harrison, W. B., III, and Barnes, D. A., 1988, Lower and Middle Ordovician sequence stratigraphy and lithofacies, Michigan Basin, U.S.A. [abs.], in Abstracts, Lower Paleozoic of the Michigan Basin Symposium, Kalamazoo, Michigan, December 9, 1988: Western Michigan University Core Research Laboratory and Michigan Basin Geological Society.

Hinze, W. J., and Merritt, D. W., 1969, Basement rocks of the Southern Peninsula of Michigan: Michigan Basin Geological Society Annual Field Excursion Guidebook, p. 28–59.

Hinze, W. J., Kellogg, R. L., and O'Hara, N. W., 1975, Geophysical studies of basement geology of Southern Peninsula of Michigan: American Association of Petroleum Geologists Bulletin, v. 59, p. 1562–1584.

Houghton, D., 1841, Fourth annual report of the state geologist: Michigan House of Representatives, Document 27, p. 3–89.

Janssens, A., 1973, Stratigraphy of the Cambrian and Lower Ordovician rocks in Ohio: Ohio Geological Survey Bulletin 64, 197 p.

Kalliokoski, J., 1982, Jacobsville Sandstone, in Wold, R. J., and Hinze, W. J., eds., Geology and tectonics of the Lake Superior Basin: Geological Society of America Memoir 156, p. 147–155.

——, 1988, Jacobsville Sandstone; An update, in Wollensak, M. S., ed., Upper Keweenawan rift fill sequence, Mid-Continent rift, Michigan: Michigan Basin Geological Society Annual Field Excursion p. 127–136.

Kellogg, R. L., 1971, An aeromagnetic investigation of the Southern Peninsula of Michigan [Ph.D. thesis]: East Lansing, Michigan State University, 128 p.

Lidiak, E. G., Marvin, R. F., Thomas, H. H., and Bass, M. N., 1966, Geochronology of the midcontinent region, United States; Part 4, Eastern area: Journal of Geophysical Research, v. 71, p. 5427–5438.

Lilienthal, R. T., 1978, Stratigraphic cross-sections of the Michigan Basin: Michigan Geological Survey Report of Investigation 19, 38 p.

Lundgren, C. E., Jr., and Barnes, D. A., 1989, Influence of depositional environment on diagenesis in St. Peter Sandstone, Michigan Basin [abs]: American Association of Petroleum Geologists Bulletin, v. 73, p. 384.

Michigan Basin Stratigraphic Committee, 1969, Stratigraphic cross-sections, Michigan Basin: Michigan Basin Geological Society, 22 p.

Newcombe, R. B., 1933, Oil and gas fields of Michigan: Michigan Geological Survey Publication 37, Geological Series 32, 293 p.

North American Commission on Stratigraphic Nomenclature, 1983, North American stratigraphic code: American Association of Petroleum Geologists Bulletin, v. 67, p. 841–875.

Ojakangas, R. W., and Morey, G. B., 1982, Keweenawan sedimentary rocks of the Lake Superior region; A summary, in Wold, R. J., and Hinze, W. J., eds., Geology and tectonics of the Lake Superior Basin: Geological Society of America Memoir 156, p. 157–164.

Repetski, J., and Harris, A., 1981, Report on referred fossils, Lower and Middle Ordovician, Ogemaw County, Michigan: U.S. Geological Survey Open-File Report 81-39, 13 p.

Reynolds, D. J., 1984, Structural and dimensional repetition in continental rifts [M.S. thesis]: Durham, North Carolina, Duke University, 175 p.

Shaw, T. H., 1988, The conodont biostratigraphy of the Prairie du Chien Group in its type area; Implications for petroleum exploration in the Michigan and Illinois Basins [abs.], in Abstracts, Lower Paleozoic of the Michigan Basin Symposium, Kalamazoo, Michigan, December 9, 1988: Western Michigan University Core Research Laboratory and Michigan Basin Geological Society.

Sloss, L. L., 1963, Sequences in the cratonic interior of NOrth America: Geological Society of America Bulletin, v. 74, p. 93–114.

Van der Voo, R., and Watts, D. R., 1978, Paleomagnetic results from igneous and sedimentary rocks from the Michigan Basin borehole: Journal of Geophysical Research, v. 83, p. 5844–5848.

——, 1980, Tectonic implications from the paleomagnetism of the Upper Keweenawan sediments [abs]: EOS Transactions of the American Geophysical Union, v. 61, p. 1195.

Van Schmus, W. R., and Hinze, W. J., 1985, The Midcontinent rift system: Annual Review of Earth and Planetary Sciences, v. 13, p. 345–383.

Wheeler, C. T., Jr., 1987, Subsurface study of Lower Ordovician Prairie du Chien Group and underlying Cambrian formations and their relation to pre-Glenwood unconformity in southern peninsula of Michigan [abs]: American Association of Petroleum Geologists Bulletin, v. 71, p. 1112.

MANUSCRIPT ACCEPTED BY THE SOCIETY JUNE 1, 1990

Printed in U.S.A.

# Diagenetic history of the Trenton and Black River Formations in the Michigan Basin

**Joyce M. Budai and James L. Wilson***
*Department of Geological Sciences, University of Michigan, Ann Arbor, Michigan 48109-1063*

## ABSTRACT

The Trenton and Black River Formations of the Michigan Basin have been diagenetically altered by a complex sequence of events related to both the stratigraphic and structural history of the basin. The physical distribution and chemical composition of dolomite in the Trenton and Black River Formations are variable and suggest multiple episodes of dolomitization. The most extensive diagenetic alteration of both Trenton and Black River limestones has occurred in fracture-controlled hydrocarbon reservoirs. Within reservoirs several stages of dolomitization were followed by carbonate and sulfate cementation, and sulfide mineralization. Although the general patterns of reservoir alteration have been recognized for some time, possible causes of such alteration have not been adequately addressed.

Several lines of evidence indicate that mineralization and hydrocarbon migration are related and occurred in the late Paleozoic, perhaps in response to compressional deformation caused by Appalachian tectonism. During such episodes, fluids were mobilized and channeled vertically through preexisting fracture zones. This fluid migration also served to drive maturing hydrocarbons out of Trenton–Black River source beds and into previously dolomitized, high-porosity intervals. This general mechanism could be applied to other fracture-related reservoirs in the Michigan Basin area based on the regional distribution of Mississippi Valley–type (MVT) reservoir alteration and compressional stress fabrics.

## INTRODUCTION

The Michigan Basin is almost circular in map view and contains nearly 5 km of Phanerozoic sediments in the basin center. Secondary structures within the basin are dominated by NW-SE–trending anticlines and faulted ridges, generally displaying structural relief of less than 60 m (Fisher and others, 1988). Based on magnetic and gravity anomaly data and limited well-core information, the Precambrian basement in the Southern Peninsula of Michigan is thought to consist of four provinces: the western margin of the Grenville province in southeastern Michigan; the Central province in the southwest; the Penokean province in the north; and the Keweenawan rift zone, which extends from the northwest to the southeast (Fig. 1; Hinze and others, 1975). The southern part of the Bowling Green fault is in Ohio and is coincident with the contact between the Grenville and Central Precambrian provinces (Fisher and others, 1988). The Howell anticline and the northern extension of the Lucas-Monroe faults parallel the structural trend of the rift zone. The Albion-Scipio and many other oil and gas fields in the Southern Peninsula are also oriented parallel to this NW-SE trend (Prouty, 1989). The Michigan Basin is separated from the Appalachian Basin to the east by the Cincinnati, Findlay, and Algonquin arches and from the Illinois Basin to the southwest by the Kankakee arch. The Wisconsin arch defines the western margin of the Michigan Basin (Fig. 1).

The Michigan Basin contains Cambrian through Pennsylvanian-age rocks and a thin Jurassic sequence in the basin center. The Trenton and Black River Formations in the Michigan Basin constitute a continuous marine carbonate unit correlative with

---

*Present address: New Braunfels, Texas 78130.

Budai, J. M., and Wilson, J. L., 1991, Diagenetic history of the Trenton and Black River Formations in the Michigan Basin, *in* Catacosinos, P. A., and Daniels, P. A., Jr., eds., Early sedimentary evolution of the Michigan Basin: Geological Society of America Special Paper 256.

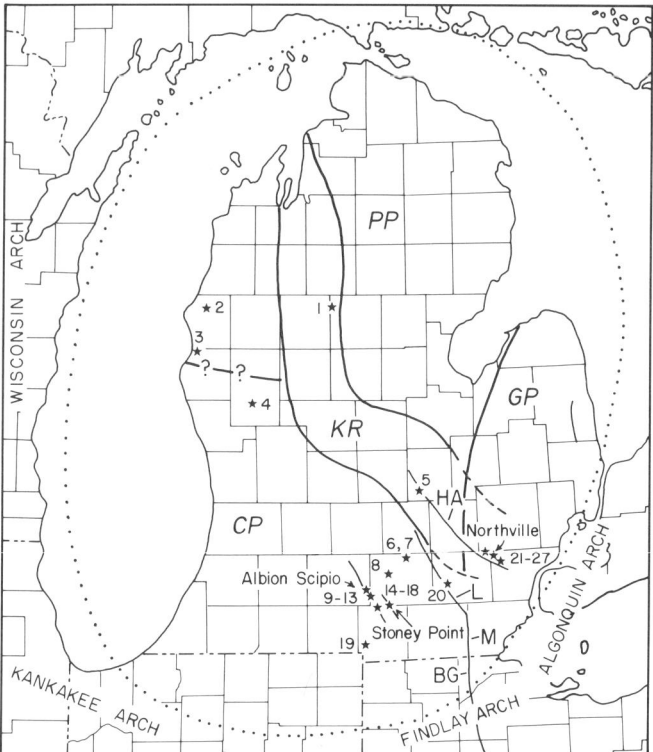

Figure 1. Michigan Basin and surrounding region showing basement provinces, major structures, and well core locations (stars). PP = Penokean province, CP = Central province, KR = Keweenawan rift zone, GP = Grenville province, HA = Howell anticline, L = Lucas fault, M = Monroe fault, BG = Bowling Green fault. Dotted ellipse outlines the edge of the Michigan Basin. Complete information on wells is listed in Table 1.

the host carbonates by mineralization typically associated with MVT deposits (Shaw, 1975; Ardrey, 1978). Similar mineralization has been reported in Trenton and Black River reservoirs in Ohio (Haefner and Mancuso, 1986; Wickstrom and Gray, 1989).

## METHODS

Twenty-seven well cores were examined for this study (Fig. 1; Table 1). Sample surfaces were finely ground and stained with Alizarin Red S and potassium ferricyanide to evaluate percent limestone and ferroan content in limestones and dolomite, respectively. Polished thin sections were examined under both cathodoluminescence and ultraviolet fluorescence. Mineralogical identification of late-stage phases was made optically and with scanning electron microscopy (SEM) in backscattered electron mode (BSE). Stable isotope analyses for this study were collected in the Stable Isotope Laboratory at the University of Michigan. Sampling and analytical techniques were the same as those described in Budai and others (1987). No acid fractionation correction was applied to dolomite analyses. All data are reported in delta per mil notation ($\delta^{18}O$, $\delta^{13}C^0/_{00}$) with respect to the PDB standard.

## DEPOSITIONAL HISTORY

The Trenton and Black River Formations record subtidal marine sedimentation throughout the basin. The Trenton Limestone consists of skeletal wackestones and packstones (biomicrites and poorly washed biosparites) with occasional small bryozoan concentrations (Fig. 3A). There is little vertical change in facies within the Trenton Formation and no evidence of peritidal or restricted facies at the top of the formation in Michigan or in surrounding states (Keith, 1985, 1988; Wickstrom and Gray, 1989). Based on paleobathymetric and tectonic subsidence modeling, together with detailed textural analysis, water depths are thought to have been shallower during Black River deposition and to have gradually deepened during Trenton deposition

the Galena-Decorah-Plattville in Wisconsin; the Trenton and Black River Formations in Ohio, the Lexington–Black River in Kentucky; and the Trenton and Black River Groups in Indiana, New York, and southern Ontario. All of the Black River and most of the Trenton Formation are probably Middle Ordovician in the Michigan Basin, but the uppermost Trenton Formation may belong in the Late Ordovician Cincinnatian series (Fig. 2; Sweet, 1984). The Trenton and Black River Formations represent an important time interval in the basin for several reasons. The underlying Cambrian and Ordovician sections are composed of very shallow marine sandstones and carbonates. Deposition of the Black River and Trenton limestones marks the beginning of deeper-water, open-marine sedimentation within the basin, suggesting some change in subsidence rates or the regional structural setting. There are significant economic reasons for understanding the diagenetic history of Middle Ordovician rocks: some of the largest hydrocarbon deposits in the Michigan Basin and adjacent regions are found in Trenton–Black River reservoirs. In southern Michigan, the Albion-Scipio, Stoney Point, and Northville fields have exceptional porosity development due to solution and extensive dolomitization along linear fracture zones. One of the interesting aspects of these reservoirs is the diagenetic alteration of

| System | | North American Series/Stages | | Michigan Rock Units Lower Peninsula | MA |
|---|---|---|---|---|---|
| Silurian | U | Cayugan | | Bass Islands Salina Group | 408 |
| | L | Niagaran | | Niagaran Clinton | |
| | | Alexandrian | | Cataract Group | 438 |
| Ordovician | U | Cincinnatian | Richmondian | Undifferentiated | |
| | | | Maysvillian | | |
| | | | Edenian | Utica | |
| | M | Champlainian | Shermanian | Trenton | |
| | | | Kirkfieldian | | |
| | | | Rocklandian | Black River | |
| | | | Blackriverian | | |
| | | | Chazyan | Glenwood | |
| | | | Whiterockian | St. Peter | |
| | L | | Canadian | Prairie du Chien | 478 |
| Cambrian | U | St. Croixian | Trempealeauan | Trempealeau | 505 |
| | | | Franconian | Munising Group | |
| | | | Dresbachian | Mount Simon | |

Figure 2. Early Paleozoic stratigraphy of the Michigan Basin, adapted from Fisher and others (1988).

## TABLE 1. WELL CORE LOCATIONS AND INTERVALS

| Well Name and Number on Figure 1 | Location (Sec. T.–R) | Interval (Ft) | Comment* |
|---|---|---|---|
| 1. Hunt Winterfield Deep A-1 | 30–20N–6W | 9,223–10,289 | no C |
| 2. Miller Carnagle 2-30 | 30–20N–17W | 5,216–5,497 | no C |
| 3. Carter Lauber #12 | 6–16N–17W | 4,947–5,240 | C?(100%) |
| 4. Sun Bradley #4 | 11–12N–13W | 5,960–6,460 | C(<2') |
| 5. Jelinik Ferris #1 | 5–5N–2E | 6,110–6,190 | no C |
| 6. Total Faist 2-12 | 12–1S–1W | 4,872–5,255 | C(11') |
| 7. Total Luck 2-12 | 12–1S–1W | 4,870–5,134 | C?(7') |
| 8. Texaco Konkol #1 | 15–2S–2W | 4,420–4,600 | no C |
| 9. Humble Riley #2 | 26–3S–4W | 4,071–4,262 | C?(14') |
| 10. Humble Hartung #1 | 1–4S–4W | 3,985–4,187 | C?(100%) |
| 11. Humble Kryst #1 | 7–4S–3W | 3,966–4,112 | C?(100%) |
| 12. Humble Wilson #6 | 7–4S–3W | 4,148–4,274 | C?(100%) |
| 13. Davis Wooden #4 | 3–5S–3W | 3,850–3,978 | C?(15') |
| 14. Jem Casler 1-30 | 30–4S–2W | 3,803–3,862 | NA |
| 15. Jem Casler 5-30 | 30–4S–2W | 4,081–4,191 | C?(30') |
| 16. Arco Conklin 1-31 | 31–4S–2W | 3,695–3,898 | C(100%) |
| 17. Jem Tolle 1-33 | 33–4S–2W | 4,240–4,270 | C?(4') |
| 18. Jem Hall 1-13 | 13–4S–2W | 4,307–4,361 | NA |
| 19. Anderson Whitaker #2 | 29–7S–4W | 3,042–3,265 | C?(100%) |
| 20. Sun Buss Haab #1 | 8–3S–4E | 3,760–4,530 | C(4') |
| 21. CPC #210 | 2–1S–7E | 3,980–4,510 | C(1.5') |
| 22. Torosian Nerreter #1 | 2–1S–7E | 4,311–4,551 | C?(100%) |
| 23. CPC #204 | 1–1S–7E | 4,123–4,572 | C(100%) |
| 24. CPC #219 | 17–1S–8E | 3,880–4,326 | C(100%) |
| 25. CPC Burroughs #4 | 24–1S–8E | 3,716–4,076 | C?(8') |
| 26. CPC Zittel #2 | 25–1S–8E | 3,675–4,042 | C(12') |
| 27. CPC Holman #1 | 30–1S–9E | 3,600–3,958 | C(7') |

*C = cap dolomite present (estimated cap thickness; 100% indicates that upper interval is all dolomite and the lower boundary of cap dolomite cannot be distinguished).
C? = core begins below the Trenton–Utica contact, but drillers' reports describe dolomite at the top of the section (same as in C).
NA = information not available.

(Howell and Budai, 1989). The central part of the basin during Trenton deposition may have been as much as 100 m deep. The Trenton limestones in the central basin are dark, have lower faunal content and diversity, and may record starved-basin accumulation. The Collingwood Shale, a thin, organic-rich shale, occurs as a member of the upper Trenton Formation in the northern part of the basin (Hiatt and Nordeng, 1985). Throughout the basin the Trenton Formation grades abruptly into the overlying Utica Shale, which is the beginning of a thick sequence of Late Ordovician calcareous marine shales (Fig. 2).

The Black River Formation consists of open-marine lime mudstones, wackestones, and packstones (micrites, biomicrites, and poorly washed biosparites) similar to the Trenton Formation. Lime mudstones and wackestones are more common than in the Trenton Formation and the Black River Formation is more argillaceous (Fig. 3B). A thin bentonite layer near the top of the formation, the Black River Shale, imparts a distinctive gamma ray signal on electric logs and is used for correlation across the basin (Lilienthal, 1978; Huff and others, 1988). On the southern and western margins of the basin, the base of the Trenton and most of the Black River Formations contain large (2 to 10 cm) tan chert nodules. These chert nodules appear to be absent in the basin center. The Black River Formation overlies a thick sequence of Early and Middle Ordovician sandstones and carbonates (Fig. 2). The lower contact between the Black River Formation and the Glenwood Shale is generally gradational. Controversy exists over the proper stratigraphic nomenclature and age of the units below the Glenwood Shale due to the probable lack of regional unconformities in the lower Paleozoic section of the Michigan Basin (Bricker and others, 1983; Fisher and Barratt, 1985; Harrison, 1986; Barnes and Harrison, 1988). For simplicity, we have shown the previously accepted St. Peter Sandstone and Prairie du Chien Group for the units below the Glenwood (Fig. 2).

 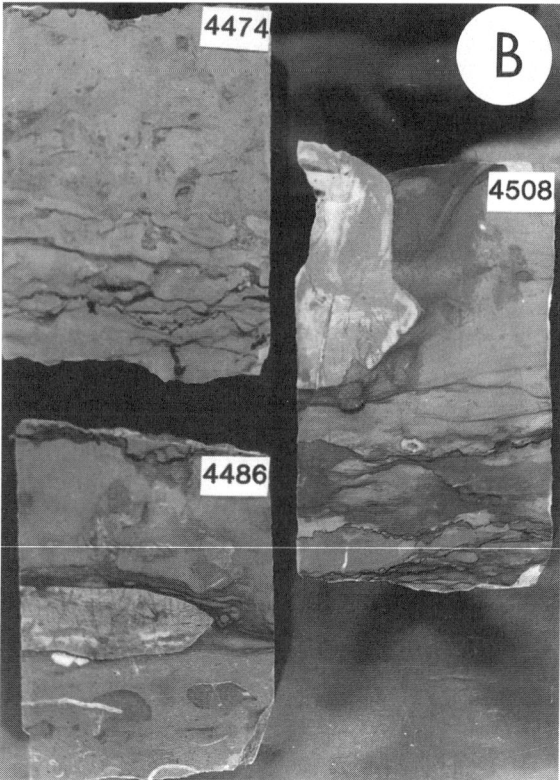

Figure 3. Photographs of Trenton and Black River Formation depositional facies. A: 4290, skeletal packstone from the middle Trenton Formation; 4307, alternating argillaceous wackestone and skeletal packstone from the middle Trenton Formation; 4310, wackestone from the middle Trenton Formation; 4335, packstone/wackestone with shaly layers and coarse skeletal debris. B: 4474, mudstone/wackestone with fine stylolites from the upper Black River Formation; 4486 and 4508, mudstones with large chert nodules typical of the upper Black River Formation. All samples are limestone from location 21 (Fig. 1; Table 1). Footage tag is 2.7 cm long for scale on rock samples.

The Trenton and Black River Formations each range in thickness from approximately 60 m in the northwest part of the basin to 150 m in the southeast (Figs. 4 and 5). In the southeastern quadrant of the basin the Trenton and Black River Formations consist of gray to brown skeletal packstones interbedded with bryozoan-rich wackestones and packstones. Red algae are locally common, but generally rare. Brachiopod, crinoid, and trilobite debris are common throughout the section. Wilson and Sengupta (1985) describe this region as a carbonate platform deposit, in part because both formations attain their greatest thickness here. As they observe, the textures and faunal content of the Trenton Formation in this region are indicative of moderate water depths and are distinct from the dark, finer-grained limestones typical in cores from near the basin center. In the center of the basin, the Trenton and Black River Formations are considerably thinner (Figs. 4 and 5), finer grained, and argillaceous. On the western margin of the Michigan Basin the Trenton and Black River Formations are thin, but exhibit depositional textures similar to the main carbonate platform on the southeast side of the basin. The Trenton and Black River Formations thin to the south in Indiana, where the Trenton Formation appears to record higher energy, shallower water conditions (Keith, 1985). Both formations are thinner in southern Ontario, where the Black River is very shallow subtidal to intertidal and the Trenton records shallow-subtidal, higher energy conditions than those in the Michigan Basin (Brookfield, 1988).

The nature of the Trenton-Utica contact has important implications for both the diagenetic history of the underlying formations and the structural/stratigraphic development of the surrounding region. Several workers have concluded that at least part of the basin was exposed at the end of Trenton deposition to account for a variety of diagenetic features in the Trenton Formation (Rooney, 1966; DeHaas and Jones, 1984; Churcher, 1986; Taylor and Sibley, 1986). Alternatively, Keith (1985, 1988) and Wilson and Sengupta (1985) suggest that alteration at the contact is related to submarine hardground development rather than subaerial exposure. There are well-documented marine hardgrounds reported in Middle Ordovician limestones equivalent to the Trenton Formation in Ontario (Wilkinson and others, 1982) and Indiana (Keith, 1985) that are identical to the mineralized upper surface of the Trenton in the Michigan Basin. Furthermore, there is no textural, faunal, or paleobathymetric evidence for shallowing at the top of the Trenton Formation within the basin (Howell and van der Pluijm, 1990). Isopach patterns for the Utica Shale indicate that following Trenton Formation deposition, subsidence in the basin became linked to that in the Appalachian Basin, which suggests the basin subsided continuously during Middle to Late Ordovician time.

## DOLOMITIZATION IN THE TRENTON AND BLACK RIVER FORMATIONS

The dolomitization history of the Trenton Formation has been the subject of considerable work, and the regional distribution of various dolomite types has been described in a number of studies (Keith, 1985, 1988; Taylor and Sibley, 1986; DeHaas and Jones, 1989; Fara and Keith, 1989; Wickstrom and Gray, 1989). The Trenton and Black River Formations have been extensively replaced by dolomite on the southwest side of Michigan, and contain a variable amount of dolomite in the rest of the basin. Both formations are pervasively dolomitized within fracture-related oil fields in southern Michigan, but may be only slightly dolomitic a short distance away from producing zones. Three distinct types of dolomite have been recognized in most studies and are generally referred to as regional, cap, and fracture dolomite (Taylor and Sibley, 1986; Keith, 1988; Wickstrom and Gray, 1989). All three have been defined, at least in part, by their geographic and stratigraphic distribution within the Trenton and Black River Formations.

Figure 5. Isopach map of the Black River Formation. Contour interval is 30 m.

### Regional dolomite

Regional dolomite, as described by Taylor and Sibley (1986), occurs on the west and southwest sides of Michigan where the Trenton and Black River Formations have been extensively dolomitized. Farther west in Wisconsin and in northern Indiana and Illinois, the Trenton is nearly completely dolomitized. The regional dolomite in Michigan has been distinguished from the cap dolomite by its geographic location and nonferroan composition (Taylor and Sibley, 1986).

Well-core samples of regional dolomite from Oceana and Mason Counties, Michigan, were examined for this study (locations 2 and 3 on Fig. 1; Table 1). Taken together, the Trenton and Black River Formations are approximately 85 percent dolomite in Oceana County and 65 percent dolomite in Mason County. For comparison, both formations are 10 to 20 percent dolomite in the Michigan Basin east of the regional dolomite area (e.g., location 4, Fig. 1).

Viewed in thin section, regional dolomite consists of fine- to medium-grained (50 $\mu m \mu m$ to 0.5 mm), anhedral to euhedral dolomite. Crinoidal fragments are commonly replaced by coarser-grained dolomite. Medium-grained dolomite has cloudy cores and clear rims (Fig. 6). Coarse (0.75 to 1.25 mm), clear dolomite cement lines intercrystalline pores, fractures, and vugs. Although regional dolomite is nonferroan when compared to cap dolomite, it does contain enough ferrous iron ($Fe^{+2}$) to stain blue with potassium ferricyanide, particularly in the upper part of the Trenton Formation. All descriptions of ferroan content in dolomite presented in this chapter are based on intensity of staining and

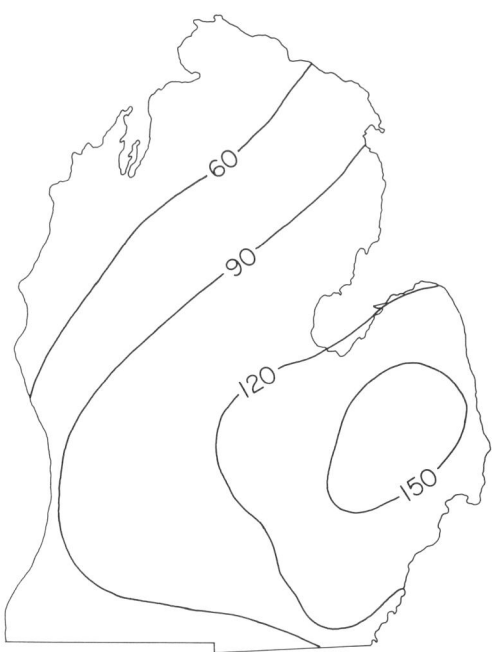

Figure 4. Isopach map of the Trenton Formation. Contour interval is 30 m (modified from Wilson and Sengupta, 1985).

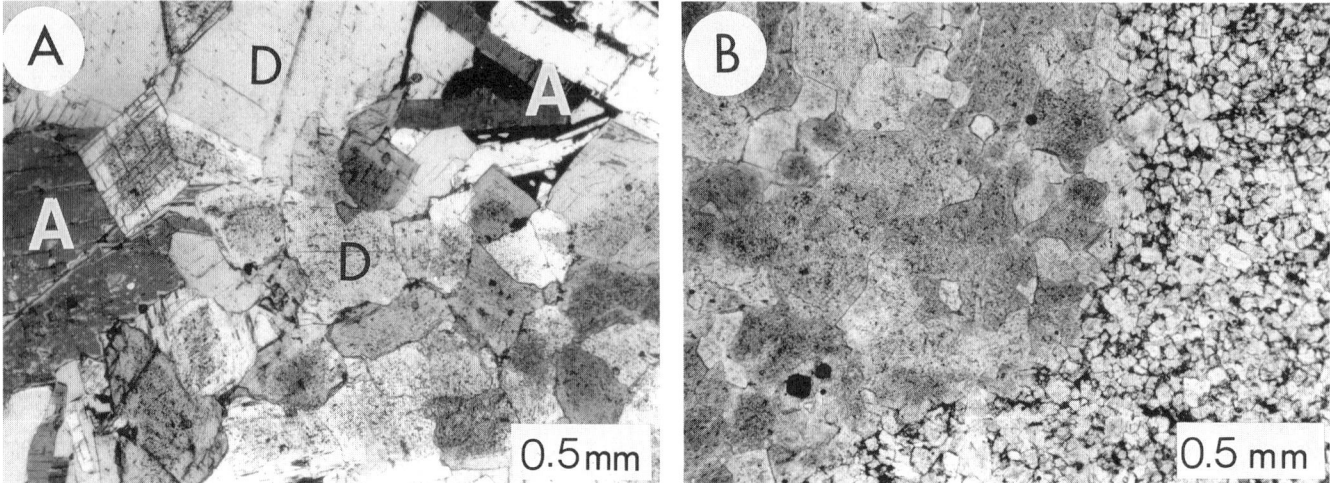

Figure 6. Photomicrographs of regional dolomite textures. A: Upper Trenton Formation sample that contains coarse dolomite (D) and anhydrite (A) lining skeletal mold. B: Regional dolomite in the Black River Formation. Sample contains both fine and medium to coarse dolomite. Both samples from location 3 (Fig. 1; Table 1).

hence are only qualitative estimates of chemical composition. Clear dolomite cement, some coarse-grained skeletal replacement, and clear rims on medium-grained dolomite stain light to medium blue with potassium ferricyanide, indicating the later-stage dolomite is moderately ferroan. Ferroan dolomite cement occurs throughout the Trenton and Black River sections. Following the final stage of dolomite cementation, vugs and fractures are commonly filled with coarse anhydrite and less commonly with calcite cement.

Under cathodoluminescence, fine- and medium-grained dolomite is dull orange with brighter orange patchy intergrowths. The cloudy part of each crystal contains the brightest orange dolomite component. Clear rims on cloudy dolomite and clear cements are dark or quenched under cathodoluminescence, which is what would be expected given their somewhat ferroan composition. The presence of iron in carbonate minerals is thought to reduce or quench luminescence, while manganese is considered an activator of luminescence (Pierson, 1981; Fairchild, 1983; ten Have and Heijnen, 1985). Under UV fluorescence, regional dolomite appears uniformly dull.

The stable isotopic composition of dolomite samples from regional dolomite locations exhibit a broad range in oxygen (–5.0 to –11.7 $\delta^{18}O$) and a somewhat smaller carbon range (1.2 to –2.1 $\delta^{13}C$; Fig. 7). The scatter in isotopic compositions is consistent with the variable luminescent character of regional dolomite. Ferroan dolomite cement in regional dolomite samples has an isotopic range similar to replacive dolomite (Fig. 7). Calcite crinoids from limestone intervals of the regional dolomite area have the same carbon and oxygen as that predicted for Late Ordovician marine calcite (Fig. 7; Lohmann, 1988). Due to a high magnesium composition and single crystal structure, echinodermal debris is often the site of early cementation (Bathurst, 1975). Assuming that crinoid fragments in the Trenton and Black River behave similarly, their isotopic composition should record the earliest diagenetic exchange and implies Late Ordovician marine water was an early diagenetic fluid in the regional dolomite area. Several regional dolomite samples have oxygen and carbon more enriched than that of Middle Ordovician marine calcite, which further indicates that some regional dolomite formed in contact with either Late Ordovician or Silurian waters.

The petrographic, cathodoluminescent, and isotopic characteristics of dolomite in the Trenton and Black River Formation from the west side of the Michigan Basin suggest that regional dolomite was not formed by one pervasive episode of dolomitiza-

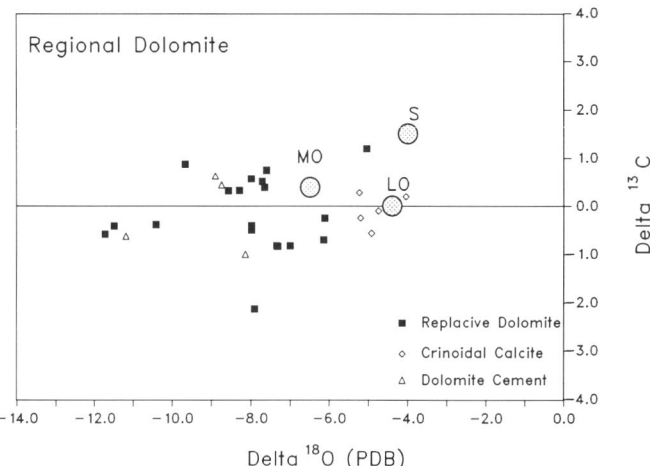

Figure 7. Stable isotopic composition of regional dolomite (solid squares) from locations 2 and 3 (Fig. 1; Table 1). Ferroan dolomite cements from the same location shown as open triangles. Calcitic crinoids shown as open diamonds. Large circles with grid pattern are the estimated values for calcite precipitated from Middle Ordovician (MO), Late Ordovician (LO), and Silurian (S) marine waters (Lohmann, 1988). All values are in per mil relative to the PDB standard.

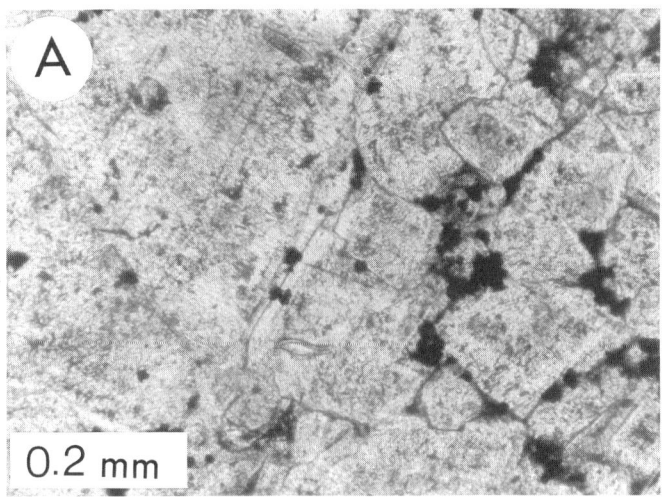
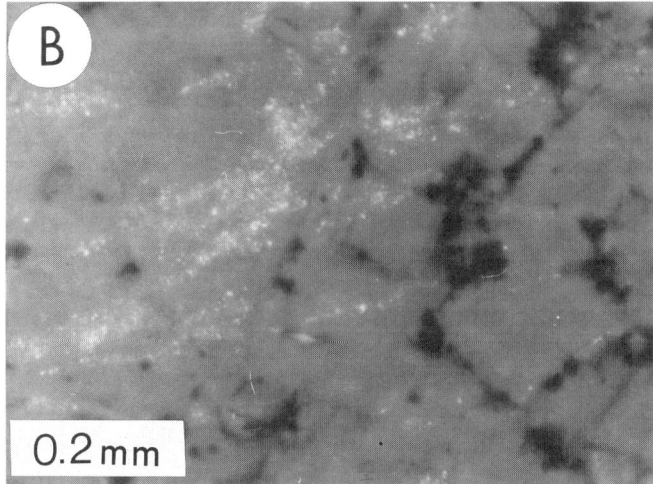

Figure 8. Photomicrographs of cap dolomite. A: Cap dolomite under plane polarized light. Black patches are pyrite. B: Same view under UV fluorescence. Note bright (white) lines of liquid hydrocarbon inclusions. Sample is from location 23 (Fig. 1; Table 1).

tion, but rather a series of replacements. Ferroan dolomite is the last stage of replacement and cementation.

## Cap dolomite

A second kind of dolomite in the Trenton Formation has commonly been referred to as cap dolomite. Cap dolomite occurs at the top of the Trenton section, ranges in thickness from 0.3 to 15 m, is moderately to highly ferroan, and is generally calcian (Taylor and Sibley, 1986). It has been reported in the southern, southwestern and northern parts of Michigan (Taylor and Sibley, 1986; Hiatt and Nordeng, 1985). Cap dolomite also occurs in northwestern Ohio (Wickstrom and Gray, 1989) and in northern Indiana, where it is thought to be the earliest stage of dolomitization (Fara and Keith, 1989). The top of the Trenton Formation in the center of the basin contains highly ferroan dolomite (Taylor and Sibley, 1986), but there is not a completely dolomitized layer as commonly occurs on the basin margins.

Cap dolomite was examined from 9 well cores in the southern and southwestern portions of the Michigan Basin for this study (Fig. 1; Table 1). The top of the Trenton is commonly dolomitic, but within fracture-controlled oil fields and on the west side of Michigan the whole section can be dolomite, which makes identification of cap dolomite difficult. In general, dolomite at the top of the Trenton Formation is more ferroan than dolomite lower in the section and can be distinguished through staining with potassium ferricyanide. One of the reported differences between cap dolomite and other dolomites in the Trenton or Black River is its highly ferroan composition. In fact, it has a highly variable composition (2 to 12 mol% $FeCO_3$ from one location) and generally contains a significant amount of manganese as well as iron (1,800 to 7,500 ppm Mn; Taylor and Sibley, 1986). The contact between highly ferroan cap dolomite and underlying fracture dolomite within oil fields is gradational and extremely difficult to define. In extensively dolomitized locations, it is not clear, based on staining and petrography, that cap dolomite is genetically distinct from fine- and medium-grained fracture dolomite.

Dolomite from the cap is generally cloudy, fine- to medium-grained and subhedral (Fig. 8a). Coarse-grained, clear dolomite lines vugs and fractures. Potassium ferricyanide-stained thin sections of cap dolomite from several locations reveal that the ferroan dolomite component of cap dolomite is distributed differently in different places. In some samples, cap dolomite consists of discrete crystals of ferroan and non-ferroan dolomite. In other samples there are distinct ferroan zones or ferroan cores within non-ferroan dolomite crystals. Under cathodoluminescence, cap dolomite is quite variable. In the Northville field (Fig. 1), dolomite samples from the top of the Trenton range from completely quenched to moderately bright, zoned crystals indicating considerable variability in the minor-element composition of the dolomite. In central Washtenaw County, Michigan, (location 20, Fig. 1; Table 1), cap dolomite is brightly zoned under cathodoluminescence. In a core from southern Hillsdale County (location 19, Fig. 1; Table 1), cap dolomite ranges from quenched to dully luminescent. Unfortunately, samples from cap dolomite in the Albion-Scipio field were not available for comparison to that in the Northville field. However, cap dolomite from the nearby Stoney Point field (location 16, Fig. 1, Table 1) is quenched under cathodoluminescence and contains brightly luminescent calcite veins. Although Taylor and Sibley (1986) do not specifically include the regional dolomite area in their cap dolomite region, the top of the Trenton in Oceana County (location 3, Fig. 1; Table 1) is similar to cap dolomite elsewhere. The top of the Trenton Formation is moderately ferroan dolomite and is very dull under cathodoluminescence.

In some samples, cap dolomite contains liquid hydrocarbon inclusions visible under UV fluorescence (Fig. 8B). These inclusions generally occur in linear arrays cutting across dolomite crystals, which strongly suggests they are secondary in origin.

The isotopic composition of cap dolomite ranges from –6.7 to –9.3 $\delta^{18}O$ and –0.2 to –1.1 $\delta^{13}C$ (Fig. 9). Based on preliminary analyses of cap dolomite from 7 wells and 3 sites, its isotopic composition does not vary significantly with vertical position relative to the overlying Utica Shale and varies slightly with geographic location. In contrast to regional dolomite, the carbon and oxygen isotopes in cap dolomite exhibit a roughly colinear variation, with the highest ratios approaching Middle Ordovician marine compositions (Fig. 9).

In Michigan, the cap dolomite has been interpreted to have formed sometime after the regional dolomite, based on textural relations between nonferroan regional dolomite and secondary ferroan pore-lining cement, implying that most ferroan dolomite in the Trenton Formation has the same origin as cap dolomite (Taylor and Sibley, 1986). However, the isotopic composition of ferroan dolomite cements in regional dolomite locations is more depleted with respect to oxygen than cap dolomite and does not support this interpretation (Figs. 7 and 9). Due to the geographic separation between the occurrence of cap and regional dolomite, it is not possible to determine their relative time of formation based on petrographic criteria. Based on cross-cutting relations, cap dolomite is clearly older than coarse, fracture-filling dolomite cements associated with alteration of reservoir rocks.

Petrographic and luminescent characteristics of cap dolomite suggest that it is a complex mixture of dolomites with variable compositions and perhaps origins. The physical restriction of cap dolomite to the top of the Trenton Formation and its ferroan composition indicate that the overlying Utica Shale played a role in cap dolomitization (Taylor and Sibley, 1986). However, most cap dolomite samples described here and in other studies are from cores taken in fracture-controlled oil fields. It is reasonable to assume that the extensive dolomitization and mineralization associated with these locations may have modified or contributed to cap dolomite and could account for its compositional and petrographic heterogeneity. Cap dolomite is minor to absent in cores located away from producing zones, and the most well-developed cap dolomite occurs within and near Trenton-hosted oil fields (Prouty, 1989, Fig. 22). Cap dolomitization of upper Trenton limestones may be related to both the proximity of the overlying Utica Shale and the location of major fracture zones that could serve as conduits for fluid movement.

*Fracture dolomite*

The Trenton and Black River Formations have been extensively dolomitized within linear fracture or fault zones located in southern Michigan, northern Ohio, and western Ontario. Major oil deposits are commonly associated with these zones, and dolomite from these locations has been referred to as fracture dolomite. Samples from the Albion-Scipio, Stoney Point, Northville, and several smaller fields were examined for this study (Fig. 1; Table 1). The distribution of dolomite within such oil fields is highly irregular and difficult to predict. While it is true that some wells penetrate a completely dolomitized Trenton and Black River section, nearby wells can be nearly all limestone. This erratic dolomite pattern suggests that fluid movement through the reservoir was highly channelized rather than diffuse. In many wells the coarsest dolomite and best porosity development occur in the lower Trenton and upper Black River Formations. The section may be dolomitized above and below this interval, but porosity is low and dolomite is finer grained. There are exceptions to this observation because porosity, like dolomite distributions, is erratic and oil is sometimes produced from limestone intervals as well.

Fracture dolomite in both the Trenton and Black River Formations is coarse-grained and generally compositionally zoned based on staining and cathodoluminescent petrography (Fig. 10). There are several generations of dolomite that occur within high-porosity fracture zones. Wall rock adjacent to small fractures has been completely replaced by coarse, anhedral to euhedral, gray to white dolomite. This dolomite commonly has cloudy cores and clear rims in plane light. The final rim on this dolomite is generally ferroan, based on staining with potassium ferricyanide. Intercrystalline porosity is high in wall-rock dolomite, and there is evidence of dissolution along fracture margins and within adjacent wall rock before the final stages of dolomite cementation. In several samples, extensive pyritization also occurred before fractures were lined with very coarse dolomite cement. In these intervals, pyrite not only replaced matrix dolomite, but also coated fracture walls. Lining both vertical and horizontal fractures is a very coarse-grained (3 to 6 mm) dolomite cement that texturally appears to be later than the dolomite within the wall rock. This dolomite ranges from normal, rhombic-shaped crystals to baroque or saddle-shaped crystals

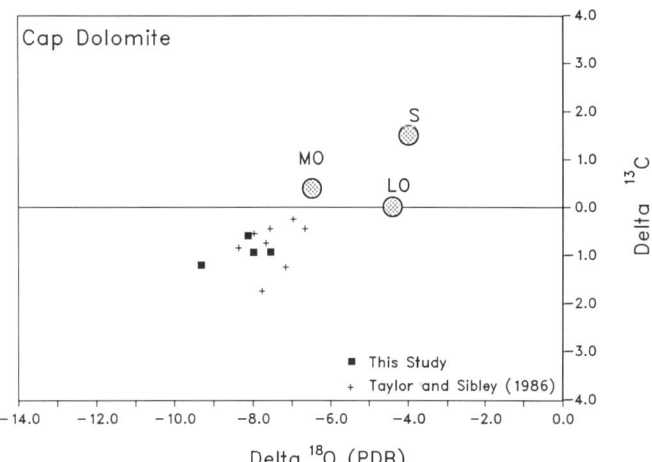

Figure 9. Stable isotopic composition of cap dolomite. Data collected for this study (solid squares) are from locations 20, 24, 26, and 27 (Fig. 1; Table 1). Symbols (+) are data from the Northville field reported by Taylor and Sibley (1986). Large circles with grid pattern are the estimated values for calcite precipitated from Middle Ordovician (MO), Late Ordovician (LO), and Silurian (S) marine waters (Lohmann, 1988). All values are in per mil relative to the PDB standard.

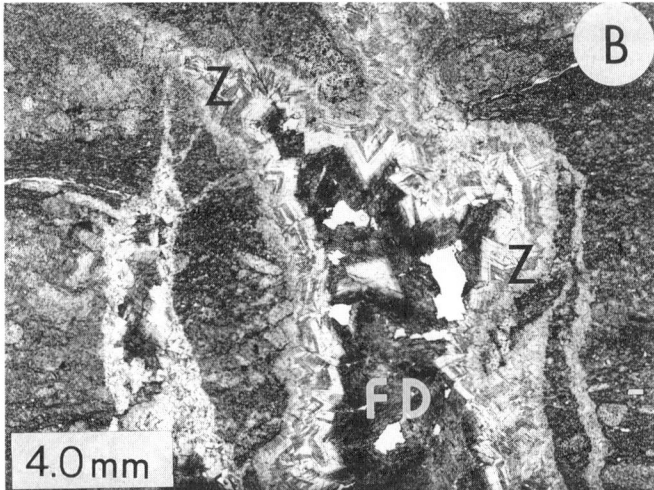

Figure 10. Photographs of fracture dolomite. A: Two samples from the Trenton Formation with very coarse, white, baroque dolomite cements and excellent fracture porosity. B: Photograph of a thin section stained with potassium ferricyanide. Light, zoned dolomite cements (Z) line large fractures. Fractures are partially filled with dark, ferroan dolomite cement (FD). All samples from location 23 (Fig. 1; Table 1).

with undulose extinction. There are sometimes several generations of dolomite cement. The first cement is non-ferroan. Where there is a second or third generation of dolomite cement, it is compositionally zoned and moderately ferroan (Fig. 10b). The final stage of fracture dolomite cement is sometimes coated with solid hydrocarbon. A variety of minerals have filled fractures lined with fracture dolomite. Calcite and anhydrite are common late-stage cements; barite, fluorite and sphalerite are rare in fractures.

There are many fine- to medium-grained dolomite intervals that also occur within fracture-controlled oil-field cores. In these intervals, dolomite has non-ferroan cores and ferroan rims. Coarse dolomite cement fills small, rare fractures and vugs. Based on petrographic relations, fine-grained dolomite pre-dates coarse dolomite in wall rock and very coarse dolomite cements.

Under cathodoluminescence, dolomite in wall rock is dull to moderately bright, the first dolomite cement is more brightly luminescent and zoned, and the latest stage ferroan dolomite cements are quenched. Liquid hydrocarbon inclusions, visible under UV fluorescence, occur as linear arrays cutting across wall rock and as patches of inclusions within coarse dolomite and calcite cements lining fractures. Based on textural relations in samples from the Stoney Point and Northville fields, it is clear that cap dolomite pre-dates coarse dolomite cement and all other late cements.

Compared to cap or regional dolomite, fracture dolomite exhibits greater isotopic variability (–6.5 to –13.5 $\delta^{18}O$, 1.5 to –2.0 $\delta^{13}C$; Fig. 11). Although data are not available from all locations in the basin, there appears to be some geographic control on $\delta^{18}O$ of fracture dolomites (Fig. 12). Dolomite in Albion-Scipio, Northville, and northern Jackson County (locations 6 and 7, Fig. 1) has a similar oxygen range, but fracture dolomite from locations on the southern and western margins of the basin has a more depleted range in oxygen and a broader range in carbon values. The clear association of fracture dolomite with hydrocarbon deposits and later-stage mineralization strongly suggests a burial origin and places its timing no earlier than mid-Carboniferous, based on oil maturation and burial-history studies of the Michigan Basin (Cercone, 1984).

## POST-DOLOMITE MINERALIZATION

Dolomitization in fracture-controlled oil fields is followed by a variety of carbonate, sulfate, and sulfide phases. In general, these late-stage minerals are more common in the Black River and lower Trenton, and rare high in the Trenton section. Volumetrically, the final stages of reservoir mineralization are less significant than fracture dolomite, but suggest that fluids moving through the Trenton and Black River Formations had changed during and following oil emplacement.

Pyrite and to a lesser extent pyrrhotite are locally abundant in the Trenton and Black River Formations throughout the basin. In most samples, pyrite occurs as a replacement of ferroan dolomite or skeletal fragments, or as framboids scattered throughout the matrix of the rock (Fig. 13A). Pyrrhotite occurs as fine laths, often in association with replacive dolomite and K-feldspar (Fig. 13B). Within fracture-controlled oil fields, pyrite also occurs as a fracture-lining cement, and its timing is intermediate between wall-rock dolomitization and coarse dolomite cements that line fractures.

Potassium feldspar (K-feldspar) is a common authigenic phase within dolomitized fracture reservoirs. K-feldspar occurs less commonly in other locations. K-feldspar is primarily a replacement of wall-rock dolomite and never fills fractures or replaces the latest stage of dolomite cement. It generally appears

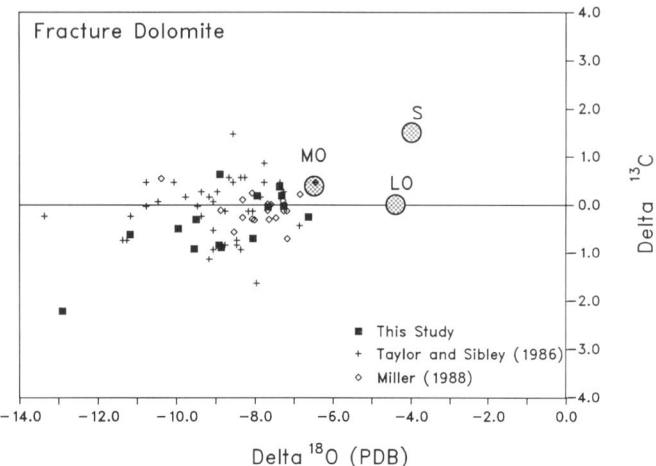
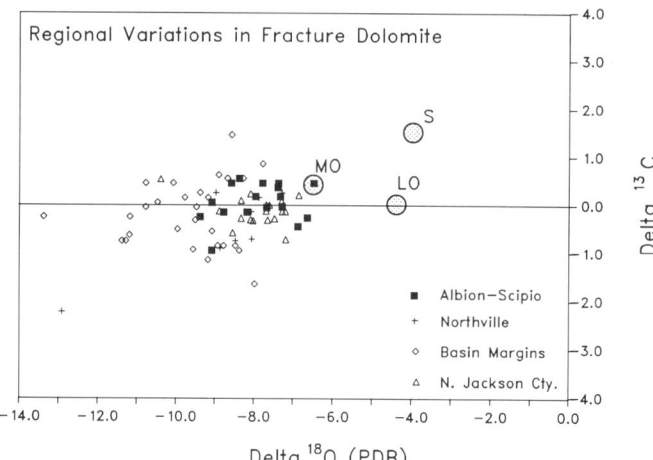

Figure 11. Stable isotopic composition of fracture dolomite from oil fields. Analyses collected for this study (solid squares) are from locations 9, 12, 19, 21, and 22 (Fig. 1; Table 1). Symbols (+) are data from Taylor and Sibley (1986). Open diamonds are data from Miller (1988). Large circles with grid pattern are the estimated values for calcite precipitated from Middle Ordovician (MO), Late Ordovician (LO), and Silurian (S) marine waters (Lohmann, 1988). All values are in per mil relative to the PDB standard.

Figure 12. Regional variation in stable isotopic composition of fracture dolomite. Data are shown by location as follows: Albion-Scipio field (solid squares); Northville field (+); western and southern basin margins (open diamonds); and interior basin (locations 6 and 7, Fig. 1; open triangles). Data sources are the same as those listed in Figure 11. Large circles with grid pattern are the estimated values for calcite precipitated from Middle Ordovician (MO), Late Ordovician (LO), and Silurian (S) marine waters (Lohmann, 1988). All values are in per mil relative to the PDB standard.

bright blue under cathodoluminescence, but is difficult to identify under plane-polarized light because it is fine-grained (10 to 20 μm) and has low-order birefringence compared to surrounding dolomite. K-feldspar is easiest to identify on an SEM in BSE mode (Fig. 13A and B).

Calcite and anhydrite are by far the most common late cements, and their occurrence is such that their relative timing is synchronous. Within fracture-controlled oil fields, vugs and fractures lined with dolomite cement are filled by either calcite or anhydrite. Calcite is intergrown with barite cement within large fractures in the Northville field (Fig. 14A), and barite is intergrown with anhydrite in a number of locations. Calcite partially fills fractures also containing fluorite and sphalerite (locations 9 and 6, respectively, Fig. 1; Fig. 14B). In many cases, the fractures that have been filled with later-stage minerals are oriented at high angles to bedding and they truncate fractures containing only dolomite. Calcite and anhydrite are common late-stage cements in the Trenton and Black River Formations throughout the basin, but are less abundant away from oil fields and generally do not occur with other sulfides and sulfates like those found in oil fields.

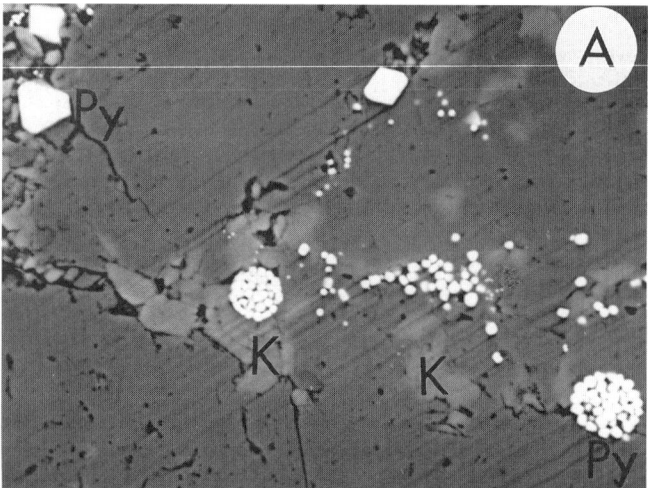
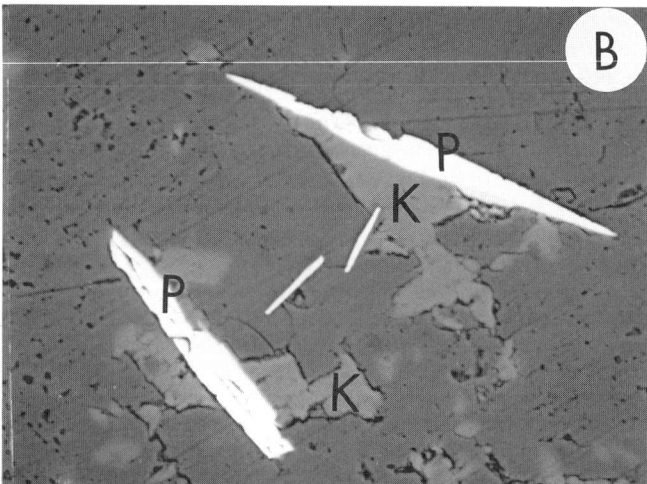

Figure 13. A: SEM micrograph in BSE mode of pyrite framboids (py) and K-feldspar (K) replacing fracture dolomite. Framboids are approximately 6 μm in diameter. B: SEM micrograph in BSE mode of pyrrhotite laths (P) and K-feldspar (K) replacing fracture dolomite. Longest lath is 60 μm. Both photographs from the lower Trenton Formation at location 23 (Fig. 1; Table 1).

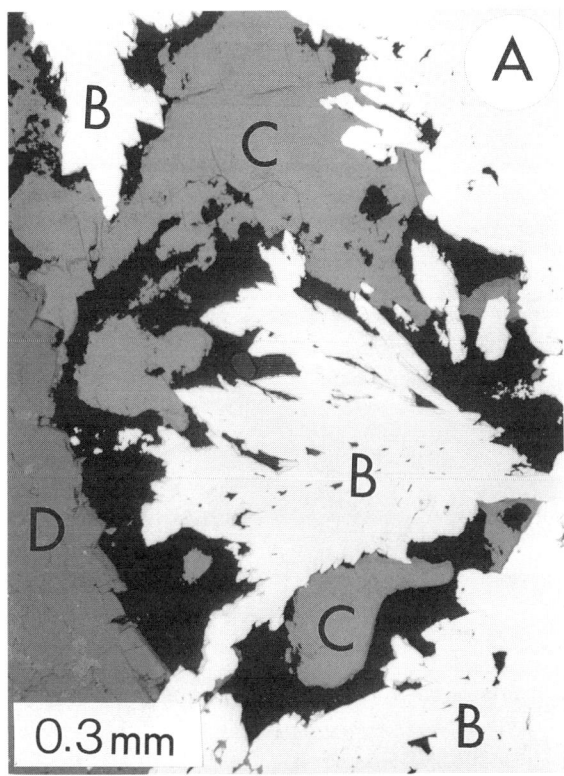
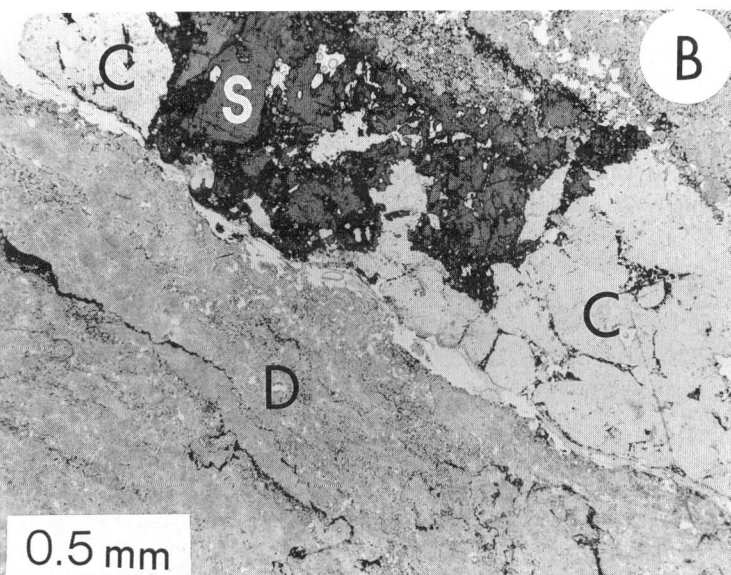

Figure 14. A: SEM micrograph in BSE mode of a barite (B) and calcite (C) vein cutting across fracture dolomite (D). Black areas are epoxy. Sample is from the lower Trenton Formation at location 23 (Fig. 1; Table 1). B: Photograph of a thin section containing a calcite (C) and sphalerite (S) vein cutting through fine-grained fracture dolomite (D). Sample is from the lower Trenton at location 6 (Fig. 1; Table 1).

Calcite in fractures is generally brightly luminescent and commonly consists of a bright yellow and a duller orange-yellow phase. The two phases are intergrown in an irregular pattern where both types occur in the same vein. Under UV fluorescence, fracture calcite contains liquid hydrocarbon inclusions that appear primary, based on petrographic criteria.

Fracture-lining calcite exhibits a range of oxygen and carbon isotopes similar to that in fracture dolomite and is strongly depleted relative to Trenton Limestone values (Figs. 11 and 15). Although it is clear that fracture calcite is later than fracture dolomite, its isotopic composition suggests that it formed from similar fluids.

To summarize, the paragenetic sequence that most commonly occurs in fracture-controlled oil fields is the following: (1a) cap dolomitization; (1b) pervasive fine- to medium-grained dolomitization; (2) coarse-grained dolomitization adjacent to highly fractured intervals; (3) extensive dissolution around fractures; (4) K-feldspar and pyrite replacement of dolomite wall rock; (5) very coarse grained dolomite cement fills or lines fractures and solution-enlarged vugs; and (6) calcite ± anhydrite, barite, fluorite, and sphalerite fill existing fractures and also a later fracture set (Fig. 16). This sequence is nearly identical to that described in Trenton-hosted fracture oil fields in Ohio (Wickstrom and Gray, 1989).

## STRUCTURAL CONSIDERATIONS

At both a regional scale and at a fine petrographic scale, it is clear that tectonic events have contributed significantly to the diagenetic history of the Trenton and Black River Formations. Although the Michigan Basin has traditionally been considered structurally uncomplicated, large hydrocarbon reservoirs in southern Michigan have formed within major fracture zones of uncertain origin. The Northville field is located astride the faulted Howell anticline, the largest structure affecting the Paleozoic section in the basin. The Albion–Scipio field has long been referred to as a fault zone, although the nature and timing of fault movement has been debated (Ells, 1962; Fisher and others, 1988; Prouty, 1989). Fisher and others (1988) suggest that movement along fault-bounded blocks within the Precambrian basement produced long-lived, intermittently active fault zones in the overlying Paleozoic section, and the fracture-controlled oil fields in southern Michigan may be evidence of such movements.

There are a number of features commonly present in well cores that indicate the basin has undergone some amount of compression. Late-stage stylolites and fractures in fracture-controlled oil fields are oriented normal or at high angle to bedding, truncating all earlier fabrics. The orientation and relative timing of these features indicate that a late-stage compressional event has deformed at least part of the lower Paleozoic section. The high-angle fractures contain dolomite, calcite, anhydrite, barite, sphalerite, and fluorite; i.e., typical MVT minerals. The carbonate and sulfate cements within these fractures contain liquid hydrocarbon inclusions. Taken together, the late-stage compression and MVT mineralization within related structures suggest that basinal brine migration through the Trenton and Black River Formation was driven by some tectonic event. In related studies,

84                                                                                                    J. M. Budai and J. L. Wilson

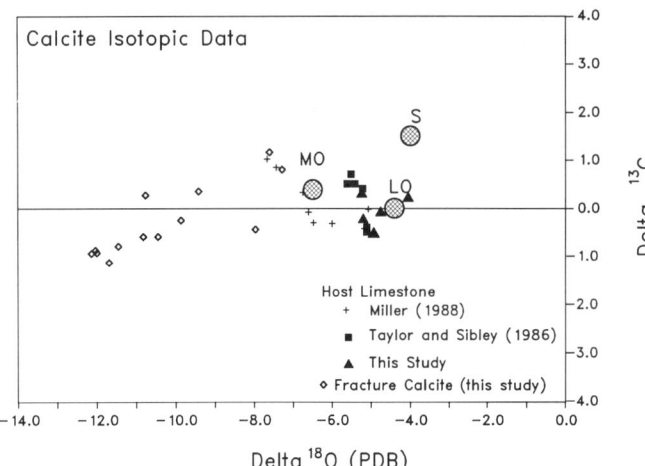

Figure 15. Stable isotopic composition of Trenton Limestone and fracture-lining calcite cements. Skeletal calcite analyses (solid triangles) are from locations 2 and 3. Fracture calcite analyses (open diamonds) are from locations 4, 6, 9, 20, and 23 (Fig. 1; Table 1). Solid squares are data from Taylor and Sibley (1986); symbols (+) are from Miller (1988). Large circles with grid pattern are the estimated value for calcites precipitated from Middle Ordovician (MO), Late Ordovician (LO), and Silurian (S) marine waters (Lohmann, 1988). All values in per mil relative to the PDB standard.

modern environments, it has been estimated that dolomite should be enriched by 2 to 6 per mil $\delta^{18}O$ relative to calcite formed from the same fluid at 25°C (Land, 1980, 1985). At higher temperatures, the isotopic fractionation between carbonate minerals and water decreases so that the $\delta^{18}O$ of calcite or dolomite formed from hot fluids is generally more negative. Therefore, as temperature increases, the difference ($\Delta$) in $\delta^{18}O$ between dolomite and calcite should decrease. Due to these relations, it is generally helpful to consider isotopic data for dolomite in a context defined by established compositions of calcite that have formed under similar diagenetic conditions.

The isotopic compositions of brachiopods, crinoids, and whole-rock limestone from various locations in the Michigan Basin provide a range in $\delta^{18}O$ and $\delta^{13}O$ that reflect marine and subsequent diagenetic values for the Trenton and Black River Formations (Fig. 15). For comparison, the expected composition of marine calcite in the Middle and Late Ordovician and Silurian are plotted with the calcite data (Lohmann, 1988; Fig. 15). Based on these data, it appears possible that, in addition to Middle Ordovician waters, both Late Ordovician and Silurian waters may have altered Trenton Limestone. The diagenetic waters most likely to affect the Trenton and Black River Formations immediately following deposition are Middle and Late Ordovician marine and meteoric water. However, the higher $^{13}C$ of some host-rock calcite analyses are difficult to explain without the admixture of a more $^{13}C$-enriched fluid, e.g., Silurian marine or meteoric water (Fig. 15).

Regional dolomite in the Trenton and Black River Formations exhibits fairly complex petrographic characteristics, which suggest a complicated diagenetic history. Although preliminary, the range in oxygen and carbon isotopic data indicate that Middle

paleomagnetic analysis and regional strain measurements of the Trenton and Black River Formations indicate that remagnetization and pervasive compressional deformation occurred throughout northeastern North America in the late Paleozoic (Suk and others, 1989; Craddock and van der Pluijm, 1989). The Trenton and Black River Formations in the Stoney Point field exhibit strong magnetic signatures, which consist of the present-day geomagnetic field direction as well as a Permian direction (Suk and others, 1989). Remagnetizations in sedimentary rocks are thought to be related to chemical alteration of detrital magnetite or hematite and authigenic growth of magnetite (Kilgore and Elmore, 1989). There is a common association of such authigenic magnetite with both solid hydrocarbons and producing hydrocarbon deposits (Donovan and others, 1979; Elmore and others, 1987; McCabe and others, 1987). The paleomagnetic results from the Trenton and Black River Formations in the Stoney Point field may indicate that Late Permian remagnetization is coeval with other aspects of reservoir rock alteration and hydrocarbon migration.

## DISCUSSION

Stable isotopic data collected from diagenetic dolomite are difficult to interpret due to problems of low-temperature dolomite formation. Whereas the calcite-water isotopic fractionation factor ($\alpha$) is well known over a large range of temperatures, the $\alpha$ for dolomite-water has not been determined below 100°C (Friedman and O'Neil, 1977). Based on extrapolation from higher-temperature and empirical observations of dolomite in

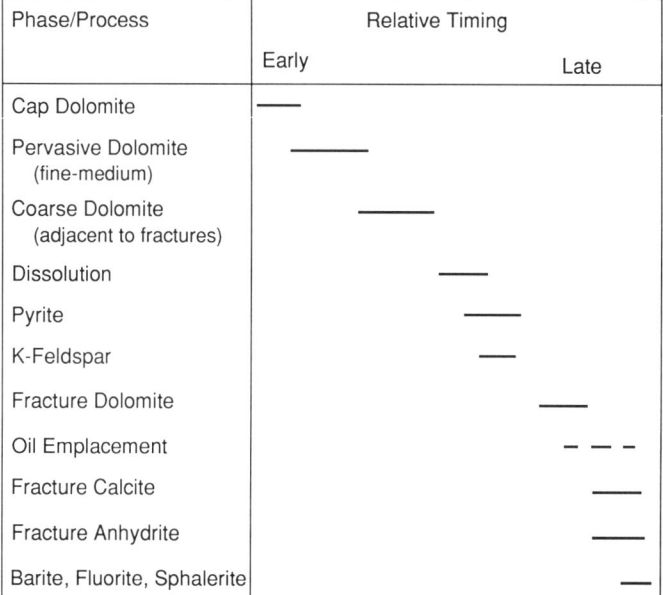

Figure 16. Paragenetic sequence most commonly observed in Trenton and Black River Formation hosted oil fields.

Ordovician, Late Ordovician, and Silurian marine waters, in addition to meteoric waters, may all have contributed to the formation of regional dolomite. There is a strong suggestion that Late Ordovician marine waters exchanged with Trenton limestones, based on calcite crinoid and other limestone analyses (Figs. 7 and 15). During the Middle Ordovician, the Michigan Basin was approximately 16°S of the equator (Van der Voo, 1988), and most of the continental interior was covered by shallow-marine water. By Late Silurian time, the Michigan Basin had moved to 25°S (Van der Voo, 1988) and was isolated from surrounding normal marine seas, as evidenced by its thick accumulation of Late Silurian evaporites. Modern meteoric water in the low-latitude, marine-dominated regions has an oxygen composition depleted by 2 to 4 per mil relative to marine water (Dansgaard, 1964). If early Paleozoic hydrologic systems behaved similarly to those today, then calcite formed from Ordovician or Silurian meteoric waters would have been 2 to 4 per mil depleted relative to marine calcite, and dolomite formed from meteoric water would have had a composition similar to marine calcite due to the expected 2 to 6 per mil enrichment associated with isotopic fractionation between calcite and dolomite at near-surface temperatures. However, a number of regional dolomite analyses in this study are more depleted with respect to oxygen than would be predicted from a mixture of marine and meteoric water at 25°C, suggesting higher formation temperatures during part of regional dolomitization. Because the range in oxygen compositions of regional dolomite (-5.0 to -11.7 $\delta^{18}O$; Fig. 7) extends from near-Silurian meteoric calcite to highly depleted relative to Middle Ordovician meteoric calcite, it is likely that regional dolomite began forming shortly after deposition of the Trenton limestones and continued, perhaps intermittently, through Late Silurian time.

The observed depletion in carbon and oxygen relative to expected marine compositions suggests that a mixture of marine and meteoric water caused regional dolomitization. Sources for marine water include marine water trapped within Trenton and Black River sediments shortly after deposition and water expressed from the Utica Shale as the overlying section was compacted. Local sources of meteoric water are more difficult to identify if there was no exposure at the end of Trenton deposition. Meteoric recharge through the Wisconsin arch has been called upon to explain the distribution of dolomitized Ordovician limestones in Wisconsin. This may also be a viable source of meteoric water for regional dolomite in the Trenton and Black River Formations due to the proximity of dolomitized locations in western Michigan to the Wisconsin arch (Badiozamani, 1973). Because regional dolomite is geographically confined to an area adjacent to the Wisconsin arch, it seems probable that groundwater flow paths related to the recharge area enabled meteoric water to mix with marine water within Trenton and Black River sediments, leading to gradual dolomitization of the section. Later-stage dolomite cements in regionally dolomitized areas are ferroan, have a more depleted $\delta^{18}O$ signature than most surrounding regional dolomite, and are physically associated with anhydrite and calcite cements. Taken together, these observations suggest that late ferroan cements and the most depleted replacive dolomite formed under different conditions, possibly from the same fluids associated with fracture dolomite within fracture oil fields.

Any explanation for the origin of cap dolomite must take into account its location at the top of the Trenton and its unusually high minor-element content. It seems clear that the Utica Shale played an important role as a source for the marine water necessary for dolomitization as well as reduced iron and manganese. Taylor and Sibley (1986) suggest that the cap dolomite occurs only in southern Michigan, due to a number of interdependent geochemical factors, but primarily to exposure of the southern margin of the basin at the end of Trenton Formation deposition and the availability of reduced iron in pore waters during cap dolomite formation. However, there is no convincing evidence for exposure of the Michigan Basin following Trenton deposition, and the distribution of cap dolomite appears to be controlled in part by the distribution of well control, which reflects known hydrocarbon occurrences. Not all wells in southern Michigan contain cap dolomite, and cap dolomite has also been reported from northern Michigan (Hiatt and Nordeng, 1985). Cap dolomite is best developed over and around fracture-controlled oil fields, suggesting more than a coincidental relation. Furthermore, the isotopic composition of cap dolomite defines a continuum between expected Middle Ordovician marine values and a more depleted end member. In contrast to regional dolomite, which displays isotopic evidence of multiple surficial fluid sources, cap dolomite appears to have formed from Middle Ordovician marine water, modified by burial temperature or isotopically depleted fluids.

It is likely that the cap dolomite is ferroan due to input from the overlying Utica Shale. As first suggested by Landes (1946), the upper Trenton Formation is commonly dolomitized because the Utica Shale provided an impermeable seal that forced ponding of upward-moving fluids throughout the burial history of the Middle Ordovician section. In areas distant from major fracture zones, most of the Trenton and Black River section is slightly dolomitic limestone, and the upper part of the Trenton may contain ferroan dolomite due to interaction with fluid expressed from the Utica Shale. Adjacent to fracture zones, the Trenton and Black River Formations have been extensively dolomitized, and the upper Trenton Limestone is always replaced by ferroan dolomite. These observations, together with the petrographic and chemical characteristics of cap dolomite, strongly suggest that it is the diagenetic product of complex fluid/rock interactions, which were most extensive over and around major fracture zones.

The regional isotopic variation observed in fracture dolomite may reflect differences in fluid sources, migration paths, or filling temperatures. Fracture dolomite with very negative oxygen compositions may reflect either formation from isotopically depleted fluids, e.g., meteoric water, or formation at higher temperatures. Because fracture dolomite on the margins of the Michigan Basin has the most depleted oxygen composition and exhibits the broadest range in carbon values, while fracture dolomite with the highest $\delta^{18}O$ occurs in the southern interior part of the basin, it is possible that meteoric recharge down through the basin margins was the source of fluids during late Paleozoic fracture dolomitization. Those locations more proximal to recharge areas would contain the most isotopically depleted fracture dolomite. If this suggestion is correct, meteoric fluids would have exchanged with carbonate rocks during deep circulation, leading to oxygen and carbon enrichment of the fluid and resulting fracture dolomite farther down flow paths. This effect would be modified by higher temperatures deeper in the basin so that actual differences in the isotopic composition of fracture dolomite between recharge and distal areas would be minimized. Alternatively, deeply sourced fluids may have been episodically expelled upward through fracture systems in response to compressional stresses related to Appalachian tectonism. The association of coarse dolomite cements with liquid hydrocarbon inclusions and typical MVT mineralization argues for basinal brine as a fluid source. However, such an interpretation does not adequately explain geographic differences in oxygen ratios observed in fracture dolomites. Most aspects of fracture dolomite and its associated mineralization can be better explained if mixing of ascending basinal brines with meteoric water occurred in recharge areas on the basin margins. Whatever combination of hydrologic scenarios is applied, Trenton and Black River limestones in fracture-controlled oil fields record complicated multiple stages of replacement, solution, and precipitation that cannot be explained by a single pervasive event or fluid source. In areas away from fracture-controlled oil fields, such alteration is rare, but present. This suggests that mineralizing fluids were present throughout the basin, though concentrated in zones where fluid migration was enhanced due to basement-controlled faulting. Furthermore, the migration of liquid hydrocarbons is closely related in time to the final stages of fracture dolomite and calcite precipitation. This implies that movement of latest carbonate-rich fluids served to drive the migration of maturing hydrocarbons out of source beds and into previously dolomitized, high-porosity zones in the reservoir. If basement structure and extrabasinal tectonics have caused episodic, large-scale, fluid-migration events, then other fracture-related hydrocarbon reservoirs in the Michigan Basin with stratigraphic relations and mineralization analogous to those in Trenton–Black River reservoirs may have formed in a similar manner.

## CONCLUSIONS

1. Regional dolomitization was driven by the mixture of marine and meteoric waters within Trenton and Black River sediments adjacent to the Wisconsin arch.

2. Stable isotopic compositions of regional dolomite indicate that Middle and Late Ordovician marine and meteoric waters as well as Silurian meteoric water were active dolomitizing fluids.

3. Cap dolomite formed primarily from Middle Ordovician marine water, modified by a more depleted water.

4. Cap dolomite is best developed over fracture oil fields, suggesting major fracture zones served as conduits for fluid movement.

5. Fracture dolomite is closely associated with MVT alteration and mineralization of major hydrocarbon reservoirs.

6. The latest stage of fracture dolomite cement is coincident with liquid hydrocarbon movement into the reservoirs and places its timing no earlier than mid-Carboniferous.

7. Regional variation in the stable isotopic composition of fracture dolomite suggests meteoric recharge on the margins of the Michigan Basin circulated down through the Middle Ordovician section, mixing with deep-sourced burial brines during dolomite precipitation.

## ACKNOWLEDGMENTS

Support for parts of this study was provided by a grant from MASERA Corporation to Wilson and Budai; NSF grant EAR8417132 to Budai; and a grant from the Michigan Research Excellence and Economic Development Program to S. E. Kesler and others. Well cores were provided by Marathon Oil Company, Arco Oil and Gas Company, the Core Research Laboratory at Western Michigan University, and the Subsurface Laboratory at the University of Michigan. Paul D. Howell and Eric J. Essene provided helpful early reviews of the manuscript. Brian D. Keith and Lawrence H. Wickstrom are gratefully acknowledged for their thoughtful reviews of the chapter.

# REFERENCES CITED

Ardrey, R. H., 1978, Diagenesis of the Middle Ordovician Trenton Formation in southern Michigan [M.A. thesis]: Ann Arbor, University of Michigan, 52 p.

Badiozamani, K., 1973, The Dorag dolomitization model; Application to the Middle Ordovician of Wisconsin: Journal of Sedimentary Petrology, v. 43, p. 965–984.

Barnes, D. A., and Harrison, W. B., 1988, Carbonate to siliciclastics-dominated sedimentation in the Lower and Middle Ordovician of the Michigan Basin; Influence of sequence boundary, paleogeography, and depositional environment [abs.]: Society of Economic Paleontologists and Mineralogists Annual Midyear Meeting, v. 5, p. 4–5.

Bathurst, R.G.C., 1975, Carbonate sediments and their diagenesis, 2nd ed.; Developments in sedimentology, v. 12, New York, Elsevier, 608 p.

Bricker, M. D., Milstein, R. L., and Reszka, C. R., Jr., 1983, Selected studies of Cambro–Ordovician sediments within the Michigan Basin: Michigan Department of Natural Resources Report of Investigation 26, 54 p.

Brookfield, M. E., 1988, A mid-Ordovician temperate carbonate shelf; The Black River and Trenton Groups of southern Ontario, Canada: Sedimentary Geology, v. 60, p. 137–153.

Budai, J. M., Lohmann, K. C., and Wilson, J. L., 1987, Dolomitization of the Madison Group, Wyoming and Utah overthrust belt: American Association of Petroleum Geologists, v. 71, p. 909–924.

Cercone, K. R., 1984, Thermal history of Michigan Basin: American Association of Petroleum Geologists Bulletin, v. 68, p. 130–136.

Churcher, P. L., 1986, Middle Ordovician (Trenton) disconformity in Michigan and Illinois Basin; Subaerial or submarine origin? [abs.]: American Association of Petroleum Geologists Bulletin, v. 70, p. 1064.

Craddock, J. P., and van der Pluijm, B. A., 1989, Late Paleozoic deformation of cratonic carbonate cover of eastern North America: Geology, v. 17, p. 416–419.

Dansgaard, W., 1964, Stable isotopes in precipitation: Tellus, v. 16, p. 436–468.

DeHaas, R. J., and Jones, M. W., 1989, Cave levels of the Trenton–Black River in central southern Michigan, in Keith, B. D., ed., The Trenton Group (Upper Ordovician Series) of eastern North America: American Association of Petroleum Geologists Studies in Geology no. 29, p. 237–266.

Donovan, T. J., Forgey, R. J., and Roberts, A. A., 1979, Aeromagnetic detection of diagenetic magnetite over oil fields: American Association of Petroleum Geologists Bulletin, v. 63, p. 245–248.

Ells, G. D., 1962, Structures associated with the Albion–Scipio oil field trend: Michigan Geological Survey Publication, 86 p.

Elmore, R. D., and 5 others, 1987, Evidence for a relationship between hydrocarbons and authigenic magnetite: Nature, v. 325, p. 428–430.

Fairchild, I. J., 1983, Chemical controls of cathodoluminescence of natural dolomites and calcites; New data and review: Sedimentology, v. 30, p. 579–583.

Fara, D. R., and Keith, B. D., 1989, Depositional facies and diagenetic history of the Trenton Limestone in northern Indiana, in Keith, B. D., ed., The Trenton Group (Upper Ordovician Series) of eastern North America: American Association of Petroleum Geologists Studies in Geology no. 29, p. 277–298.

Fisher, J. H., and Barratt, M. W., 1985, Exploration in Ordovician of central Michigan Basin: American Association of Petroleum Geologists Bulletin, v. 69, p. 2065–2076.

Fisher, J. H., Barratt, M. W., Droste, J. B., and Shaver, R. H., 1988, Michigan Basin, in Sloss, L. L., ed., Sedimentary cover–North American craton, U.S.: Geological Society of America, The Geology of North America, v. D-2, p. 361–381.

Friedman, I., and O'Neil, J. R., 1977, Compilation of stable isotope fractionation factors of geochemical interest: U.S. Geological Survey Professional Paper 440-KK, 108 p.

Haefner, R. J., and Mancuso, J. J., 1986, Mississippi Valley Type mineralization and dolomitization in the Trenton Formation, Wyandot County, Ohio [abs.]: American Association of Petroleum Geologists Bulletin, v. 70, p. 1066.

Harrison, W. B., III, 1986, Stratigraphy and depositional environments of Glenwood Formation and St. Peter Sandstone in Michigan Basin [abs.]: Association of Petroleum Geologists Bulletin, v. 70, p. 1067.

Hiatt, C. R., and Nordeng, S., 1985, A petrographic and well log analysis of five wells in the Trenton–Utica transition in the northern Michigan Basin, in Cercone, K. R., and Budai, J. M., eds., Ordovician and Silurian rocks of Michigan Basin: Michigan Basin Geological Society Special Paper 4, p. 33–43.

Hinze, W. J., Kellogg, R. L., and O'Hara, N. W., 1975, Geophysical studies of basement geology of Southern Peninsula of Michigan: American Association of Petroleum Geologists Bulletin, v. 59, p. 1562–1584.

Howell, P. D., and Budai, J. M., 1989, Quantitative paleobathymetric analysis from subsidence data; Example from Middle Ordovician Michigan Basin [abs.]: American Association of Petroleum Geologists Bulletin, v. 73, p. 365.

Howell, P. D., and van der Pluijm, B. A., 1990, Early history of the Michigan basin; Subsidence and Appalachian tectonics: Geology (in press).

Huff, W. D., Kolata, D. R., Frost, J. K., and Trevail, R. A., 1988, Correlation of the upper Ordovician Deicke K-bentonite bed from the Upper Mississippi Valley to the St. Lawrence Valley, Canada: Geological Society of America Abstracts with Programs, v. 20, p. 121.

Keith, B. D., 1985, Facies, diagenesis and the upper contact of the Trenton Limestone of northern Indiana, in Cercone, K. R., and Budai, J. M., eds., Ordovician and Silurian rocks of Michigan Basin: Michigan Basin Geological Society Special Paper 4, p. 15–32.

—— , 1988, Reservoirs resulting from facies-independent dolomitization; Case histories from the Trenton and Black River carbonate rocks of the Great Lakes area: Carbonates and Evaporites, v. 1, p. 74–82.

Kilgore, B., and Elmore, R. D., 1989, A study of the relationship between hydrocarbon migration and the precipitation of authigenic magnetic minerals in the Triassic Chugwater Formation, southern Montana: Geological Society of America Bulletin, v. 101, p. 1280–1288.

Land, L. S., 1980, The isotopic and trace element geochemistry of dolomite; The state of the art, in Zenger, D. H., Dunham, J. B., and Ethington, R. L., eds., Concepts and models of dolomitization: Society of Economic Paleontologists and Mineralogists Special Publication 28, p. 87–110.

—— , 1985, The origin of massive dolomite: Journal of Geological Education, v. 33, p. 112–125.

Landes, K. K., 1946, Porosity through dolomitization: American Association of Petroleum Geologists Bulletin, v. 30, p. 305–318.

Lilienthal, R. T., 1978, Stratigraphic cross sections of the Michigan Basin: Michigan Geological Survey Report of Investigations 19, 36 p.

Lohmann, K. C., 1988, Geochemical patterns of meteoric diagenetic systems and their application to studies of paleokarst, in James, N. P., and Choquette, P. W., eds. Paleokarst, New York, Springer-Verlag, p. 58–80.

McCabe, C., Sassen, R., and Saffer, B., 1987, Occurrence of secondary magnetite within biodegraded oil: Geology, v. 16, p. 7–10.

Miller, M. A., 1988, Dolomitization and porosity evolution, [Ph.D. thesis]: East Lansing, Michigan State University, 168 p.

Pierson, B. J., 1981, The control of cathodoluminescence in dolomite by iron and manganese: Sedimentology, v. 28, p. 601–610.

Prouty, C. E., 1989, Trenton exploration and wrench tectonics; Michigan Basin and environs, in Keith, B. D., ed., The Trenton Group (Upper Ordovician series) of eastern North America: American Association of Petroleum Geologists Studies in Geology 29, p. 207–236.

Rooney, L. F., 1966, Evidence of unconformity at top of Trenton Limestone in Indiana and adjacent states: American Association of Petroleum Geologists Bulletin, v. 50, p. 533–546.

Shaw, B., 1975, Geology of the Albion–Scipio trend, southern Michigan [M.S. thesis]: Ann Arbor, University of Michigan, 64 p.

Suk, D., Van der Voo, R., and Peacor, D. P., 1989, Late Paleozoic remagnetization of the Trenton Limestone in Michigan Basin [abs.]: EOS Transactions of the American Geophysical Union, v. 70, p. 310.

Sweet, W. C., 1984, Graphic correlation of upper Middle and Upper Ordovician rocks, North American midcontinent province, *in* Bruton, D. L., ed., Aspects of the Ordovician system: Palaeontological Contributions from the University of Oslo 295, p. 23–35.

Taylor, T. R., and Sibley, D. F., 1986, Ferroan dolomite in the Trenton Formation, Ordovician Michigan Basin: Sedimentology, v. 33, p. 61–86.

ten Have, T., and Heijnen, W., 1985, Cathodoluminescence activations and zonation in carbonate rocks; An experimental approach: Geologie en Mijnbouw, v. 64, p. 297–310.

Van der Voo, R., 1988, Paleozoic paleogeography of North America, Gondwana, and intervening displaced terranes; Comparisons of paleomagnetism with paleoclimatology and biogeographical patterns: Geological Society of America Bulletin, v. 100, p. 311–324.

Wickstrom, L. H., and Gray, J. D., 1989, Geology of the Trenton Limestone in northwestern Ohio, *in* Keith, B. D., ed., The Trenton Group (Upper Ordovician series) of eastern North America: American Association of Petroleum Geologists Studies in Geology 29, p. 159–167.

Wilkinson, B. H., Janecke, S. V., and Brett, C. E., 1982, Low Mg calcite marine cement in Middle Ordovician hardgrounds from Kirkfield, Ontario: Journal of Sedimentary Petrology, v. 52, p. 47–57.

Wilson, J. L., and Sengupta, A., 1985, The Trenton Formation in the Michigan Basin and environs; Pertinent questions about its stratigraphy and diagenesis, *in* Cercone, K. R., and Budai, J. M., eds., Ordovician and Silurian rocks of Michigan Basin: Michigan Basin Geological Society Special Paper 4, p. 1–13.

MANUSCRIPT ACCEPTED BY THE SOCIETY JUNE 1, 1990

Geological Society of America
Special Paper 256
1991

# Late Silurian pinnacle reefs of the Michigan Basin

**Gerald M. Friedman and David C. Kopaska-Merkel*

*Department of Geology, Brooklyn College, City University of New York, Brooklyn, New York 11210 and Northeastern Science Foundation, Inc., affiliated with Brooklyn College, Rensselaer Center of Applied Geology, 15 Third Street, P.O. Box 746, Troy, New York 12181-0746*

## ABSTRACT

The pinnacle reefs of the Michigan Basin form small, isolated hydrocarbon reservoirs encased in impermeable evaporites and mudstones, and account for most of Michigan's hydrocarbon reserves. The temporal relations between the reef sequence and the evaporites are still in dispute, but in the currently favored model, deposition of reefs and evaporites follow each other closely in a cyclic manner but are not synchronous. Pinnacle development occurred in four stages and included periods of subaerial exposure, which enhanced reef porosity and permeability through leaching and dolomitization. Subsequent evaporite precipitation filled much of this porosity; many reefs are completely salt plugged and impermeable. Sea-level history and depositional environments of Salina evaporites are disputed, but a model of shallow-water evaporite deposition in the basin is favored.

Regional trends have been recognized across the pinnacle-reef belt, and these predict increased salt plugging of the reefs basinward, increased dolomitization and preserved secondary porosity toward the basin margin, and in the northern trend, zones of production that pass from gas to oil and finally to water toward the basin margin. The producing reefs have porosities that range from 3 to 37 percent (average 6 percent) and average permeabilities of 11 to 12 mD (ranges to more than 1 D).

## INTRODUCTION

The Michigan Basin is a shallow (4,999 ± m; 16,000 ± ft), circular, intracratonic sag covering 316,000 km² (122,000 mi²) in the northern central United States and southern Canada (Fig. 1). During Middle and Late Silurian time, the basin was the site of extensive carbonate deposition and precipitation of basin-filling evaporites. The carbonate deposits include basin-margin barrier reef or bank complexes (including back-barrier facies), pinnacle reefs that formed on the shelf offshore from the barrier reefs, and micrites deposited in the center of the basin (Fig. 2). The major development of carbonates is overlain by extensive cyclical evaporite and carbonate deposits of Late Silurian age.

Most of the Silurian oil and gas reserves in the Michigan Basin are found within numerous but small pinnacle reefs. The first well to produce commercial hydrocarbons from a Niagaran reef in the Michigan Basin was drilled in Ontario in 1889, and estimated primary recoverable reserves in the reefs are about 350 million barrels of oil and 4 trillion ft³ of gas (Mantek, 1976; Caughlin and others, 1976; Yelling and Tek, 1976; Gill, 1985). Modern exploration for these reefs depends on sophisticated seismic processing and acquisition techniques, but success rates for reef-identification are high (McClintock, 1977; Lee and Budros, 1982; Gill, 1985).

Michigan Basin pinnacle reefs may be partially or entirely dolomitized. They commonly have good porosity and permeability as a result of periods of possible subaerial exposure, leaching, and early dolomitization during the Silurian. The pinnacles are sealed on their flanks by evaporites and lime mudstones, and on their tops by evaporites; in some reefs, evaporites completely plug porosity. Multiple factors influencing porosity creation, preservation, alteration, and destruction have produced complex porosity distributions in these reefs.

---

*Present address: Geological Survey of Alabama, 420 Hackberry Lane, P.O. Box O, Tuscaloosa, Alabama 35486-9780.

Friedman, G. M., and Kopaska-Merkel, D. C., 1991, Late Silurian pinnacle reefs of the Michigan Basin, *in* Catacosinos, P. A., and Daniels, P. A., Jr., eds., Early sedimentary evolution of the Michigan Basin: Geological Society of America Special Paper 256.

89

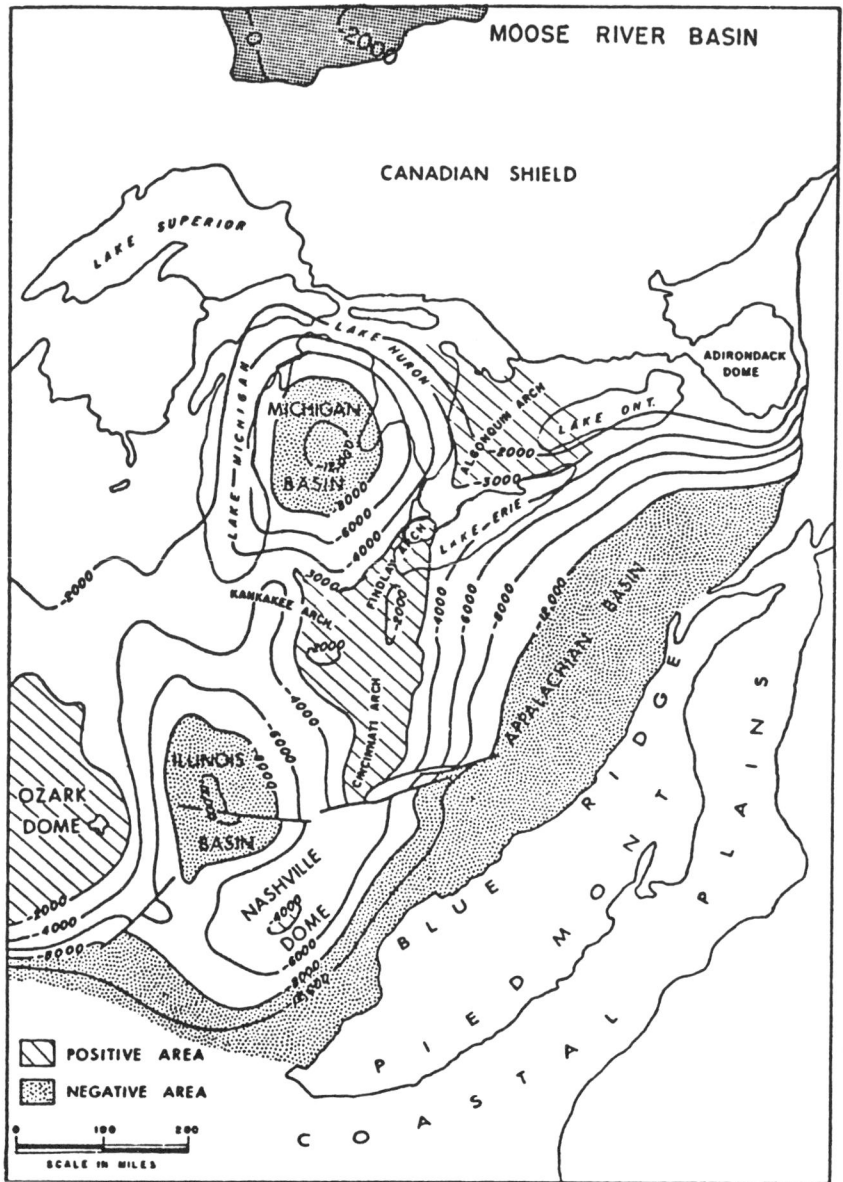

Figure 1. Structure contour map of regional basement showing the location of the Michigan Basin. Contour lines represent feet below sea level (from Nurmi, 1975, modified after Brigham, 1971).

## GEOLOGIC SETTING

The Michigan Basin began subsiding in the Precambrian and contains as much as 4,270 m (14,000 ft) of slightly deformed Paleozoic strata. Maximum subsidence occurred during the Late Silurian and Middle Devonian (Cohee and Landes, 1958).

During the Niagaran (Middle Silurian) the Michigan Basin was situated near to and south of the equator (less than 30°S; Scotese and others, 1979) and filled by a warm, shallow sea that provided ideal conditions for reef growth (Briggs and Briggs, 1974; Bentley, 1979a). During the Niagaran, the Michigan Basin could be divided into three depositional settings: (1) the shallow, broad, basin-edge carbonate bank or barrier reef with its reef limestone, back-reef lagoonal deposits, patch reefs, and fore-reef lime mudstones and lime sandstones; (2) the gently sloping shelf, which was the site of pinnacle-reef development and the deposition of interreef micritic crinoidal limestones and nodular limestones; and (3) the deep central basin with its deposits of dense, micritic, argillaceous limestones (Mantek, 1973). The stratigraphic relations between Middle and Upper Silurian rock units in the Michigan Basin, according to Gill (1985), including some equivalent oil industry names, are illustrated in Figure 3. The reefs occur in the upper Niagaran Guelph Formation (Brown Niagaran of oil industry usage) and overlying Maumee Algal Stromatolite of the basal Salina Group (lateral equivalent of the Ruff Formation or A-1 Carbonate). The Engadine Dolomite of

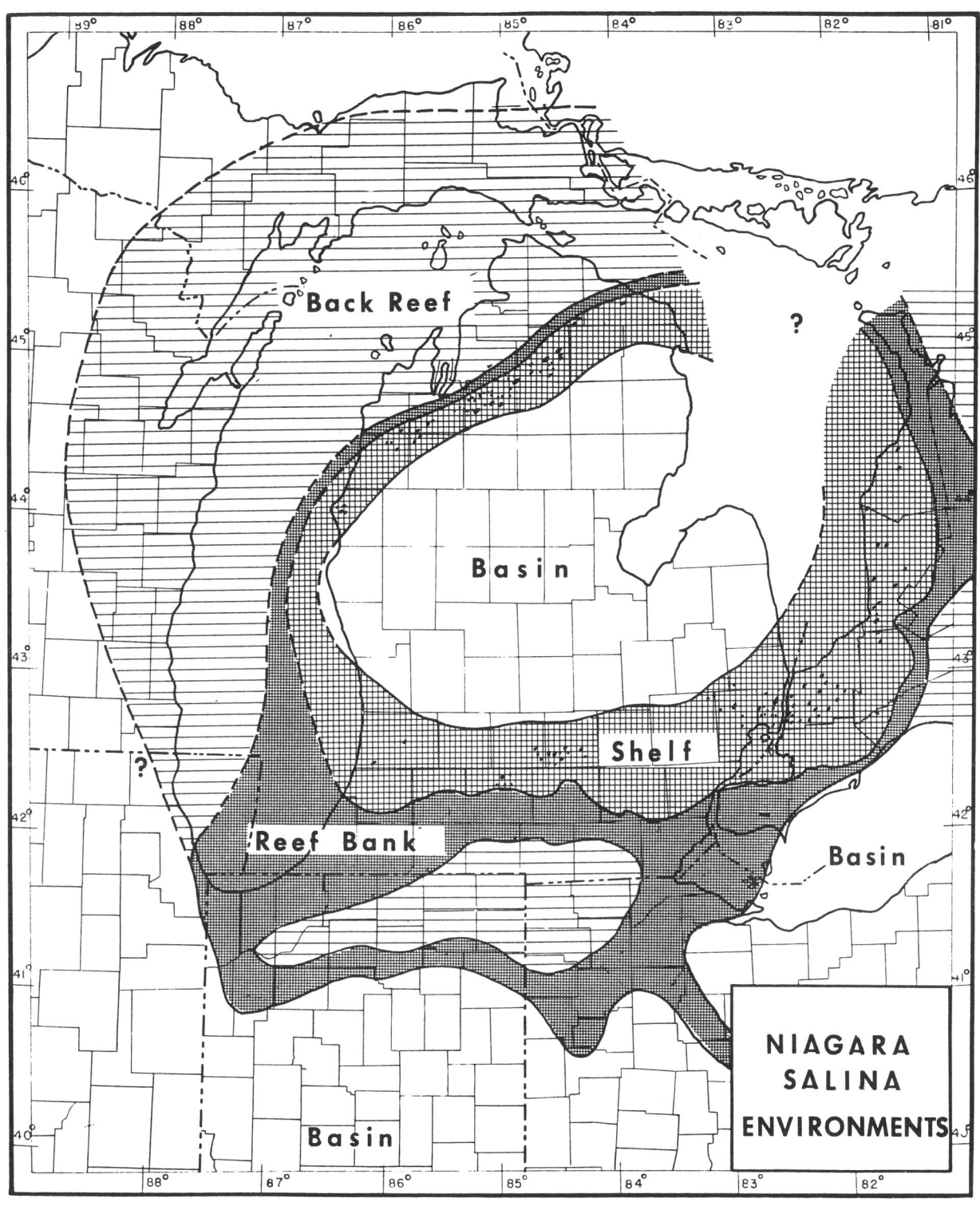

Figure 2. Middle and Late Silurian (Niagara-Salina) environments in the Michigan Basin. The pinnacle reefs are located along the shelf zone (indicated by dots within the shelf zone) (from Briggs and Briggs, 1974, modified after Ulteig, 1964; Sanford, 1969; Brigham, 1971; Shaver and others, 1971; Mantek, 1973; Meloy, 1974).

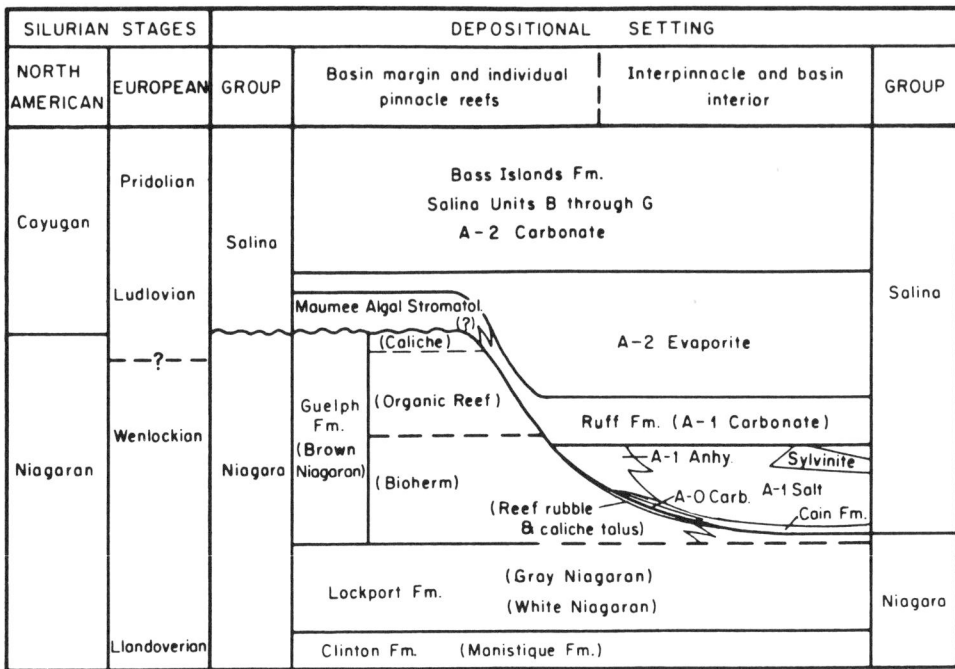

Figure 3. Stratigraphic relations and nomenclature of Middle and Upper Silurian formations in the Michigan Basin (from Gill, 1985). A-1 Carbonate and Maumee Algal Stromatolite are time-equivalent units.

Sharma (1966) is equivalent to the Lockport-Guelph interval and therefore encompasses both the lower parts of the pinnacle reefs and the subjacent pre-reef strata. A cross section through the basin margin shows the relations between the basin-edge shallowing-upward barrier-reef sequence, the shelf sequence with pinnacle reefs, and the evaporites of the overlying Salina Group in northern Michigan (Fig. 4). Early studies of the Michigan Basin Silurian reefs include Lowenstam (1950), Felber (1964), Sharma (1961, 1966), and Ells (1969).

The pinnacle reefs developed on the basin-rimming shelf in a belt transitional between the stable carbonate platform and the deep interior basin. Depositional conditions have been interpreted as uniformly subsiding (Sears and Lucia, 1979) or unstable differential subsidence (Briggs and Briggs, 1974; Gill, 1977, 1979). The pinnacles average 0.5 km² (0.2 mi²) in area and range in thickness from 90 m (300 ft) near the reef bank to 210 m (700 ft) offshore in the northwestern part of the basin (Mantek, 1976). Maximum thickness is about 130 m (430 ft) on the southern flank (Brigham, 1971). Thousands of individual pinnacle reefs with varied shapes and sizes occur in a belt that extends around the Michigan Basin (Fisher, 1973; Shaver, 1977).

At the end of Niagaran time, the Michigan Basin was barred by extensive lateral development of the basin-rimming barrier-reef sequence, and with continued evaporative conditions became hypersaline (Gill, 1977; Droste and Shaver, 1985). During this period of time, extensive evaporites were deposited, and the pinnacle reefs were subjected to one or more periods of subaerial exposure and fresh-water leaching (Balogh, 1981; Bay, 1983; Gill, 1985; Cercone and Lohmann, 1986; Cercone, 1988). By the end of Salina time, the Michigan Basin was filled by evaporites and intercalated thin limestone units, which represent alternating conditions of low and high sea level (Nurmi, 1975; Nurmi and Friedman, 1977; Droste and Shaver, 1985). These cycles can be recognized in the primarily carbonate sediments on the Wabash platform to the south, where reefs began nucleating at about the same time as in the Michigan Basin, but persisted through Salina time (Shaver, 1974; Droste and Shaver, 1977).

## MODELS FOR TIMING OF REEF GROWTH

Much controversy exists in the interpretation of the relation between pinnacle-reef development and the precipitation of evaporites. Three schools of thought, summarized by Mesolella and others (1974), relate relative timing of reef development and evaporite precipitation.

*Model 1.* The pinnacle reefs (including stromatolitic caps) developed in their entirety during Niagaran time and are postdated by the deposition of Salina carbonates and evaporites (Cain Formation and younger; see Fig. 5). The model is supported by Budros and Briggs (1977), Cercone and Lohmann (1986), and Cercone (1988).

*Model 2.* The development of the pinnacles occurred in part at the same time as the precipitation of the surrounding Salina evaporites (Droste and Shaver, 1977). This model is improbable because reef-dwelling/building organisms cannot survive under hypersaline conditions.

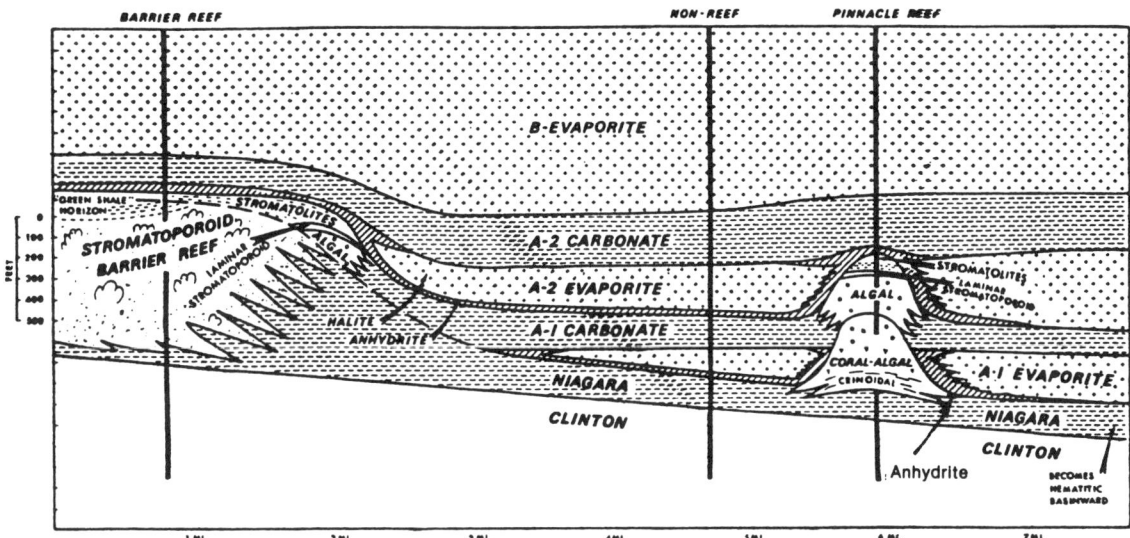

Figure 4. Schematic cross section of Middle and Upper Silurian rocks in the Michigan Basin showing the relation between the basin-margin reefs and shelf pinnacle reefs (after Mesolella and others, 1974).

*Model 3.* Periods of reef development were followed by periods of hypersalinity and precipitation of evaporites. In this model, the reefs and evaporites follow each other closely in time through several cycles of deposition, but are not synchronous (Figs. 3 and 4). At least one episode of evaporite deposition before the end of reef growth in the early Cayugan is implied by this model, but details vary among proponents.

*Evidence supporting model 3*

Models 1 and 3 are both supported by the mutually exclusive relation between reefs and evaporites: reef organisms prefer not to be pickled in brine (Friedman and Sanders, 1978)! In a recent study of the middle Miocene reefs and evaporites of the Gulf of Suez and Red Sea, Monty and others (1987) concluded, on the basis of field relations of these well-preserved and well-exposed sequences, that the basinal evaporites were entirely younger than the associated carbonates. The transition from normal marine to hypersaline conditions was marked by the extinction of the coral reefs, and their replacement with stromatolitic "reefs," followed by the elimination of even the euryhaline stromatolite-forming taxa and the initiation of evaporite deposition. This model corresponds either to model 1 or a single cycle of model 3 as outlined above. Model 3 is the most popular today (e.g., Balogh, 1981; Bay, 1983; Droste and Shaver, 1985), because it is supported by detailed lithostratigraphic (e.g., Gill, 1985) and biostratigraphic analysis (e.g., Droste and Shaver, 1985, 1987).

In a detailed discussion of the depositional history of some Michigan Basin Niagaran pinnacle reefs, Gill (1985) correlated the stromatolitic reef cap with the A-1 Carbonate (Ruff Formation) and Maumee Algal Stromatolite of Ohio (Kahle, 1974, 1978). The correlation was made on the basis of stratigraphic and petrographic similarity to the Maumee. A caliche-pebble conglomerate at the base of the Cain (A-0 Carbonate), which contains clasts derived from the caliche horizon of the coralgal reef below the stromatolitic cap (Gill, 1985, p. 124; Figs. 3 and 4), confirms the post-coralgal reef age of the Cain and supports model 3. Pentameracean brachiopods, conodonts, graptolites, acritarchs, and ostracodes were used by Droste and Shaver (1985, 1987) to support the inference of stratigraphic equivalency of the A-1 Carbonate and the stromatolitic reef-capping facies. It should be noted that all reefs are different; some may even lack the A-1 Carbonate-equivalent cap. In some reefs, the A-1 Carbonate-equivalent consists of a brachipod biostrome, rather than a stromatolite "reef" (Cercone, personal communication, 1988).

## THE PINNACLE REEFS

The development of an individual pinnacle reef has been divided into several stages, but the interpretation of these stages differs depending on the reef/evaporite depositional model that is upheld. In general, the pinnacle reefs overlie crinoidal and bryozoan wackestones of the upper Lockport Formation. The actual pinnacle development has been divided into four stages (Briggs and Briggs, 1974; 1974; Huh and others, 1977; Bentley, 1979a, b; Gonzales, 1981, 1982).

Stage 1 was the initial development on the shelf of mound-like bodies of carbonate mud mixed with the skeletons of bryozoans and crinoids (Fig. 6), and rare corals (Bay, 1983). The deposition of these mounds began below wave base, probably in relatively quiet water (Gill, 1985). The mounds may have been cemented on the sea floor and therefore possibly were wave resistant (Pratt, 1982). Mounds probably nucleated on local paleotopographic highs on which water clarity and light penetration

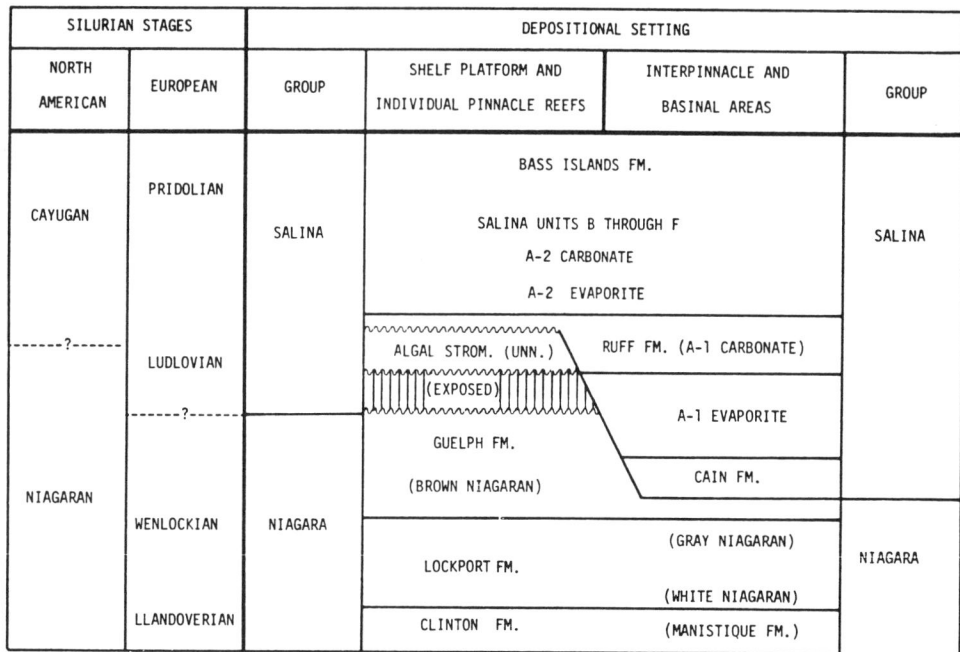

Figure 5. Model 1 stratigraphic relations and nomenclature of Middle and Upper Silurian formations in the Michigan Basin (from Gill, 1979). A-1 Carbonate, A-1 Evaporite, and Cain Formation entirely postdate algal stromatolite.

were greater than neighboring paleotopographic lows. Reefs expanded upward and outward as mound/reef growth outpaced deposition of detrital interreef carbonates (Droste and Shaver, 1985, 1987). As the mounds grew, they were colonized by corals and stromatoporoids and, finally, when they reached the high-energy wave zone, frame-building organisms became dominant and the mounds passed into their reefal stage (autogenic succession). A similar effect can result from relative sea-level change, but allogenic control of reef development has not been documented for Niagaran pinnacle reefs in the Michigan Basin.

Stage 2 of pinnacle growth was characterized by the development of a wave-resistant organic reef or boundstone (Fig. 6). The reef fauna included frame-building stromatoporoids, tabulate corals, sediment-binding algae, crinoids, bryozoans, and brachiopods. Changes in the dominant organisms in the reef assemblages occurred with reef growth, and there was a progression from stromatoporoids to corals and finally to LLH-type stromatolitic algae (Kahle, 1974, 1978; Gill, 1985). Reef growth kept pace with relative sea-level rise, resulting in a vertical pinnacle-reef morphology. At the climax of growth, the pinnacles may have stood from 90 to 210 m (300 to 700 ft) above the sea floor (130 m or 430 ft on the southern flank). However, Droste and Shaver (1987) suggested that slightly younger Illinois Basin reefs may have had small depositional relief due to simultaneous reef and interreef deposition. Michigan Basin Niagaran pinnacle reefs may also have had less depositional relief than commonly supposed (Jodry, 1969).

Stage 3 of pinnacle development is interpreted as a period of subaerial exposure when sea level fell an estimated 9 to 15 m (30 to 50 ft). During this time the reef core underwent apparent fresh-water leaching, with the development of vugs, internal sediment, vadose pisolites, calcrete, and laminar calcite.

During Stage 4, the top of the pinnacles formed supratidal islands characterized by the development of intertidal stromatolites, burrowed mudstone, peloidal wackestone, and flat-pebble conglomerate (Huh and others, 1977; Gill, 1985). In some places, a brachiopod biostromal facies formed during stage 4. During this stage, sea level fluctuated, and the reefs alternated between the supratidal, intertidal, and subtidal zones (Fig. 6). With the redevelopment of hypersaline conditions and fluctuating sea levels, the interreef areas were filled by carbonates and evaporites, and the reefs themselves were finally overlain by a thick layer of anhydrite (e.g., Gill, 1985, especially Fig. 7-5). Development of slightly younger Silurian reefs in the Illinois Basin seems to have involved only two principal reef generations due to greater environmental stability in that basin, which lacks evidence of regional evaporite deposition (Droste and Shaver, 1987).

The history of pinnacle-reef growth and resultant stratigraphy are complex, and the preceding summary is greatly simplified. Some pinnacle reefs apparently consist of multiple reservoirs, separated by subaerial exposure surfaces (emersion surfaces), paleowater tables, or other surfaces that may accumulate mud-size detrital material and/or become coated with cemented layers (such as caliche) and form barriers to vertical permeability. Thus, the details of reef stratigraphy, which depend in part on reef growth history, may strongly influence the distribution of hydrocarbon pools.

Relevant recent references on reef development include:

Balogh, 1981; Davidson, 1981; Gonzales, 1981, 1982; Griest and Shaver, 1981; Aminian, 1982; Hartsell, 1982; Bay, 1983; Leonard, 1983; Orr, 1984; Bourque and others, 1986.

## SEA-LEVEL CHANGE, EVAPORITE DEPOSITION, AND REEF DIAGENESIS

There have traditionally been two schools of thought in regard to sea-level change in the Late Silurian Michigan Basin. According to one model, the Salina evaporites are primarily shallow-water facies, deposited in a dry or highly evaporitic basin. The other model supposes that the Michigan Basin remained filled or mostly filled with water throughout the Cayugan, and that the Salina evaporites were deposited in deep water.

According to the dry-basin theory, leaching and other vadose diagenetic processes affected the Niagaran pinnacle reefs during an extended period of subaerial exposure, while sulfates, halite, sylvite, and borates were deposited in the interior basin (Nurmi and Friedman, 1977; Gill, 1985). It is unlikely that the entire Michigan Basin dried out during this time, but the occurrence of the sylvinite facies, as well as sedimentary structures and other features indicative of sabkha environments in the Salina evaporites in the basin center seem to confute a deep-water origin for the evaporites (Matthews, 1970; Mesolella and others, 1974; Nurmi and Friedman, 1977; Briggs and others, 1980).

The deep-water theory infers that the apparent vadose features of the pinnacle reefs formed subaqueously (Droste and Shaver, 1977, 1987) by analogy with similar rock units in other basins that have been interpreted in this way (e.g., Scholle and Kinsman, 1974; Esteban and Pray, 1975). In this scenario, chemical changes in Michigan Basin water associated with increased restriction caused syndepositional diagenetic alteration of pinnacle-reef rocks. This theory was developed to account for the occurrence on the Wabash platform south of the Michigan Basin of apparently continuous marine-reef growth coeval with topographically lower Salina evaporite deposition in the Michigan Basin (Shaver, 1974, 1977; Droste and Shaver, 1977, 1985, 1987).

A model of evaporative drawdown in an isolated Michigan Basin was recently articulated (Cercone, 1988). This model combines elements of preexisting models in an attempt to resolve the seeming inconsistency of a continuous Late Silurian marine record on the topographically high Wabash platform with evidence of prolonged subaerial exposure of the Michigan Basin pinnacle reefs and of shallow-water evaporite deposition in the Michigan Basin center. In this model, the barrier-reef and bank complexes surrounding the Michigan Basin are interpreted to be capped with supratidal deposits, which formed a continuous ring 50 to 150 km (31 to 93 mi) wide around the basin. This supratidal barrier partially isolated the basin so that evaporation exceeded inflow, and rapid evaporative drawdown and precipitation of shallow-water evaporites occurred. At the same time, leaching occurred in the exposed reefs ringing the basin, while reefs farther to the south

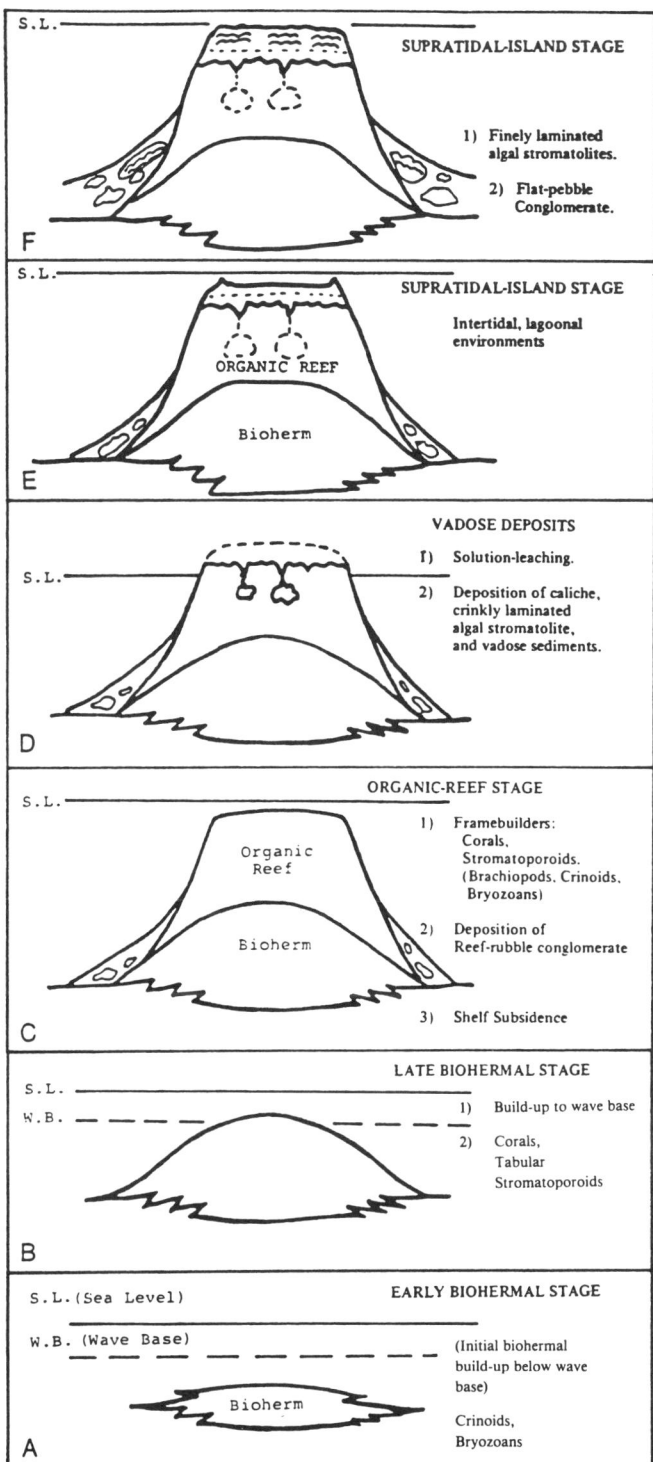

Figure 6. The postulated development of pinnacle reefs from the bioherm to the supratidal island stage (from Huh and others, 1977).

on the Wabash platform continued to grow in normal marine water (Cercone, 1988, Fig. 3).

Most previous workers have either dwelt on the Niagaran-Salina stratigraphy of the Michigan Basin itself while ignoring the contemporaneous deposits of the neighboring Wabash platform (e.g., Gill, 1977, 1979, 1985), or concentrated on the platform and non-evaporitic Illinois Basin to the south to the exclusion of the Michigan Basin (e.g., Droste and Shaver, 1987). The former have marshalled impressive evidence for subaerial exposure and evaporative drawdown in the basin and on its fringes, whereas the latter have equally convincingly documented relatively continuous marine-reef growth throughout the Late Silurian. Droste and Shaver have seriously addressed the reconciliation of these separate bodies of evidence in recent papers (Shaver, 1977; Droste and Shaver, 1977, 1985, 1987) but continue to support continuity between Illinois Basin, Wabash platform, and Michigan Basin water bodies during the Niagaran and Cayugan, thereby rendering evaporative drawdown in the Michigan Basin and prolonged subaerial exposure of the pinnacle reefs unlikely (Droste and Shaver, 1977).

The isolated basin model requires that the Michigan Basin pinnacle and barrier reefs have been exposed above sea level while at the same time the Wabash platform was continuously submerged. The detailed and comprehensive stratigraphic and structural studies needed to test this point have not yet been done. It also has not been established whether or not the pinnacle reefs were subaerially exposed during stage 3 of their development, nor has the shallow-water or deep-water origin of the Salina evaporites in the Michigan Basin been conclusively demonstrated. Accordingly, consensus has not been reached on any of these points. For simplicity, we shall in the following discussion adopt the position that subaerial exposure of the pinnacle reefs did occur, following the majority viewpoint.

## DIAGENESIS AND RESERVOIR OCCURRENCE

The pinnacle reefs were affected by fresh-water leaching, dolomitization, and the precipitation of void-filling evaporite minerals during their early history, which has had a great influence on subsequent porosity and permeability (e.g., Cercone and Lohmann, 1986, and references therein; Cercone, 1988).

During periods of subaerial exposure, either prior to or during the formation of the Salina evaporites, porosity of the pinnacles was enhanced by fresh-water leaching and the creation of vadose fractures, vugs, and channels in both the biohermal and organic reef facies (Mesolella and others, 1974; Gill, 1985; Cercone and Lohmann, 1986). Early dolomitization occurred in mixed waters, either schizohaline or fresh/marine water, which enhanced porosity especially in the biohermal skeletal micrites (Petta, 1980; Sears and Lucia, 1980; Cercone, 1984; Cercone and Lohmann, 1985, 1986). The model of early dolomitization preferred depends in part on the model for reef depositional history. Model 1 (evaporites entirely postdate reef development) appears to require schizohaline dolomitization because pervasive early dolomitization at the bases of some northern flank reefs would require very large-scale dilute phreatic systems under depositional model 1, which would be incompatible with the extremely arid climate (Cercone and Lohmann, 1986). Model 3 allows fresh/marine water dolomitization. Porosity occlusion during exposure was the result of introduction of internal sediment and calcite cements (Bentley, 1979b; Cercone and Lohmann, 1983).

A second episode of dolomitization occurred during deposition of the A-2 evaporite. This has been interpreted to have occurred in northwestern Michigan by brine reflux from the A-2 evaporite down through the organic-rich A-1 carbonate (Sears and Lucia, 1980; Cercone and Lohmann, 1986), one inferred source of the hydrocarbons now trapped in the pinnacle reefs (Lee and Budros, 1982; Gill, 1985; Cercone and Lohmann, 1986). However, Jodry (1969) suggested that expulsion of connate water from the A-1 Carbonate through the reefs after the start of A-2 Evaporite deposition caused dolomitization of the reefs, based on evidence from the southern flank of the basin.

During stages of evaporite precipitation, the formation waters within the pinnacles became hypersaline, and in some, available porosity was plugged by anhydrite and halite. In some reefs, a thick salt-plugged zone developed as relative sea level rose and the reefs were buried in evaporites. The tendency in some reefs is an increase in the amount of salt plugging toward the top of the reef (Petta, 1980). This effect is greater downdip, where the evaporite sequence is thicker and purer. Some salt-filled vugs contain halite crystals with oil inclusions, suggesting that some salt plugging may postdate some hydrocarbon migration.

Regional deep-burial diagenesis of the pinnacle reefs occurred mainly before the end of the Jurassic, and included formation of geopetal diagenetic "sediment," pyrite, pyrobitumen, rhombic dolomite cement, and equant calcite spar in fractures, stylolites, and late solution vugs (Cercone and Lohmann, 1987). The diagenetic fluids were basinal brines that accessed the reefs through regional aquifers (Cercone and Lohmann, 1984, 1987) such as the Lockport (Gray Niagaran; Gill, 1979). The effect of deep-burial diagenesis on reservoir quality of the pinnacle reefs has been minimal (Cercone and Lohmann, 1986, 1987).

Within producing reefs, intercrystalline and vuggy porosity occurs in the biohermal and reef-core facies. Porosities range from 3 to 37 percent with an average of 6 percent (Gill, 1979). Average permeability is about 11 to 12 mD, but locally is much higher.

Large-scale trends in diagenesis and production have been noted in the northern pinnacle-reef belt (Gill, 1979; Cercone and Lohmann, 1983). The plugging of the reefs with halite and anhydrite appears to increase basinward, whereas the degree of dolomitization and the amount of secondary porosity increase toward the shore. The trend in dolomitization results in reefs that are composed completely of dolomite on the updip edge of the pinnacle belt, reefs of interbedded limestone and dolomite within the belt, and reefs completely composed of limestone on the basinward edge of the shelf. These trends apparently result from availability and chemistry of diagenetic fluids, not from primary lithologic differences (Cercone and Lohmann, 1983, 1986). Reefs in the southern belt are entirely dolomite.

The distribution of reservoir fluids for the northern basin flank is shown in both plan and section in Figure 7. On the basinward margin of the shelf, the reefs tend to be barren or salt plugged. Moving in an updip direction, the reefs produce gas, then oil, and finally closest to the shore they are barren or produce water. This distribution has been explained as the result of updip migration and differential entrapment of fluids in the reefs, which are hydraulically connected through the underlying Lockport Formation (Gray Niagaran; Gill, 1979), which is in full agreement with the classic theory of Gussow (1954, 1968). Likely source rocks for Niagaran pinnacle-reef oil are the Cain Formation (top part of Brown Niagaran in interreef sequences; Gill and others, 1978; Gill, 1979), which overlies the porous and permeable Lockport Formation, and the A-1 Carbonate (Gill,

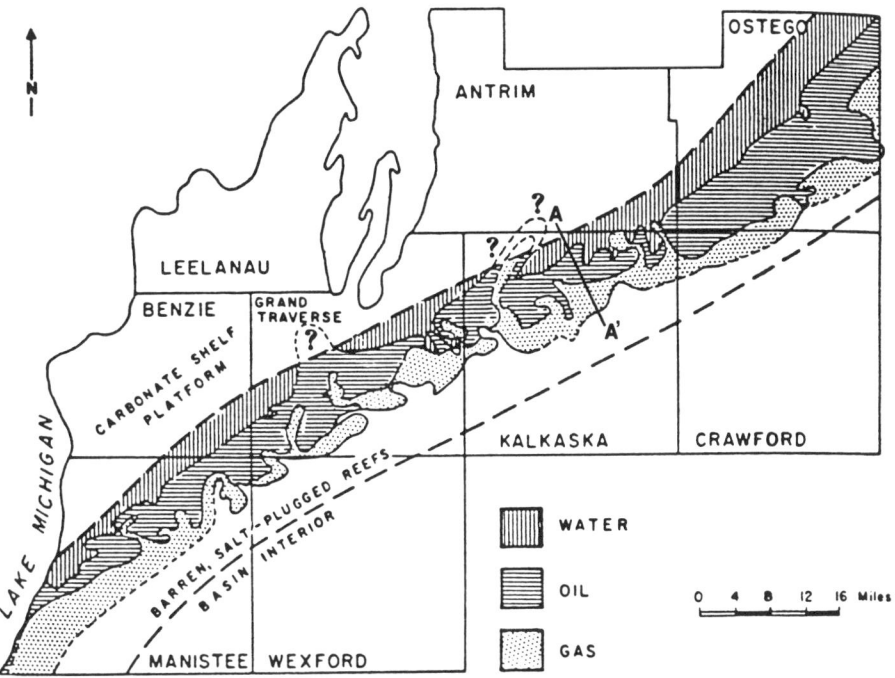

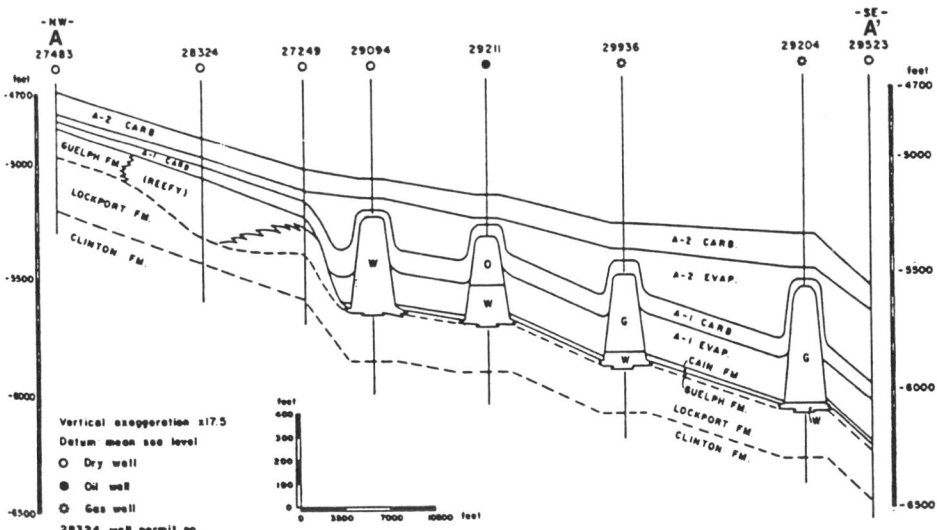

Figure 7. Map showing the distribution of salt-plugged, and gas-, oil-, and water-bearing reefs in the northern pinnacle-reef belt of the Michigan Basin. The cross-section (A-A′) shows the segregation of fluids within the reefs (from Gill, 1979).

1985). The Cain is preferred as the primary source because the A-1 Carbonate is relatively impermeable (Gill, 1979). The basal argillaceous mudstones of the Cain contain on average 0.30 percent TOC (Gill, 1979), and this formation is very similar to other preevaporite sapropelic bituminous carbonates inferred to be excellent hydrocarbon source beds (Gill, 1979). Hydrocarbons from the Cain would probably have migrated to the reefs through the Lockport, whereas those from the A-1 Carbonate could have entered the reefs directly. Migration of oil into Michigan Basin pinnacle reefs could not have occurred before the Carboniferous, as indicated by source-rock maturity studies (Cercone, 1984), but the precise timing is unknown.

## CONCLUSIONS

The pinnacle reefs of the Michigan Basin form small, isolated hydrocarbon reservoirs encased in impermeable evaporites and carbonates. The temporal relations between the reef sequence and the evaporites are still in dispute, but many writers seem to be leaning toward model 3 (e.g., Gill, 1985 versus Gill, 1979). Pinnacle development occurred in four stages and included periods of subaerial exposure that enhanced reef porosity and permeability through leaching and dolomitization. The occurrence of subaerial conditions during stage 3 of reef growth is also disputed, as is deposition of Salina evaporites in shallow water, but both are currently favored. Subsequent evaporite precipitation filled much of this porosity, producing many reefs that are completely salt plugged and impermeable.

Regional trends have been recognized across the northern pinnacle-reef belt, and these predict increased salt-plugging of the reefs basinward, increased dolomitization and preserved secondary porosity toward the shore, and zones of production that pass updip from gas to oil and finally to water. The producing reefs have porosities that range from 3 to 37 percent (average 6 percent) and permeabilities that range up to more than 1 D, but average 11 to 12 mD. Stratigraphy and diagenesis of Michigan Basin pinnacle reefs are complex and variable, both regionally and locally; despite the existence of established trends, unsuspected types of hydrocarbon traps probably exist.

## ACKNOWLEDGMENTS

Brian Pratt, Bruce Levell, Paul A. Daniels, Karen Rose Cercone, William Mantek, and one anonymous reviewer critically evaluated earlier drafts of this chapter.

## REFERENCES CITED

Aminian, K., 1982, The synergetic study of Silurian-Niagaran pinnacle reef belt around the Michigan Basin for exploration and production of oil and gas [Ph.D. thesis]: Ann Arbor, University of Michigan, 568 p.

Balogh, R. J., 1981, A study of the Middle Silurian Belle River Mills, Peters, and Ray pinnacle reefs from the Michigan Basin: Geological Society of America Abstracts with Programs, v. 13, p. 402.

Bay, T. A., 1983, The Silurian of the northern Michigan Basin, in Harris, P. M., ed., Carbonate buildups; A core workshop (SEPM Core Workshop No. 4): Tulsa, Oklahoma, Society of Economic Paleontologists and Mineralogists, p. 53–72.

Bentley, B. P., 1979a, Carbonate lithofacies and diagenetic features of the Guelph Formation (Middle Silurian) in the Amoco Production Berg-Berg 1-21 unit well, Presque Isle County, Michigan [M.S. thesis]: Troy, New York, Rensselaer Polytechnic Institute, 147 p.

——, 1979b, Carbonate lithofacies and diagenetic features of the Guelph Formation (Middle Silurian) in the Amoco Production Berg-Berg 1-21 unit well, Presque Isle County, Michigan: The Compass of Sigma Gamma Epsilon, v. 57, p. 16–26.

Bourque, P. A., and 6 others, 1986, Silurian and Lower Devonian reef and carbonate complexes of the Gaspé Basin, Quebec; A summary: Bulletin of Candian Petroleum Geology, v. 34, p. 452–489.

Briggs, L. I., and Briggs, D., 1974, Niagara-Salina relationships in the Michigan Basin, in Kesling, R. V., ed., Silurian reef-evaporite relationships: Michigan Basin Geological Society Field Conference, p. 1–23.

Briggs, L. I., Gill, D., Briggs, D. Z., and Elmore, R. D., 1980, Transition from open marine to evaporite deposition in the Silurian Michigan Basin, in Nissenbaum, A., ed., Hypersaline and evaporite environments: Amsterdam, Elsevier Scientific Publishing Co., p. 253–270.

Brigham, R. J., 1971, Structural geology of southwestern Ontario and southeastern Michigan: Ontario Department of Mines and Ministry of Northern Affairs, Petroleum Resources Section, Paper 71-2, 110 p.

Budros, R. and Briggs, L. I., 1977, Depositional environment of Ruff Formation (Upper Silurian) in southeastern Michigan, in Fisher, J. H., ed., Reefs and evaporites; Concepts and depositional models: American Association of Petroleum Geologists Studies in Geology 5, p. 53–71.

Caughlin, W. C., Lucia, F. J., and McIver, N. L., 1976, The detection and development of Silurian reefs in northern Michigan: Geophysics, v. 41, p. 646–658.

Cercone, K. R., 1984, Diagenesis of Niagaran (Middle Silurian) pinnacle reefs, northwest Michigan [Ph.D. thesis]: Ann Arbor, University of Michigan, 382 p.

——, 1988, Evaporative sea-level drawndown in the Silurian Michigan basin [sic]: Geology, v. 16, p. 387–390.

Cercone, K. R., and Lohmann, K. C., 1983, Regional aspects of diagenesis in Niagaran pinnacle reefs, northwest Michigan; Evidence for differential fluid migration [abs.]: American Association of Petroleum Geologists Bulletin, v. 67, p. 438.

——, 1984, Tracking diagenetic fluids; Isotope and fluid inclusion evidence from Silurian reefs of northern Michigan: Geological Society of America Abstracts with Programs, v. 16, p. 466.

——, 1985, Early diagenesis of Middle Silurian pinnacle reefs, northern Michigan, in Cercone, K. R., and Budai, J. M., eds., Ordovician and Silurian rocks of the Michigan Basin and its margins: Michigan Basin Geological Society Special Paper 4, p. 109–130.

——, 1986, Diagenetic history of the Union 8 pinnacle reef (Middle Silurian), northern Michigan, USA, in Schroeder, J. H., and Purser, B. H., eds., Reef diagenesis: Berlin, Springer-Verlag, p. 381–398.

——, 1987, Late burial diagenesis of Niagaran (Middle Silurian) pinnacle reefs in Michigan Basin: American Association of Petroleum Geologists Bulletin, v. 71, p. 156–166.

Cohee, G. V., and Landes, K. K., 1958, Oil in the Michigan Basin, in Weeks, L. G., ed., Habitat in oil: Tulsa, Oklahoma, American Association of Petroleum Geologists, p. 473–493.

Davidson, R. H., 1981, Petrology of pinnacle reefs in the Guelph Formation

(Niagaran), northern Michigan [M.S. thesis]: Nacogdoches, Texas, Stephen F. Austin State University, 93 p.

Droste, J. B., and Shaver, R. H., 1977, Synchronization of deposition; Silurian reef-bearing rocks on Wabash platform with cyclic evaporites of Michigan Basin, in Fisher, J. H., ed., evaporites; Concepts and depositional models: American Association of Petroleum Geologists Studies in Geology 5, p. 93–109.

——, 1985, Comparative stratigraphic framework for Silurian reefs; Michigan Basin to surrounding platforms, in Cercone, K. R., and Budai, J. M., eds., Ordovician and Silurian rocks of the Michigan Basin and its margins: Michigan Basin Geological Society Special Paper 4, p. 73–93.

——, 1987, Upper Silurian and Lower Devonian stratigraphy of the central Illinois Basin: Indiana Geological Survey Special Report 39, 29 p.

Ells, G. D., 1969, Architecture of the Michigan Basin: Michigan Basin Geological Society Annual Field Excursion, p. 60–88.

Esteban, M., and Pray, L. C., 1975, Subaqueous, syndepositional growth of in-place pisolite, Capitan reef complex (Permian) Guadalupe Mountains, New Mexico and West Texas: Geological Society of American Abstracts with Programs, v. 7, p. 1068–1069.

Felber, B. E., 1964, Silurian reefs of southeastern Michigan [Ph.D. thesis]: Evanston, Illinois, Northwestern University, 194 p.

Fisher, J. H., 1973, Petroleum occurrence in the Silurian reefs of Michigan: Ontario Petroleum Institute, 12th Annual Conference, Paper 9, 10 p.

Friedman, G. M., and Sanders, J. E., 1978, Principles of Sedimentology: New York, John Wiley and Sons, 792 p.

Gill, D., 1977, Salina A-1 sabkha cycles and the Late Silurian paleogeography of the Michigan Basin: Journal of Sedimentary Petrology, v. 47, p. 979–1017.

——, 1979, Differential entrapment of oil and gas in Niagaran pinnacle-reef belt of northern Michigan: American Association of Petroleum Geologists Bulletin, v. 63, p. 608–620.

——, 1985, Depositional facies of Middle Silurian (Niagaran) pinnacle reefs, Belle River Mills gas field, Michigan Basin, southeastern Michigan, in Roehl, P. O., and Choquette, P. W., eds., Carbonate petroleum reservoirs: New York, Springer-Verlag, p. 121–139.

Gill, D., Briggs, L. I., and Briggs, D., 1978, The Cain Formation; A transitional succession from open marine carbonates to evaporites in a deep water basin, Silurian, Michigan Basin [abs.], in Abstracts 10th International Congress on Sedimentology: International Association of Sedimentologists, v. 1, p. 244–245.

Gonzales, B. R., 1981, Facies analysis of pinnacle reefs of the Guelph Formation (Middle Silurian), northern Michigan [M.S. thesis]: Nacogdoches, Texas, Stephen F. Austin State University, 150 p.

——, 1982, Facies analysis of pinnacle reefs of the Guelph Formation (Middle Silurian), northern Michigan [abs.]: American Association of Petroleum Geologists Bulletin, v. 66, p. 573.

Griest, S. D., and Shaver, R. H., 1981, Geometric and paleocologic analysis of Silurian reefs near Celina, Ohio: Proceedings of the Indiana Academy of Science, v. 91, p. 373–390.

Gussow, W. C., 1954, Differential entrapment of oil and gas; A fundamental principle: American Association of Petroleum Geologists Bulletin, v. 38, p. 816–853.

——, 1968, Migration of reservoir fluids: Journal of Petroleum Technology, v. 20, p. 353–363.

Hartsell, M. Y., 1982, Niagara pinnacle reefs of western Michigan [M.S. thesis]: East Lansing, Michigan State University, 109 p.

Huh, J. M., Briggs, L. I., and Gill, D., 1977, Depositional environments of pinnacle reefs, Niagara and Salina Groups, northern shelf, Michigan Basin, in Fisher, J. H., ed., Reefs and evaporites; Concepts and depositional models: American Association of Petroleum Geologists Studies in Geology 5, p. 1–21.

Jodry, R. L., 1969, Growth and dolomitization of Silurian reefs, St. Clair Co., Michigan: American Association of Petroleum Geologists Bulletin, v. 53, p. 957–981.

Kahle, C. F., 1974, Nature and significance of Silurian rocks at Maumee quarry, Ohio, in Silurian reef-evaporite relationships: Michigan Basin Geological Society Field Conference, p. 31–54.

——, 1978, Patch reef development and effects of repeated subaerial exposure in Silurian shelf carbonates, Maumee, Ohio, in Kesling, R. V., ed., Guidebook to Field Excursions: Geological Society of America North-central Section, p. 63–115.

Lee, J., and Budros, R., 1982, Reef exploration in Michigan Basin; Problems and solutions [abs.]: American Association of Petroleum Geologists Bulletin, v. 66, p. 593.

Leonard, B. J., 1983, Environment of deposition of the A-1 carbonate, Salina Group, Michigan Basin [M.S. thesis]: Kalamazoo, Western Michigan University, 83 p.

Lowenstam, H. A., 1950, Niagaran reefs of the Great Lakes area: Journal of Geology, v. 58, p. 430–487.

Mantek, W., 1973, Niagaran pinnacle reefs in Michigan: Michigan Basin Geological Society Annual Field Conference Guidebook, p. 35–46.

——, 1976, Recent exploration activity in Michigan, in 15th Annual Conference Proceedings: Ontario Petroleum Institute, 29 p.

Matthews, D., 1970, The distribution of Silurian potash in the Michigan Basin, in 6th Forum on Geology of Industrial Minerals: Lansing, Michigan Geological Survey, p. 20–33.

McClintock, P. L., 1977, Seismic data processing techniques in exploration for reefs, northern Michigan, in Fisher, J. H., ed., Reefs and evaporites; Concepts and depositional models: American Association of Petroleum Geologists Studies in Geology 5, p. 111–124.

Meloy, D. L., 1974, Depositional history of the Silurian northern carbonate bank of the Michigan Basin [M.S. thesis]: Ann Arbor, University of Michigan, 78 p.

Mesolella, K. J., Robinson, J. D., McCormick, L. M., and Ormiston, A. R., 1974, Cyclic deposition of Silurian carbonates and evaporites in Michigan Basin: American Association of Petroleum Geologists Bulletin, v. 58, p. 34–62.

Monty, C.L.V., Rouchy, J. M., Maurin, A., Bernet-Rollande, M. C., and Perthuisot, J. P., 1987, Reef-stromatolites-evaporites facies relationships from middle Miocene examples of the Gulf of Suez and the Red Sea, in Peryt, T. M., ed., Evaporite basins; Lecture notes in earth sciences 13: New York, Springer-Verlag, p. 133–188.

Nurmi, R. D., 1975, Stratigraphy and sedimentology of the Lower Salina Group (Upper Silurian) in the Michigan Basin [Ph.D. thesis]: Troy, New York, Rensselaer Polytechnic Institute, 261 p.

Nurmi, R. D., and Friedman, G. M., 1977, Sedimentology and depositional environments of basin-center evaporites, Lower Salina Group (Upper Silurian), Michigan Basin, in Fisher, J. H., ed., Reefs and evaporites; Concepts and depositional models: American Association of Petroleum Geologists Studies in Geology 5, p. 23–52.

Orr, G. D., 1984, Niagaran reefs, northwestern Michigan [M.S. thesis]: East Lansing, Michigan State University, 41 p.

Petta, T. J., 1980, Silurian pinnacle reef diagenesis; Northern Michigan, in Halley, R. B., and Loucks, R. G., eds., Carbonate reservoir rocks; Notes for SEPM core workshop 1: Society of Economic Paleontologists and Mineralogists, p. 32–42.

Pratt, B. R., 1982, Stromatolitic framework of carbonate mudmounds: Journal of Sedimentary Petrology, v. 52, p. 1203–1227.

Sanford, B. V., 1969, Silurian of southwestern Ontario: Ontario Petroleum Institute 8th Annual Conference, Paper 5, 44 p.

Scholle, P. A., and Kinsman, D.J.J., 1974, Aragonitic and high-Mg calcite caliche from the Persian Gulf; A modern analog for the Permian of Texas and New Mexico: Journal of Sedimentary Petrology, v. 44, p. 904–916.

Scotese, C., Bambach, R. K., Barton, C., Van der Voo, R., and Ziegler, A., 1979, Paleozoic base maps: Journal of Geology, v. 87, p. 217–277.

Sears, S. O. and Lucia, F. J., 1979, Reef-growth model for Silurian pinnacle reefs, northern Michigan reef trend: Geology, v. 7, p. 299–302.

——, 1980, Dolomitization of northern Michigan Niagaran reefs by brine refluxion and freshwater/seawater mixing, in Mazzullo, S. J., and Zenger, D. R., eds., Concepts and models of dolomitization: Society of Economic Paleontologists and Mineralogists Special Publication 28, p. 215–236.

Sharma, G. D., 1961, Geology of the Peters field, St. Clair Co., Michigan [Ph.D. thesis]: Ann Arbor, University of Michigan.

—— , 1966, Geology of the Peters reef, St. Clair Co., Michigan: American Association of Petroleum Geologists Bulletin, v. 50, p. 327–350.

Shaver, R. H., 1974, Structural evolution of northern Indiana during Silurian time, *in* Silurian reef-evaporite relationships: Michigan Basin Geological Society Field Conference, p. 55–77.

—— , 1977, Silurian reef geometry; New dimensions to explore: Journal of Sedimentary Petrology, v. 47, p. 1409–1424.

Shaver, R. H., and 6 others, 1971, Silurian and Middle Devonian stratigraphy of the Michigan Basin; A view from the southwest flank, *in* Forsyth, J. L., Geology of the Lake Erie islands and adjacent shores: Michigan Basin Geological Society Guidebook, p. 37–59.

Ulteig, J. R., 1964, Upper Niagaran and Cayugan stratigraphy of northeastern Ohio and adjacent areas: Ohio Geological Survey Report of Investigations 51, 48 p.

Yelling, W. F., Jr., and Tek, M. R., 1976, Prospects for oil and gas from Silurian-Niagaran trend in Michigan: Ann Arbor, University of Michigan Institute of Science and Technology, 35 p.

MANUSCRIPT ACCEPTED BY THE SOCIETY JUNE 1, 1990

Geological Society of America
Special Paper 256
1991

# A history of study of Silurian reefs in the Michigan Basin environs

**Robert H. Shaver**
*Indiana Geological Survey, 611 North Walnut Grove, Bloomington, Indiana 47405, and*
*Indiana University Department of Geological Sciences, 1005 East Tenth Street, Bloomington, Indiana 47405*

## ABSTRACT

The first Silurian carbonate buildup in North America to be correctly identified as a reef was in eastern Wisconsin, in 1862. In the ensuing 125 years, many hundreds of studies were presented on other Silurian reefs in the Michigan Basin and environs. Successive trends in both scientific and economic interests have characterized the quest to learn what riches these reefs may yield. Three periods are recognizable: (1) An early period of discovery, extending until 1926, saw more incorrect ideas to explain the once-mysterious reefs than correct ideas, e.g., ideas of volcanism and upheavals. James Hall, who advanced both incorrect and correct ideas, best typified this period, but T. C. Chamberlin should be credited most for his lasting insight. (2) A middle period of enlightenment, 1927 to 1960, saw a convincing reef proof and a systematic set of biologic reef parameters set forth by (especially) E. R. Cumings, R. R. Shrock, and H. A. Lowenstam. (3) A modern period of integration, 1961 to present, could have been designated as one of proliferation, so numerous were the new reef models and ideas concerning reef geometry, distribution, diagenesis, evaporite relations, deep- versus shallow-water environments, basin-to-shelf differences, cyclicity in deposition, sea-level changes, tectonism, and hydrocarbon accumulation. Many of these ideas conflict; thus, I choose the regionally broad stratigraphic integration that developed as the most significant key to the modern period and the several debates that yet require reckoning against the modern stratigraphic framework. The stratigraphic relations favored in this chapter depart from tradition, but they suggest several kinds of studies that need to be undertaken.

## INTRODUCTION

The Silurian reefs that are exposed in the Niagara Escarpment along the far reaches of the Michigan Basin have been recognized and figured for more than a century and are a part of classical fossil-reef study. The many hundreds and probably thousands of Silurian reefs that lie buried within the Michigan Basin are a much later discovery. The early discoveries of Silurian reef-associated hydrocarbons within the basin occurred about 100 years ago in Ontario and about 60 years ago in Michigan. Even earlier discovery occurred outside the basin in western Indiana, but few producers or the geological profession then understood the connection between reefs and hydrocarbons. Major Silurian exploration in Michigan did not begin until 1952. By that time much pertinent history of Silurian reef study had passed, as recorded in the abbreviated perspective below:

*1841:* Sir Charles Lyell first applied the term "reef" to the previously mysterious Silurian buildups in England.

*1862:* In Wisconsin, James Hall first recognized a North American Silurian reef as such.

*1865:* The first Silurian reef-associated hydrocarbons in North America were discovered in Indiana but were not produced commercially until 23 or 24 years later.

*1901 to 1913:* Amadeus Grabau asserted that the Silurian reef-depositional and salt-depositional episodes were wholly successive events everywhere in the Great Lakes area.

*1927:* Edgar Cumings and Robert Shrock began to publish their milestone studies of reefs along the far southwestern border of the Michigan Basin in Indiana, studies that essentially ended an 87-year history of controversy in favor of an organic origin for the midwestern Silurian carbonate buildups.

*1950:* Heinz Lowenstam added important refinements to the

Shaver, R. H., 1991, A history of study of Silurian reefs in the Michigan Basin environs, *in* Catacosinos, P. A., and Daniels, P. A., Jr., eds., Early sedimentary evolution of the Michigan Basin: Geological Society of America Special Paper 256.

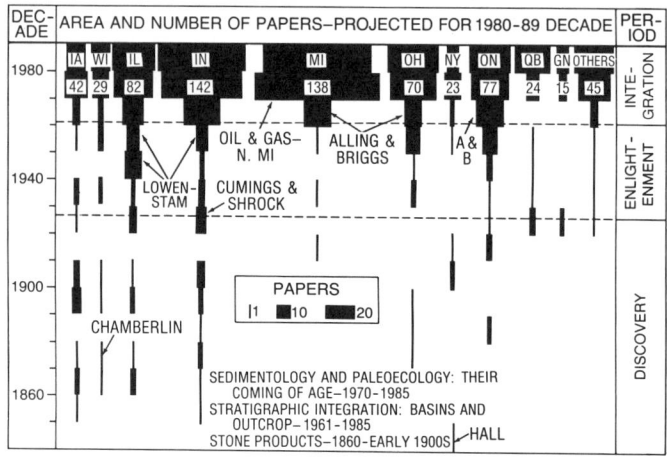

Figure 1. A 539-paper history of Silurian reef study arranged by geographic area and projected from 1985 through the 1980–89 decade. Numbers at top are cumulative for the entire history. Sources of information are the Pugh (1950) reef bibliography, Geo-Ref (American Geologic Institute), and my personal library.

Silurian reef concept, including for hydrocarbon-yielding reefs in the Illinois Basin.

*1955 and 1961:* Studies by Thomas Bolton and Bruce Liberty and by Harold Alling and Louis Briggs, respectively, suggested more intimate reef-evaporite relations than had Grabau's (above).

*1974:* Hydrocarbon production had become so successful in Michigan, particularly in the northern Silurian reef trend, that Briggs and Briggs (1974a) predicted that the northern trend would amount to a giant oil field (500,000,000 bbl or more).

*1980 to 1989:* Silurian reef literature for Michigan became more voluminous than that for any other state or province during this decade.

In North America, only the Canadian Arctic and North Greenland provinces have had major Silurian reef studies that began after major study of the Michigan Basin. Until 1985, Indiana had led other areas slightly in cumulative numbers of papers (Fig. 1). T. C. Chamberlin, Edgar Cumings, Robert Shrock, and even Heinz Lowenstam as late as the 1950s, were somewhat lonely pioneers for their times. There was a great burgeoning of reef study during the 15 years preceding 1985 (Figs. 1 and 2).

In these years, sedimentology dominated more traditional emphases in geology as the primary preoccupation in Silurian-reef study. Hydrocarbon interest was also strong. The period from 1975 to 1985 saw the appearance of new paleoecological studies on Silurian reefs and associated strata that were significant additions to the founding work of (especially) Cumings and Shrock (1928a, b) and Lowenstam (1948–1957).

Despite the long history of Silurian-reef study prior to Michigan Basin studies, the struggle to understand Silurian reefs in general seems only to have intensified to the present time. Silurian workers have yet to achieve unanimity on several questions, especially: (1) ecologic (organic, framework) reefs versus buildups having something less than reef status and known variably as carbonate-mud mound, bioherm, or clinothem deposits; (2) deepwater (aphotic) versus shallow-water (photic) reef environments; (3) wholly successive reef and evaporite histories versus overall contemporaneity, even though reciprocal in some places, within the whole of the midwestern reef province; (4) in relation to item 3, shallow-water (or even sabkha) evaporites versus deep-water evaporites; (5) modes and timing of diagenetic processes and their textural, structural, and geometric effects on the reefs; and (6) general regional or global control (epeirogenesis?, tectonism?) of reef and evaporite sedimentation versus local, independent control (tectonism?, evaporative drawdown?, autocyclic sedimentation?).

Much of the intensity of debate on these questions has been evoked through lately evinced Michigan Basin interests that are mostly economic in nature. The basin-based literature focuses narrowly on what is a small segment of the entire Silurian reef archipelago but does not answer all the questions posed above. This assertion is made because of what has been learned recently of wide-ranging control of Silurian sea levels and sedimentational events and because newer regional Silurian stratigraphic studies now link depositional histories, basin and interbasin alike, throughout the Great Lakes area.

To me, for example, the larger, preevaporite parts of the buried Michigan Basin reefs are the middle generation of several within the larger basin environs (Figs. 3 and 4); their particular diagenetic and preservational history is one of a variety of such histories; their stratigraphic relations, narrow as they are (Fig. 3), are not the most critical key to understanding the overall Silurian history of reef development, and they are only one key to understanding the overall relations of reefs and evaporites and their depositional environments.

I believe, therefore, that much of the Michigan Basin evaporite-dominated view of Silurian geology, including associated reef development, needs to be reexamined, particularly in

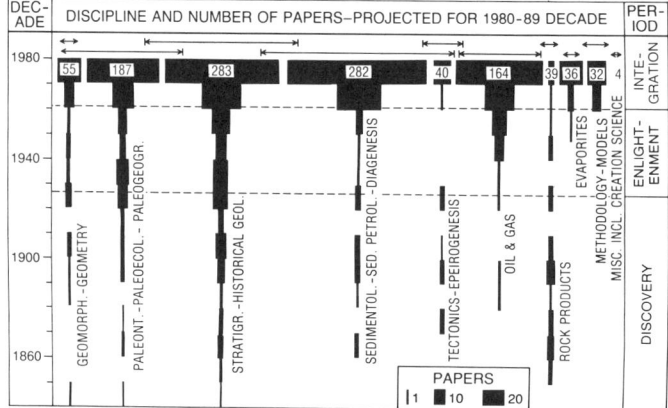

Figure 2. A 539-paper history of Silurian reef study arranged by subdisciplinary emphases and projected from 1985 through the 1980–89 decade. Numbers at top are cumulative for the entire history. Sources of information are the same as for Figure 1.

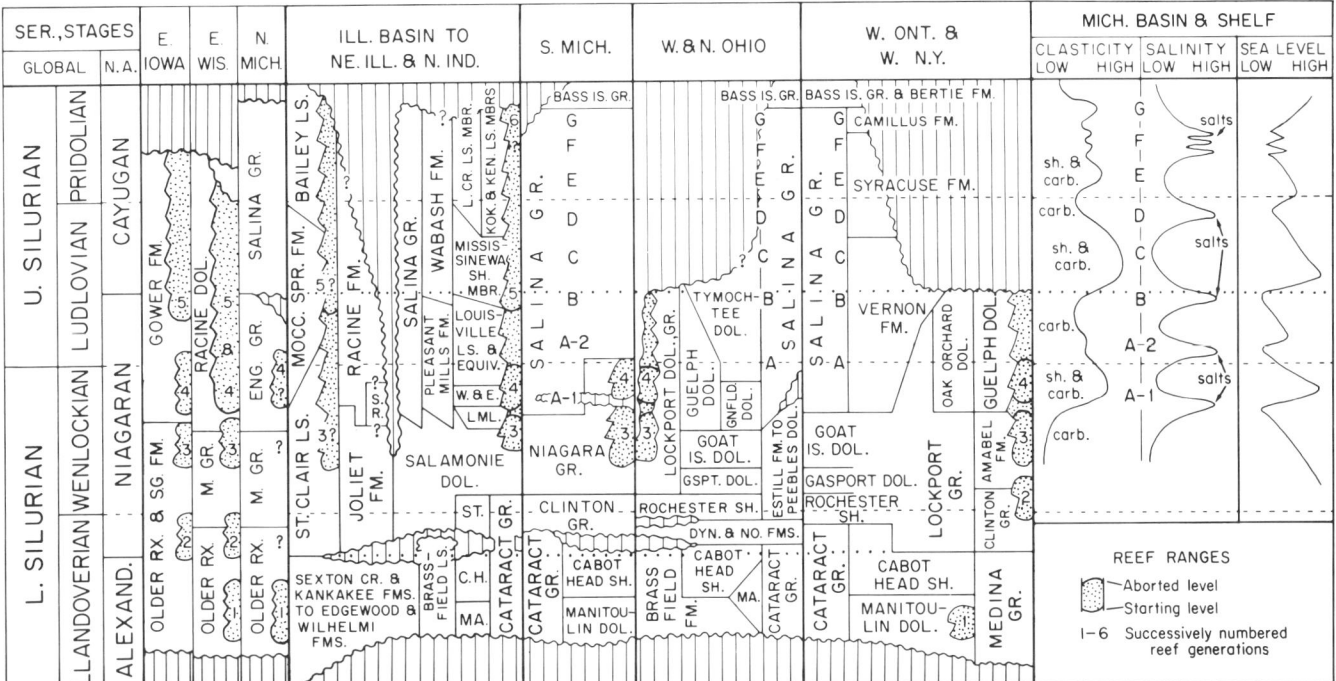

Figure 3. Chart showing selected correlations of Silurian rocks in the Great Lakes region together with cyclical factors in sedimentation, including reef generation and abortion, for the Michigan Basin and environs. Adapted from Shaver and others (1983), Droste and Shaver (1987a, b), and many earlier reports. Letters A through G represent informally designated units of the Salina Group. Stratigraphic abbreviations: C. H., Cabot Head Member; ENG., Engadine; IS., Island; KEN., Kenneth; KOK., Kokomo; L. CR., Liston Creek; LML., Limberlost Dolomite Member; M., Manistuque; MA., Manitoulin Dolomite Member; MOCC, SPR., Moccasin Springs; S.G., Scotch Grove; ST., Stroh Member; S.R., Sugar Run Formation; and W. & E., Waldron Shale and equivalents. "Reef generation" here refers to a group of reefs that were closely coordinated for extrinsic reasons in their starting levels, although at least generation 1 is a loosely collective group. Not all reefs of a given generation were aborted at the same time; however, some were for extrinsic reasons. For example, some reefs of generation 3 were quickly aborted in coordinated style, whereas some others survived even to the end of Silurian time.

the light of newer regional stratigraphic studies. Similarly, the unsettled questions on the fundamental nature of Silurian buildups noted above need further study in the same light.

This history, therefore, extends well beyond the Michigan Basin proper and provides a broad backdrop through which a better understanding of the Michigan Basin reefs may be gained. But this history is more than that: it is a mirror of the geological profession. It records many of the grand names in geology that figure in other histories of study; it reflects the changing economic impetus that fueled geological acumen in general; and it embodies the waves of major scientific emphases ranging from discovery, description, and historical geology through the burgeoning period of sedimentology and global tectonics.

Particularly in the Great Lakes area, the vast Silurian reef archipelago of northeastern North America (Fig. 5, a time-composited portrayal) now belongs to classic study, and the ideas on its development have influenced geological study well beyond the direct measure of its part of the geologic record.

It is one of the world's great fossil-reef systems. Nevertheless, it is partly conjectural as portrayed (Fig. 5), both as to numbers and locations of discrete reef masses. If the smaller reefs that are measured in mere meters were counted, probably many millions of discrete reefs dotted this archipelago. Moreover, the archipelago was active for several millions of years, and some large complexes reached 500 m, more and less, in aggregate thickness.

Study of these reefs is conveniently characterized by three historical phases: (1) an early period (1843 to 1927) and its romance of discovery, (2) a middle period (1928 to 1961) of scientific enlightenment, and (3) a modern period (from 1961) of proliferation of ideas but, more significantly, also of stratigraphic integration that sorts out the ideas and indicates those by which further enlightenment is yet to come.

## HISTORIAN'S ROLE AND ACKNOWLEDGMENTS

To characterize these periods, many facts and an extensive bibliography have been compiled. This history, however, goes beyond such obvious tasks of annotation and includes analysis and interpretation in keeping with my view as to what a historian's role should be. It suggests, therefore, some of the backdrops,

Figure 4. Map of the western and central Great Lakes area showing locations of major structural features and most reef sites named in the text and figures.

motivations, and states of knowledge against which developments in Silurian reef study occurred. In presenting my judgment as to what developments are most significant and what answers to problems are most likely to be correct, and why, for the several ongoing modern-period debates. Such judgment is conditioned especially by my assessment of what I consider to have been my own and my many Indiana-based associates' contribution to Silurian reef study: a broadly based integrative regional stratigraphy.

Some of the record of study is presented by illustrations that are individually focused on a development or a debate and that give pertinent bibliographic citations. Placing given author names on one side or another of a debate featured in a figure necessarily results in oversimplification. My best judgment has been used, nonetheless, and I offer regrets to any worker who believes he has been misinterpreted.

The many reviewers and editors to be acknowledged with thanks for critiquing the whole or selected parts of this history are: Curtis Ault and Stanley Keller, Indiana Geological Survey, Bloomington; Albert Carozzi, University of Illinois, Champaign; Paul Catacosinos, Delta College, Michigan; Paul Daniels, Southlake, Texas; Thomas Evans, Wisconsin Geological and Natural History Survey; Howard Feldman, Kansas Geological Survey; Joseph Gregory, University of California, Berkeley; William Harrison, Western Michigan University, Kalamazoo; Jeffrey Packard, Geological Survey of Canada, Calgary, Alberta; James E. McGovney, Exxon USA, Denver, Colorado; Randall Milstein, Michigan Geological Survey, Lansing; Charles Kahle, Bowling Green State University, Bowling Green, Ohio; William Roth, McClure Oil Company, Alma, Michigan; David Stith, Ohio Geological Survey, Columbus; Daniel Textoris, University of North Carolina, Chapel Hill; and Steven Whitaker, Illinois State Geological Survey, Champaign.

## DISCOVERY PERIOD (1843 to 1927)

### Those mysterious reefs

The earliest notes of Silurian reefs in the Silurian System in Europe (Fig. 6) preceded those of North America by only a few years, but the first observers, including the great R. I. Murchison (1839), referred to the reefs as mysterious crystallizations into concretions. The full impact of Charles Darwin's and others' ideas on the theory of modern reefs was yet to be felt, and it is fitting that in addition to Murchison, another great geologist of his time, Sir Charles Lyell (1841), was probably the first to say "reef" in an organic sense for the Silurian buildups.

In North America, the prolific James Hall (1843, 1862) figured prominently in the earliest attempts to understand what once were altogether enigmatic sedimentary structures (Fig. 7), as he apparently was the first North American geologist to be both incorrect (in western New York) and correct (in eastern Wisconsin) in his interpretations. Not all early geologists, however, were necessarily correct or incorrect; they simply described the Silurian buildups with minimal or no speculation. The preeminent Canadian geologist, Sir William Logan (1863), was one of these. He puzzled over the buildups on Anticosti Island in eastern Canada

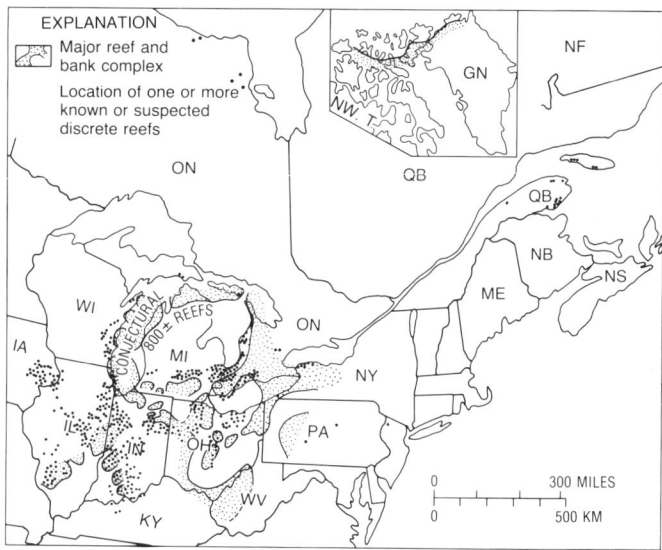

Figure 5. Map of part of northeastern North America, including Greenland and part of the Canadian Northwest Territories (inset), showing a time-composited portrayal of the Silurian reef archipelago. Locations of smaller reef masses in the Great Lakes area are not shown; other North American locations not shown are in central Nevada, west Texas, and central Manitoba.

## EARLIEST NOTES AND INTERPRETATION IN EUROPE

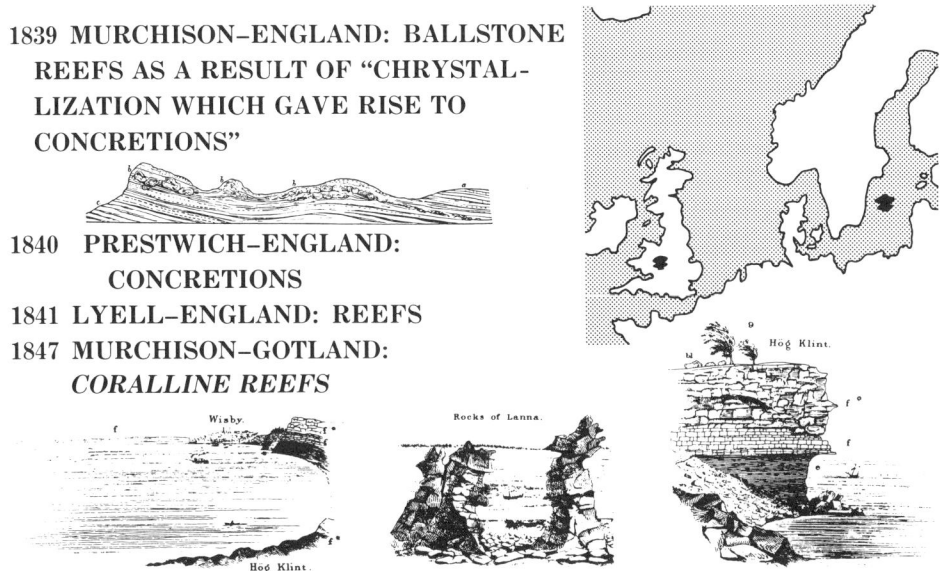

1839 MURCHISON–ENGLAND: BALLSTONE REEFS AS A RESULT OF "CHRYSTALLIZATION WHICH GAVE RISE TO CONCRETIONS"

1840 PRESTWICH–ENGLAND: CONCRETIONS

1841 LYELL–ENGLAND: REEFS

1847 MURCHISON–GOTLAND: *CORALLINE REEFS*

Figure 6. Notes on the earliest scientific mention of Silurian reefs in Europe. Reef symbols shown here in England and Gotland; similar symbols in other figures indicate the locales to which notes and bibliographic references apply.

and those in western Ontario at the eastern edge of the Michigan Basin with these words:

Dome shaped masses; the rocks . . . appear to be somewhat disturbed; the dip and strike are irregular, and the inclination amounts . . . to twelve degrees. Masses of coral . . . aspect of having been disturbed . . . Forty-seven feet of rock . . . constitute a single massive bed without divisional planes.

Much of the early interest was fueled by the economic needs of a rapidly developing industrial society, and the early state and federal geological surveys in the United States and Canada figured prominently in the study of Silurian reefs by furthering their more direct purpose to gather knowledge of mineral resources (Fig. 8).

Science was enough of an inspiration, however, and more than a few geologists were dramatically, even romantically, inspired by the thrill of their discoveries. Consider these words:

The whole expanse must have been one broad blue ocean, with its infinity of coral groves . . . the crinoideans reared their beautiful and gorgeous heads, and above them shone the bright tropical sun (Hall, 1843, in New York).

The eon of air and sunlight on the young continent . . . was short. Atlas tired, and the earth sank into the waters . . . nearly all over were dropped precipitates from the distant Atlantis [and] the corals built reefs in shoal waters (McGee, 1891, in Iowa).

Their enduring mounds, mocking at time and change, still rise in mighty klintar and hurl defiance at the elements (Cumings and Shrock, 1928a, in Indiana).

Whatever the poetic license taken with this kind of writing, the advantage hardly exceeded the importance of the discoveries.

Whatever the economic inspiration and romance of discovery, the early period was beset with an 87-year history of misconceptions (Fig. 9). From volcanism to deltaic mud lumps, these misconceptions did not leave much to the imagination. One interpretation, that the reefs simply represented cross or false bedding or cleavage, became common in the southern Great Lakes area because typical 3 to 10 m (10 to 30 ft) natural exposures and quarry sections were so small as to reveal rarely anything but small, nondiagnostic portions of the reefs; for example, only unidirectionally dipping flank beds truncated at the top along either the pre–Middle Devonian unconformity or by the multiple Quaternary unconformity.

In eastern Iowa, Calvin (1896a, Plate I; and 1896b as reported by Norton, 1899, p. 426–427) said:

The Niagara waters were shallow, and strong currents . . . swept calcareous material in various directions and [piled] it up in eddies of lenticular shape or in obliquely bedded masses, and [this produced] false bedding or oblique bedding, such as to be seen wherever in sea, lake, or river, a current is extending a bar by dumping sand or silt over its edge, thus adding layer after layer at angles more or less inclined.

The buried Silurian reefs of the Michigan Basin did not figure in the early debates, but those reefs exposed in the Niagara Escarpment and elsewhere along the basin fringes did. Gorby (1886) and Thompson (1889, 1892) were sure that the exposed reefs extending from Ohio, across northern Indiana and northeastern Illinois, through the escarpment in eastern Wisconsin, and

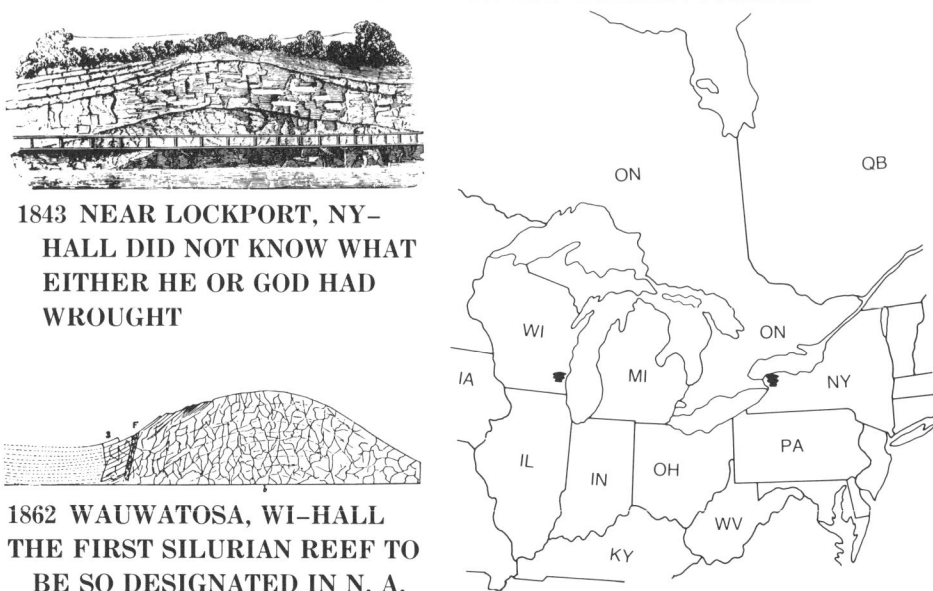

Figure 7. Notes on the earliest figured Silurian reefs in North America.

Figure 8. A reminder from a bygone era of the early partnership between economic and scientific interests that furthered development of knowledge of Silurian reefs. Photos supplied through courtesy of Markes Johnson, Williams College, Massachusetts.

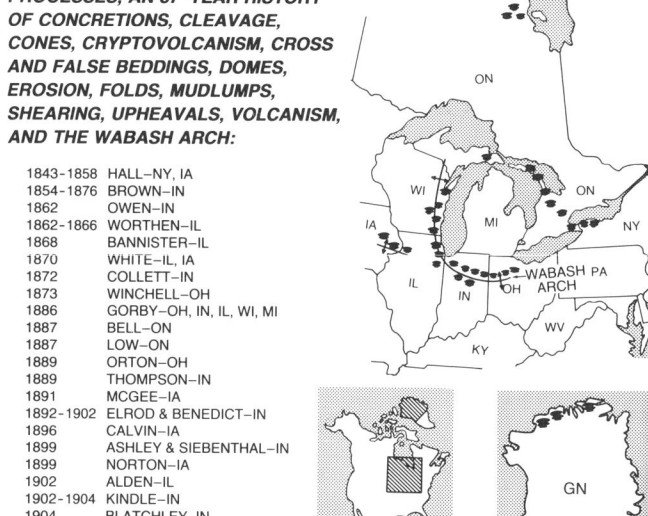

Figure 9. Bibliographic and geographic records of nearly a century of early-period thought on Silurian reefs as products of nonorganic processes.

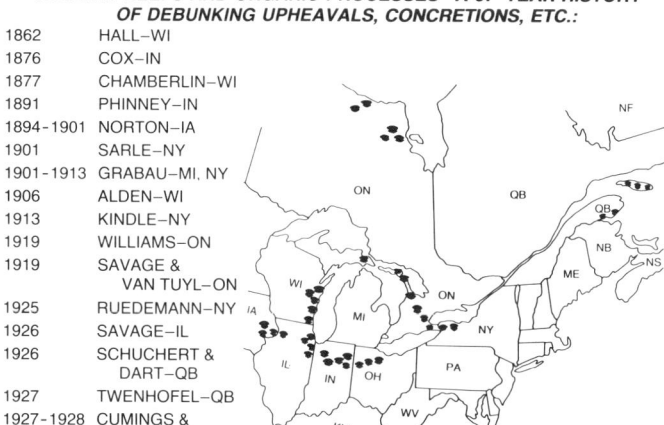

Figure 10. Bibliographic and geographic record of early-period thought on Silurian reefs as products of organic processes.

into "the volcanic regions of Lake Superior" represented a structurally upheaved arch (the so-called "Wabash Arch"). (See its approximate axis, depicted in the Great Lakes–area map of Fig. 9.)

As Thompson (1892) proposed:

The arch formed by this upheaval consists of a vast series of low bubbles or cones . . . a range of rudimentary mountains whose highest peaks are only a few feet in altitude.

It was the reefs that were the so-called "rudimentary mountains," in part the same ones on which Chamberlin (1877) 15 years earlier had published his organic views of the reefs.

The several kinds of early misconceptions, applied repeatedly from Iowa to North Greenland, are understandable, given the state of the then-fledgling geological science and given the highly altered, mostly dolomitized (particularly in the Great Lakes area), structurally deformed, and then-poorly-exposed reefs. Some 80 years were needed before the nonorganic ideas were all but dispelled from new increments of an ever-growing Silurian reef literature.

In those decades, organic ideas became dominant, as they were advanced by several prominent persons in sedimentary geology. Some of these persons, as shown by Figures 10 and 11, crossed from the nonorganic to the organic side of the reef debate. Chamberlin (1877; see Fig. 12) led the way with his still-valid statements of Silurian reef principles based on the exposed reefs in Wisconsin at the far western Michigan Basin edge. Norton (1894–1901), Phinney (1891), and Sarle (1901) (see Fig. 10) are not so well known, and history has not yet given them the recognition they deserve for their early enlightened ideas on organic processes and stratigraphic relations applicable to reefs distributed from Iowa to New York.

In the Canadian Arctic and North Greenland, organic-process ideas scarcely gained momentum during the early period. Koch's 1920s expeditions by dog sled in North Greenland led to his ideas that the spectacular Silurian reefs exposed there originated as erosional structures (Fig. 13). Virtually all the meaningful scientific studies that favored organic-buildup origins awaited the modern period of study (Fig. 13).

## Summary

The early period of discovery is perhaps typified best by James Hall, who was the first to be both incorrect and correct in his interpretations (Fig. 7). But history most needs to acknowledge T. C. Chamberlin (1877) for his careful description of the westernmost exposures of Michigan Basin reefs and his setting forth still-valid Silurian-reef criteria (Fig. 12) for such buildups that "stood as . . . reefs in the depositing sea [and whose preservation] is fatal to any theory of . . . violent action."

## ENLIGHTENMENT PERIOD (1928 to 1961)

### Silurian reef criteria

The enlightenment period began in 1928 with the painstaking work of Edgar R. Cumings and Robert R. Shrock. Their (1928a, 1928b) advancement of a definitive Silurian-reef concept (Fig. 14) some 60 years ago effectively ended the earlier controversy. The northern Indiana reefs near the southern Michigan Basin edge (Fig. 4) figured most prominently in their definition, but Cumings (1930) also described Silurian reefs in northwestern Ohio. Even today, I feel there is still almost nothing in this reef definition that should be disputed, although there is dispute as detailed farther on. The dispute focuses on the heart of the Silurian-reef concept itself and includes the question of whether

"The Attawapiskat coral reef is thought to correspond in time to some part of the Niagaran epoch."

Figure 11. Notes on Silurian reefs of the Hudson Bay region as an early-period example of correct interpretation of reefs as results of organic processes.

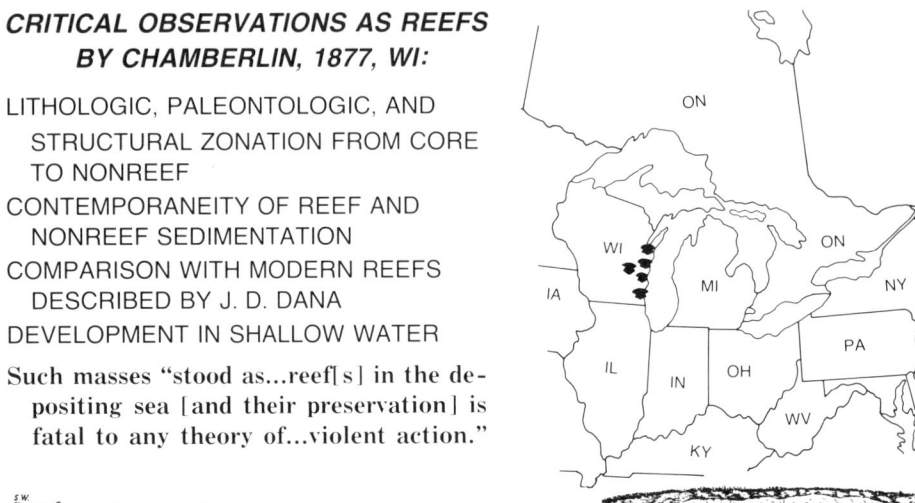

Figure 12. Notes on T. C. Chamberlin's early-period views on Silurian reefs, the most important statement for their time.

any large ecologic Silurian reef (in the sense of the most common modern definition of "reef") exists at all. During the enlightenment period of study, the seeds of such doubt were revived, especially by Marius LeCompte (1938) through his then-new idea that the midwestern Silurian buildups were of deep-water origin and were not, therefore, "actual recifs."

Heinz Lowenstam (1948–1957) figured prominently; he added some new ideas and refinements to the Chamberlin (1877) and Cumings and Shrock (1927, 1928a, b) theses, which championed the proposition that the large Silurian buildups are true ecologic reefs (Fig. 15). Further, Lowenstam is to be credited for his early stimulation of interest in the relation between Silurian reefs and economically exploitable hydrocarbons.

History records that Lowenstam (1948) pioneered in reef paleoecology in relation to paleogeography through a paper published in the Illinois State Museum of Science Papers series. That paper, which has scarcely been noted by the modern body of Silurian workers, addressed the paleoecological differences between reef and nonreef faunas. Lowenstam was able to show that some long-held paleogeographic views for North America were incorrect—views that had become standard information in the most popular historical geology textbooks of that time.

Specifically, the idea of a northern, or Arctic, Silurian sea and fauna and a southern Silurian sea and fauna separated by the so-called "Isthmus of Indiana" was discarded. Lowenstam pointed out that, because of fortuitous outcrop relations, the northern fauna was largely a reef fauna in contrast to the different and almost wholly nonreef southern fauna. (For further analysis of the differences between Silurian reef and nonreef faunas and their stratigraphically keyed changes in the southern Michigan Basin environs, see Shaver, 1974a, b).

One of Lowenstam's generalized reef tenets, however, on conceptualized growth from initially deep water into eventually shallow water, cannot be applied rigorously to all Silurian reefs; rather, it must be reconciled with present evidence that midwestern reefs variably had their early stages in both shallow water (photic zone) and deep water (aphotic zone). Still another of Lowenstam's tenets, that of wave resistance as a primary reef criterion, has been so rigorously interpreted by much of the modern generation of workers as to become a special problem in the classification and identification of reefs. (See discussions under "Stratigraphic Integration Period.")

One important note on both Lowenstam's (1948) and Cumings and Shrock's (1927, 1928a, b) work is that they showed remarkable insight into the origin and distribution of Silurian reef sediment, including its relations from core into flank rocks and thence to interreef rocks. Further, they understood that diagenetic processes wrought internal and external changes in structure or geometry of the reefs. Nevertheless, some modern workers seem to have discounted these observations, as recorded farther on and as detailed by Shaver and Sunderman (1989).

*Great reef-evaporite debate*

As noted above, the principal debate generated by Silurian reef study during the early and middle periods addressed the fundamental nature (organic versus nonorganic) of the Silurian buildups, and it extended to the question of water depth. A newer kind of controversy became focal during the middle period, however, although its basis had been laid earlier by no less prominent geologists than Amadeus Grabau (1901–1913) and W. H. Scherzer (Grabau and Scherzer, 1910). In their view, the Silurian reef-depositional episode was to be entirely separated from and wholly preceded the Silurian evaporite-depositional episode (Fig.

16). Indeed, they proposed that the reef episode terminated by the end of Niagaran, or Middle Silurian, time in the classical scheme of threefold division of the Silurian Period.

Moreover, in their view, the Silurian sea shrank to isolated basinal-desert and playalike locales and there received salts that were derived from erosion of the reefs and associated Silurian rocks. The salts were brought to the great playas by rivers wending their ways through gorges in the then-exposed reef tracts and on into the Late Silurian desert. Here too was the idea of a great regional Niagaran-Cayugan unconformity (or Middle to Late Silurian unconformity), a concept that became synonymous with "Niagara-Salina unconformity," "Lockport-Salina unconformity," "reef-evaporite unconformity," and "postreef unconformity."

Such a grand concept and its corollary ideas continued through and gained much acceptance during the enlightenment period (e.g., Cumings and Shrock, 1928a; Schuchert, 1943; Lowenstam, 1949, 1950). Such acceptance continued popularly into the modern period (e.g., Gill, 1977a, b; Nurmi, 1974), despite doubts expressed early on a stratigraphic basis as to the validity of these ideas when applied in two basins (e.g., by Alling, 1928; Caley, 1940; and Shaw, 1937; see Fig. 16). Perhaps as many geologists today follow the Grabau thesis as follow the opposed school that gained its greatest momentum during the next period of study. (See further discussions under "Stratigraphic Integration Period.")

### Summary

The enlightenment period owes much to Cumings and Shrock (1927, 1928a, b) because they set forth and abundantly illustrated most of the still-valid Silurian reef criteria. History also owes much to Lowenstam (1948–1957) for strengthening and adding to the criteria, for his early stimulation of interest in reef-hydrocarbon potential, and for showing that reef paleoecology has important bearing on Silurian paleogeography.

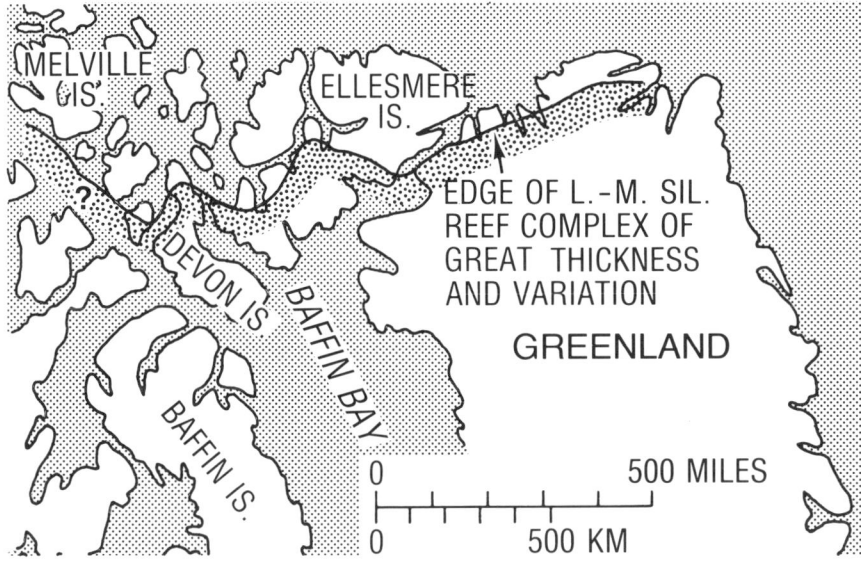

**EROSIONAL STRUCTURES:**
1920-1929–KOCH

**ORGANIC BUILDUPS**
1966-1976–DAWES
1967–KERR
1972–NORFORD
1976–MAYR
1979–LANE & THOMAS
1980–HURST
1982–PACKARD & DIXON
1984–HURST & SURLYK

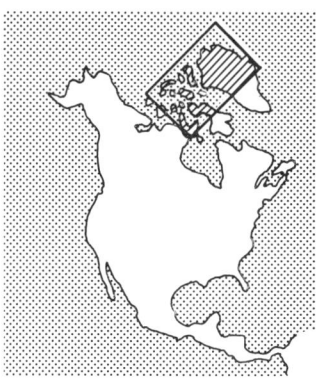

Figure 13. Bibliographic and geographic record of thought on Silurian reefs of Greenland and the Canadian Arctic variably as products of nonorganic and organic processes. (Photographs are from Dawes, 1976.)

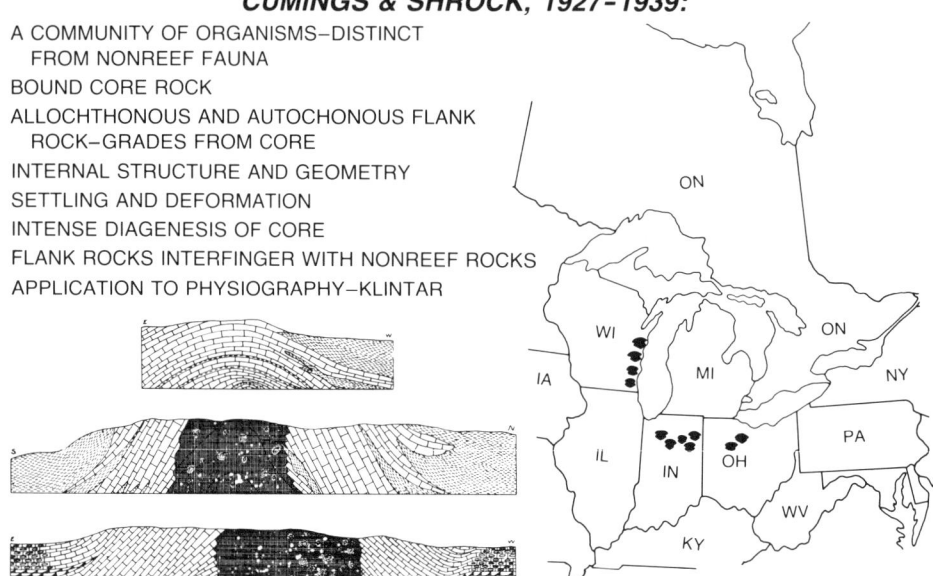

Figure 14. Notes on Edgar Cumings' and Robert Shrock's definitive views on Silurian reefs, the keystone to the middle period of enlightenment.

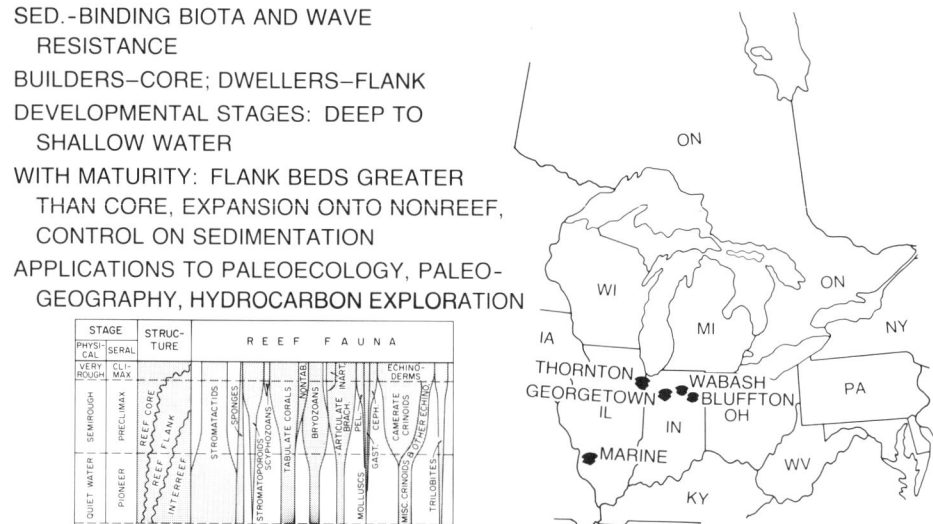

Figure 15. Notes on Heinz Lowenstam's contribution to the concept of Silurian reefs.

*A history of study of Silurian reefs, Michigan Basin environs* 111

SHADES OF GRABAU, 1913: "The close of the Guelph epoch marks the period of destruction of the animal life of the Siluric sea...Niagaran strata which underlay and surrounded the salt basins suffered extensive erosion during this period."

| SUCCESSIVE EPISODES AND A GENERAL UNCONFORMITY: | | REGIONAL FACIES AND/OR NO GENERAL UNCONFORMITY: | |
|---|---|---|---|
| 1913-1928 | GRABAU, WILLIAMS, CUMINGS & SHROCK–E. N. A. | 1927 | CARMAN–OH |
| 1943 | SCHUCHERT–NE. N. A. | 1937-1940 | CALEY, SHAW–ON |
| 1950 | EVANS–ON | 1940-1958 | ELLS, SLOSS–MI BASIN |
| 1962 | EHLERS & KESLING–MI, IN | 1956-1971 | LIBERTY & BOLTON–ON |
| 1969 | SANFORD–ON | 1961-1978 | ALLING & BRIGGS, MESOLELLA, ET AL.–MI & AP. BASINS |
| 1973-1980 | BRIGGS, FRIEDMAN, GILL, KAHLE, NURMI, HUH, ET AL.–MI, ON, OH, IN | 1964-1977 | JANSSENS, ULTEIG–OH |
| | | 1969-1974 | JODRY, MATTHEWS & EGLESON–MI |
| | | 1969-1985 | RICKARD, DROSTE & SHAVER ET AL.–NE. N. A. |

Figure 16. Notes on the continuing controversy on the relationship between Silurian reef and evaporite depositional histories.

## STRATIGRAPHIC INTEGRATION PERIOD (1961 TO PRESENT)

### Stratigraphic base for reef study

An improved stratigraphic base (Figs. 3 and 17) for the many kinds of reef studies is singled out as a key development during the modern period. This base forms a solution, whole or partial, for several current controversies, the most critical of which concerns reef and evaporite relations. To be sure, however, this stratigraphic solution has not been fully accepted by some modern Silurian workers, including those noted on the left side of Figure 16. The key to the selected beginning date of 1961 is a widely cited paper by Harold Alling and Louis Briggs. It states that a complete facies relationship exists on a regional scale between reef-bearing and evaporite-bearing rocks in the Great Lakes area (to be more precisely explained later on the basis of both space and time). As Alling and Briggs (1961) said,

The stratigraphic relations also substantiate the concept that the evaporites were deposited in reef-ringed basins surrounded by normal marine seas [Fig. 18].

Credit for such ideas, however, must be shared especially with Bolton and Liberty, who as early as 1955 (Fig. 19) stated similar concepts. Indeed, the beginning date for the modern period could as well be stated as 1955, an observation supported by the fact that Louis Briggs and his associates seem to have subsequently disavowed much of this concept or otherwise, in the writer's opinion, did not come to a fully satisfactory accommodation with its regional scope as their studies continued. (See Briggs' citations on both sides of Fig. 16.)

The stratigraphic integration provided by Lawrence Rickard (e.g., 1969, 1975) for the Appalachian and Michigan Basins is also focal, partly because he dealt with the type sections of the standard provincial series that are used for Silurian correlation in North America. One of his important conclusions was that the original type Niagaran rocks overlap type Cayugan rocks in their time values, which meant that a new, unambiguous series boundary needed definition (Rickard, 1975). This was done by restricting the redefined type Cayugan Series to those original Cayugan rocks that do not overlap Niagaran rocks in time value. The New York–Ontario standard is the one used in the Michigan Basin both traditionally and to the present time, but basin workers have continued to overlook Rickard's redefinition and its need. This means, therefore, that a residual semantic element exists in the reef-evaporite controversy noted above.

Although the provincial New York series (epoch) terms recommended by Rickard were eschewed by Berry and Boucot (1970) for North American use, they are often used, nevertheless, in the Michigan Basin environs along with the more recently adopted European series terms (see Fig. 3). The Berry and Boucot correlations of rock units in the basin environs aided in the modern integration of separately developed stratigraphies, but many newer advances are now evident for the past 20 years. (Compare Fig. 3 herein and Shaver and others, 1984, with Berry and Boucot, 1970).

Similar interpretations (to those of Rickard) of stratigraphic overlap (time and facies senses) between rocks called the Lockport Group (or Lockport-Guelph) and the Salina Group, traditionally assigned respectively to the Niagaran and Cayugan Series, were made by Ulteig (1964), Janssens (1974, 1977), and by still others for the Appalachian Basin and the Findlay Arch area bounding the southeastern Michigan Basin. Janssens also noted such relations for the Michigan Basin in northwestern Ohio, as did an Indiana group of workers in several papers (e.g., Droste and Shaver, 1982, 1985).

Other stratigraphic studies in the Appalachian and Illinois Basins were conducted during the modern period that have a bearing on problem solving in the Michigan Basin. In the Appalachian Basin, Mesolella (1978) and a West Virginia group (e.g., Patchen and Smosna, 1975; Smosna and Patchen, 1978, 1980;

| UNIFYING REEF STRATIGRAPHIES: | | | |
|---|---|---|---|
| 1961 | ALLING & BRIGGS–MI & AP. BASINS | 1972 | PHILCOX–IA |
| 1964 | PINSAK & SHAVER–IN | 1974-1978 | MESOLELLA ET AL.–MI & AP. BASINS |
| 1964 | ULTEIG–OH | | |
| 1967 | KERR–GN | 1974-1985 | DROSTE, REXROAD, SHAVER ET AL.–G. LAKES, IL BASIN |
| 1969 | SANFORD–ON | | |
| 1969-1975 | RICKARD–G. LAKES | 1976-1980 | PATCHEN, SMOSNA–WV, OH |
| 1970-1981 | BOURQUE & LESPÉRANCE ET AL.–QB | 1977 | JANSSENS–OH |
| | | 1980-1984 | HURST, SURLYK–GN |
| 1971 | LIBERTY, BOLTON–ON, QB | 1981-1985 | WITZKE ET AL.–IA |
| 1971-1976 | DAWES, MAYR, NORFORD–GN | | |

| ARE KEYS TO (WITH EXAMPLES): | |
|---|---|
| PALEOECOLOGY–BOURQUE, WITZKE | TECTONICS–HURST & SURLYK; SANFORD (1985, ON) |
| PALEOGEOGRAPHY–DROSTE, REXROAD, & SHAVER | SEA LEVEL CHANGES–SHAVER |
| REEF GEOMETRY–SHAVER | ECONOMIC EXPLOITATION–AULT (1975, IN); SANFORD |
| ORIGIN OF EVAPORITES–MESOLELLA ET AL. | |

Figure 17. Notes on development and significance of integrative Silurian stratigraphies.

Smosna and others, 1989) collectively presented reef- and evaporite-stratigraphic studies for the eastern part of the basin that helped to join basins, not just parts of one basin, in an interbasin history. Some of their findings appear to support the advocacy of Rickard (1969, 1975) and, longer ago, of Alling (1928), who denied the classic concept of Niagaran-Cayugan unconformity and even claimed time overlap between the original type series and the rock units for which the series were named. Mesolella, however, did not follow Rickard's 1975 recommendation (as he should have, in my opinion) because he did not assign the partially Lockport-equivalent A-carbonate rocks (Salina Group) to the Niagaran Series. Possibly, neither has any Appalachian Basin or Michigan Basin worker followed the Rickard expedient as he or she should to avoid semantic problems if the North American provincial series terms are to continue in use.

Newer findings in the Illinois Basin also attested to a broad regionality of reef-related events. The Illinois Basin was traditionally said to have no Cayugan rocks (e.g., Lowenstam, 1948–1957; Willman, 1971, 1973). Even though various Illinois State Geological Survey reports long afterward remained ambiguous as to the presence of normal-marine Cayugan rocks in the basin (e.g., Willman and others, 1975), other Illinois reports and several reports generated by Indiana workers presented evidence of hundreds of feet of Cayugan rocks in that basin. For example, Collinson and others (1967) and Ross (1962) reported early Late Silurian and late Late Silurian ("Late" used here as in Fig. 3) graptolites and conodonts from the Moccasin Springs and Bailey Formations, respectively, and reports by Bristol (1974), Becker (1974), Becker and Droste (1978), Droste and Shaver (1980, 1987a), and Schalb (1975, as reported by Droste and Shaver, 1980) presented new geophysical log, sample log, and biostratigraphic data that have twofold significance.

First, thick (as much as 300 m; 1,000 ft) Illinois Basin reefs are encased in contemporaneously accreted Silurian sediments well toward their tops. These reefs, therefore, were not surrounded by water nearly as deep as the reef was thick, as one would have to conclude from older stratigraphic interpretation. Second, at least two principal Illinois Basin reef generations are recognized and have coordination with other reefs (e.g., Moccasin Springs reefs with Mississinewa reefs; Fig. 3).

Such information effectively undermines the concept of a great Niagaran-Cayugan unconformity. Further, considering that hundreds of feet of Silurian carbonate rocks, formerly assigned to the Devonian System, have been reassigned to the reef-bearing sequence, serious doubt exists as to whether any of Lowenstam's (1948–1957) reefs, as much as 800 or even 1,000 feet thick, required water depths of many hundreds of feet for its development. (See the depth provision in Lowenstam's reef postulates in Fig. 15.) They did not require such depths, in my opinion, because interreef sedimentation and basin subsidence and/or sea-level rise obviated the need for great depths of water.

The larger Great Lakes reef province was extended to Kentucky during the modern period (Anderson, 1980; Seale, 1985; Smosna and others, 1989), but generation 4(?) reefs in the Louis-

*A BEGINNING TO REGIONAL STRATIGRAPHIC SENSE BY ALLING & BRIGGS IN 1961:*

"These studies recognize the conformity of deposition that links Niagaran and Cayugan rocks."

"The contemporaneity of reef growth and salt deposition is illustrated by the growth of a large algal reef...."

"The stratigraphic relations also substantiate the concept that the evaporites were deposited in reef-fringed basins surrounded by normal marine seas which covered Wisconsin, Illinois, Indiana, Kentucky, West Virginia, Maryland, and Virginia."

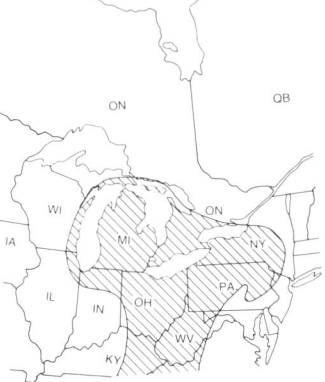

Figure 18. Notes on the beginning keystone to the modern period of Silurian reef-stratigraphic integration, a paper by Alling and Briggs (1961). Area within curved line is that area dealt with by Alling and Briggs. (These notes are intended as a tribute to the late Lou Briggs.)

ville Limestone (see Fig. 3) of western Kentucky have yet to be assessed with respect to a possible coordinated history with northern reefs. Similarly, the eastern Kentucky Lockport reefs of Smosna and others (1989) need evaluation with respect to possible positioning within the scheme of Figure 3. They were said, however, to have developed under the cyclical influence of sea-level fluctuations, which suggests probable coordination in the scheme of Figure 3.

A modern series of intense reef-stratigraphic and paleoecological studies in eastern Canada (e.g., Bolton, 1981; Bourque, 1979–1982; Bourque and others, 1977–1986; Barnes and others, 1981) have stratigraphic import for interpretations in the Michigan area, including the fact of very Late Silurian reef growth. When this information is considered together with modern knowledge of real or potential fluctuation in global sea level during Silurian time, it suggests greater plausibility for the modern idea of very Late Silurian reef growth in the Great Lakes area. As already noted, such an idea contrasts with strong traditional interpretations. Perhaps more importantly, the eastern Canada studies hold paleoecological lessons from that dominantly limestone province that scarcely have been applied to reef studies in the Great Lakes dolomite province.

With doubts expressed by Shaver (1962) and Pinsak and Shaver (1964) as to the accuracy of the traditional Silurian stratigraphic views for the Michigan Basin environs, a long-continuing group of studies at the Indiana Geological Survey and Indiana University fostered the idea of a broad regionality for events composing midwestern Silurian stratigraphy. Geophysical-log tracing, along with supporting sample and core work (e.g., Becker, 1974; Becker and Droste, 1978; Droste and Shaver, 1982, 1987a), resulted in long-distance physical correlations of single rock bodies known by several different names, e.g., the Moccasin Springs, Mississinewa, C shale, and part of the Vernon

*IDEAS BEFORE THEIR TIME (?) BY BOLTON & LIBERTY, 1955-1971, ON REEF-EVAPORITE RELATIONS*

"Indeed one might suspect that the Salina was initiated, not as a superjacent formation, but as an extension of late Guelph interreefal facies. This suspicion must be carried further when one finds Guelph reefs with sufficient relief that the Salina A unit is cut out. These phenomena cannot be explained by calling a disconformity at the [Guelph-Salina] contact."

"The evaporite series...would be initiated contemporaneously with maximum reef growth on the... marginal area.... The evaporite seas would oscillate [resulting in] the intimate relations observed between the reefal and evaporitic facies...."

Figure 19. Notes on the Bolton and Liberty (1955) and Liberty and Bolton (1971) interpretations of relations between Silurian reefs and evaporites. Credit is shared with Alling and Briggs (1961) for ushering in the modern period of study.

Formation. (See Illinois to New York columns, Fig. 3.) Indiana outcrop terminology became readily applicable to western Ohio carbonate sections (e.g., Droste and Shaver, 1976; Griest and Shaver, 1982) in the distal southern reaches of the Michigan Basin to where Janssens (1974, 1977) and the Indiana University Paleontology Seminar (1976) had already brought the standard New York and Ontario terminology in relation to reef stratigraphy.

Such Indiana-based studies were found to be mutually supporting with some minority stratigraphic views for the Michigan Basin proper, views that spoke of at least partial, in part reciprocal, equivalency of rocks called, respectively, the Niagara and Salina Groups. Such published views as those by Ells (1958), Mesolella and others (1974), Budros and Briggs (1977), and Droste and Shaver (1982) were joined by some unpublished views from the petroleum industry to show that a substantial part of the great so-called carbonate-bank system of southern Michigan (e.g., as depicted by Fisher, 1973, and Fincham, 1975) consists not of reeflike, light-colored carboante-bank (skeletally derived) sediments, but of dark-colored, in part laminated rocks of one or more of the A-carbonate units of the Salina Group. This assignment of dark laminated Salina rocks to the Niagara carbonate bank has been one of convenience to geophysical well loggers for areas distal from the edge of the A-1 evaporite (Salina Group; Fig. 3), but such assignment has added greatly to the problems attending an unambiguous presentation of sedimentational history. These circumstances temper the often-made statement that the A-1 carbonate does not extend into, through, or past the carbonate bank (so-called "massive reef") in southern Michigan. Equivalent rocks, whatever their name, do exist beyond the evaporite margin, in the Ohio and Indiana parts of the Michigan Basin and very likely also in the carbonate bank of southern Michigan as that bank has sometimes been mapped. Here, then, has been further basis for paleoenvironmental and stratigraphic misinterpretation. Specifically, the very question is raised as to how much the so-called carbonate bank in southern Michigan is truly carbonate bank (nonreef), how much consists of coalesced reef structures, and how much consists of the dark-colored laminated Salina rocks.

These distal Michigan Basin findings appear, therefore, to offer serious problems for the more traditionally taken positions of Gill (1973), Nurmi (1974), and Nurmi and Friedman (1975, 1977) on reef-evaporite relations. Where had the great Niagaran-Cayugan unconformity gone?

The physically based and the geophysical log–based findings of the Indiana group (1960s to present time) were accompanied by far-ranging biostratigraphic studies. Conodont work, directed mostly by Carl Rexroad, Indiana Geological Survey, was focal, but these studies also integrated some brachiopod, graptolite, and ostracod zones. The unpublished ostracod studies of Tollefson (1978) and the conodont studies have repeatedly revealed that the highest standard ostracod and conodont zones occur in northern Indiana reefs and associated strata that are correlated with very young Salina rocks in the Michigan Basin (Fig. 3). The biozonal evidences were summarized by Shaver and others (1971), Pollock and Rexroad (1973), Droste and Shaver (1977, 1985), and Rexroad and others (1985).

The modern Indiana-based stratigraphic work will be refined in time, as gaps still remain in that segment of stratigraphic knowledge. In my opinion, however, the present stratigraphic backdrop (Fig. 3) provides a crucial test for several kinds of current propositions bearing on the Michigan Basin–area reefs.

### Age of proliferation?

So numerous were the new and revived proposals related to Michigan Basin–area Silurian reefs made in recent decades (Fig. 2) that the modern period of study could have been designated the "Period of Proliferation," the "Era of Publish or Perish," or more seriously, the "Age of Sedimentology." Several kinds of new proposals, or simply advances in knowledge, are taken up below, complete with exemplary literature citations and annotations. Selection from among many sources has been guided both by import and to highlight disagreements yet needing resolution.

***Deep-water buildups revisited.*** Pray (1976), Lehmann (1978), McGovney (1978), and McGovney and others (1982), in full-length papers (including theses) and in abstracts, proposed deep-water origins, especially for well-known reefs at and just beyond the southwestern Michigan Basin edge Their proposals are a revival of sorts of the ideas of Marius LeCompte (1938). Their subject reefs, the Thornton Reef south of Chicago, Illinois, and the Pipe Creek Jr Reef in central northern Indiana (Fig. 4), were believed to have grown in several hundred feet of water, and, "Why not 1,000 to 2,000 feet?" (Pray, 1976, p. 28). This idea is not a repeat from Lowenstam (1950–1957; Fig. 15 herein), because the latter allowed that the reefs did grow up into surf level.

The one permissive shallow-water episode was said by a part of the renaissance deep-water group to be a eustatic lowering of sea level. This event, they believed, resulted in development of megabreccias in the Thornton Reef, and it ushered in A-1 evaporite (Salina Group) deposition in the Michigan Basin concomitantly with the break in development of the buried pinnacle reefs in Michigan (see southern Michigan column in Fig. 3). This lowering was said to be a Pridolian (Late Silurian) event. Apart from the timing of this event (not Pridolian, the Indiana group, e.g., Droste and Shaver, 1985, has repeatedly claimed), the stratigraphic scheme of this group agrees closely with the scheme of Figure 3.

Further, these workers reasoned that virtually all reef sediment was autochthonous and was cemented in place penecontemporaneously with its production, which meant also to them that reef geometry changed little with diagenesis. That is, steeply inclined flank beds (40°, more and less) remained virtually with the original depositional inclination. These views discount the core-flank relations that were detailed by Cumings and Shrock (1928a, b) and Lowenstam (1948–1957). According to Shaver and Sunderman (1989), they also necessarily discount the generally accepted Silurian reef concept and much of what is known of Silurian paleoecology in general. Nevertheless, much valuable sedimentologic and diagenetic data were presented, including data on marine cementation and porosity development.

*Clinothem buildups revisited.* In considerable conflict with the deep-water ideas of Pray and his associates (above), a series of abstracts and papers by Wilkinson (1982) and Devaney and others (1983, 1986) proposed that the same reefs and many others in the Midwest are not reefs at all in the common sense. Rather, they are very shallow, high-energy clinothem deposits, genetically similar to marl deposits in some respects, that were derived from wave-swept benches. These modern clinothem ideas are reminiscent of very similar ideas advanced during the 19th century by Calvin (1896a, b), Collett (1872), and Worthen (1866). (See discussion of false bedding and cleavage for the early period of study.)

The glacial pavement at the Pipe Creek Jr Reef in Indiana, which developed after pre–Middle Devonian erosion and even after later, pre-Quaternary erosion, was thought to nearly coincide with Silurian sea level during clinothem deposition. Further, nearly all the buildup sediment was said to be allochthonous, and much flank-bed rotation occurred as a diagenetic effect that led to oversteepening of beds by as much as 15° or more. This group also denied the significance of any core-flank relation and of the common, ecologic-reef concept, which denial seems to have been the only significant point they had in common with views of the Pray group (above).

*Biblical deluge revisited.* According to D'Armond (1982; Fig. 20 herein), the intensely studied Thornton Reef near Chicago is much more likely to have resulted from the biblical deluge, being transported from the Hudson Bay region, than it is to be a coral reef formed in situ. Could it be that creation science, growing weary of (or taking advantage of?) the arguments between deep-water, low-energy reef theorists and shallow-water, high-energy reef theorists, has proposed the ultimate compromise, the deepest but highest energy water that has yet been proposed?

Figure 20. Cover of September 1980 *Creation Research Society Quarterly* slightly modified (drafting) for purposes of readability at reduced scale and of showing author's name on cover. A compromise between deep-water and shallow-water proponents?

*Ecologic buildups (reefs) revisited.* In this history, the terms "reef," "ecologic reef," "organic reef," "framework reef," and even "true reef" are used mostly interchangably but in the collective genetic senses discussed by Heckel (1974): topographic relief, evidence of potential wave resistance or turbulent water, mainly organically built, organic binding or framework, and some control over surrounding sedimentation. In Heckel's somewhat broad sense, even those midwestern Silurian buildups that represent water below wave base conceivably could qualify as reefs, as could those buildups that are bound inorganically. This somewhat broad definition has some advantage when applied to midwestern Silurian carbonate buildups, many of which have not been, or cannot be, classified in each detail presented by Heckel. Further, many midwestern buildups represent immature stages, where observed on fortuitous outcrop, of what became (in my opinion) mature, shallow-water, ecologic reefs.

## BIOLOGIC AND LITHOLOGIC ZONATION OF REEFS:

1963 INGELS–IL
1964 TEXTORIS & CAROZZI–IN
1968 SEPKOSKI–IN
1980 IND. U. PALEO. SEM.–IN

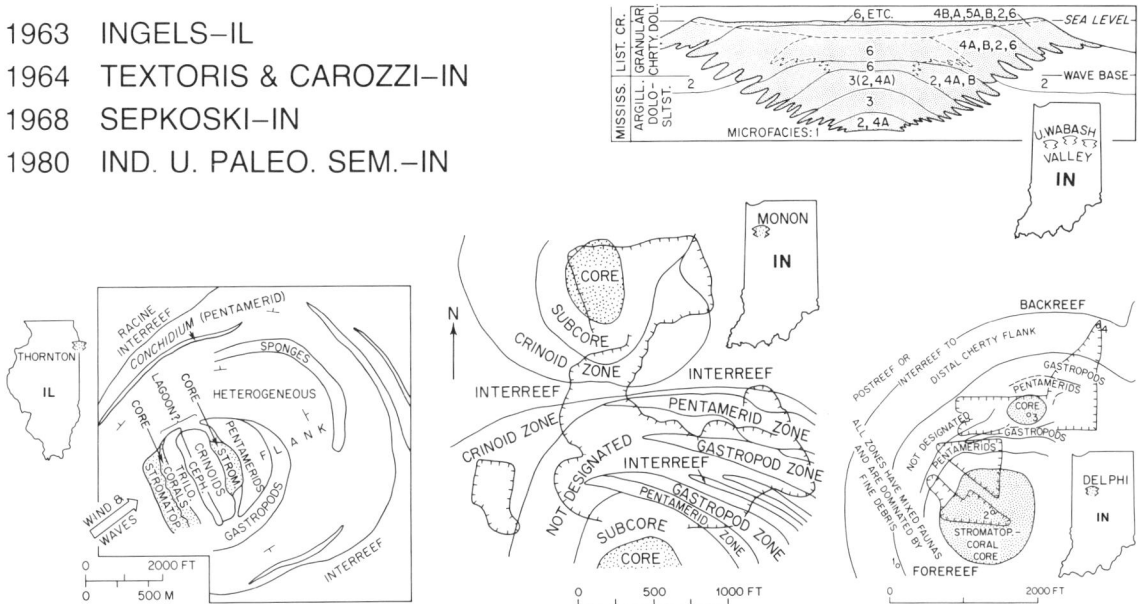

Figure 21. Graphic and bibliographic notes attesting to the validity of Silurian carbonate buildups as ecologic reefs. Abbreviations of stratigraphic names: Mississ., Mississinewa Shale Member; List. Cr., Liston Creek Limestone Member. Both are of the Wabash Formation.

The ecologic-reef principles advanced by Chamberlin, Cumings and Shrock, and Lowenstam during the discovery and enlightenment periods of study (Figs. 12, 14, and 15) received much support and refinement during the stratigraphic integration period, although as noted above, contradictions also were offered. Examples of refinements and extensions include (categorized in the five items that follow):

(1) Small and large buildups on outcrop were shown to be zoned both biologically and lithologically in degrees that went beyond the early-founded concept of intergradation of core and flank rocks (e.g., Ingels, 1963, Thornton Reef near Chicago; Sepkoski, 1968, and as further reported for Sepkoski by Shaver, 1977, Monon Reef, northwestern Indiana; Indiana University Paleontology Seminar, 1976, 1980, reefs at Rockford, western Ohio, and the Delphi Reef, north-central Indiana; Sunderman and others, 1982; Shanty Falls Reef, upper Wabash Valley, Indiana; Suchomel, 1975, and Shaver and Sunverman, 1982, Pipe Creek Jr and Delphi Reefs, north-central Indiana; Yoder, 1982, Francesville Reef, northwestern Indiana; Winzeler, 1974, Buckland Reef, western Ohio; and Carozzi and Zadnik, 1959, and Textoris and Carozzi, 1964, upper Wabash Valley reefs, Indiana; see Figs. 4 and 21 herein). Especially the first and last cited sources established lithologic criteria for Silurian reef zonation, some of it in relation to changing depth of water as the reefs grew.

Some of these studies showed that, although there is a general vertical succession of zones, a lateral zonation also exists. That is, a given zone is diachronous.

(2) Many hundreds of so-called pinnacle reefs were discovered in the subsurface Michigan Basin proper during the modern period, and several studies concluded that these buildups qualify, for at least part of their phases of growth, as true ecologic reefs complete with framework, boundstone, etc., and associated bioclastic sediments (e.g., Felber, 1963; Sharma, 1966; Gill, 1973; Mesolella and others, 1974; Sears and Lucia, 1979). For the most part, these studies emphasized only the vertical relations among the reef zones, an expectable result of vertical core drilling. Therefore, the growth histories and attendant reef-environmental parameters are not yet well understood because direct knowledge of the lateral relations is mostly lacking; it may be inferred, however, from some studies of exposed reefs in which given zones were shown to be diachronous (e.g., Indiana University Paleontology Seminar, 1976). Nevertheless, many workers agreed that a general vertical succession consists of: (a) an initial crinoid-rich biohermal phase (variably competing with a carbonate-mud phase), followed by (b) a coral-stromatoporoid (true-reef) phase, and thence by (c) an algal-dominated phase (e.g., Jodry, 1969; Huh and others, 1977; Sears and Lucia, 1979; Bay, 1983). Reef abortion and rejuvenation generally occurred between the last two phases according to many workers, but not according to Sears and Lucia (1980) in the example of buried northern Michigan reefs.

Several studies of small reefs exposed around the far margins of the Michigan Basin also attest to vertical succession of reef communities. Among them are studies by Wahlman (1974,

Montpelier Reef, northern Indiana, late Early Silurian) and Grawberger (1978, Manitowaning patch reef, Manitoulin Island, Ontario, early Early Silurian). (See locations in Fig. 4.)

(3) Application of the basic reef tenets to many newly discovered reefs in the Michigan Basin environs and beyond was accompanied by closely related studies that measured organic evolution of the reef community itself in a time-stratigraphic sense, not simply in a reef-environmental sense. These and related studies also furthered the 1948 work of Lowenstam by clearly distinguishing the differences between reef faunas and contemporaneous nonreef faunas (e.g., Hinman, 1968, Philcox, 1970, and Witzke, 1983, in Iowa; Shaver, 1974a, b, in Indiana; and Indiana University Paleontology Seminar, 1976, and Griest and Shaver, 1982, in Ohio).

(4) Several modern investigations resulted in improved geometric and structural knowledge of reefs from both external and internal perspectives. New observations that flank rocks (or the clinothems of Devaney and others, 1986), dominated by bioclastic debris, abut against and gradually merge with core rocks, not merely lean against the core, pass over the core, or exist in the absence of core, have repeatedly shown the correctness of the Cumings and Shrock (1928a, b; Fig. 21 herein) thesis. (For example, see Crowley, 1973, western New York; Indiana University Paleontology Seminar, 1976, 1980, Ohio and Indiana; and, Textoris and Carozzi, 1964, Indiana.) Nevertheless, these depictions of internal structure were contradicted during the modern period by Devaney and others (1986, Fig. 30, Pipe Creek Jr Reef, Indiana) and McGoveny (in Shaver and others, 1983, Fig. 39, Thornton Reef, Illinois). Such contradictions in turn were contradicted (Shaver and Sunderman, 1989, see especially Fig. 4).

Silurian reefs were shown during the modern period to have wide ranges in size, geometric form, and gradation to nonreef deposits (e.g., Shaver, 1977; Shaver and others, 1978). A most critical geometric consideration, however, is that few midwestern Silurian reefs are known to have flat bottoms, contrary to the great majority of published cross section depictions of haystack-like reefs even for the modern period (e.g., Kahle, 1978; Gill, 1977b). The flat-bottomed depictions by Gill (1977b) of the Belle River Mills Reef (Fig. 4) in southern Michigan are questioned because very few wells used in his synthesis penetrate the full reef thickness as portrayed. A flat bottom is not necessarily the conservative interpretation.

Probably most large reefs began growth from small-area sources and/or from coalescence from several small-area or point sources as the fledgling reefs grew upward and spread laterally over contemporaneously accreting interreef sediments. (In addition to sources cited above, see Bristol, 1974; Suchomel, 1975; Shaver and others, 1983; Indiana Paleontology Seminar, 1980; and Griest and Shaver, 1982, who collectively showed or suggested typical multiple-point sources for a common, Great Lakes–wide reef generation, 3 of Fig. 3.) This geometric consideration has much bearing on how reef-environmental parameters (including depth of water) are interpreted and, therefore, on how growth histories are reconstructed.

In example, these item-4 observations hardly support the contention of Gill (e.g., 1973, 1977b, and other papers) and of some others that virtually all of the Michigan Basin pinnacle-reef growth was accomplished while virtually no offreef sediment accumulated; rather, these geometric considerations favor the alternatives of either synchronous or reciprocal reef and offreef sedimentation variably proposed by Sloss (1969), Sarg (1982), and others. Further, they do not support any interpretation drawn from Lowenstam's (Fig. 15 herein) idea of reef growth from deep to shallow water that necessarily holds that water around a reef was ever as deep as the reef is thick. Also, they temper the deep-water theses of Pray (1976), Lehmann (1978), and McGovney (1978). It is not clear from Pray, for example, that the thickness of such large reefs as Thornton in Illinois (Fig. 4) is not a direct measure of minimum water depth surrounding a reef during its growth.

The idea that external and internal geometry and structure of reefs changed postdepositionally through various diagenetic processes arose during the enlightenment period of study (e.g., Cumings and Shrock, 1928a, b). Such changes are manifest in peripheral and subreef sagging, oversteepening of upper flank beds, and in suprareef drape as often has been reported during the modern period (e.g., Shaver and others, 1978; Indiana University Paleontology Seminar, 1976, 1980; Droste and Shaver, 1980; and Devaney and others, 1986). Nevertheless, the advocates of deep-water buildups (cited variably above) have largely dissented. (See pertinent discussions in Shaver and Sunderman, 1982 and 1989, and in Devaney and others, 1986.)

(5) The problem of enigmatic structural cores for so many buildups in the highly dolomitized midwestern Silurian reef province has long plagued investigators as they sought a consensus on the basic nature of the buildups (Fig. 22). Cumings (1932) coined the all-embracing term "bioherm" as a result. Several modern workers, studying one or a few buildups, sometimes very small and immature buildups or highly dolomitized buildups, have concluded that they dealt merely with so-called carbonate-mud mounds or with bioherms lacking ecologic-reef status. Some pronouncements were accompanied by observations that framework was not continuous within the structural core of the reef (e.g., Kahle, 1974, Maumee Reef, southeastern Michigan Basin in Ohio, Fig. 4 herein); some spoke of discontinuous framework and, therefore, also of nonresistance to waves (e.g., Philcox, 1970, Palisades complex in eastern Iowa); and others spoke of blue-gray carbonate-mud cores lacking any ready evidence of framework, and resembling, in their opinion, modern-day carbonate-mud accumulations in appreciable depths of water.

Nevertheless, several modern studies suggest that these kinds of conservative positions have serious shortcomings whenever the nonreef labels are intended to characterize the Silurian buildups in general. These studies include the following.

(1) Indiana University Paleontology Seminar (1976) on

TOWARD A SOLUTION OF THE ENIGMA OF CARBONATE-
MUD CORES — FITS AND STARTS:

| DISCOUNTING THE REEF IDEA | SUPPORTING THE REEF IDEA |
|---|---|
| 1938-LECOMPTE, IN–Stromatactis is inorganic? Lack of algae | 1928-CUMINGS & SHROCK, IN–Dissolution of framework |
| 1972-GILL, MI–Stromatactis and karstic diagenesis | 1950-LOWENSTAM, IL, IN–Stromatactis is organic |
| 1974-KAHLE, OH–Discontinuous framework | 1964-TEXTORIS & CAROZZI, IN–Stromatactis and fistuliporids |
| 1976-PRAY, IL–Emphasis on cementation at expense of stromatactis | 1973 & 1982-CROWLEY, NY, AND GRIEST & SHAVER, OH–Dolomitization and destruction of framework |
| | 1974 & 1981-CORON & TEXTORIS, IN, AND BOURQUE ET AL., QB–Algal role |
| | 1983-BOURQUE & GIGNAC, QB–Sponge role |
| | 1986-ARCHER & FELDMAN, IN–Algal role; diagenetic destruction of framework |

IS THIS THE ORIGINAL LOOK (BEFORE DIAGENESIS) OF THE ENIGMATIC BLUE-GRAY CORE ROCKS OF MANY FOSSIL REEFS?

Figure 22. Notes on a continuing disagreement on the appropriateness of the term "reef" for Silurian carbonate buildups. The sketch is from Archer and Feldman (1986).

reefs in the Rockford Quarry complex, western Ohio, where gradational diagenetic destruction of coralline framework from flank into core may be observed as well as many resilient (to blasting), relatively uncompactible blue-gray mud cores that grade laterally into coarse-grained bioclastic flank rocks; these cores had an obvious role in postdepositional structural changes as relatively unchanging bulwarks against which other changes occurred (e.g., compaction and flank-bed rotation);

(2) Griest and Shaver (1982) and Ganley (1984) on reefs in the Celina complex, western Ohio, where dozens of blue-gray carbonate-mud cores on causal inspection appear to lack any framebuilders but that with painstaking observations exhibit a convincing array of colonial corals and stromatoporoids, albeit a nearly destroyed array in this highly altered, nearly pure (99 percent) dolomite complex;

(3) Shaver (1974c), Winzeler (1974), and Shaver and others (1983) on a modest-sized carbonate buildup at Buckland, Ohio, that is inverted-cone-shaped in cross section, has a coral and stromatoporoid framework, and that has other attributes hardly debatable as to ecologic reef status; at the cited times of study, the bottom of the downward narrowing reef was not exposed, but in 1989 the very small bottom part was exposed and there consisted of blue-gray dense carbonate mud (dolomitized) in which only a few nearly obliterated framework fossils (the colonial coral *Syringopora*) could be found; such findings of superposed carbonate-mud and typical reef attributes in one exposure should dispel much of the doubt that has attached to many less well-exposed and incomplete buildups and to the significance of the carbonate-mud structures in general;

(4) Textoris and Carozzi (1964), on upper Wabash Valley reefs, Indiana, and Sears and Lucia (1979), on buried pinnacle reefs in Michigan, who concluded that many carbonate-mud cores were the early, immature stages of reefs that originated below wave base but that do exhibit framework and did attain ecologic-reef status in their later stages, reflecting the zone of wave action;

(5) Archer and Feldman (1986) on so-called microbioherms in the Waldron Shale of Indiana, who proposed that a prediagenetic coralline-algal framework, illustrated here in Figure 22, possibly applied to many of the enigmatic blue-gray carbonate-mud cores throughout the Great Lakes area; and, very importantly,

(6) Several modern reports on eastern Canada reefs (e.g., citations in Fig. 23 herein), which in that dominantly and relatively unaltered limestone province, record an abundant algal, sponge, and other frame-binding organic roles within carbonate-mud cores (see Fig. 23).

These studies and others seem to me to obviate further need for doubts of ecologic-reef status, as were summarized for Silurian reefs in general during the modern period on largely philosophical and semantic bases by Braithwaite (1973) and Stanton (1967). These studies also suggest that the conservative modern-period claims by some persons of nonresistance to waves for given buildups (see criterion in Fig. 15) are questionable, and therefore, that designations by the noncommittal term "bioherm" have been grossly overworked.

The question may be asked, "Nonresistance to what waves? Silurian inland-sea waves or modern North Atlantic gale-driven, deep-water-generated, long-fetch waves?" Even the most obvious rigid fossil and modern reefs might not wholly resist the latter kind, although some Silurian reefs of North Greenland and the Canadian Arctic that were in geosyncline-marginal settings (Hurst and Surlyk, 1984) might survive with minimal damage. The most appropriate wave-resistance test for midwestern Silurian reefs ought to be in direct relation to construction above or below wave base, and even then the potential-resistance factor comes into play (see Heckel, 1974). Those Silurian buildups exhibiting framework and core-to-flank relations would not be seen today if they had failed the Silurian wave test. As Shaver and

ENLIGHTENED INFORMATION ON:
ORIGIN OF CARB.-MUD CORES, STROMATACTIS, CYCLIC REEF GENERATION THROUGHOUT SILURIAN, REEF COMMUNITIES, ECOLOGIC-REEF CONCEPT, PHYSICAL REEF PARAMETERS, AND TECTONIC AND SEDIMENTOLOGIC CONTROLS

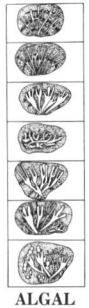 
1979-1982 BOURQUE
1980 BOURQUE & LACHAMBRE
1981 BOLTON
1981 BOURQUE, MAMET, & ROUX
1983 BOURQUE & GIGNAC
1986 BOURQUE & OTHERS

ALGAL BOUND-STONES

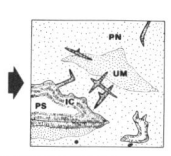

DIAGENESIS OF SPONGE-CONSTRUCTED MOUNDS

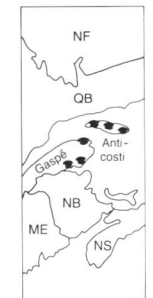

Figure 23. Notes on significant advances in knowledge of Silurian reefs through paleoecological studies in eastern Canada.

others (1978) said, the wave-resistance factor has been so zealously guarded by some investigators as to define Silurian reefs out of existence and to negate an extremely useful sedimentologic-organic concept. (e.g., of reefs).

***Reefs versus carbonate banks.*** The terms "bank," "carbonate bank," "carbonate platform," "carbonate sand bank," "barrier," "barrier reef," "barrier reef," "barrier bar," "crinoid barrier bar," "blanket reef," "massive reef," "reef bank," transgressing "littoral-lagoonal-barrier reef association," and "complexes" thereof, have been used for various kinds of rock bodies and concepts beginning in the early period of study and continuing to the present time. Such uses, therefore, have often been confused, as exemplified in the following four items:

(1) Not all persons have agreed as to the differences between reefs and banks as outlined by Heckel (1974) and Hoffman and Narkiewicz (1977), two examples of definitive recommendations.

(2) A few persons have identified reefs in the field as banks even though they recognized framework cores, and they have added "bioherm" to their nomenclature, forgetting that this term is the all-embracing one for both reefs and banks (Cumings original definition, 1932) and limiting it only to moundlike structures lacking ecologic-reef status (e.g., compare treatment in Shaver, 1974c, with Kahle, 1974 and 1978, for Buckland and Maumee, Ohio).

(3) Stratigraphic relations of the carbonate banks discussed here are not all agreed on; for example, not their relations with certain reef and evaporite rocks.

(4) Not all persons agree that each major carbonate-bank system, as depicted in Figure 5, necessarily exists.

Nevertheless, great strides were made during the modern period to understand carbonate banks in the Michigan Basin environs. The discussion that follows is limited to the large-scale carbonate-bank trends as depicted in Figure 5 herein and in paleogeographic maps of the Great Lakes area, Wenlockian through Pridolian times, that were originated by Droste and others (1980) and later published by Shaver and others (1983) and Droste and Shaver (1983, 1985). Crude, but nevertheless the first, paleobathymetric maps reflecting these carbonate banks were presented by Droste and Shaver (1987b). The proposition that such major sedimentary features exist and are to be distinguished from other major sedimentary systems is supported by Briggs and others (1980). Their graphic depiction (Briggs and others, 1980, Fig. 17-1) fairly well agrees with the geography outlined here, but their term is "carbonate platform."

The discussion below is based on seven groups of investigations, selected from among others and arranged by geographic area, that have yielded data adding up to a perhaps fair, but incomplete conception of Michigan-area carbonate banks and their reef relations. That conception, following the seven groups of bibliographic citations, is mine; it does not necessarily have agreement in all or any parts from each of the cited authors, some of whom did not directly address carbonate banks as such. The seven investigative groups are as follows.

(1) Buried carbonate-bank system within the Michigan Basin proper—Ells (1958, 1967), Pounder (1962), Burgess and Benson (1982), Meloy (1974), Mesolella and others (1974), and Sarg (1982), among others, collectively clarified many sedimentologic, paleoecologic, diagenetic, and stratigraphic questions, but not to everyone's satisfaction. Some of these workers recognized that the carbonate bank ("massive reef" of some reports), represented mostly by the Niagara Group (see Fig. 3) and surrounding much of the basin, is not mostly reeflike but, as the host structure, has intimate relations with many framework reefs that grade into bank rocks.

(2) Far southern Michigan Basin outcrop, at Buckland, Rockford, and Celina in western Ohio, and Pleasant Mills and New Corydon in adjacent Indiana (Fig. 4)—Shaver (1974c), Droste and Shaver (1976), Griest and Shaver (1982), Shaver and others (1983), and Ganley (1984) described light-colored, thick-bedded, pure-carbonate biostromal Lockport and Salamonie deposits becoming cross bedded upward and intimately hosting the beginning of many ecologic reefs, although not all descriptions were to the direct purpose of carbonate-bank interpretation. These deposits have been interpreted by some workers as distal expressions of the buried Niagara carbonate bank, or so-called "massive reef" of southern Michigan, or simply the "reef" covering thousands of midwestern square miles (e.g., Kahle, (1978; Sears and Lucia, 1980; Gill, 1977a).

(3) Buried to outcropping relations of a carbonate bank ("barrier reef" of Beards, 1967), eastern Michigan Basin and environs in western Ontario and western New York—Bolton (1957), Zenger (1965), Beards (1967), Sanford (1969), Bolton and Liberty (1955), Liberty and Bolton (1971), Crowley (1973), and Bailey (1986) collectively presented several carbonate-bank features variably in the Amabel and Guelph Formations that are reminiscent of those known elsewhere but are not all in full agreement or directed to the carbonate-bank concept.

(4) Buried to outcropping relations of the Fort Wayne Bank, distal Michigan Basin in northern Indiana—Sunderman and Mathews (1975, p. 4-6, 31-33), Droste and Shaver (1982, 1985, 1987b), and Shaver and others (1983) collectively detailed many aspects, including local relations among reeflike, banklike, and other rocks, and the relation of this carbonate bank to an expanded Michigan Basin and to evaporite deposition younger than the A-1 evaporite (see Fig. 3). This interpretation is in concert with Sonnenfeld's (1985) proposals for evaporite deposition. This bank system, mostly younger than the inner basin system, was assigned erroneously, in my opinion, to the Illinois Basin regime by Briggs and Briggs (1974b).

(5) Illinois Basin—Droste and Shaver (1980, 1987a) presented geographic, geometric, sedimentologic, stratigraphic, and paleoenvironmental interpretations of the conceptual Terre Haute Bank and its relation to various discrete reef types. This bank-like system ranges stratigraphically from the St. Clair Limestone into the Bailey Limestone.

(6) Buried to outcropping relations of carbonate-bank segments, eastern Findlay Arch and Cincinnati Arch area, central-

northern and southwestern Ohio—Cumings (1930), Mallin (1950), Ulteig (1964), Janssens (1974, 1977), Shaver and others (1978), Lehle (1980), Tomassetti (1981), Shaver (1985), and Kahle (1988) presented descriptive and interpretive data on very illustrative carbonate-bank segments, but much of it was disconnected from the bank concept and much study is yet needed. Nevertheless, these studies show much in common between this bank, or banks, including intimate relationships between massive biostromal deposits and ecologic reefs, and other banks noted above. The segments variably range upward from the Peebles and Goat Island Dolomites into the upward-extended part of the Lockport Group of Janssens (1977; see Fig. 3 herein) that is the equivalent of lower Salina rocks.

(7) Perhaps the most conjectural of barrierlike carbonate banks in the Michigan Basin environs is that proposed by the Indiana group (1980–1987 citations above) for the outcrop and near-outcrop areas of eastern Wisconsin, northern Michigan, and Manitoulin Island, Ontario, distal Michigan Basin. It is said to be most conjectural because of scarcity of subsurface data and modern published geologic reports. Many descriptions of banklike rocks, some in intimate association with true reef structures, include those by Shrock (1939), Mikulic (1977, p. A9–A14), and Shaver and others (1978, Fig. 18 and p. 25) in the Racine Formation; Ehlers (1973) and Johnson and others (1979) in the upper part of the Engadine Group; and Shelden (1963) in the Amabel Formation (only). Both the Racine and upper Engadine faunas are younger than the brachiopod Zone of *Pentamerus oblongus*, the upper Racine fauna probably much younger (see Droste and Shaver, 1985, Fig. 3; Shaver and others, 1984). Therefore, at least the upper Racine and upper Engadine carbonate-bank segments (if so qualified) are younger than all or most of the buried bank system of the inner Michigan Basin. This bank qualifies, along with the Fort Wayne Bank (item 4 above), as the conceptualized younger barrier of Sonnenfeld (1985) and Droste and Shaver (1987b) that was thought necessary to explain the second and later phases of evaporite deposition in the Michigan Basin.

The longer and better understood of these seven interpretive examples of carbonate banks in the Michigan Basin and adjoining regions seem to have a common stratigraphic origin in the upper parts of partially equivalent rock units known as the Niagara Group (inner Michigan Basin), Amabel Formation (distal eastern Michigan Basin), Lockport Group (Appalachian Basin), Joliet Formation and Salamonie Dolomite (distal southwestern Michigan Basin), and Salamonie Dolomite and St. Clair Limestone (Illinois Basin). (See Figs. 3 and 4.)

Some of the bank examples noted here are overlain by dark-colored fine-grained laminated carbonate rocks of the Salina Group, variably the A-1 carbonate (= Greenfield Dolomite?) or higher Salina carbonate units, including the Tymochtee Dolomite (Fig. 3). This relation in some places was affected by pre-Salina erosion (e.g., in southwestern Ohio according to Kleffner, 1988a, b), but in some places the relation is due to a time-transgressive carbonate-bank event. That is, the top of a given bank is depositionally diachronous, and a single major unconformity along the top is not apparent. These circumstances are thought to apply somewhere in all three basins of concern (opinion stated here but seemingly disputed by Kahle, 1988, for the carbonate bank along the Findlay Arch in western Ohio).

From the discussion immediately above, it follows that upward growth of banks into younger rock units was not everywhere continued into Late Silurian time. In some places where growth did continue it was set back to locales more distant from basin centers than was early growth. Later growth is now represented by parts of these rock units: Guelph Dolomite (parts of Michigan and Appalachian Basins); Louisville Limestone and Wabash Formation (northern Indiana); Engadine Dolomite and Racine Dolomite (eastern Wisconsin and northern Michigan); and Moccasin Springs, Bailey, and Wabash Formations (Illinois Basin). (See Figs. 3 and 4.)

The concept presented here of these features provides for thickened, narrowly to broadly linear interbasin platform-to-basin edge deposits that extend (with interruptions) as many as hundreds of miles, cover as many as thousands of square miles, have very gentle topographic relief effected over many miles, and consist in large part, if not everywhere dominantly, of fairly pure biostromal to cross-bedded organic-carbonate sediments (see discussions and concepts in Heckel, 1974). Thinning phases of these banks commonly extend both into the basin and well out into interbasin areas so that any defined boundary is arbitrary in such areas.

Within and at the top of these thickened biostromal to biohermal features are many ecologic-reef masses complete with framework and quaquaversal structure. Indeed, it is the upper lower part of the carbonate-bank system singled out above that is host to the beginning stages of growth for reefs in the probably greatest single Silurian reef generation in the Great Lakes area (generation 3, Fig. 3). Nonreef carbonate-bank sediments vastly exceed the reef rocks in volume.

As already noted, not all students of Silurian reefs have agreed with some conceptualizations presented here. For example, the very low ratio of thickness to lateral extent and the many interruptions in pure bank rocks do not fit their concept of a carbonate bank well enough. The interpretations presented by the Indiana group, however, provide not only for nonreef faunas to have contributed importantly to large carbonate banks but also for reef-shed sediments to have travelled miles from their sources (e.g., Lowenstam, 1949) and to have changed both lithologic character and thicknesses of such basically nonreef formations as the Bailey Limestone (Illinois Basin) and the Liston Creek Limestone Member (northern Indiana) in proximity to reef trends (Droste and Shaver, 1980, 1987a; Shaver, 1974a).

A special problem attaches to recognition of the buried carbonate bank (upper part of Niagara Group) in Michigan because, in areas distal from the A-1 evaporite margin, many stratigraphers have lumped the dark-colored A-carbonate rocks (Salina Group) with the lighter-colored and very different bank rocks below. Some such interpretations have been based on geophysical logging alone. The problem then becomes one of either different bank concepts among different workers or of incorrect stratigraphic and environmental interpretations or both. (Stratigraphic specifics of this problem are given elsewhere.)

*Cyclical reef generation.* Some workers from the discovery and enlightenment periods of study knew that various Michigan Basin–area reefs were of differing ages, as described from incomplete exposures; that some reefs were aborted at an early stage; and possibly even that some reefs had earlier origins than others (e.g., Shrock, 1939, eastern Wisconsin reefs). Nevertheless, it was not until the modern period and the opening of large quarries and use of extensive core drilling that a modest number of workers accepted the idea that reef growth continued in some places well past Wenlockian and Niagaran time and even until the end of Silurian time. During the modern period, stratigraphic identification became more precise with respect to reef generation and abortion. Indeed, that some reefs flourished, aborted, and then were rejuvenated on the one spot, with or without intervening nonreef strata, was documented in several places.

Eleven examples of these differing phenomena, derived during the modern period, follow.

(1) Eastern Iowa buildups—Philcox (1972), Witzke (1981a, b, 1983), Bunker and others (1985), Shaver and others (1978), and Shaver (1985) recorded generative and abortive histories at varying stratigraphic levels, including vertical stacking in some places (e.g., reefs exposed in the Palisades complex, Iowa, Philcox, 1970, and in quarries at LeClaire, Iowa, and Port Byron, adjacent Illinois, Shaver, 1985; see Fig. 4).

(2) Delphi Reef, Indiana—the Indiana University Paleontology Seminar (1980) interpreted a cyclic expansive and restrictive history of growth, possibly without a prolonged abortive period or without recognizable unconformity.

(3) Bluffton Reef, Indiana—Shaver (1976) and Johnson (1981) depicted successive reef generation, abortion, and regeneration on one spot, but the two reef intervals are separated by tens of feet of nonreef strata.

(4) Buried Michigan Basin pinnacle reefs (Peters, Marine City, Big Hand, and others; Fig. 4)—Sharma (1966) (Jodry (1969), Mesolella and others (1974), and others furnished many examples of two-generation reefs in their views, the regenerative phase thought by some persons to be separated from the first phase by unconformity and dominated by algal stromatolites.

(5) Maumee Reef, northwestern Ohio exposure—Kahle (1974, 1978); reported cyclicity of development much like that of item 4 above but not necessarily representing the same stratigraphic interval if Janssens' stratigraphy (1974, 1977) is correct.

(6) Several eastern Indiana and western Ohio exposures, far southern margin of Michigan Basin—Shaver (1974c), Winzeler (1974), Droste and Shaver (1976), and Shaver and others (1983) described many examples that range from: (a) simple generative and abortive history for a reef, to (b) multiple generative and abortive histories at one place (vertical stacking) with intervening nonreef strata but no apparent major unconformity, and to (c) multiple history for one reef, the regenerative reef phase being dominated by algal stromatolites and separated from the first reef phase by a localized, reef-top unconformity (my interpretation). Such examples are believed by some persons to represent a penecontemporaneous miniature replica of the buried Michigan Basin pinnacle-reef history. The approximate 1:5 ratio in respective thicknesses of reef sections is thought here to represent the difference in subsidence rate from inner to outermost basin.

(7) Bruce Peninsula exposures, Ontario, far eastern Michigan Basin—collected data from several authors, including Liberty and Bolton (1971), and still available exposures suggest close parallels to phenomena recorded in item 6 above.

(8) Reef exposures at Valders, eastern Wisconsin, at the far western Michigan Basin edge—Shaver and others (1978) proposed two phases of normal-marine reef growth, vertically stacked but separated by tens of feet of nonreef strata, this being an interpretation differing from that of Shrock (1939).

(9) Reef exposures at West Milgrove, north-central Ohio along east flank of Findlay arch—Shaver and others (1978) and Lehle (1980) depicted multiple generative and abortive histories for normal-marine reefs below and, in places, algal-stromatolitic reefs above, the two phases being separated variably by minor, reef-top unconformities only (my opinion) and several feet of straight-algal-laminated rocks. According to Janssens' (1977) stratigraphy, this section represents younger cyclic events than those described in items 4, 6, and 7 above.

(10) Buried and exposed reefs in southwestern Ontario and western New York, Michigan Basin to Appalachian Basin—Pounder (1962, 1963a, b), Hadley (1970), and Crowley (1973) described multiple generative and abortive histories like that of item 4 above (Michigan Basin) but also included an earlier generative phase in their interpretation (generation 2 of Fig. 3). In southwestern Ontario, Hadley (1970) interpreted vertical stacking of reefs separated by nonreef sediments and claimed not only Guelph reefing, but also A-1 and A-2 (Salina) carbonate reefing, which episodes would be reciprocal with evaporite deposition. Hutt (1974) expressed similar views for southwestern Ontario reefs, and as late as 1986, Bailey agreed with the proposition of Guelph, A-1, and A-2 reefing in southwestern Ontario, although some other geologists had disavowed this proposition for the eastern Michigan Basin (e.g., Koepke and Sanford, 1966; Gill, 1977; Lee and others, 1976). At least three reef generations, therefore, were recognized by this group that have not only vertical (stratigraphic) distinctiveness, but also some geographic differentiation by generation (see generations 2 through 4, Fig. 3).

(11) Buried Illinois Basin reefs—Droste and Shaver (1980, 1987a) emphasized that at least two generations of reefing (or periods of reef stabilization) occurred in the Illinois Basin, one arising from the St. Clair Limestone and Salamonie Dolomite (generation 2 or 3 or both 2 and 3?, Fig. 3) and another with the advent of Moccasin Springs and Mississinewa deposition; these two generations closely correspond with generations 3 (Salamonie) and 5 (Mississinewa) in northern Indiana.

Not all reefs of a given generation were aborted at the same approximate time. For example, the reef at Montpelier, Indiana (e.g., Wahlman, 1974; Shaver and others, 1983) arose in common with many reefs of generation 3 (Fig. 3) that are exposed farther east in eastern Indiana and western Ohio and that were quickly aborted. (See Shaver, 1974c; Droste and Shaver, 1976; Winzeler, 1974; Griest and Shaver, 1982; Ganley, 1984.) But the Montpelier Reef continued to grow through a shallowing-water period amidst oolite deposits, which same period of shallowing resulted in reef abortion farther east except in the example of algal stromatolites at the top of some reefs (e.g., Buckland Reef, Ohio).

With these aforementioned modern-period evidences of cyclic reef generation and abortion from around the Michigan Basin and beyond, and with their own integrative stratigraphy, the Indiana group (e.g., Droste and Shaver, 1977, 1985; Shaver and others, 1978; Shaver and others, 1983) proposed that coordinated cyclic reef generation and abortion had occurred throughout the Silurian Period and throughout the Great Lakes area. This added a new dimension to the already current ideas of such cyclic phenomena involving reefs within the Michigan Basin proper that were largely limited stratigraphically to the section ranging upward from the upper part of the Niagara Group into the A-1 carbonate of the Salina Group (Fig. 3). Further, they proposed the specifics of interrelated regional-scope changes in sea level, salinity, and terrigenous clasticity (right side of Fig. 3), not that such principles were wholly new with them.

An enabling factor for the proposition of widespread reef coordination was the fact that two Late Silurian reef generations (5 and 6 of Fig. 3) seem to be indigenous to the southern Great Lakes area and nowhere else, except in Iowa and adjacent Illinois where generation 5 is also present (Shaver, 1985). Generation 5 includes most of the outcropping, classic upper Wabash Valley reefs and many buried reefs, which were found in ever larger quarries and by an extensive coring program of the Indiana Geological Survey to bottom out at about the Louisville-Mississinewa contact (Fig. 3). The corresponding Great Lakes–wide horizon is the approximate bottom of the C shale (Salina Group) and of other shaly units, including the Moccasin Springs Formation in the Illinois Basin. This formation had its own new (at that depositional time) reef generation (e.g., Droste and Shaver, 1980, 1987a).

This generation seems to have been fostered by a return of midwestern Silurian seas to near-normal salinity, probably as a result of deepening (sea-level rise?) and improved circulation that also brought about cessation of evaporite depositions in the salt basins. In Iowa, the C-datum position falls within the LeClaire Member (LeClaire Facies) of the Gower Formation of eastern Iowa (see nomenclature of Witzke, 1981a, b, 1983; Bunker and others, 1985; Shaver 1985) and within the Racine Formation of adjacent northwestern Illinois. Bioherms (including true reefs?) of that area were part of classic studies (e.g., Hall and Whitney, 1858, and Worthen 1862, Iowa; Savage, 1926, Illinois) and modern ones (Willman, 1973, Illinois).

Until the Indiana-originated proposals were made (Shaver and others, 1978; and later, Shaver, 1985), these buildups, having intimate, interfingering relations with dense, sublithographic algal(?)-laminated carbonate rocks representing a restricted, high-salinity(?) environment, were considered to be no younger than Niagaran in age, an error verified also by the Iowa group (e.g., Witzke, 1983). Here, then, in an area far from the Michigan Basin, is further testimony—albeit circumstantial in the lack of Iowa evaporites—to a broad regionality for reef and evaporite coordination. (See relations between reef starts, sea levels, and evaporite nondeposition shown in Fig. 3.)

Reef generation 6 of Figure 3, identified so far only on northwestern Indiana outcrop in the Kenneth Limestone Member and limited to small, poor examples of reefs, is very likely Late Silurian (Pridolian) in age as verified both by conodont and ostracod zonation (e.g., see summaries in Droste and Shaver, 1977, 1985). Curiously, one of these Late Silurian buildups in the Kenneth, presumed to project disconformably with 27.4 m (90 ft) of erosional relief through laminated Salina rocks (Kokomo Limestone Member of Fig. 3), was a main basis for Cumings and Shrock's (1928a) support of Grabau's (1901-1913) idea of a grand Niagaran-Cayugan unconformity.

The unconformity in that area was disproved by the Indiana program of core drilling of reefs, which showed that Kenneth buildups both overlay and interfingered with the Late Silurian laminated Salina rocks (e.g., Pinsak and Shaver, 1964; Shaver and others, 1983).

These proposals of a Great Lakes–wide coordinated history of reef generation and abortion, which extended until the end of Silurian time, have found additional support in principle by the work of Markes Johnson and associates (e.g., Johnson, 1987; Johnson and Campbell, 1980; Colville and Johnson, 1982), which provides for similar cyclicity in Early Silurian sea levels that affected the entire area from Iowa to Michigan and western Ontario and on to eastern Canada. These proposals, nevertheless, offered as-yet-unresolved problems for some Michigan Basin workers who have been preoccupied with the narrow stratigraphic section ranging upward from upper rocks of the Niagara Group through the A carbonates of the Salina Group, a range that embraces only the two-stage reefs of generations 3 and 4 (Fig. 3).

These proposals do not seem to fit well with ideas of hundreds of feet of sea-level drawdown in the Michigan Basin and of major regional unconformity, at least not those ideas that failed in reconciliation with a continuing history of sedimentation, including reef growth, outside the basin proper. Cercone's (1986, 1988) papers seem to be rare in this regard, in that she allowed for a possible basin excursion from eustacy as she advocated major drawdown and yet reconciled this idea with the extra-basin history.

These proposals for cyclicity in reef development also offer problems to some recently advanced ideas pertinent to the Illinois Basin reefs: specifically, the ramp theories of Coburn (1986) and Whitaker (1988) to explain, in their views, the reef distribution in that basin. Similarly, Lowenstam's (1948-1957) conception of clastic belts to explain reef distribution needs ot be reevaluated in the light of new stratigraphic information. The clastic belts of Lowenstam are not simply geographically related, but also stratigraphically related.

*Reefs and depth of water.* The modern Indiana group and some others proposed that the cyclicity in Silurian reef and evaporite stratigraphy noted above was effected by modest fluctuations of sea level, perhaps a few tens of feet at most (e.g., Droste and Shaver, 1987b; Matthews and Egleson, 1974). Johnson (1987) suggested Llandoverian water depths in the continental interior ranging from 0 to 60 m in the Michigan Basin to as much as 90 m in Iowa and as little as 30 m maximum in the Williston Basin. These ranges, if valid, place limits on the range of fluctuation in eustatic sea level, and in these shallow ranges they are in keeping with those suggested by Droste and Shaver (1987b). These propositions, taken together with the facts of existence of many very shallow water reef phases of growth in the Michigan Basin environs, some including algal-stromatolitic and oolitic associations, suggest that the great majority, if not all, the Great Lakes area buildups commonly called "reefs" grew in relatively shallow waters ranging from not far below wave base upward into the surf zone. This circumstance seems to apply generally to Silurian reefs whatever their stratigraphic generation. For example, the some 300 collective-generation-1 bioherms (reefs?) now known on Manitoulin Island, western Ontario, are thought to be shallow-water products (Copper, 1978; Mielczarek and Copper, 1986). The same conclusion has been reached for the middle-generation buried pinnacle reefs of Sears and Lucia (1979) and Bailey (1986) and the generation-6 Kenneth reefs in northern Indiana (Shaver and others, 1983).

Modest fluctuations in sea level would provide for the delicate balances between reef generation and abortion and would explain why some individual reefs experienced an expansive-restrictive history of growth without apparent major intervening periods of abortion. Some reefs would originate in an area where water depth was increasing to a reef-critical level. At other times, reefs would originate in an area where water depth was decreasing to the critical level. These circumstances would effect differences between groups of reefs. Moreover, explanation would be provided for the described differences between those buildups commonly called carbonate-mud mounds that lack ready evidence of framework (possibly represent water below wave base) and those that are interpreted either to have grown through wave base or to have originated above wave base. (For example, see upper Wabash Valley generation 5 reefs, Textoris and Carozzi, 1964; eastern Indiana and adjacent Ohio reefs of generation 3, citations given above; and Manitoulin Island reefs of collective generation 1 of Fig. 3 herein, Grawberger, 1978.) These reef differences also would find explanation through the ramp theories of Sears and Lucia (1979, Michigan Basin) and Whitaker (1988, Illinois Basin), by which reefs could have originated both above and below wave base, and some could have grown through it.

Even the thickest reefs in the Midwest, ranging from 122 to 305 m (400 to 1,000 ft), may have originated in shallow water, above wave base or not far below, because they originated amidst rather pure bioclastic, crinoid-rich, biostromal and bank-like carbonate rocks of the St. Clair Limestone and Salamonie Dolomite (Fig. 3). These reefs appear to belong mostly to generation 3, many of whose small reefs, where seen on outcrop, are known to have originated in shallow water. (See also Bailey's 1986 ideas, and Lee and others' 1976 ideas on Michigan Basin subsidence to accommodate substantial thicknesses of generally shallow water pinnacle-reef growth.)

If these interpretations are correct, great reef thicknesses were accommodated in large part by basin and shelf subsidence (or sea-level rise?), not solely by initial depths of water, whatever modifications in reef growth and character were imposed by modest fluctuations in eustatic sea level. The further implication is that any strict interpretation of Lowenstam's (1950, 1957) proposal of growth from initial depths (equal to eventual reef thickness) below wave base to depths near sea level (Fig. 15) may be incorrect when applied to large reefs. Further, modest depths of water in the cyclic theme favored here would tend to promote restricted environments and reef abortion at times (for discussion of which, see section entitled "Reefs in relation to evaporites").

Cyclically fluctuating water depths would also explain the cyclically appearing influxes of terrigenously derived sediments into the Midwest, inasmuch as greater depth and improved circulation would foster long-distance transport of clay and silt as are found in the Waldron, Moccasin Springs, Mississinewa, C shale (Salina Group), and Racine units (Fig. 3). The coincidence of new reef generations occurring at those same depositional times is interpreted as just that—coincidence. Return to normal circulation fostered reef growth as well as terrigenous clastic influxing, which in the amount that took place in the Midwest was not seriously detrimental to reef growth, as were the greater amounts of clastics that came into the eastern Appalachian Basin after Wenlockian time.

Negating this idea of deepening water in relation to clastic influxes, Owens (1981) proposed that the Mississinewa clasitcs in a small part of northern Indiana were possibly wind derived, but this thesis needs testing in the scale of the huge clastic rock body that accumulated during middle Salina-depositional time throughout much of the larger Great Lakes area.

*Diagenesis of reefs.* Although some enlightenment-period workers (e.g., Cumings and Shrock (1928a, b) mentioned or identified diagenetic and other processes that changed the reef character, such as dolomitization, compaction, structural deformation, boring, and pisolitization, it was during the modern period of Silurian reef study that newly acquired sedimentologic expertise helped to bring about an explosion in reef studies (Figs. 1 and 2). The eight examples that follow are by no means exhaustive:

(1) Jodry (1969) discussed dolomitization of buried southern Michigan Basin reefs by connate waters and did not necessarily support the idea of refluxion of brines during evaporite deposition that was reciprocal with reef growth.

(2) Sears and Lucia (1980) favored the idea of dolomitization of buried northern Michigan Basin reefs by brine refluxion and fresh- and sea-water mixing. They did not support the idea of reciprocal reef and evaporite deposition or the idea of unconformity between the coral-stromatoporoid and algal portions of

reefs, this being in disagreement with Mesolella and others (1974).

(3) Gill (1972, 1973, 1977a, b), Huh (1973), and Huh and others (1977) favored especially vadose-water processes to account for solution features, leaching, caliche, and pisolites in the buried pinnacle reefs in the northern and southern Michigan Basin. Unlike Sears and Lucia (1980), they interpreted intrareef unconformity, but opposing Jodry (1969) and agreeing with Sears and Lucia (1980), they interpreted no reciprocity in reef and evaporite deposition.

(4) Berry and Boucot (1970) proposed that the generally dolomitized suite of midwestern Silurian rocks, including reefs in the southern Michigan Basin environs, became dolomitized as normal-marine sea water moved over the interbasin platform areas and into the evaporite basins. As this occurred, these waters became concentrated in magnesium through evaporation and because lime-secreting organisms, reef and level-bottom organisms alike, removed calcium from the water. This idea of northward-moving current finds support in the independently reported fore- and backreef relations of many platform-situated reefs south of the Michigan Basin (see Shaver and others, 1978).

(5) Cercone and Lohman (1985) proposed processes of early diagenesis of northern Michigan Basin pinnacle reefs as interpreted through isotopic geochemistry in relation to position of reefs on a ramp. They minimized the idea of effect of exposure to ground water during early subaerial exposure.

(6) Kahle (1974, 1978, 1988) described diagenetic processes altering the exposed Maumee Reef, other northwestern Ohio reefs, and the carbonate-bank and reef deposits along the eastern flank of the Findlay Arch, especially as interpreted as a result of paleokarstification processes during Silurian subaerial exposure. His karst evidence includes solution features, brecciation, micritization, dolomitization, and pisolitization, but his interpreted timing of events in the Maumee Reef history may not agree with Janssens' (1974, 1977) subsurface stratigraphy or with that of Figure 3 (herein); it does agree in part with Gill's (1975, 1977a) interpretation of pinnacle-reef history in the Michigan Basin. Therefore, the collective Gill-Kahle interpretation apparently calls for total reef and evaporite independence, development of a major regional unconformity, and denial of any contemporaneity between reef-bearing rocks and evaporite-bearing rocks (see Kahle, 1988, Fig. 11.2).

(7) Pray (1976), Lehmann (1978), and McGovney (1978), studying the exposed Pipe Creek Jr and Thornton Reefs, northern Indiana and northeastern Illinois (Fig. 4), especially emphasized processes of early marine cementation and porosity development. Their interpretations are related to ideas of deep-water origin, autochthonous sedimentation, and unchanging angles of flank-bed inclination.

(8) Devaney and others (1986), interpreting the Pipe Creek Jr Reef, related diagenesis especially to shallow-water origin, allochthonous sedimentation, compaction, solution features, and much-changing angles of flank-bed inclination. Much of their interpretation is a contradiction to major conclusions of the deep-water school (item 7 above).

***Reefs in relation to evaporites.*** The question that is implicit in this title has literally plagued the Michigan Basin workers during the modern period of study. The stratigraphic range (so to speak) of the problem is exemplified by Gill (1973, 1975, 1977a) and associates versus Hadley (1970) and Bailey (1986). At one extreme, the former group interpreted no post-Guelph, post-A-1 evaporite reefing in buried Michigan pinnacle reefs; at the other extreme, the latter group thought that even A-2 carbonate reefing had occurred in buried pinnacle reefs in the vicinity of Clinton, southwestern Ontario. Such disagreement has not been settled on the basis of different geographic and sedimentational settings for the disparate interpretations.

The question that remains requires for its solution reef evidence that is discussed in other sections above, especially on basic stratigraphy (both local and regional), reef paleoecology, cyclicity of reef generation, and depth of water. Of course, the solution also requires sedimentologic and geochemical evidence directly pertinent to evaporites.

The modern period of reef study saw three most common kinds of the compromise that is needed from among these kinds of evidence and that has been applied especially to the buried pinnacle reefs within the Michigan Basin proper. They range from one extreme of entirely successive reef and evaporite relations, through the middle ground of integrated but reciprocal reef and evaporite deposition, and to another extreme of exactly contemporaneous reef and evaporite deposition. These three basic kinds of proposals have been repeatedly detailed in the basin literature, so that repetition need not be continued here except for bibliographic citation of some reports that explain the three hypotheses or some variations thereof: Mesolella (1972), Kahle (1974), Gill (1975), Johnson and others (1979), and Cercone (1988).

Many proposals, although in partial disagreement with one another, have in common the idea that evaporites were deposited concomitantly with a lowered sea level (as depicted in Fig. 3), whether the lowering was of modest extent (e.g., few tens of feet) or of much greater extent (e.g., as much as 122 m (400 ft). Such lowering of sea level would, in the view of many persons, tend to bring about reef abortion or a restricted growth phase (Fig. 3). In contrast, a raised sea level would encourage new reef growth and a replacement of evaporite deposition by carbonate deposition.

Nevertheless, a rather novel reversal of these relations was proposed by Straw (1985): "the evaporites may have been derived from waters of marine origin [during high sea level stand and periods of reef growth], and the A-1 carbonate from meteoric waters draining from a broad carbonate shelf into a land-locked bolson—the dessicated Michigan Basin [while reefs were necessarily aborted]."

It seems that during the past two decades a preponderance of Michigan Basin-oriented literature has debated the different possibilities noted above for reef-evaporite relations. A final answer is not given here, but the probable trend in thinking is suggested along with my personal view. An increasing number (still a minority?) of persons have come to accept the basic tenet of Figure 3. In the Great Lakes–wide area there is a general, regional-scale facies relation between reef-bearing rocks and evaporite-bearing rocks. This means, in some views, that reefs continued to grow somewhere during the same time that evaporites were being deposited in the Michigan and Appalachian Basins. For example, the reviews cited above to Cercone and Straw were couched in agreement with at least a part of the here-claimed basic truth of Figure 3.

Some other workers, however, have misinterpreted what is meant by the Indiana-based statement of "regional facies relation between reef-bearing rocks and evaporite-bearing rocks." Some have apparently assumed that it meant normal-marine organic-reef growth contemporaneous with exactly juxtaposed evaporite deposition, and so they rejected the general facies idea. The Johnson and others (1979) paper noted above is an example. It rejected the basic tenets of Figure 3 and spoke of the unacceptability of the idea that Silurian reef builders could survive prolonged immersion to pickling brine, which is a misstatement of what others have said. So did Kahle (1988), in his study of reef-associated unconformities (diastems?, paraconformities?) in western Ohio, apparently believe that such unconformities, many (or all?) of them being minor, obviate the sense of the Indiana-based interpretation.

Neither do some actual compromises with Figure 3 offer satisfactory solution. The paper by Briggs and Briggs (1974b) recognized the evidence of a regional facies, minor unconformities or not, well exemplified in southern Michigan and northern Indiana, but it avoided the ramifications by assigning to the Illinois Basin the Indiana stratigraphic evidence that possibly would require change in sedimentologically based interpretation in the Michigan Basin. Straw's (1985) solution is hardly better. One of his premises was part of the truth the writer asserts for Figure 3, but he then remade critical parts of the stratigraphy on the basis of his novel conceptualization of evaporite and related carbonate deposition, not on the facts of physical stratigraphy and biostratigraphy. These facts, I believe, contradict his conceptualization.

Straw explained, nevertheless, perhaps better than did the Indiana workers and Kahle (1988), that any unconformity, of minor or greater magnitude, between reef-bearing rocks (Niagara Group in Michigan) and evaporite-bearing rocks (Salina Group) ought to be diachronous. Exactly! And this idea still preserves the idea of overall regional equivalency (time sense) between the two groups of rocks, that is, "a general regional facies relation." The Indiana group repeatedly has advocated minor unconformities, especially on reef tops, at differing stratigraphic levels from place to place. In Indiana, where primary bedded Silurian evaporites are unknown, the dark-colored fine-grained carbonate rocks (intervening between evaporites in Michigan) become interbedded, actually, with reef-bearing rocks, including reefs themselves. All this occurs within one named group of rocks, the Salina Group as designated in Indiana. There it is impossible to continue the simple vertical-stack (Niagara-Salina) nomenclature and its lithologic connotations used in Michigan (Fig. 3).

Cercone's (1988) compromise is much more to the point, because it basically accepted the regional facies relation as its statement by the Indiana group intended, and further, it recognized another fairly well-supported idea: that a sea-level lowering was necessary to explain evaporite deposition. This is the direction that basin workers must take, I believe, if they are to solve critical problems attending the reef-evaporite controversy.

The question that follows from Cercone's proposal, however, is how much sea-level lowering is needed, a modest amount (calling for evaporite deposition in modest depths of water) or an amount (as much as 122 m; 400 ft) needed to desiccate the Michigan Basin (calling for deposition of sabkha evaporites). Cercone favored the latter answer as an independent basin excursion from regional eustacy because she apparently thought that potash salts and other features in the A-1 evaporite necessarily required playalike settings. This requirement(?), however, only raises the spectre of diametrically opposed views among evaporite geochemists and sedimentologists, some of whom believe that they have no problem explaining potash salts as resulting from metasomatic processes. Discussion of the relative merits of the strictly geochemical views peculiar to this group is beyond the scope of this chapter.

Nevertheless, the stratigraphy and reef history developed during the modern period and presented here decidedly favor the deeper water evaporite group (e.g., Rickard, 1966; Matthews and Egleson, 1974; Schmalz, 1969; Sloss, 1969 and 1977; Sonnenfeld, 1985). Sonnenfeld's overall proposal calls variably on modest sea-level changes and a barrier mechanism, with sills or passes, including an actively building reef barrier somewhere—back to Alling and Briggs (1961) and Bolton and Liberty (1955)! To be sure, his proposal fits well with the reef-and-carbonate-bank-barrier stratigraphy of Droste and Shaver (1987b, and earlier papers).

The decided favor by me comes, not because I necessarily reject Cercone's final answer, but because nearly all other proposers of sabkha evaporites in a deeply drawn down, desiccated basin seem to have rested their cases on regional projections (great unconformity, end of reef growth, etc.) that simply are denied by stratigraphic facts, e.g., Gill (1973, 1977a), Nurmi (1974), and Nurmi and Friedman (1977). Gill's propositions seem to be least supportable for added reasons beyond his no allowance at all for reciprocity or equivalency in reef and evaporite deposition anywhere, neither within the basin nor between the basin and far-removed interbasin areas. His view, followed by some others (e.g., Lee and others, 1976), that pinnacle-reef growth began and ended while offreef deposition remained virtually at zero, would make the basin pinnacles truly unique, including in their geometric character. Elsewhere in the basin environs and throughout the Midwest, such Silurian reefs are virtually unknown as indicated from many modern-period obsevations of local reef stratigraphy.

Further, the problem with his idea of totally independent

reef and evaporite histories and its regional projection was compounded by semantics. There are no time-rock type sections in the Michigan Basin, and hardly any trusted biozones (time indicators) have been delineated, but clearly, Gill defined his rock units and his time-rock units (series) circularly on the basis of local lithology. This semantic problem carries over into much of the Michigan Basin literature, albeit harmlessly enough in some instances (e.g., Cercone, 1988). The Niagara Group of the Michigan Basin is not also the Niagaran Group, nor is it time equivalent or synonymous with "Niagaran Series."

*Grand postreef unconformity resurrected.* In 1960, Sangree's doctoral thesis on Silurian reefs and other rocks in northern Indiana appeared. The date is nearly coincident with 1961, the year of Alling and Briggs' prophetic paper that was selected here to symbolize the beginning of the modern period of Silurian reef study. The Sangree paper showed that dark-colored partly laminated micritic rocks, which would have been identified as Salina carbonate rocks if in Michigan, extended as far south as Indianapolis. These rocks are at different stratigraphic levels and are intimately sandwiched and intertongued with normal-marine rocks, reefs and all. If published, this thesis would have signaled to the Silurian fraternity that something was badly wrong with the classic concept of a grand postreef (Niagara-Salina, Niagaran-Cayugan, or reef-evaporite) unconformity.

At the same time, the modern group of Indiana-based workers began to question the classic concept (e.g., Shaver, 1962; Pinsak and Shaver, 1964.) Later, joined by Adriaan Janssens and Bruce Liberty, they threw out the classic concept altogether from their interpretation with such aggressive titles as "The mythical Niagaran-Cayugan unconformity, southern and eastern Michigan Basin area" (Droste and others, 1975), and even by claiming that the Silurian System in Indiana or parts of it are keys to settling the reef and evaporite controversy (Droste and Shaver, 1976; Droste and others, 1980).

The unconformity that they eschewed was the one having several kinds of historic to modern expressions, as exemplified below:

(1) Grabau and Sherzer's (1910) ideas of a grand reef-evaporite unconformity necessitated that all reef growth ceased by the end of Niagaran time. The reefs and their host rocks were then eroded and leached of their salts that eventually found their way to the great salt playas via rivers wending their way through the exposed and deadened reef barriers and thence into the Cayugan deserts.

(2) Cumings and Shrock (1928a, p. 131–33) interpreted a Niagaran-Cayugan unconformity as having nearly 30 m (100 ft) of erosional relief near Logansport, Indiana, where reef or reeflike rocks were thought to project through later-deposited Salina algal laminates (Kokomo of Fig. 3) lying as far as 27 m (90 ft) below. Later core drilling by Pinsak and Shaver (1964) showed a Kenneth (Fig. 3) reef or reeflike body interbedded at its base with a solid 30 m (100 ft) of Salina algal laminates below. A nearly identical section is repeated at Bunker Hill, Indiana (Shaver and others, 1983).

(3) Schuchert (1943) recorded his standard Niagaran-Cayugan unconformity in his stratigraphy textbook as, in ascending order, Middle Silurian (middle, or upper Niagaran), Disconformity, Upper Silurian (Cayugan or Salinan) for five midwestern Great Lakes states. His ideas, of course, extended into the Niagaran and Cayugan type areas of western New York and adjacent Ontario.

(4) Gill (1973, 1977a), Nurmi (1974), and Nurmi and Friedman's (1977) reef-evaporite unconformity in the deep Michigan Basin was translated by Gill, via strictly lithologic definitions, into a Niagaran-Cayugan unconformity; this unconformity, when extended by Gill to Indiana, was said to be manifested by so-called "river gorges" through the Fort Wayne Bank of northern Indiana, that is, the river gorges of Grabau 1901–1913. Nevertheless, most of the Fort Wayne Bank in northeastern Indiana depositionally postdates the time that Gill needed for his A-1 evaporite event (Droste and Shaver, 1982; Fig. 3).

(5) Johnson and others (1979) adhered to the classically conceived Engadine-Salina unconformity in northern Michigan, and they seemed to apply it throughout the Great Lakes area as a Niagaran-Cayugan unconformity (Johnson and others, 1979, Fig. 7 and p. 6).

(6) Kahle (1974) repeated Gill's Michigan Basin-based thesis of a Niagara-Salina (Niagaran-Cayugan; Guelph-Salina, postreef) unconformity for the Maumee Reef site, northwestern Ohio. By 1978, this unconformity became three unconformities (Kahle, 1978), one of which was picked as the Niagaran-Cayugan boundary without benefit of age indicators but this time including some Salina-like rocks (algal stromatolites) in the Niagaran Series and, unlike Gill, including some reeflike rocks (algal-stromatolite reef) in the Cayugan Series (Kahle, 1978, Fig. 4 and accompanying text).

(7) Kahle (1988) presented a somewhat modified view of the grand postreef unconformity along the Cincinnati and Findlay Arches in western Ohio, which he characterized as follows:

The Lockport/Peebles and Greenfield contact in western Ohio is a major, regional, subaerial unconformity.... The contact between the Lockport Dolomite and the Greenfield Dolomite is not a facies contact as depicted by some workers [citations from Janssens, 1977; Shaver, 1977; Shaver and others, 1978; Droste and Shaver, 1982]....

[The] author believes that the evidence ... for a regional subaerial unconformity at or near the top of the reef-bearing Silurian strata in Michigan and Ohio countermands the viewpoint which opposes ... the possibility of such an unconformity.

This statement was accompanied by Kahle's Figure 11.2, a correlation chart showing an essentially isochronous upper Lockport contact (as the chart is constructed) and an isochronous lower Greenfield contact, respectively below and above a possible Niagaran-Cayugan series (epoch) boundary.

The classic concept of a Niagaran-Cayugan unconformity should have become passé by 1988, in my opinion, if for no other reason because continued use of this cliché only furthers the semantic part of the problem among those who have some real disagreement on stratigraphic correlation. The semantic part ought to be eliminated quite apart from the possibility that many Michigan Basin and Ohio workers may be correct that an unconformity or unconformities of some magnitude exist within or over the top of the buried basin pinnacle reefs or within or over the top of the reefs along the Findlay Arch. Beginning in 1970, three at least guiding, if not quite authoritative sources pointed up the folly of continuing the turn-of-the-century tradition in Silurian time-rock classification:

(1) In 1970, Berry and Boucot, in their worldwide Silurian correlation efforts, refused to use the provincial North American series terms, so ambiguous that usage had become with respect to real stratigraphic facts, particularly the usage of Niagaran and Cayugan nomenclature.

(2) In 1975, Rickard, who earlier recognized (e.g., 1969) the lack of unconformity between the rocks of these two series in the New York and Ontario type areas, redefined the relation between what should be accepted as the North American standards for these two series if workers are to continue use of these terms.

(3) In the late 1970s and early 1980s the American Association of Petroleum Geologists, in cooperation with Canadian and Mexican geologists, constructed its series of COSUNA (Correlation of Stratigraphic Units of North America) charts. An internationally composed committee set up both the global chronostratigraphic scale and the North American provincial scale that were to be used on these charts uniformly throughout the United States. The midwestern chart (Shaver and others, 1984), which includes the Michigan Basin and Findlay Arch area, shows no Niagaran-Cayugan unconformity for the provincial scale, as it heeds both the Rickard (1975) and the COSUNA (e.g., Shaver and others, 1984) provisions.

One may properly ask the question: By what good reason do midwestern stratigraphers, without reference to standard age indicators or other direct consideration of the age relations in New York and Ontario, continue to use local lithology and a conceptual single midwestern unconformity as a basis for proclamation of the grand Niagaran-Cayugan (Niagara-Salina, postreef, reef-evaporite) unconformity of Grabau and Sherzer (1910) and of the Schuchert (1943) textbook and other textbooks?

For many years prior to 1988, the Indiana group of Silurian workers spoke of reef-top unconformities, probably at different stratigraphic levels and of which some appear to die out downdip. It suggested many as-yet-unrecognizable bedding-plane unconformities, and as did Straw (1985), it spoke of a regional relation between reef-bearing rocks and evaporite-bearing rocks that should be a time-transgressive relation, whether the actual boundary in places is conformable or unconformable. An overall "facies relation" would apply either way in the sense of partial to nearly complete age overlap for the Great Lakes region.

Parts of Kahle's (1988) study are to the point of the Indiana-based view:

(1) The Lockport-Greenfield contact is a paraconformity, Kahle proposed.

(2) Multiple unconformities (diastems?), vertically stacked, occur at some single localities as also was observed by Kahle's associates (e.g., Lehle, 1980).

(3) Many of the localities are reef sites or otherwise are massive carbonate-bank sites, which expectably were deposited to near sea level, thence experienced radical change with minor sea-level movements.

(4) Some locales exhibit unconformities that were fault controlled independently of Kahle's major regional unconformity.

(5) No age indicators and no detailed subsurface stratigraphy were presented to assure the singleness of major unconformity or to assure that the appearance of unchanging age relations along the unconformity (see Kahle's Fig. 11.2) is apt.

(6) Kahle (p. 251) agreed that his contact ("regional unconformity") may not represent the Niagaran-Cayugan boundary in the Great Lakes region.

Further perspective is needed. The Kahle view of regional scope and major hiatus does find some support in the far southwestern Ohio type area of both the Peebles Dolomite and the Greenfield Dolomite. There, new conodont-based work by Kleffner (1988a, b) suggested that a number of standard Silurian conodont zones are missing from between the Peebles and the overlying Greenfield. The latter unit, in fact, is represented by the youngest standard Silurian zone, *eosteinhornensis* (Kleffner, 1988a, b).

Immediately on the western side of the Ordovician exposures along the Cincinnati Arch in Indiana and Kentucky, however, the missing zones begin to appear, in the Louisville Limestone (Rexroad and others, 1978), which together with the Waldron Shale lie next above the Peebles-equivalent Salamonie Dolomite. Further, an unconformity can be observed atop the Peebles equivalent and below the Limberlost Dolomite in one or more easternmost Indiana quarries just north of the Ordovician exposures near Richmond, Indiana, and structurally high on the arch (see Fig. 3). Here the unconformity is minor, within the Zone of *Pentamerus oblongus* and hundreds of feet stratigraphically below the level of the *eosteinhornensis* Zone. Still farther north at Celina, westernmost Ohio, no unconformity has been found at that level (Shaver, 1974c; Griest and Shaver, 1982), whence the section climbs well up into the A-carbonate-equivalent section (Salina Group) below the Quaternary erosion surface.

Still farther north and northwestward in Indiana and into the distal Michigan Basin, still higher conodont zones appear, including that of *eosteinhornensis,* which are collated with two major episodes of Salina algal-laminate deposition as well as with reef rocks to the top of the Silurian System there (F-Salina equivalent?). (See biostratigraphic summaries in Shaver and others, 1971; Droste and Shaver, 1977, 1985; Rexroad and others, 1985).

A problem remains as to just what the Greenfield represents

along its full western Ohio extent and as to just how high the reef-bearing rocks extend in northern Ohio (as high as in Indiana?). In its collective use, the Greenfield is a wastebasket term, and so also are the terms Lockport, Guelph, and Cedarville Dolomites in some of their past use in western Ohio. Apparently a semantic factor will remain in the debate over the assumed Niagaran-Cayugan unconformity (or some possible major unconformity) until such time as the classicists truly weigh the biostratigraphic and physical subsurface data that are available and that suggest something other than a simple pervasive unconformity.

Differential tectonism seems evident from south to north along the axis of the Cincinnati Arch and thence the axes of the Findlay (Ohio) and Kankakee (Indiana) Arches. Considering the subsurface Appalachian Basin stratigraphy of Ulteig (1964), the full ramifications of Kahle (1988) and Janssens' (1974, 1977) evidence, and the relations described just above for the arch areas northward in Ohio, Indiana, and Michigan, I conclude that Walther's Law works beautifully in the southern Michigan Basin region and is all but proved to do the same in the western Appalachian Basin region, central northern Ohio. This idea hardly resembles the classic idea of Niagaran-Cayugan (Niagara-Guelph, reef-evaporite, or, simply, postreef) unconformity.

**Numbers and sizes of reefs in relation to stratigraphy and tectonism.** Very little appears in the Silurian-reef literature for any period of study that testifies directly to the relations suggested in the title above, but the proliferation in modern-period studies resulted in many data that are ready for analysis. As to the quantity of reefs in given stratigraphic and geographic situations, relatively huge numbers seem to apply expectably only to small reefs measured in a few meters, but such numbers can apparently occur at any or most of the reef-generation starting levels depicted in Figure 3.

For example, the small spongiostromatid(?) reefs described by Soderman and Carozzi (1963) in the Nazbro Quarry near Knowles, Wisconsin, surely numbered a thousand or more in the quarry alone (Shaver and others, 1978). These reefs, which belong to collective-generation 1 and were quickly aborted, are stratigraphically low. In further example, the very small to small-medium normal-marine generation-3 reefs described by Griest and Shaver (1982) in the Karch Quarry at Celina, Ohio, once numbered (before quarrying) about 250 in a 60-acre site. This number applies near starting level 3 (stratigraphically intermediate), but after upward growth, through about 18 stratigraphic meters (60 ft), the number was reduced by one-third or more of small to small-medium reefs. Similar phenomena apply to the Rockford complex in Ohio, a 50-acre site, where after about 12 m (40 ft) of additional upward growth, only one medium-size reef and a few smaller reefs survived the competitive process of coalescence (Indiana University Paleontology Seminar, 1976). These larger numbers of small reefs, figured from relatively small sites, stagger the imagination when projected to square miles along an expectable trend for a given generation.

Medium-size reefs such as generation-5 reefs, as they happen to be exposed in the upper Wabash Valley, northern Indiana, average 0.3 km (0.2 mi) or less in diameter and are little more than 30 m (100 ft) thick below their eroded tops. They are many fewer in number, two or three per square mile (Shaver and others, 1978, Fig. 20). Even so, this projects to more than 1,000 such reefs in a single northern Indiana county.

The still-larger buried pinnacle reefs of the Michigan Basin (generations 3 and 4), many of them about 122 m (400 ft) thick and covering about half a square mile each, were known in 1988 to number about 1,000 in the northern reef trend. At this size, therefore, still fewer reefs are present. The thickest pinnacle reefs known in the Midwest are about 305 m (1,000 ft) thick and are in the Illinois Basin. Extremely few are known at that size. Some appear to have originated low in the St. Clair and to have grown upward to the top of Late Silurian rocks (Fig. 3).

Two modern-period studies have addressed these apparently interrelated phenomena of sizes, spacing, and numbers of reefs in given situations: (1) Griest and Shaver (1982), small generation-3 reefs at Celina, Ohio; and (2) Whiteman and Gardner (1975), middle-size generation-5 reefs in the upper Wabash Valley, Indiana. Both studies concluded that the distribution of these reefs is nonrandom and that in this way the reef communities behaved in accord with what is known about the nonrandom distribution of modern ecologic systems. These studies also suggested that a mathematical predictability applies to numbers, survival rates, spacing, and sizes of any fossil-reef system, e.g., the buried pinnacle-reef system in the Michigan Basin.

From the data above, it has become apparent that stratigraphic range has much to do with sizes and numbers of Silurian reefs in given situations. For example, no very thick more or less discrete reef in the Great Lakes area has a short stratigraphic range. Perhaps the buried pinnacle reefs in Michigan nearly fulfill such a circumstance in that their range is thought here to be only from the upper part of the Niagara Group into the lower part of the Salina Group (Fig. 3).

Some other representatives of these aborted generation-3 and -4 reefs are 23 m (75 ft), more or less, in thickness where exposed within the same stratigraphic level at the southern edge of the Michigan Basin in western Ohio and adjacent Indiana. The ratio of 23/122 m (75/400 ft) is thought here to represent the ratio of relative subsidence of basin-platform edge to basin proper. (See Droste and Shaver, 1985, p. 76 and Fig. 2.) In carrying this idea farther, Droste and Shaver (1980) explained the difference between tall, areally modest pinnacle-like reefs basinward from the Terre Haute Bank, Illinois Basin, and less thick but areally broad coalesced reef masses platformward.

These ideas, however, were challenged by two proponents of the carbonate ramp theory Coburn (1986) and Whitaker (1988) to essentially describe the Silurian proto-Illinois Basin from Tennessee northward all the way to the axis of the present Cincinnati and Kankakee Arches separating the Illinois and Michigan Basins. This theory, if applicable, would modify somewhat the expectations from reef distributions in specified size ranges that attach to the more conventional theory that invokes a hingeline, however broad, and some differentially acting Silurian tectonism.

Although Coburn claimed that discontinuous reef trends are expectable on a carbonate ramp, some actually existing linear reef trends were insufficiently integrated with the facts of physical stratigraphy and cyclical sedimentation. This is to say that the carbonate-ramp theory as applied has oversimplified the real sequence of events. I believe that sequence appears to require some differential regional tectonism. For example, basin development in a conventional sense may be required to explain the known reef phenomena; global sea-level changes acting in concert with only a starting tectonic (structural) circumstance, which remained neutral throughout Silurian time, may be insufficient explanation. Given some active tectonism and the fact that basin-shelf hingelines were not extremely prominent slope breaks in this example, the need to invoke the extremes of either a carbonate ramp or a conventional differentially subsiding basin becomes a moot point.

The thickest of the reefs discussed here, as much as 305 m (1,000 ft) thick?, have a great stratigraphic age range, from St. Clair depositional time to latest Silurian time (Fig. 4), suggesting as much as 10 m.y. in the reef making. The Michigan Basin pinnacles would appear to have grown at much more rapid rates but nevertheless at rates much less than those known for some Pleistocene and Holocene reefs. Compaction and other diagenetic processes acting on Silurian reefs are not enough explanation for these differences, nor is major(?) unconformity in some reefs sufficient answer for all reefs. New paleoecologic and sedimentologic studies are needed to assess, for example, the full significance of so many bedding planes and repetitious sets of graded bedding within reef-flank deposits and growth planes within the reef cores. Furthermore, the potential difference between the reef capacities of Silurian and modern marine communities needs to be considered.

## SUMMARY

The discussions above in their collectivism are characterized by two main themes. One theme considers discrete buildups and how they fulfill rather well most of the ecologic-reef criteria listed in Figures 14 and 15 (my opinion). An extension of this theme is that the different geographic groups of reefs and different stratigraphically based generations of reefs have both common characters and characters tending toward separateness. This circumstance explains the range in discrete reef (buildup) types, such as carbonate-mud mound, incipient reef, patch reef, pinnacle reef, coalesced reef, and even reef-bank complex, that the profession has defined during the modern period of study.

A second theme deals with the relation of discrete reef masses with other rocks, whether called oolite bars, algal stromatolites, evaporites, carbonate banks, or simply interreef strata. My account, weighted by both my perspective and that of others, provides for active reefs having had much to do with these other rocks.

All these characters, seemingly so diverse and disordered, suggest the impossibility of defining a single reef model for the Michigan Basin environs. But as stated by Shaver and others (1978) in concurrence with Johnson and others (1979), perhaps the majority of modern-period workers have come to agree that:

This seeming disorder . . . lends character to the model. Silurian reefs, like all reefs, first of all were responses to a set of physical-chemical conditions that were both permissive and limiting. Within the limitations, the reef community and the reef body itself demonstrated great versastility, using all survival potentials in whatever combinations were necessary, to exploit each possible niche and opportunity. An ordered array of form, sizes, etc., that we have yet to understand fully was the result. This array helped to reset its own limitations, so that in much of the Great Lakes area, development of the Silurian System increasingly partook of organically controlled responses. The model [still] has much to tell.

Surely, the modern period of study saw adequate fulfillment of a final criterion demanded of Silurian and other reefs (Figs. 14 and 15) that was not discussed directly above, the criterion specifying reef influence or control on other sedimentation. The question of how much control, whether in relation to evaporite sedimentation or to nonreef carbonate sedimentation, needs additional assessment.

## REEF ECONOMICS

### Stone products

Silurian reefs have played and continue to play an important and variable role in the North American industrial complex. During the early period of study, stone products were the main economic interest. They remain very important today but share that interest with the petroleum industry.

Burned lime was a principal reef-related resource during the early period of reef study (Fig. 8), and quarries developed for this purpose typically had 6- to 9-m (20- to 30-ft) exposures. Vertical reef sections of greater magnitude that could be examined were rare. The reef-based part of the lime industry declined to virtually zero during the middle and modern periods of reef study. In 1988, however, lime-burning kilns remained active near Knowles, Wisconsin (Thomas Evans, written communication, September 6, 1988), far western Michigan Basin, where the quarried materials included thousands of small algal-dominated reefs. These Early Silurian reefs were described by Shrock (1939), Soderman and Carozzi (1963), and Shaver and others (1978, p. 12).

The early-period literature for Indiana (and for some other Great Lakes states) records the production of foundation stone and flagstones from Silurian rocks. Especially the Liston Creek Limestone Member (Fig. 3), which contains much reef-shed detritus and itself rises into reefs and becomes reef-flank rock, was used. The reef-based part of this industry, never large, is nearly defunct in Indiana, but one flatstone quarry was in operation during 1989 (Carr and others, p. 4) and produced from flank rocks of the Yorktown Reef (Fig. 4).

As the reef-based burned lime and foundation-stone industries declined, a still-growing crushed stone and aggregate industry based on Silurian reefs became established. During the 1970s in Indiana, for example, Silurian reefs yielded as much as 16 percent annually of the total crushed stone product. By 1973 in Illinois, 40 percent of the state's crushed stone product of 66.5 million tons valued at more than $144 million came from quarries containing Silurian reefs. As much as 10 percent came from the Thornton Reef (Fig. 4) alone. In Ohio in 1987, about 7 percent (3.04 million tons) of the state's total production (45.05 million tons) of limestone and dolomite came from three quarries that are developed in the Lockport carbonate-bank and reef complex exposed at Carey along the eastern flank of the Findlay Arch. The value of sales from the three Carey pits was about $10.4 million, or about 6.3 percent of the state's total sales.

The Silurian-reef product has expanded to include agricultural lime, fluxes, and high-calcium limestone. The latter finds use in food extenders, mineral animal feed, athletic field marker, whiteners, and pharmaceutical, deodorizing, and desulfurization materials. (The information on stone products from reefs was furnished by Curtis H. Ault, Indiana Geological Survey, and David Stith, Ohio Geological Survey.)

Much of the current research on removing sulfur from coal involves the use of high-calcium limestone as a desulfurization material. The current concern for acid rain in the environment emphasizes the need for such research, and the potential for increased use of high-calcium limestone for desulfurization is very high. The search for high-calcium limestone in reefs has received attention in Indiana for many years (Ault and Carr, 1981), although only one quarry in the state now produces high-calcium limestone from a Silurian reef, the Pipe Creek Jr Reef (Fig. 4). A second use of high-calcium limestone, as a high-purity, high-brightness product for filler or whiting applications, has also received increasing attention (Shaffer and others, 1982). Silurian reef limestone also holds promise for this use.

As a result of modern-period quarrying operations, Silurian reef exposures in some Great Lakes states have become very extensive, nearly 90 m (300 ft) high in some quarries and more than 1.6 km (1 mi) in lateral extent. Silurian reef science has benefited immensely from direct access to these reef exposures, and much of the latest knowledge of reef geometry, internal structure and composition, stratigraphic relations, and diagenesis rests on investigations of these exposures.

As yet, reef science has returned only a modest amount of benefit to the industry; for example, in quality-control programs and in locating new quarry sites as older operations completely removed some reefs (see the reef-location maps for Indiana by Ault and others, 1976, and in preparation). The industry, however, has scarcely begun to realize other potential benefits, e.g., in planning quarry and underground-mining operations in accord with scientifically developed information on size, composition, geometry, and interreef relations and as was advocated by Ault (1975) for northern Indiana reefs.

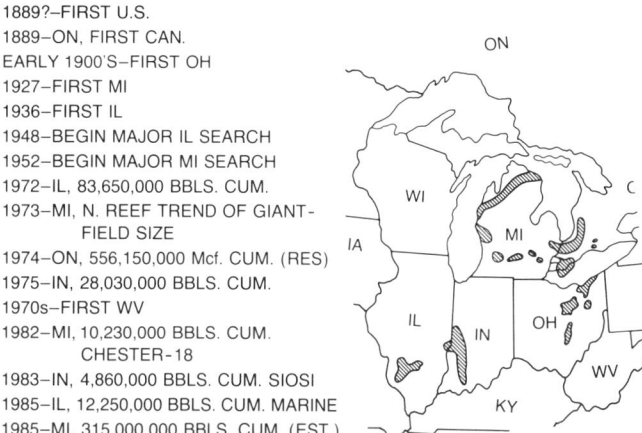

Figure 24. Notes on some highlights in the impetus given to Silurian reef study by the petroleum industry. Patterned areas have yielded significant quantities of Silurian reef-related hydrocarbons. "Chester-18," "Siosi," and "Marine" are reef-associated oil fields; RES., reserves.

## Hydrocarbon production

What may have been the earliest North American drilling discovery of oil in Silurian reefs or reef-induced structures occurred at Terre Haute, Indiana, in 1865 (Scovell, 1897, p. 517). This discovery occurred shortly after the first commercial oil well in the United States was completed in Pennsylvania, but according to Scovell (1897, p. 527), commercial quantities of oil at Terre Haute were not produced until 1888; Rarick (1980) stated that the date was 1889 (Fig. 24).

In the same year, 1889, the first Silurian reef-associated fields in Canada were brought in. Little or no connection between reefs and hydrocarbons was recognized during these early-period ventures.

In the Michigan Basin proper, production began near the turn of the century from pinnacle reefs and reef-associated structures in southwestern Ontario; in Michigan, such production began in the 1920s (see Fig. 24; Quillian, 1966; Hadley, 1970; Fisher, 1973). The major Michigan search began in 1952. By 1973 the newly discovered northern pinnacle reef trend became a major play. In the 1970s, the western Ontario reef-based gas fields, which were discovered during the early period, remained the largest gas reserves in that province. The estimated 1987 cumulative production of about 350,000,000 barrels from reef-associated fields in Michigan seems to bear out the 1974 prediction of Briggs and Briggs (1974a) that the northern reef trend, now numbering nearly 1,000 discovered reefs, would eventually amount to giant oil-field size (500,000,000 barrels) in its collectivism.

Reef-associated hydrocarbon production from the Illinois and Appalachian Basins has been much less than from the Michigan Basin despite the fact that production from those basins began early and major searches began earlier in Illinois than in Michigan (Fig. 24). In the Illinois Basin, the Marine Reef (Fig. 4) began its hydrocarbon yield in 1943 (Lowenstam and DuBois, 1946) and is still in the active stage of exploitation. Its yield is almost entirely from Silurian reef rock, and its 1985 cumulative yield of well over 12,000,000 barrels suggests that it is still the all-time champion among individual Silurian-reef fields. Although reef-associated production in Illinois began in 1936, the Marine Reef was the first documented subsurface reef in Illinois. Its discovery set off the reef hunt in Illinois.

Good single-reef fields in the Illinois Basin are expected to yield a few million barrels of oil. In Illinois, less than half the production has come directly from reefs; the remainder has come from reef-induced reservoirs above the reefs. In Indiana, all the production has come from reef-induced structures (see Bristol, 1974; Becker and Keller, 1976). Reef production in the Michigan Basin comes both from the reefs and from reef-induced structures (Michigan Geological Survey Division, 1984). Some good individual fields, like those in the Illinois Basin, are expected to yield a few million barrels, but most Michigan Basin reefs are probably expected to yield less than the better Illinois Basin reef fields. The larger Michigan Basin reefs are generally smaller than the larger Illinoss Basin reefs, but the Michigan Basin reefs seem to be more numerous.

Accurate figures on production of hydrocarbons from Silurian reefs in the Appalachian Basin are difficult, if not impossible to obtain, but the number of fields in the Newburg Sand (drillers' term, Lockport Group) has increased slightly since the report by Floto (1955) appeared. At least some Newburg production is still believed to come from Silurian reefs. (Ohio information on hydrocarbons is from David Stith, personal communication, September 14, 1988).

The petrolem industry has been of increasing benefit, beginning during the middle period, to Silurian reef science and is now of immense value. In return, reef science has given immense returns to the petroleum industry. Today the partnership between science and industry acommodated by Silurian reefs is perhaps one of the finest examples of its kind. The partnership with the stone-products industry has yet to achieve this level, but the value of reef research in this field nevertheless increased greatly during the past decade.

## SOME NEEDED STUDIES

Much has yet to be learned from the Silurian reefs in the Michigan Basin and environs. The following suggestions are both predictions and pleas:

1. A program of close coring across one or more Michigan Basin pinnacle reefs, such coring to extend through the bottom of the reef and well into offreef strata (a plea more idealistic than practical?). This would lead to much-needed better understanding of internal structure and zonation, environmental parameters, growth history, and interreef relations, including those with evaporites.

2. Conodont-based biostratigraphic studies of basin pinnacle reefs and offreef strata would aid physical stratigraphers in finding exact niches for the reefs in the regional stratigraphic history.

3. Intensified sample and core studies, using available materials and collated with geophysical log signatures, are still needed to delineate the buried carbonate bank ("massive reef" of Fisher, 1973) of southern Michigan. Its mass, as commonly represented, consists of both typical bioclastic carbonate-bank sediments and dark-colored micritic laminated lower Salina carbonate rocks.

4. Similar studies, extending to outcrop of the carbonate bank and associated reef structures along the east flank of and atop the Findlay and Cincinnati Arches in Ohio, combined with conodont biostratigraphic investigations, should parallel those in Michigan. They would ascertain further the degree of applicability (inapplicability?) of Walther's Law for understanding a regionally transgressive facies relation (or a transgressive regional unconformity?) between restricted-basin facies of rocks and more normal marine interbasin facies. Conversely, such conodont-based studies would test the applicability of a single major regional unconformity with apparently unchanging time value throughout that area as depicted by Kahle (1988, Fig. 11.2).

5. New reef and reef-stratigraphic studies in several outcrop locales, especially in eastern Wisconsin and in the northern Michigan extension of this Niagara Escarpment region, need to be carried out. Conodont and other biozonal studies should help to ascertain whether the upper Racine and upper Engadine rocks and the restricted-phase algal?-laminates of the Waubakee Formation of Wisconsin (see Klug, 1977; Shaver and others, 1984) qualify as a younger expression of the relation between carbonate-bank and laminated Salina rocks described within the basin.

6. The lessons learned from such intense paleoecological studies of reef and interreef strata as have been carried out during the past 10 years in Iowa and eastern Canada need increased application and emulation in the Michigan Basin environs.

7. As a special ramification from item 6 above, paleoecological and sedimentological studies as outlined under the heading "Numbers and sizes of reefs" needs to be directed toward reef-growth rates and the reef-building capacity of Silurian communities.

8. Further stratigraphic testing of the concept of cyclic reef generation and abortion and related clastic and evaporite sedimentation of regional or greater scope is needed for better understanding of this phenomenon in relation to such potentially causative factors as basin tectonism, global sea-level cyclicity, oceanic sea-floor spreading, and Silurian glaciation. According to Figure 3 and Shaver and others (1985), sea-level lowerings (water shallowings) were slow events and sea-level risings were short events, which proposal contradicts the sea-level behavior curve of Vail and others (1977) as originally proposed.

9. In partial relation to item 8 above, basin tectonism or plate tectonics in general should be increasingly considered in potentially causative relation with genesis, distribution, and configuration of Silurian reefs, as suggested recently in various ramifications for Michigan Basin and Illinois Basin reefs and for reefs along the east flank of the Findlay Arch (e.g., Sandford and others, 1985; Droste and Shaver, 1980; Coburn, 1986; Whitaker, 1988; Kahle, 1988). Such consideration should extend to appraisal of basins as describing carbonate ramps and as having fault-controlled configurations in part, e.g., in the sense of having marginally transverse faulting.

10. The petroleum industry has already assessed many of the factors discussed above in relation to reef distribution and to discovery of potential hydrocarbon reservoirs, but this assessment should never be considered finished. During the modern period of study, such papers as those by Droste and Shaver (1980, 1987a), Shaver (1977), and Whitaker (1988) for the Illinois Basin, and Sanford and others (1985) and Bailey (1986) for the Michigan Basin, suggested many new subtleties on reef size and shape, reef distribution (both stratigraphic and geographic), potential for reservoir cap rock, and erosional, sedimentational, burial, and structural controls in Silurian reefs that are not yet well assessed.

In the Illinois Basin, for example, reassessment needs to be done to whatever extent Lowenstam's (1948–1957) mapping of high-, medium-, and low-clastic zones, each with its own prediction for reef distribution, has directly guided reef exploration. The clastics are real enough, but most use of the clastic map has ignored post-Silurian erosion effects, hundreds of feet of Late Silurian carbonate rocks, and the vertical distribution of clastics within the entire Silurian section. Therefore, a given geographic clastic zone can also be other than what it is labelled, depending on the stratigraphic interval in question. Clastics may be an indirect key to the reef information sought, but in the Illinois Basin they are dependent themselves on the actual factors that controlled reef distribution.

Should the fact of continued sedimentational cyclicity, including in reef growth, beyond the Michigan Basin and after A-evaporite depositional time encourage exploration in the Michigan Basin for reef trends, geographically and stratigraphically offset from known trends; for example, in the C shale (middle Salina) interval? After all, that depositional time was a time of magnificent reef growth on the southern interbasin platform and in the Illinois Basin, but the concept of an expanding, carbonate-bank and reef-restricted basin suggests that a new reef trend will not be discovered in the C shale of Michigan. Reefs of C-shale age moved, so to speak, beyond Michigan's borders. Are there buried reefs of pre-Niagara age within the central Michigan Basin (collective generation 1 of Fig. 3)? No answer is given here, except that, theoretically, reefs are not barred from that area.

## CONCLUSION

The concluding moral for this history relates especially to the fact of so many debates and otherwise opposed viewpoints on various reef facets. The discussion under "Diagenesis of reefs" and other headings illustrates this. Hardly any reef interpretation that was based mostly or wholly on modern sedimentological and geochemical principles did not provoke serious disagreement on the same or other bases. Some theoretical conceptualizations, so based, were then projected widely, both to reefs in general and to an assumed regional stratigraphy. In some ways this was backwards, and in some examples a modern stratigraphic base was present but ignored or outright denied. And that is certainly backwards! Stratigraphy, for both local reef details and the regional setting, should be a yardstick against which sedimentologic and geochemical interpretations can be tested. Indeed, the Doctrine of Uniformitarianism and the lack of some needed modern analogs cannot be the exclusive tests for Silurian reefs and their relations.

I suggest, therefore, that modern geological training and research interests have possibly strayed too far from basic stratigraphic tenets, which still are the very foundation of geologic science. Silurian reef sedimentologists and geochemists, armed with such a persuasion, however, probably will maintain clear consciences in the knowledge that practicality has required most studies to proceed on too independent and both geographically and stratigraphically isolated bases. Further, too many of the older stratigraphic and paleontologic pronouncements have proved to be inaccurate because the impact of facies relations in a dominantly carbonate and evaporite province (in the Silurian example) was not understood, neither as a part of physical stratigraphy nor as a paleoecological factor affecting so-called index fossils.

Perhaps the best that we can hope is, given that the special incentives geologists have for studying Silurian reefs have been frustrated far more than we should like and than often need be, Silurian reef students will accept the idea that the stratigraphic base is everyone's obligation to build and/or to use, not just to conceptualize without use of existing conventional evidence.

# REFERENCES CITED

Not all the citations given below are found as identical listings in the text and figures; rather, a few are from such collective entries in the figures as 1970–81 Bourque and Lespérance and others.

Alden, W. C., 1902, Chicago folio, Illinois–Indiana: U.S. Geological Survey Geologic Atlas Folio 81, 41 p.

——, 1906, Milwaukee special folio, Wisconsin: U.S. Geological Survey Geologic Atlas Folio 140, 16 p.

Alling, H. R., 1928, The geology and origin of the Silurian salt of New York state: New York State Museum Bulletin 275, 139 p.

Alling, H. R., and Briggs, L. I., 1961, Stratigraphy of the Upper Silurian Cayugan evaporites: American Association of Petroleum Geologists Bulletin, v. 45, p. 515–547.

Anderson, R. L., 1980, Reef structures in the Louisville Limestone (Silurian) in Bullitt County, Kentucky [M.S. thesis]: Richmond, Eastern Kentucky University, 55 p.

Archer, A. W., and Feldman, H. R., 1986, Microbioherms of the Waldron Shale (Silurian, Indiana); Implications for organic framework in Silurian reefs of the Great Lakes area: Palaios, v. 1, p. 133–140.

Ashley, G. H., and Siebenthal, C. E., 1899, The coal deposits of Indiana: Indiana Department of Geology and Natural Resources Annual Report 23, 1428 p.

Ault, C. H., 1975, Origin, geometry, and distribution and industrial significance of Silurian reefs in northern Indiana: Geological Society of America Abstracts with Programs, v. 7, p. 712–713.

Ault, C. H., and Carr, D. D., 1981, Search for high-calcium limestone in Silurian reefs of northern Indiana: Geological Society of America Bulletin, v. 92, Part 1, p. 641–647.

Ault, C. H., and 4 others, 1976, Map of Indiana showing thickness of Silurian rocks and location of reefs and reef-induced structures: Indiana Geological Survey Miscellaneous Map 22.

Bailey, S.B.M., 1986, A new look at the development, configuration, and trapping mechanisms of the Silurian Guelph reefs of southwestern Ontario: 25th Annual Conference Paper 13, Ontario Petroleum Institute, Inc., 28 p.

Bannister, H. M., 1868, Geology of Cook County, in Worthen, A. H., Geology and paleontology: Illinois Geological Survey, v. 3, p. 239–256.

Barnes, C. R., and others, 1981, Field meeting, Anticosti–Gaspé Quebéc, 1981: Montréal, Québec, Université de Montréal Departement de Géologie and International Union of Geological Sciences Subcommission on Silurian Stratigraphy, Ordovician–Silurian Working Group, v. 1, Guidebook, 56 p.

Bay, T. A., 1983, The Silurian of the northern Michigan Basin, in Harris, P. M., ed., Carbonate buildups; A core workshop: Society of Economic Paleontologists and Mineralogists Core Workshop 4, p. 53–60.

Beards, R. J., 1967, Guide to subsurface Palaeozoic stratigraphy in southern Ontario: Ontario Energy and Resources Management Paper 67-2, 19 p.

Becker, L. E., 1974, Silurian and Devonian rocks in Indiana southwest of the Cincinnati arch: Indiana Geological Survey Bulletin 50, 83 p.

Becker, L. E., and Droste, J. B., 1978, Late Silurian and Early Devonian sedimentologic history of southwestern Indiana: Indiana Geological Survey Occasional Paper 24, 14 p.

Becker, L. E., and Keller, S. J., 1976, Silurian reefs in southwestern Indiana and their relation to petroleum accumulation: Indiana Geological Survey Occasional Paper 19, 11 p.

Bell, R., 1987, Report on an exploration of portions of the Attawapishkat and Albany Rivers, Lonely Lake to James Bay: Canada Geological Survey Annual Report 1886, pt. G, p. 1G–38G.

Berry, W.B.N., and Boucot, A. J., 1970, Correlation of the North American Silurian rocks: Geological Society of America Special Paper 102, 289 p.

Blatchley, W. S., 1904, The lime industry in Indiana: Indiana Department of Geology and Natural Resources Annual Report 28, p. 211–257.

Bolton, T. E., 1957, Silurian stratigraphy and paleontology of the Niagara Escarpment in Ontario: Geological Survey of Canada Memoir 289, 145 p.

——, 1981, Ordovician and Silurian biostratigraphy, Anticosti Island, Québec, in Lespérance, P. J., ed., Stratigraphy and paleontology: Subcommission on Silurian Stratigraphy, Ordovician–Silurian Boundary Working Group, Field Meeting, Anticosti-Gaspé, Québec 1981, v. 2, p. 41–57.

Bolton, T. E., and Liberty, B. A., 1955, Silurian stratigraphy of the Niagara Escarpment, Ontario, in The Niagara Escarpment of peninsular Ontario, Canada: Michigan Basin Geological Society Guidebook, p. 19–38.

Bourque, P.-A., 1979, Facies of the Silurian West Point reef complex, Baie des Chaleurs, Gaspésie, Québec: Geological Association of Canada Guidebook B-2, 29 p.

——, 1980, The upper reef complex of the West Point carbonate buildup—The Pointe de l'Ouest to Point de l'Indien coastal section, Port Daniel, Québec, in Pickerill, R. K., ed., Ordovician, Silurian, and Devonian strata of northern New Brunswick and southern Gaspé: Fredericton, New Brunswick, Canadian Paleontology and Biostratigraphy Seminar Guidebook, p. 30–49.

——, 1982, An Upper Silurian reefal limestone platform; From supratidal plain to marginal slope, in Hesse, R., Middleton, G. V., and Rust, B. R., eds., Paleozoic continental margin sedimentation in the Québec Appalachians: Hamilton, Ontario, International Association of Sedimentologists, 11th International Congress, Guidebook to Field Excursion 7B, p. 87–106.

Bourque, P.-A., and Gignac, H., 1983, Sponge-constructed stromatactis mud mounds, Silurian of Gaspé, Québec: Journal of Sedimentary Petrology, v. 53, p. 521–532.

Bourque, P.-A., and Lachambre, G., 1980, Stratigraphie du Silurien et du Devonien basal du Sud de la Gaspésie: Québec Ministére de l'Energie et des Ressources, Direction générale de la Recherche géologique et minérale ES-30, 123 p.

Bourque, P.-A., and Lespérance, P. J., 1977, The Silurian–Devonian boundary in northeastern Gaspé Peninsula, Québec, in Martinsson, A., ed., The Silurian–Devonian boundary: Stuttgart, West Germany, International Union of Geological Sciences, Ser. A, no. 5, p. 245–255.

Bourque, P.-A., Mamet, B. L., and Roux, A., 1981, Algues calcaires siluriennes du synclinorium de la Baie des Chaleurs, Québec, Canada: Revue de Micropaléontologie, v. 24, p. 83–126.

Bourque, P.-A., and 5 others, 1986 Silurian and Lower Devonian reef and carbonate complexes of the Gaspé Basin, Québec; A summary: Bulletin of Canadian Petroleum Geology, v. 34, p. 452–489.

Braithwaite, C.J.R., 1973, Reefs; Just a problem of semantics?: American Association of Petroleum Geologists Bulletin, v. 57, p. 1100–1116.

Briggs, L. I., and Briggs, D., 1974a, Reefs and evaporites in the Michigan Basin: Geotimes, v. 19, no. 8, p. 31.

——, 1974b, Niagara–Salina relationships in the Michigan Basin, in Kesling, R. V., ed., Silurian reef-evaporite relationships: Michigan Basin Geological Society Guidebook, p. 1–30.

Briggs, L. I., Gill, D., Briggs, D. Z., and Elmore, R. D., 1980, Transition from open marine to evaporite deposition in the Silurian Michigan Basin, in Nissenbaum, A., ed., Hypersaline brines and evaporitic environments: Amsterdam, Netherlands, Elsevier Scientific Publishing Company, p. 253–270.

Bristol, H. M., 1974, Silurian pinnacle reefs and related oil production in southern Illinois: Illinois Geological Survey Illinois Petroleum 102, 98 p.

Brown, R. T., 1854, Geological survey of the State of Indiana: Indiana State Board of Agriculture Annual Report 3, p. 299–332.

——, 1876, Geology, topography, etc., illustrated atlas; State of Indiana: Chicago, Illinois, Baskin, Forster and Co., p. 199–200.

Budros, R., and Briggs, L. I., 1977, Depositional environment of Ruff Formation (Upper Silurian) in southeastern Michigan, in Fisher, J. H., Reefs and evaporites; Concepts and depositional models: American Association of Petroleum Geologists Studies in Geology 5, p. 53–71.

Bunker, B. J., Ludvigson, G. A., and Witzke, B. J., 1985, The Plum River Fault Zone and the structural and stratigraphic framework of eastern Iowa: Iowa Geological Survey Technical Information Series 13, 41 p.

Burgess, R. J., and Benson, A. L., 1969, Exploration for Niagaran reefs in northern Michigan: Ontario Petroleum Institute, Inc., Eighth Annual Conference, Paper 1, 30 p.

Caley, J. F., 1940, Paleozoic geology of the Toronto–Hamilton area: Canada Geological Survey Memoir 224, 284 p.

Calvin, S., 1896a, Geology of Jones County: Iowa Geological Survey Annual Report for 1895, v. 5, p. 33–112.

——, 1896b (title unknown): Iowa Academy of Science Proceedings, v. 3. (Article not seen but cited as reported by W. H. Norton, 1899.)

Carman, J. E., 1927, The Monroe Division of rocks in Ohio: Journal of Geology, v. 35, p. 481–506.

Carozzi, A. V., and Zadnick, V. E., 1959, Microfacies of Wabash Reef, Wabash, Indiana: Journal of Sedimentary Petrology, v. 29, p. 164–171.

Carr, D. D., Ault, C. H., and Brown, M. A., 1989, Directory of dimension stone quarries in Indiana: Indiana Geological Survey Directory Series, 15 p.

Cercone, K. R., 1986, Sedimentation in Michigan Basin during earliest Salina; Evidence for an excursion from eustacy [abs.]: American Association of Petroleum Geologists Bulletin, v. 70, p. 1063–1064.

——, 1988, Evaporative sea-level drawdown in the Silurian Michigan Basin: Geology, v. 16, p. 387–390.

Cercone, K. R., and Lohmann, K. C., 1985, Early diagenesis of Middle Silurian pinnacle reefs, northern Michigan, in Cercone, K. R., and Budai, J. M., eds., Ordovician and Silurian rocks of the Michigan Basin and its margins: Michigan Basin Geological Society Special Paper 4, p. 109–117.

Chamberlin, T. C., 1877, Geology of Wisconsin: Wisconsin Geological Survey, v. 2, p. 93–405.

Coburn, G. W., 1986, Silurian of the Illinois Basin; A carbonate ramp: Oil and Gas Journal, October 6, 1986, p. 96–100.

Collett, J., 1872, Geological reconnaissance of Jasper, White, Carroll, Cass, Miami, Wabash, and Howard Counties: Indiana Geological Survey Annual Reports 3 and 4, p. 289–337.

Collinson, C., and 4 others, 1967, Devonian of the north-central region, United States, in International symposium on the Devonian System: Alberta Society of Petroleum Geologists, v. 1, p. 933–939.

Colville, V. R., and Johnson, M. E., 1982, Correlation of sea-level curves for the lower Silurian of the Bruce Peninsula and Lake Temiskaming District (Ontario): Canadian Journal of Earth Sciences, v. 19, p. 962–974.

Copper, P., 1978, Paleoenvironments and paleocommunities in the Ordovician–Silurian sequence of Manitoulin Island: Michigan Basin Geological Society Special Papers 3, p. 47–61.

Coron, C. R., and Textoris, D. A., 1974, Non-calcareous algae in Silurian carbonate mud mound, Indiana: Journal of Sedimentary Petrology, v. 44, p. 1248–1250.

Cox, E. T., 1876, Huntington County: Indiana Geological Survey Annual Report 7, p. 116–133.

Crowley, D. J., 1973, Middle Silurian patch reefs in Gasport Member (Lockport Formation), New York: American Association of Petroleum Geologists Bulletin, v. 57, p. 283–300.

Cumings, E. R., 1922, Nomenclature and description of the geological formations of Indiana, in Logan, W. N., and others, Handbook of Indiana geology: Indiana Department of Conservation Publication 21, p. 403–570.

——, 1930, Silurian reefs near Tiffin, Carey, and Marseilles, Ohio: Indiana Academy of Sciences Proceedings, v. 39, p. 199–204.

——, 1932, Reefs or bioherms?: Geological Society of America Bulletin, v. 43, p. 331–352.

Cumings, E. R., and Shrock, R. R., 1927, The Silurian coral reefs of northern Indiana and their associated strata: Indiana Academy of Science Proceedings, v. 36, p. 71–85.

——, 1928a, The geology of the Silurian rocks of northern Indiana: Indiana Department of Conservation Publication 75, 226 p.

——, 1928b, Niagaran coral reefs of Indiana and adjacent states and their stratigraphic relations: Geological Society of America Bulletin, v. 39, p. 576–620.

D'Armond, D. B., 1980, Thornton Quarry deposits; A fossil coral reef or a catastrophic flood deposit? A preliminary study: Creation Research Society Quarterly, v. 17, p. 88–105.

Dawes, P. R., 1966, Lower Paleozoic geology of the western part of the North Greenland fold belt: Grønlands Geologiske Undersøgelse Rapport, v. 11, p. 11–15.

——, 1971, The North Greenland fold belt and environs: Geological Society of Denmark Bulletin, v. 20, p. 197–239.

——, 1976, Precambrian to Tertiary of northern Greenland, in Escher, A., and Watt, W. S., eds., Geology of Greenland: Grønlands Geologiske Undersøgelse Bulletin, p. 248–303.

Devaney, K. A., Wilkinson, B. H., Van der Roo, R., and McCabe, C., 1983, Silurian reefs and carbonate clinothems; Allochthonous carbonate deposition on inclined marine slopes: Geological Society of America Abstracts with Programs, v. 15, p. 557.

Devaney, K. A., Wilkinson, B. H., and Van der Roo, R., 1986, Deposition and compaction of carbonate clinothems; The Silurian Pipe Creek Junior complex of east-central Indiana: Geological Society of America Bulletin, v. 97, p. 1367–1381.

Droste, J. B., and Shaver, 1976, The Limberlost Dolomite of Indiana; A key to the great Silurian facies in the southern Great Lakes area: Indiana Geological Survey Occasional Paper 15, 21 p.

——, 1977, Synchronization of deposition; Silurian reef-bearing rocks on Wabash Platform with cyclic evaporites of Michigan Basin, in Fisher, J. H., ed., Reefs and evaporites; Concepts and depositional models: American Association of Petroleum Geologists Studies in Geology 5, p. 93–109.

——, 1980, Recognition of buried Silurian reefs in southwestern Indiana: Journal of Geology, v. 88, p. 567–587.

——, 1982, The Salina Group (Middle and Upper Silurian) of Indiana: Indiana Geological Survey Special Report 24, 41 p.

——, 1983, Atlas of early and middle Paleozoic paleogeography of the southern Great Lakes area: Indiana Geological Survey Special Report 32, 32 p.

——, 1985, Comparative stratigraphic framework for Silurian reefs; Michigan Basin to surrounding platforms, in Cercone, K. R., and Budai, J. M., eds., Ordovician and Silurian rocks of the Michigan Basin and its margins: Michigan Basin Geological Society Special Paper 4, p. 73–93.

——, 1987a, Upper Silurian and Lower Devonian stratigraphy of the central Illinois Basin: Indiana Geological Survey Special Report 39, 29 p.

——, 1987b, Paleoceanography of Silurian seaways in the midwestern basins and arches region: Paleoceanography, v. 2, p. 213–227.

Droste, J. B., Janssens, A., Liberty, B. A., and Shaver, R. H., 1975, The mythical Niagaran–Cayugan unconformity, southern and eastern Michigan Basin area: Geological Society of America Abstracts with Programs, v. 7, p. 749–750.

Droste, J. B., Rexroad, C. B., and Shaver, R. H., 1980, The Silurian System in Indiana and environs; A key to regional paleogeography and to reef and evaporite controversies: Geological Society of America Abstracts with Programs, v. 12, p. 224.

Ehlers, G. M., 1973, Stratigraphy of the Niagaran Series of the Northern Peninsula of Michigan: Ann Arbor, University of Michigan Museum of Paleontology Papers on Paleontology 3, 200 p.

Ehlers, G. M., and Kesling, R. V., 1962, Silurian rocks of Michigan and their correlations, in Fisher, J. H., ed., Silurian rocks of the southern Lake Michigan area: Michigan Basin Geological Society Annual Field Conference 1962, p. 1–20.

Ells, G. D., 1958, Notes on the Devonian–Silurian in the subsurface of southwest Michigan: Michigan Geological Survey Progress Report 18, 55 p.

——, 1967, Michigan's Silurian oil and gas pools: Michigan Geological Survey Report of Investigations 2, 49 p.

Elrod, M. N., and Benedict, A. C., 1892, Geology of Wabash County: Indiana Department of Geology and Natural Resources Annual Report 17, p. 192–259.

——, 1902, Niagara Group unconformities in Indiana: Indiana Academy of Science Proceedings for 1901, p. 205–215.

Evans, C. S., 1950, Underground hunting in the Silurian of southwestern Ontario:

Geological Association of Canada Proceedings, v. 3, p. 55–85.

Felber, B. E., 1963, Silurian reefs of southwestern Michigan [Ph.D. thesis]: Evanston, Illinois, Northwestern University, 104 p.

Fincham, W. J., 1975, The Salina Group of the southern part of the Michigan Basin [M.S. thesis]: East Lansing, Michigan State University, 56 p.

Fisher, J. H., 1973, Petroleum occurrence in the Silurian reefs of Michigan: Ontario Petroleum Institute, Inc., Paper 9, 10 p.

Floto, B. A., 1955, The possible presence of buried Niagaran reefs in Ohio and their relation to the Newburg oil and gas zone: Ohio Geological Survey Report of Investigations 7, part 2, p. 41–58.

Ganley, M. C., 1984, Carbonate petrology of a Middle Silurian patch reef near Celina, Ohio (Karch Quarry) [M.S. thesis]: Dayton, Ohio, Wright State University, 182 p.

Gill, D., 1972, Karstic diagenesis in Niagaran Guelph reefs, Michigan, and the origin of stromatactis: Geological Society of American Abstracts with Programs, v. 4, p. 721–722.

——, 1973, Stratigraphy, facies evolution and diagenesis of productive Niagaran–Guelph reefs and Cayugan sabkha deposits, the Belle River Mills gas field, Michigan Basin [Ph.D. thesis]: Ann Arbor, University of Michigan, 275 p.

——, 1975, Cyclic deposition of Silurian carbonates and evaporites in Michigan Basin: American Association Petroleum Geologists Bulletin, v. 59, p. 535–538.

——, 1977a, Salina A-1 sabkha cycles and the Late Silurian paleogeography of the Michigan Basin: Journal of Sedimentary Petrology, v. 47, p. 979–1017.

——, 1977b, The Belle River Mills gas field; Productive Niagaran reefs encased by sabkha deposits, Michigan Basin: Michigan Basin Geological Society Special Paper 2, 187 p.

Gorby, S. S., 1886, The Wabash Arch: Indiana Department of Geology and Natural History Annual Report 15, p. 228–241.

Grabau, A. W., 1901, Guide to the geology and paleontology of Niagara Falls and vicinity: New York State Museum Bulletin 45, 284 p.

——, 1909, Physical and formal evolution of North America during Ordovic, Siluric, and early Devonic time: Journal of Geology, v. 17, p. 209–252.

——, 1913, Principles of stratigraphy: New York, A. G. Seiler and Co., 1184 p.

Grabau, A. W., and Sherzer, W. H., 1910, The Monroe Formation of southern Michigan and adjoining regions: Michigan Geological and Biological Survey Publication 2, Geologic Series 1, 248 p.

Grawberger, D. J., 1978, The Manitowaning Bioherm; An Early Silurian patch reef, in Sanford, J. T., and Mosher, R. E., eds., Geology of the Manitoulin area: Michigan Basin Geological Society Special Papers 3, p. 85–86.

Griest, S. D., and Shaver, R. H., 1982, Geometric and paleoecologic analysis of Silurian reefs near Celina, Ohio: Indiana Academy of Science Proceedings, v. 91, p. 373–390.

Hadley, C. J., 1970, Subsurface studies of reef debris will help reveal the presence and quality of pinnacle reefs in southwestern Ontario: Ontario Petroleum Institute, Inc., 9th Annual Conference, Paper 8, 15 p.

Hall, J., 1843, Geology of New York; Part 4, Survey of the Fourth Geological District: Natural History of New York, 525 p.

——, 1858, Geology of Iowa; General reconnoissance: Iowa Geological Survey, v. 1, pt. 1, Geology, p. 45–146.

——, 1862, Report on the geological survey of the State of Wisconsin: v. 1, 455 p.

Hall, J., and Whitney, J. D., 1858, Report on the geological survey of the State of Iowa: State of Iowa, v. 1, 472 p.; v. 2, p. 473–724.

Heckel, P. H., 1974, Carbonate buildups in the geologic record; A review, in Laporte, L. F., ed., Reefs in time and space, selected examples from the recent and ancient: Tulsa, Oklahoma, Society of Economic Paleontologists and Mineralogists Special Publication 18, p. 90–154.

Hinman, E. E., 1968, A biohermal facies in the Silurian of eastern Iowa: Iowa Geological Survey Report of Investigations 6, 52 p.

Hoffman, A., and Narkiewicz, M., 1977, Developmental pattern of lower to middle Paleozoic banks and reefs: Neues Jahrbuch für Geologie und Paläontologie Monatshefte, v. 5, p. 272–283.

Huh, J.M.S., 1973, Geology and diagenesis of the Salina–Niagaran pinnacle reefs in the northern shelf of the Michigan Basin [Ph.D. thesis]: Ann Arbor, Michigan, University of Michigan, 253 p.

Huh, J. M., Briggs, L. I., and Gill, D., 1977, Depositional environments of pinnacle reefs, Niagara and Salina Groups, northern shelf, Michigan Basin, in Fisher, J. H., ed., Reefs and evaporites; Concepts and depositional models: American Association of Petroleum Geologists Studies in Geology 5, p. 1–21.

Hurst, J. M., 1980, Paleogeographic and stratigraphic differentiation of Silurian carbonate buildups and biostromes of North Greenland: American Association of Petroleum Geologists Bulletin, v. 64, p. 527–548.

Hurst, J. M., and Surlyk, F., 1984, Tectonic control of Silurian carbonate-shelf margin morphology and facies, North Greenland: American Association of Petroleum Geologists Bulletin, v. 68, p. 1–17.

Hutt, R. B., 1974, Geology and hydrocarbon potential of the east shore of Lake Huron: Ontario Petroleum Institute, Inc., 13th Annual Conference, Technical Paper 6, 12 p.

Indiana University Paleontology Seminar, 1976, Constitution, growth, and significance of the Silurian reef complex at Rockford, Ohio: American Association of Petroleum Geologists Bulletin, v. 60, p. 428–451.

——, 1980, Stratigraphy, structure, and zonation of a large Silurian reef at Delphi, Indiana: American Association of Petroleum Geologists Bulletin, v. 64, p. 115–131.

Ingels, J.J.C., 1963, Geometry, paleontology, and petrography of Thornton Reef complex, Silurian of northeastern Illinois: American Association of Petroleum Geologists Bulletin, v. 47, p. 405–440.

Janssens, A., 1974, The evidence for Lockport–Salina facies changes in the subsurface of northwestern Ohio: Michigan Basin Geological Society Field Conference 1974, p. 79–88.

——, 1977, Silurian rocks in the subsurface of northwestern Ohio: Ohio Geological Survey Report of Investigations 100, 96 p.

Jodry, R. L., 1969, Growth and dolomitization of Silurian reefs, St. Clair County, Michigan: American Association of Petroleum Geologists Bulletin, v. 53, p. 957–981.

Johnson, A. M., Kesling, R. V., Lilienthal, R. T., and Sorenson, H. O., 1979, The Maple Block Knoll Reef in the Bush Bay Dolostone (Silurian, Engadine Group), Northern Peninsula of Michigan: University of Michigan Museum of Paleontology Papers on Paleontology 20, 33 p.

Johnson, M. E., 1987, Extent and bathymetry of North American platform seas in the Early Silurian: Paleoceanography, v. 2, p. 185–211.

Johnson, M. E., and Campbell, G. T., 1980, Recurrent carbonate environments in the Lower Silurian of northern Michigan and their inter-regional correlation: Journal of Paleontology, v. 54, p. 1041–1057.

Johnson, N. C., 1981, Geometry, stratigraphy, and paleoecology of a Silurian reef at Bluffton, Indiana [A.M. thesis]: Bloomington, Indiana University, 66 p.

Kahle, C. F., 1974, Nature and significance of Silurian rocks at Maumee Quarry, Ohio, in Kesling, R. V., ed., Silurian reef-evaporite relationships: Michigan Basin Geological Society Guidebook, p. 31–54.

——, 1978, Patch reef development and effects of repeated subaerial exposure in Silurian shelf carbonates, Maumee, Ohio: Michigan Basin Geological Society Guidebook, p. 63–131.

——, 1988, Surface and subsurface paleokarst, Silurian Lockport and Peebles Dolomites, western Ohio, in James N. P., and Choquette, P. W., eds., Paleokarst: New York, Springer-Verlag, Inc., p. 229–255.

Kerr, J. W., 1967, Nares submarine rift valley and the relative rotation of North Greenland: Bulletin of Canadian Petroleum Geology, v. 15, p. 483–520.

Kindle, E. M., 1902, The Niagara limestones of Hamilton Co., Ind.: American Journal of Sciences, ser. 4, v. 14, p. 221–224.

——, 1903, The Niagara domes of northern Indiana: American Journal of Science, ser. 4, v. 15, p. 459–468.

——, 1904, The stratigraphy and paleontology of the Niagara of northern Indiana: Indiana Department of Natural Resources Annual Report 28, p. 397–486.

——, 1913, Niagara folio, New York: U.S Geological Survey Geologic Atlas

Folio 190, 26 p.

Kleffner, M. A., 1988a, Taxonomy and biostratigraphic significance of Wenlockian and Ludlovian (Silurian) conodonts in the midcontinent outcrop area, North America [Ph.D. thesis]: Columbus, Ohio State University, 251 p.

—— , 1988b, Lilley Formation, Lilley-Peebles transition unit, and Peebles Dolomite (Silurian, southern Ohio); Conodont biostratigraphy and chronostratigraphy and depositional history: Geological Society of America Abstracts with Programs, v. 20, 1 p.

Klug, C. R., 1977, The Upper Silurian of Wisconsin, in Nelson, K. G., ed., Geology of southeastern Wisconsin: Milwaukee, University of Wisconsin–Milwaukee, 41st Annual Tri-State Field Conference Guidebook, p. A35–A39.

Koch, L., 1920, Stratigraphy of northwest Greenland: Dansk Geologiske Forening Meddelelser, v. 5, 78 p.

—— , 1925, The geology of North Greenland: American Journal of Science, ser. 5, v. 9, p. 271–285.

—— , 1929, Stratigraphy of Greenland: Meddelelser om Grønland, v. 73, p. 205–320.

Koepke, W. E., and Sanford, B. V., 1966, The Silurian oil and gas fields of southwestern Ontario: Geological Survey of Canada Paper 65-30, 138 p.

Lane, P. D., and Thomas, A. T., 1979, Silurian carbonate mounts in Peary Land, North Greenland: Grønlands Geologiske Undersøgelse Rapport, v. 88, p. 51–54.

LeCompte, M., 1938, Quelques types de "recifs" Siluriens et Devoniens de l'Amerique du Nord: Museum royale histoire naturel Belgique Bulletin, v. 41, no. 19, p. 1–51.

Lee, J., Budros, R., and Byar, G., 1976, Examples of geologic and seismic reef proximity indicators from the Michigan Basin: Ontario Petroleum Institute, Inc., 15th Annual Conference Paper, 52 p.

Lehle, P. F., 1980, Deposition and development of Lockport and Salina (Silurian) rocks at West Milgrove, Ohio [M.S. thesis]: Bowling Green, Ohio, Bowling Green State University, 105 p.

Lehmann, P. J., 1978, Deposition, porosity evolution, and diagenesis of the Pipe Creek Jr. Reef (Silurian), Grant County, Indiana [M.S. thesis]: Madison, University of Wisconsin, 234 p.

Lespérance, P. J., and Bourque, P.-A., 1970, Silurian and basal Devonian stratigraphy of northeastern Gaspé Peninsula, Quebec: American Association of Petroleum Geologists Bulletin, v. 54, p. 1868–1886.

Liberty, B. A., and Bolton, T. E., 1956, Early Silurian stratigraphy of Ontario, Canada: American Association of Petroleum Geologists Bulletin, v. 40, p. 162–173.

—— , 1971, Paleozoic geology of the Bruce peninsula area, Ontario: Canada Geological Survey Memoir 360, 163 p.

Logan, W. E., 1863, Report on the geology of Canada: Geological Survey Canada Report of Progress to 1863, 983 p.

Low, A. P., 1887, Preliminary report on an exploration of country between Lake Winnipeg and Hudson Bay: Canada Geological Survey Annual Report 1886, pt. F, p. 1F–24F.

Lowenstam, H. A., 1948, Biostratigraphic studies of the Niagaran interreef formations in northeastern Illinois: Illinois State Museum of Science Papers, v. 4, p. 1–146.

—— , 1949, Niagaran reefs in Illinois and their relation to oil accumulation: Illinois Geological Survey Report of Investigation 145, 36 p.

—— , 1950, Niagaran reefs of the Great Lakes area: Journal of Geology, v. 58, p. 430–487.

—— , 1957, Niagaran reefs in the Great Lakes area, in Ladd, H. S., ed., Treatise on marine ecology and paleoecology: Geological Society American Memoir 67, v. 2, p. 215–248.

Lowenstam, H. A., and DuBois, E. P., 1946, Marine pool, Madison County; A new type of oil reservoir in Illinois: Illinois Geological Survey Report of Investigations 114, 30 p.

Lyell, C., 1841, Some remarks on the Silurian strata between Aymestry and Wenlock: Geological Society Proceedings, v. 3, p. 463–465.

McGee, W. J., 1891, The Pleistocene history of northeastern Iowa: U.S. Geological Survey Annual Report 11, pt. 1, p. 187–577.

McGovney, J.E.E., 1978, Deposition, porosity evolution, and diagenesis of the Thornton Reef (Silurian), northeastern Illinois [Ph.D. thesis]: Madison, University of Wisconsin, 454 p.

McGovney, J.E.E., Lehmann, P. J., and Sarg, J. F., 1982, Eustatic sea-level control of Silurian (Niagaran) reefs Michigan Basin [abs.]: American Association of Petroleum Geologists Bulletin, v. 66, p. 604.

Mallin, J. W., 1950, The Peebles Formation, a Niagaran reef deposit [M.S. thesis]: Cincinnati, Ohio, University of Cincinnati, 36+ p.

Mantek, W., 1973, Niagaran pinnacle reefs in Michigan: Michigan Basin Geological Society Guidebook, p. 35–46.

Matthews, R. D., and Egleson, G. C., 1974, The origin and implications of a mid-basin potash facies in the Salina salt of Michigan [preprint]: Northern Ohio Geological Society, 68 p.

Mayr, U., 1976, Middle Silurian reefs in southern Peary Land, North Greenland, Bulletin of Canadian Petroleum Geology, v. 24, p. 440–449.

Meloy, D. U., 1974, Depositional history of the Silurian northern carbonate bank of the Michigan Basin [M.S. thesis]: Ann Arbor, University of Michigan, 78 p.

Mesolella, K. J., 1972, Reciprocal deposition within Niagara and early Cayugan (Silurian) carbonates and evaporites, northern Michigan Basin: Ontario Petroleum Institute, Inc., 11th Annual Conference, Paper 8, 34 p.

—— , 1978, Paleogeography of some Silurian and Devonian reef trends, central Appalachian Basin: American Association of Petroleum Geologists Bulletin, v. 62, p. 1607–1644.

Mesolella, K. J., and 3 others, 1974, Cyclic deposition of Silurian carbonates and evaporites in Michigan Basin: American Association of Petroleum Geologists Bulletin, v. 58, p. 34–62.

Mesolella, K. J., and 3 others, 1975, Reply to Comment on 'Cyclic deposition of Silurian carbonates and evaporites in Michigan Basin': American Association of Petroleum Geologists Bulletin, v. 59, p. 538–542.

Michigan Geological Survey Division, 1984, Michigan's oil and gas fields, 1982: Michigan Geological Survey Division Annual Statistical Summary 3, 52 p.

Mielczarek, W., and Copper, P., 1986, Early Silurian (Llandoverian) Leask Point and Charlton Bay bioherms, Manitoulin Island, Ontario, Canada [abs.]: American Association of Petroleum Geologists Bulletin, v. 70, 1 p.

Mikulic, D. G., 1977, Preliminary revision of the Silurian stratigraphy of southeastern Wisconsin, in Nelson, K. G., ed., Geology of southeastern Wisconsin: Milwaukee, University of Wisconsin–Milwaukee, 41st Annual Tri-State Field Conference Guidebook, p. A6–A34.

Murchison, R. I., 1839, The Silurian System: London, Murray, 768 p.

—— , 1847, On the Silurian and associated rocks in Dalecarlia, and on the succession from Lower to Upper Silurian in Smoland, Oland, and Gotland, and in Scania: Geological Society of London Quarterly Journal, v. 3, p. 1–46.

Norford, B. S., 1972, Silurian stratigraphic sections at Kap Tyson, Offley Ø. and Kap Schuchert, northwestern Greenland: Meddelelser om Gronland, v. 195, no. 2, 40 p.

Norton, W. H., 1894, Geology of Linn County: Iowa Geological Survey Annual Report 3, p. 123–195.

—— , 1899, Geology of Scott County: Iowa Geological Survey Annual Report 9, p. 391–519.

—— , 1901, Geology of Cedar County: Iowa Geological Survey Annual Report 11, p. 281–396.

Nurmi, R. D., 1974, The lower Salina (Upper Silurian) stratigraphy in a desiccated deep Michigan Basin: Ontario Petroleum Institute, Inc., Paper 14, 24 p.

Nurmi, R. D., and Friedman, G. M., 1975, Sedimentology and diagenesis of lower Salina Group (Upper Silurian) evaporites in Michigan Basin [abs.]: American Association Petroleum Geologists Bulletin v. 59, p. 1738.

—— , 1977, Sedimentology and depositional environments of basin-center evaporites, lower Salina Group (Upper Siluria), Michigan Basin, in Fisher, J. H., ed., Reefs and evaporites; Concepts and depositional models:

American Association of Petroleum Geologist Studies in Geology 5, p. 23–52.

Orton, E., 1889, The Trenton Limestone as a source of petroleum and inflammable gas in Ohio and Indiana: U.S. Geological Survey Annual Report 8, pt. 2, p. 475–662.

Owen, R., 1862, Report of a geological reconnoissance of Indiana made during the years 1859 and 1860 under the direction of the late David Dale Owen: Indianapolis, H. R. Dodd and Co., 368 p.

Owens, R. N., 1981, Petrologic analysis of the Mississinewa Member of the Wabash Formation and the effect of reef proximity on interreef sedimentation [M.S. thesis]: Muncie, Indiana, Ball State University, 83 p.

Packard, J. J., and Dixon, O. A., 1982, Carbonate buildups associated with a progradational basin-slope succession from the Upper Silurian of Arctic Canada [abs.]: Eleventh International Congress on Sedimentology, p. 117.

Patchen, D., and Smosna, R., 1975, Stratigraphy and petrology of Middle Silurian Mckenzie Formation in West Virginia: American Association of Petroleum Geologists Bulletin, v. 59, p. 2266–2287.

Philcox, M. E., 1970, Coral bioherms in the Hopkinton Formation (Silurian) of Iowa: Geological Society of America Bulletin, v. 81, p. 969–974.

—— , 1972, Burial of reefs by shallow-water carbonates, Silurian, Gower Formation, Iowa, U.S.A.: Geologisches Rundschau, v. 61, p. 686–708.

Phinney, A. J., 1891, The natural gas field of Indiana: U.S. Geological Survey Annual Report 11, pt. 1, p. 589–742.

Pinsak, A. P., and Shaver, R. H., 1964, The Silurian formations of northern Indiana: Indiana Geological Survey Bulletin 32, 87 p.

Pollock, C. A., and Rexroad, C. B., 1973, Conodonts from the Salina Formation and upper part of the Wabash Formation in north-central Indiana: Geologica et Palaeontologica, v. 7, p. 77–92.

Pounder, J. A., 1962, Guelph–Lockport Formation of southwestern Ontario: Ontario Petroleum Institute, Inc., 1st Annual Conference, Paper 5, 29 p.

—— , 1963a, Guelph–Lockport drilling should reveal more reefs: Oil and Gas Journal, v. 61, p. 144–148.

—— , 1963b, Structure, economics play key roles in Guelph–Lockport search: Oil and Gas Journal, v. 61, p. 162–164.

Pray, L. C., 1976, Guidebook for a field trip on the Thornton Reef (Silurian), northeastern Illinois: Kalamazoo, Western Michigan University Department of Geology and Geological Society of America North-Central Section, 47 p.

Prestwich, J., 1840, On the geology of Coalbrookdale: Geological Society Transactions, ser. 2, v. 5, p. 413–495.

Pugh, W. E., ed., 1950, Bibliography of organic reefs, bioherms, and biostromes: Tulsa, Oklahoma, Seismograph Service Corporation of America, 139 p.

Quillian, R. G., 1966, Geologic, reservoir, and production characteristics, Tilbury Field (offshore-Lake Erie): Ontario Petroleum Institute, Inc., Paper 10, 17 p.

Rarick, R. D., 1980, The petroleum industry; Its birth in Pennsylvania and development in Indiana: Indiana Geological Survey Occasional Paper 32, 36 p.

Rexroad, C. B., Noland, A. V., and Pollack, C. A., 1978, Conodonts from the Louisville Limestone and the Wabash Formation (Silurian) in Clark County, Indiana, and Jefferson County, Kentucky: Indiana Geological Survey Special Report 16, 17 p.

Rexroad, C. B., Droste, J. B., and Shaver, R. H., 1985, Conodonts from the Wabash Formation (Ludlovian and Pridolian Series) in northern Indiana: Report of progress: Geological Society of America Abstracts with Programs, v. 17, p. 323.

Rickard, L. V., 1966, Gamma-ray logs and the origin of salt, in Rau, J. L., and Delwig, L. F., eds., Third symposium on salt: Cleveland, Ohio, Northern Ohio Geological Society, Inc., p. 34–39.

—— , 1969, Stratigraphy of the Upper Silurian Salina Group, New York, Pennsylvania, Ohio, Ontario: New York State Museum and Science Service Geological Survey Map and Chart Services 12, 57 p.

—— , 1975, Correlation of the Silurian and Devonian rocks in New York State: New York State Museum and Science Service Map and Chart Series 24, 16 p.

Ross, C. A., 1962, Silurian monograptids from Illinois: Palaeontology, v. 5, p. 59–72.

Ruedemann, R., 1925, Some Silurian (Ontarian) faunas of New York: New York State Museum Bulletin 265, 134 p.

Sanford, B. V., 1969, Silurian of southwestern Ontario: Ontario Petroleum Institute, Inc., Paper 5, 44 p.

Sanford, B. V., Thompson, F. J., and McFall, G. H., 1985, Plate tectonics; A possible controlling mechanism in the development of hydrocarbon traps in southwestern Ontario: Bulletin of Canadian Petroleum Geology, v. 33, p. 52–71.

Sangree, J. B., 1960, Silurian of northern Indiana [Ph.D. thesis]: Evanston, Illinois, Northwestern University, 170 p.

Sarg, J. F., 1982, Off-reef Salina deposition (Silurian), southern Michigan Basin; Implications for reef genesis, in Handford, C. R., Loucks, R. G., and Davies, G. R., eds., Depositional and diagenetic spectra of evaporites; A core workshop: Tulsa, Oklahoma, Society of Economic Paleontologists and Mineralogists Core Workshop 3, p. 354–380.

Sarle, C. J., 1901, Reef structures in Clinton and Niagara strata of western New York: American Geologist, v. 28, p. 282–299.

Savage, T. E., 1926, Silurian rocks of Illinois: Geological Society of America Bulletin, v. 37, p. 513–534.

Savage, T. E., and Van Tuyl, F. M., 1919, Geology and stratigraphy of the area of Paleozoic rocks in the vicinity of Hudson and James Bays: Geological Society of America Bulletin, v. 30, p. 330–378.

Schmalz, R. F., 1969, Deep-water evaporite deposition; A genetic model: American Association of Petroleum Geologists Bulletin, v. 53, p. 776–789.

Schuchert, C., 1943, Stratigraphy of the eastern and central United States: New York, John Wiley and Sons, 1013 p.

Schuchert, C., and Dart, J. D., Stratigraphy of the Port Daniel–Gascons area of southeastern Quebec: Canada Geological Survey Museum Bulletin 44 (Contributions to Canadian Paleontology), p. 35–58.

Schwalb, H. R., 1975, Oil and gas in Butler County, Kentucky: Kentucky Geological Survey Report of Investigations 16, 65 p.

Scovell, J. T., 1897, Geology of Vigo County, Indiana: Indiana Department of Geology and Natural Resources Annual Report 21, p. 509–576.

Seale, G. L., 1985, Relationship of possible Silurian reef trend to middle Paleozoic stratigraphy and structure of the southern Illinois Basin of western Kentucky: Kentucky Geological Survey Thesis Series 3, 63 p.

Sears, S. O., and Lucia, F. J., 1979, Reef growth model for Silurian pinnacle reefs, northern Michigan reef trend: Geology, v. 7, p. 299–302.

—— , 1980, Dolomitization of northern Michigan reefs by brine refluxion and freshwater/seawater mixing: Tulsa, Oklahoma, Society of Economic Paleontologists and Mineralogists Special Publication 28, p. 215–235.

Sepkoski, J., 1968, A Niagaran reef complex near Monon, Indiana: Notre Dame Science Quarterly, November, p. 11–13.

Shaffer, N. R., Ault, C. H, and Carr, D. D., 1982, High-brightness limestones in Indiana: Indiana Academy of Science Proceedings, v. 91, p. 406–418.

Sharma, G. D., 1966, Geology of Peters Reef, St. Clair County, Michigan: American Association of Petroleum Geologists Bulletin, v. 50, p. 327–350.

Shaver, R. H., 1962, Silurian rocks at the southern edge of the Michigan Basin in Indiana, in Fisher, J. H., ed., Silurian rocks of the southern Lake Michigan area: Michigan Basin Geological Society Annual Field Conference, p. 21–29.

—— , 1974a, The Silurian reefs of northern Indiana; Reef and interreef macrofaunas: American Association of Petroleum Geologists Bulletin, v. 58, p. 934–956.

—— , 1974b, The Niagaran (Middle Silurian) macrofauna of northern Indiana; Review, appraisal and inventory: Indiana Academy of Science Proceedings, v. 83, p. 301–315.

—— , 1974c, Structural evolution of northern Indiana during Silurian time, in Kesling, R. V., ed., Silurian reef-evaporite relationships: Michigan Basin Geological Society Field Conference 1974, p. 55–77, 89–109.

—— , 1976, Silurian reefs, interreef facies, and faunal zones of northern Indiana and northwestern Illinois: Kalamazoo, Western Michigan University and North-Central Section Geological Society of America guidebook (Indiana portion), 37 p.

——, 1977, Silurian reef geometry; New dimensions to explore: Journal of Sedimentary petrology, v. 47, p. 1409–1424.

——, 1985, Field trip on Silurian sedimentary geology with special emphasis on the reefs, Great Lakes area: Bloomington, Indiana University, 75 p.

Shaver, R. H., and Sunderman, J. A., 1982, Silurian reefs at Delphi and Pipe Creek Jr Quarry, Indiana, with emphasis on the question of deep vs. shallow water: West Lafayette, Indiana, Geological Society of America North-Central Section and Purdue University Department of Geosciences Field Trip 5 Guidebook, 30 p.

——, 1989, Silurian seascapes; Water depths, reef clinothems, and reef geometry; A critical review of the Silurian reef model: Geological Society of America Bulletin, v. 101, p. 939–951.

Shaver, R. H., and 6 others, 1971, Silurian and Middle Devonian stratigraphy of the Michigan Basin; A view from the southwestn flank, in Forsyth, J. L., ed., Geology of the Lake Erie islands and adjacent shores: Michigan Basin Geological Society Guidebook, p. 37–59.

Shaver, R. H., and 9 others, 1978, The search for a Silurian reef model; Great Lakes area; Indiana Geological Survey Special Report 15, 36 p.

Shaver, R. H. and 6 others, 1983, Silurian reef and interreef strata as responses to a cyclical succession of environments, southern Great Lakes area, in Shaver, R. H., and Sunderman, J. A., eds., Field trips in midwestern Geology: Bloomington, Indiana, Geological Society of America, Indiana Geological Survey, and Indiana University Department of Geology, field trip 12, v. 1, p. 141–196.

Shaver, R. H., and 31 others, 1984, Midwestern basins and arches region correlation chart: Tulsa, Oklahoma, American Association of Petroleum Geologists Correlation of Stratigraphic Units of North America (COSUNA) project, Chart MBA.

Shaver, R. H., Droste, J. B., and Rexroad, C. B., 1985, Cyclicity in Silurian carbonate deposition, southern Great Lakes area, in relation to models for changes in global sea level: Geological Society of America Abstracts with Programs, v. 17, p. 326.

Shaw, E. W., 1937, The Guelph and Eramosa Formations of the Ontario Peninsula: Royal Canadian Institution Transactions, v. 21, p. 317–362.

Sheldon, F. D., 1963, Transgressive marginal lithotopes in Niagaran (Silurian) of northern Michigan Basin: American Association of Petroleum Geologists Bulletin, v. 47, p. 129–149.

Shrock, R. R., 1929, The klintar of the upper Wabash Valley: Journal of Geology, v. 37, p. 17–29.

——, 1939, Wisconsin Silurian bioherms (organic reefs): Geological Society of America Bulletin, v. 50, p. 529–562.

Sloss, L. L., 1947, Environments of limestone deposition: Journal of Sedimentary Petrology, v. 17, p. 109–113.

——, 1953, The significance of evaporites: Journal of Sedimentary Petrology, v. 23, p. 143–161.

——, 1969, Evaporite deposition from layered solutions: American Association of Petroleum Geologists Bulletin, v. 53, p. 776–789.

——, 1977, Reefs and evaporites; A summary, in Fisher, J. H., ed., Reefs and evaporites; Concepts and depositional models: American Association of Petroleum Geologists Studies in Geology 5, p. 189–191.

Smosna, R., and Patchen, D., 1978, Silurian evolution of central Appalachian Basin: American Association of Petroleum Geologists Bulletin, v. 62, p. 2308–2328.

——, 1980, Niagaran bioherms and interbioherm deposits of western West Virginia: American Association of Petroleum Geologists Bulletin, v. 64, p. 629–637.

Smosna, R., Conrad, J. M., and Maxwell, T., 1989, Stratigraphic traps in Silurian Lockport Dolomite of Kentucky: American Association of Petroleum Geologists Bulletin, v. 73, p. 874–886.

Soderman, J. W., and Carozzi, A. V., 1963, Petrography of algal bioherms in Burnt Bluff Group (Silurian), Wisconsin: American Association of Petroleum Geologists Bulletin, v. 47, p. 1682–1708.

Sonnenfeld, P., 1985, Flow regimes in the Silurian of the Michigan Basin, in Cercone, K. R., and Budai, J. M., eds., Ordovician and Silurian rocks of the Michigan Basin and its margins: Michigan Basin Geological Society Special Paper 4, p. 157–171.

Stanton, R. J., Jr., 1967, Factors controlling shape and internal facies of organic carbonate buildups: American Association of Petroleum Geologists Bulletin, v. 51, p. 2462–2467.

Straw, W. T., 1985, Facies and depositional setting of the A-1 carbonate (Silurian) in the Michigan Basin, in Cercone, K. R., and Budai, J. M., eds., Ordovician and Silurian rocks of the Michigan Basin and its margins: Michigan Basin Geological Society Special Paper 4, p. 143–156.

Suchomel, D. M., 1975, Paleoecology and petrology of Pipe Creek Jr Reef (Niagaran–Cayugan), Grant County, Indiana [A.M. thesis]: Bloomington, Indiana University, 40 p.

Sunderman, J. A., and Mathews, G. W., 1975, Silurian reef and interreef environments: Fort Wayne, Indiana, Society of Economic Paleontologists and Mineralogists Great Lakes Section and Indiana University–Purdue University at Fort Wayne, 94 p.

Sunderman, J. A., and 5 others, 1982, Geometry and petrology of the Silurian Shanty Falls Reef near Wabash, Indiana: Geological Society of America Abstracts with Programs, v. 14, p. 289.

Textoris, D. A., and Carozzi, A. V., 1964, Petrography and evolution of Niagaran (Silurian) reefs, Indiana: American Association of Petroleum Geologists Bulletin, v. 48, p. 397–426.

Thompson, M., 1889, The Wabash Arch: Indiana Department of Geology and Natural History Annual Report 16, p. 41–53.

——, 1892, Geological and natural history report of Carroll County: Indiana Department of Geology and Natural Resources Annual Report 17, p. 171–191.

Tollefson, L.J.S., 1978, Paleoenvironmental analysis of the Kokomo and Kenneth Limestone Members of the Salina Formation in the vicinity of Logansport, Indiana [M.S. thesis]: Urbana, University of Illinois, 173 p.

Tomassetti, J. A., 1981, Geology of Lockport (Silurian) rocks at the Ohio Lime Company quarry, Woodville, Ohio [M.S. thesis]: Bowling Green State University, 72 p.

Twenhofel, W. H., 1927, Geology of Anticosti Island: Canada Geological Survey Memoir 154, 481 p.

Ulteig, J. R., 1964, Upper Niagaran and Cayugan stratigraphy of northeastern Ohio and adjacent areas: Ohio Geological Survey Report of Investigations 51, 48 p.

Vail, P. R., Mitchum, R. M., Jr., and Thompson, S., III, 1977, Seismic stratigraphy and global changes of sea level; 4, Global cycles of relative changes of sea level, in Payton, C. E., ed., Seismic straigraphy; Applications to hydrocarbon exploration: Tulsa, Oklahoma, American Association of Petroleum Geologists Memoir 26, p. 83–97.

Wahlman, G. P., 1974, Stratigraphy, structure, paleontology, and paleoecology of the Silurian reef at Montpelier, Indiana [A.M. thesis]: Bloomington, Indiana University, 71 p.

Whitaker, S. T., 1988, Ramp-platform model for Silurian pinnacle reef distribution in the Illinois Basin: Oil and Gas Journal, May, p. 102–108.

White, C. A., 1870, Report of the Geological Survey of the State of Iowa: Geology of Iowa, v. 1, 381 p.

Whiteman, S. K., and Gardner, J. V., 1975, Near-neighbor analysis of Lagro bioherms; A first approximation, in Sunderman, J. A., and Mathews, G. W., eds., Silurian reef and intereef environments: Fort Wayne, Indiana, Society of Economic Paleontologists and Mineralogists Great Lakes Section and Indiana University–Purdue University at Fort Wayne, p. 84–90.

Wilkinson, B. R., 1982, Carbonate facies and petrology: Bloomington, Indiana University Department of Geology Colloquium and Sohio Sedimentology Seminar lectures.

Williams, M. Y., 1919, The Silurian geology and faunas of Ontario Peninsula, and Manitoulin and adjacent islands: Canada Geological Survey Memoir 111, 195 p.

Willman, H. B., 1971, Summary of the geology of the Chicago area: Illinois Geological Survey Circular 460, 77 p.

—— , 1973, Rock stratigraphy of the Silurian System in northeastern and northwestern Illinois: Illinois Geological Survey Circular 479, 55 p.

Willman, H. B., and 7 others, 1975, Handbook of Illinois stratigraphy: Illinois Geological Survey Bulletin 95, 261 p.

Winchell, A. N., 1873, Geology of Seneca and Wyandot Counties: Ohio Geological Survey, v. 1, pt. 1, Geology, p. 611–639.

Winzeler, T. J., 1974, Petrology and evaluation of Silurian reef and associated rocks, Buckland Quarry, Ohio [M.S. thesis]: Bowling Green, Ohio, Bowling Green State University, 104 p.

Witzke, B. J., 1981a, Stratigraphy, depositional environments, and diagenesis of the eastern Iowa Silurian sequence [Ph.D. thesis]: Iowa City, University of Iowa, 574 p.

—— , 1981b, Silurian stratigraphy of eastern Linn and western Jones Counties, Iowa: Iowa Geological Survey Guidebook 35, 37 p.

—— , 1983, Silurian benthic invertebrate associations of eastern Iowa and their paleoenvironmental significance: Wisconsin Academy of Sciences, Arts, and Letters, v. 1, pt. 1, p. 21–47.

Worthen, A. H., 1862, Remarks on the age of the so-called "Leclaire Limestone" and "Onondaga Salt Group of the Iowa report: American Journal of Science, ser. 2, v. 33, p. 46–48.

—— , 1866, Devonian and Silurian Systems: Illinois Geological Survey, v. 1, Geology, p. 119–152.

Yoder, G. E., 1982, Stratigraphy, structure and paleoecology of a Silurian reef at Francesville, Indiana [A.M. thesis]: Bloomington, Indiana University, 113 p.

Zenger, D. H., 1965, Stratigraphy of the Lockport Formation (Middle Silurian) in New York State: New York State Museum and Science Service Bulletin 404, 210 p.

MANUSCRIPT ACCEPTED BY THE SOCIETY JUNE 1, 1990

Printed in U.S.A.

# The Salina evaporites in the Michigan Basin

**Peter Sonnenfeld and Ihsan Al-Aasm**
*Deparment of Geology, University of Windsor, Windsor, Ontario N9B 3P4, Canada*

## ABSTRACT

Evaporitic conditions in the Michigan Basin commenced in Late Ordovician time and continued into the Early Devonian. Five major evaporite (anhydrite/halite) cycles in the Silurian Salina Group (termed "A-1," "A-2," "B," "D," and "F") can be recognized. They are further subdivided into two or more cycles of evaporite deposition by intercalations of less soluble members. Only the lowermost Salina evaporite produced a sylvinite deposit. The water surface was enlarged in subsequent evaporite cycles, which did not go beyond halite saturation, suggesting an enlargement of the water supply. Additional seawater came from the Kokomo Sea in Indiana; continental waters likely also entered at times from the Moose River Basin. The Chatham sag eventually fed Michigan Basin brines into Ohio and Pennsylvania. A progressive overstepping of one evaporite unit over the one below indicates either subsidence affecting a wider area of rising sea level, the latter suggested by the inundation of northern Ohio beginning with the "B" unit of salt deposition. In no case does brine depth appear to have exceeded some tens of meters. The present morphology of reefs is dictated as much by compaction of surrounding micrites as by renewed growth on older reef mounds where rates of subsidence remained moderate. The amount of clay influx in the upper part of the Salina suggests that humid periods became more frequent at the expense of the duration of dryer periods. Post-Salina rocks initially contain numerous short-lived evaporite cycles, and then grade to open-marine sediments, deposited before the Michigan Basin became emergent land in the late Paleozoic

## INTRODUCTION

Of the Salina evaporites, only the basal unit has been extensively discussed in various publications. Much of the material is found in Master's theses that are not readily accessible, or in company files. It is, therefore, appropriate to attempt a synthesis of what is known about the Salina evaporites in the light of present knowledge about other evaporite basins.

## PALEOGEOGRAPHIC SETTING

The outline of the Michigan Basin dates back to the late Proterozoic. Its boundaries were defined by the Kankakee arch in the (present) southwest, Wisconsin dome in the northwest, Frontenac arch in the northeast, Algonquin arch in the east, and Findlay arch in the southeast (Pirtle, 1932; Prouty, 1986; Fig. 1). Since there are no sediments preserved for more than 300 million years prior to the Cambrian transgression, it is not possible to date the start of the downwarp of the Michigan Basin.

At the onset of the Cambrian–Lower Ordovician cycle, the Fraserdale, Algonquin, and Frontenac arches were undergoing active uplift, as evidenced by the substantial thickness of siliciclastics that accumulated in immediately adjacent embayments (Sanford and others, 1984). Cambrian sediments indicate an ovate configuration of the Michigan Basin with intermittent spurts of subsidence (Catacosinos, 1973; Fisher and others, 1988); this configuration was maintained for the remainder of the Paleozoic with only minor shifts in the depocenter.

To the southwest, in Early Silurian time, the Kokomo Sea covered much of Indiana; to the southeast, a mud flat extended initially into Pennsylvania and New York; and to the north, there was land until mid-Silurian time.

Sonnenfeld, P., and Al-Aasm, I., 1991, The Salina evaporites in the Michigan Basin, *in* Catacosinos, P. A., and Daniels, P. A., Jr., eds., Early sedimentary evolution of the Michigan Basin: Geological Society of America Special Paper 256.

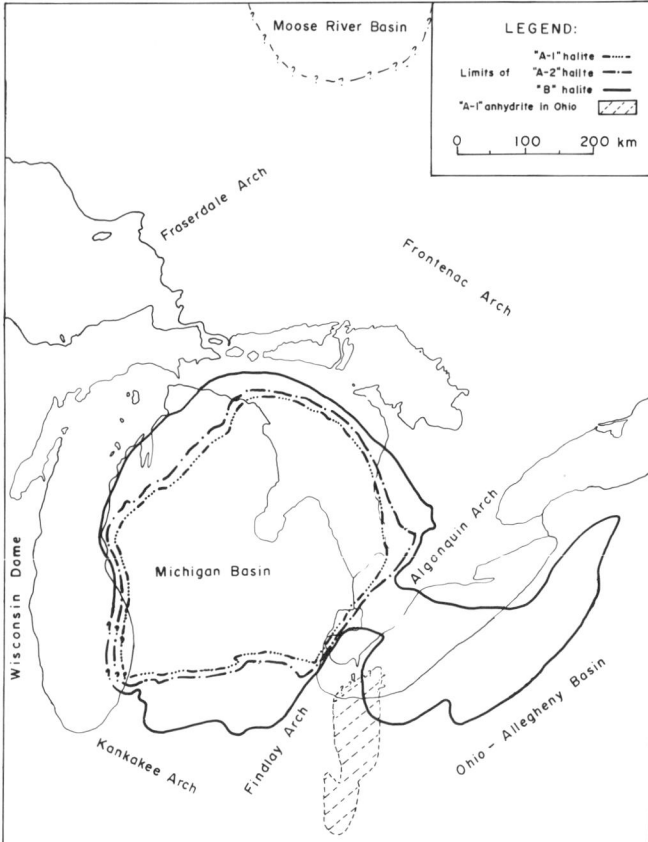

Figure 1. Overstepping of "A-1," "A-2," and "B" halites in the Michigan Basin (after Sanford, 1969; Tremper, 1973; Fincham, 1975; Mesolella, 1978; and others).

The Michigan Basin was never within the strictly equatorial belt sensu stricto, but only in low latitudes marked by warm, subtropical weather with alternating yearly wet and dry periods. During the early Paleozoic, it was situated in the southern hemisphere, slowly moving closer to the equator and reaching about 25°S in Middle Silurian time (Van der Voo, 1988) and a little under 20°S toward the end of the Salina deposition (Droste and Shaver, 1983). Because of the Coriolis effect, any surface currents would therefore be directed against the coast of its left side and then circulate clockwise, and any bottom currents counterclockwise. The direction of prevailing trade winds would have been from the (paleo-) southeast (present northeast), assisting this deflection of surface currents against the western coast.

The climate became more arid in Late Cambrian time, and an evaporative water deficit developed. The earliest anhydrite crusts and lenses occur in inadequately dated Cambro-Ordovician sediments assigned to the Munising Group (Catacosinos, 1973), indicating that in peritidal sediments the evaporative water deficit was taking its toll, concentrating the brine and the groundwater to gypsum saturation. Because basins with water deficit are prone to develop a nutrient-rich surface inflow, reefs flourished in this environment in the Ordovician. Locally, some anhydrites continue to form interbedded with these reefal carbonates. No evaporite other than anhydrite (originally gypsum) is found in rocks of Cambrian to Early Silurian (pre-Wenlockian) rocks.

Carbonate banks and reefs also fringed the Michigan Basin throughout Early to Middle Silurian time. Eventually, a chain of patch reefs of Niagaran age was established offshore, roughly parallel to the basin margin, indicating that even here water depth remained well within the photic zone. Basinward, pre-Salina carbonates gently thin without developing reefs. The Ordovician and early to mid-Silurian Michigan Basin thus subsided very gradually, and the sea remained shallow.

## THE SALINA GROUP

In mid-Wenlockian time (the base of the Salina Group) the rate of subsidence in the Michigan Basin began to fluctuate, with periods of rapid downwarp alternating with slow subsidence. Greatest downwarping of the Michigan Basin coincided with deposition of the Silurian Salina and Bass Islands Groups (Fig. 2), and the Devonian Detroit River Group (Cohee and Landes, 1958). Each interval of more rapid subsidence is marked by halite precipitation in the center of the basin skirted by anhydrite toward the margins, each phase of lesser rate of subsidence by a carbonate bank or carbonate intercalation.

Terminology of the individual units (Fig. 2) was first established by Landes (1945) and later amended by Evans (1950) and Gill (1977). Starting with an "A" unit containing two evaporite cycles "A-1" and "A-2" and three carbonate interbeds "A-0," "A-1," and "A-2," the next unit "B" is again an evaporite unit that can be subdivided into several evaporite cycles. A "C" unit of mainly dolomitic shales separates these evaporitic cycles from a "D" unit of at least two cycles of evaporite deposition, capped by an "E" unit of shales and dolomites. The "F" unit represents again several evaporite cycles and is capped by a "G" unit composed mainly of shales, but containing locally minor halite. An evaporitic unit "H" containing dolomite with some anhydrite and salt is now renamed the Bass Islands Group.

### The "A-1" unit

The first cycle of evaporite precipitation, generally known as the "A-1" unit, is essentially restricted to the Michigan Basin and is composed of a lower "A-1" evaporite and an upper "A-1" carbonate unit. It overlies a carbonate unit of Wenlockian age composed of Niagaran barrier and patch reefs as well as an off-reef facies.

A basal carbonate underlies the "A-1" evaporite (now referred to as the "A-0" carbonate or Cain Formation; Gill, 1977, 1979). This "A-0" carbonate is composed of blackish gray mudstones, containing 0.7 to 7 ppt bituminous matter and as many as five cycles of calcite/anhydrite varvites alternating with halite. The salinity of bottom waters in the basin apparently began to oscillate with a wider amplitude, repeatedly reaching halite saturation. The "A-0" carbonate is underlain by a conglomerate largely composed of reef fragments derived from bioherms circling the margin. The transition from underlying carbonates to the "A-1" evaporites is gradational over 1.5 to 3 m (Mesolella, 1972).

From anhydrites that pinch out toward the margins against reef flanks the "A-1" evaporite grades to about 145 m of halite in the basin center. This halite extends to the eastern shore of Lake Huron (Sanford, 1969), but does not extend into Ohio. The "A-1" evaporite (as anhydrite) is absent or extremely thin over reefs, and only the subsequent "A-1" carbonate drapes over them. It is the only evaporite cycle in which the concentration of the brine reached saturation for sylvite, suggesting that the inflow restriction was greatest and thus the water exchange at the entrance became more restricted than in any of the subsequent evaporite cycles.

The evaporites are, in turn, capped by about 15 to 20 m of a dark grayish brown, fetid carbonate mudstone, often with bituminous partings, that doubles in thickness within the pinnacle reef zone. These "A-1" carbonates are also referred to as the Ruff Formation (Budros, 1974; Budros and Briggs, 1977). Mesolella (1972) wondered whether they do not represent an interreef facies between rejuvenated algal reefs, as they directly abut against the reefs (Mantek, 1973). The topography then was very flat, with pinnacles protruding only about 3 m above the surrounding interreef area (Mesolella, 1972). Thin anhydrite bands within the "A-1" carbonate in the vicinity of the southern pinnacle reef belt have long been referred to as "rabbit ears" (Gill, 1973; Budros and Briggs, 1977; Lilienthal, 1978).

*The "A-2" unit*

The "A-2" unit is also divided into a lower evaporite and an upper carbonate unit. There must have been a larger basin entrance in this cycle of evaporite deposition, as the brine reached only halite, but not sylvite saturation. The "A-2" evaporite is composed of more than 145 m of halite in the center of the basin and grades laterally to anhydrite, eventually pinching out (Lilienthal, 1978). Where the halite laterally changes to anhydrite, the "A-2" evaporite thins markedly, and this facies boundary coincides approximately with the limit of dolomitization in the "A-1" unit (Tremper, 1973). The facies change and concurrent thinning likely marks the hinge line between shelf of moderate subsidence and basin of accelerated subsidence.

*The "B" unit*

The "B" unit is primarily an evaporite formation. It thins from 145 m in the center of the basin to less than 15 m on the margins, mainly due to the pinching out and changing to anhydrite of a lower halite member. None of the evaporite beds extends to the southern margin of the basin, where the unit is entirely composed of shales and dolomites.

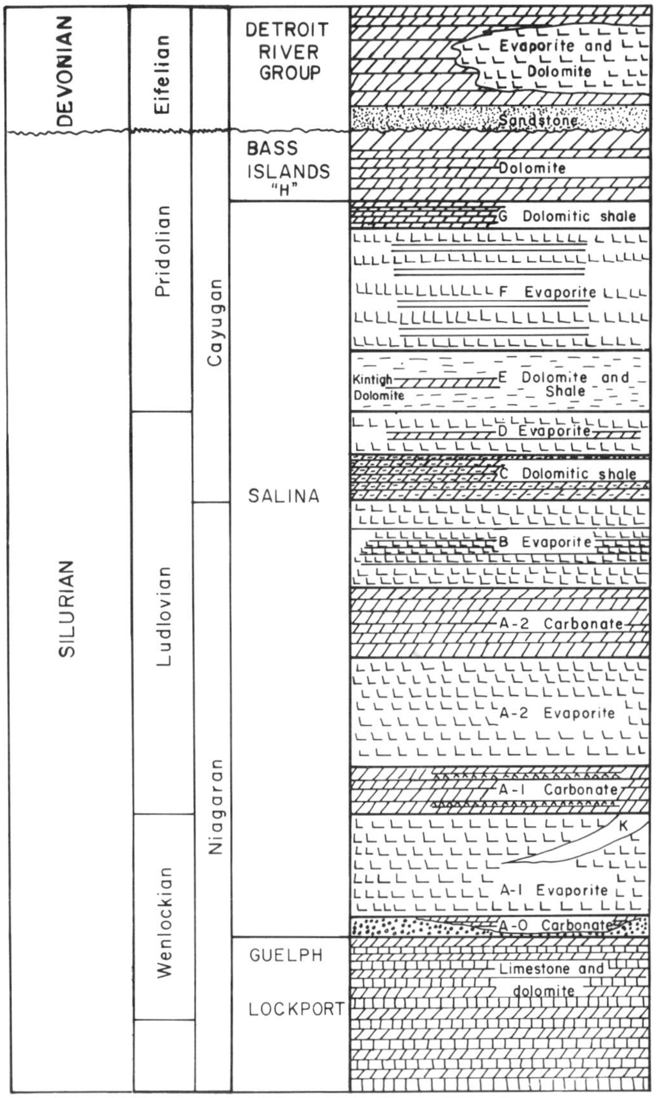

Figure 2. The stratigraphic sequence of mid-Paleozoic evaporites in the Michigan Basin (according to the Correlation Chart, Michigan Department of Conservation, 1964).

*The "C" unit*

The "C" unit is composed of gray dolomitic shale grading to argillaceous dolomite and is as much as 36 m thick. In the northern part of the basin, it thickens, wherever the "B" salt locally is thinner (Tremper, 1973). The "C" unit is present in the Michigan, Ohio, and Allegheny Basins (Lilienthal, 1978) and thus forms an excellent correlation marker into Ohio and Indiana Basins (Droste and Shaver, 1985). In its upper part it contains rounded, frosted quartz grains of apparently eolian origin, which are used as a correlation marker.

*The "D" unit*

The "D" evaporite is about 12 m thick, has the smallest areal extent in the basin center, and is split in two by a thin dolomite intercalation (Sanford, 1969). As such it continues into Ohio, and northern and western Pennsylvania (Mesolella, 1978). Toward the east and south it contains three halite units (Jacoby, 1969). Along the basin margin the halites change to thinner anhydrite and some shale (Tremper, 1973).

### The "E" unit

This unit comprises as much as 36 m of red and greenish gray shale interbedded with thin stringers of dolomite in its lower part, containing an occasional anhydrite stringer (Tremper, 1973). On the western side of the basin it also contains a very porous dolomite, the "Kintigh zone" (Burns, 1962).

### The "F" unit

The "F" unit thins southwestward from 290 m in the center of the basin; it is composed of as many as six salt cycles separated by shales, dolomites, and anhydrites. The lower salts are more widespread; the upper salts recede into the Michigan Basin and are replaced on the margins by shales. The upper contact is conformable. Dellwig and Evans (1969) observed truncation at the crests of minor folds in the "F" salt and at least two local unconformities within the salt sequence of the Detroit mine. Tremper (1973) interpreted the abrupt thinning of the salt against the basin margin as caused by redissolution.

### The "G" unit

The "G" unit comprises as much as 20 m of a sheet-like, dark gray, dolomitic shale with occasional anhydrite stringers (Tremper, 1973). The unit thins and its dolomite content increases toward the basin margins.

### Bass Islands Group

The Bass Islands Group, formerly referred to as unit "H," is the uppermost Silurian unit, composed of light-colored dolomites containing several thin anhydrite and halite beds in the center of the basin.

## DEPOSITIONAL HISTORY OF THE SALINA GROUP

The "A-1" evaporite precipitated in the deeper parts of the basin, but did not cover protruding marginal reefs. Their tops remained in surface inflow of lesser salinity. Neither the "A-1" nor the "A-2" halites extend into Ohio. Only a thin intercalation of "A-1" anhydrite is found within a carbonate sequence in western Ohio (Mesolella, 1978; Fig. 1). However, Janssens (1977) recognized in western Ohio both a 3- to 11-m-thick "A-1" anhydrite and a 1.5- to 10.5-m-thick "A-2" anhydrite. The anhydrites abut against subtidal and supratidal carbonates covering the crest of the Findlay arch (Droste and Shaver, 1987) and appear to have been deposited in a small local basin that had no connection to the Michigan Basin. The "A-1" carbonate, composed of limestone in the basin center, and of dolomite all around the basin margin, covers the "A-1" evaporite. It signals a temporary freshening of the brine.

Unlike the "A-1" evaporites, the "A-2" evaporites drape over pinnacle reefs. Here they generally consist of anhydrite and are substantially thinner than away from the reefs. Basinward, the "A-2" anhydrite encloses a 145-m-thick halite. The "A-2" evaporite is capped by the "A-2" carbonate, which is as much as 45 m thick, but increases to 84 m near pinnacle reefs. Like the "A-1" carbonate it is also more dolomitized along the basin rims; the dolomitized north rim contains a large lens of oolites, suggesting strong currents (Tremper, 1973). The "A-2" carbonate is rich in carbonaceous material in its base, and contains some anhydrite toward the basin center. Beyond the northern limits of the "A-2" evaporite, the "A-2" carbonate directly overlies the "A-1" carbonate. One or more surfaces of subaerial exposure occur within the "A-1" carbonate, but the "A-2" carbonate was also briefly exposed subaerially, before deposition of the "B" unit commenced (Bay, 1983).

The "B" salt unit oversteps the limits of the "A-1" and "A-2" units in all directions (Mesolella and others, 1974). The "B" salt is the first unit that transgresses into Ohio, Pennsylvania, and New York (Mesolella, 1978). Brines precipitating the "B" salt were able to spread through the Chatham sag, a breach in the Algonquin arch, but generally missed the area that had been the site of "A-1" anhydrite deposition (Fig. 1). The halite section above a local unconformity within the "B" salt overlaps, progressively older beds in the Allegheny Basin (Dellwig and Evans, 1969). These transgressions would not be compatible with any lowering of sea level during "B" salt deposition.

More than doubling the water surface (Droste and Shaver, 1987) during the transgression of unit "B" increased the water deficit correspondingly. If, nevertheless, no potash saturation is encountered, it does not imply that the amount of runoff and frequency of rains doubled, lowering the overall water deficit. There is no evidence of a significant increase in frequency of mud-laden flash floods. Instead, it merely suggests that the water exchange through the entrance strait increased. That the "B" unit represented a deeper brine column is indicated by its apparent cooler brine temperatures compared to those in the A-units (Miles and others, 1985); based on fluid inclusion studies, the halite precipitation in the A-units fluctuated between about 32 and 48°C (Dellwig, 1955).

The thickness of the "A-2" carbonate unit indicates massive dissolution of older halites along the northern margins of the Michigan Basin during "B" salt deposition, while such dissolution in the southern part of the basin had occurred earlier during the deposition of underlying carbonates (Mesolella and others, 1974). Thicknesses of the "B" unit in Michigan are greater where the "A-2" evaporite is thin. Along the north rim, a dolomitic marl or shaly dolomite overlying the halite is about 10 m thick, but thickens where the halite thins (Tremper, 1973). This capping

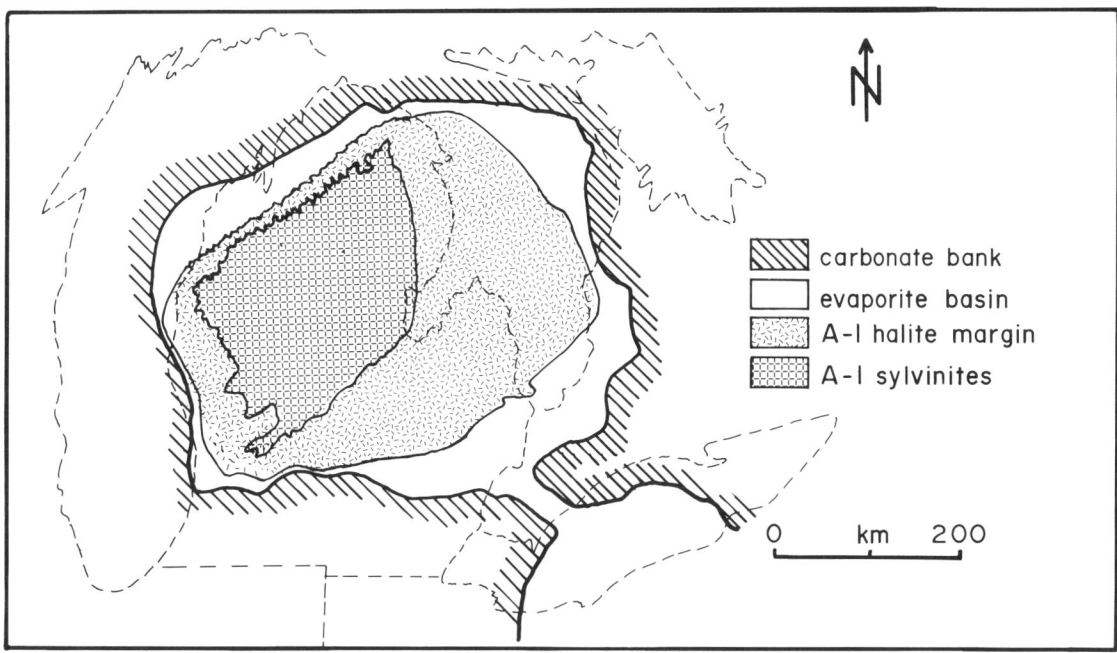

Figure 3. Distribution of "A-1" sylvinite (from Sonnenfeld, 1985).

dolomitic marl apparently has filled in after some halite dissolution.

In the Ohio and Allegheny Basins, the "B" unit contains three distinct major halite-bearing units, of which the middle one alone can be subdivided into five halite cycles (Jacoby, 1969). Crossbedding, some 30 to 60 cm thick, has been observed in halite in the Retsof mine, dipping south, away from the Allegheny Basin margin (Dellwig and Evans, 1969). This indicates a brine current swinging back and forth, i.e., oscillating in its strength, possibly with sea level oscillating through the seasons.

Starting with the "B," "D," and "F" salts and continuing into the evaporite cycles of the Detroit River Group, a small amount of offlap is noticeable on the southwestern margin of the basin (Fincham, 1975; Johnson and Gonzales, 1978). This tilting process is possibly due to more rapid subsidence in the north or to late pulses in a rise of the surrounding Kankakee and Algonquin arches (Ekblaw, 1938).

However, in the Allegheny Basin, salt sequences expanded progressively southeastward over previous shoal areas. At the same time, the thickest section or site of most rapid subsidence progressively migrated eastward. The "D" salt overstepped the "B" salt, and the "F" salt overstepped the "D" salt in all directions, especially flooding red beds in southern Pennsylvania and northern West Virginia (Mesolella, 1978). Each halite unit thus transgressed over a previously established basin margin.

Salts of the "B," "D," and "F" units covered only the northeasternmost portion of the Allegheny Basin. Contrary to Alling and Briggs (1961), the Ohio and Allegheny Basins were one broad, featureless mud flat during most of early Salina deposition, and depressions developed only in upper Salina (Cayugan) time (Treesh, 1973; Mesolella, 1978).

This apparent expansion of the original evaporite basin, i.e., the subsequent onset of more subsidence of adjacent lowlands, is typical of other evaporite basins worldwide. The only other example to be mentioned here is the expansion of the Castile Sea into the Salado Sea in the Permian Basin of West Texas (Sonnenfeld, 1984).

*The "A-1" sylvinite*

The "A-1" sylvinite (a mixture of halite and sylvite) thickens rapidly and coalesces into one thick bed in and around the northwestern reef chain (Fig. 3); it evidently precipitated on the flank of the basin (Sonnenfeld, 1985). The bromide content suggests precipitation during considerable cooling (Anderson and Egleson, 1970), when potassium-rich brines become lighter and move up the basin flanks (Sonnenfield, 1984). The expectation of finding the thickest sylvinite section in the center of the basin (Matthews, 1970; Matthews and Egleson, 1974) has not been fulfilled by later drilling. Instead it thickens toward the reef chain in the northwestern part of the basin, with individual sylvinite intercalations coalescing into a single member. Although the sylvinite unit reaches in excess of 17 m in thickness in interreef areas, it is not present beyond the reef chain. In analogy to other occurrences it may be that the sylvite originally did precipitate to within sight of the shores (compare, e.g., the Upper Kama sylvinites in Fig. 5-4 of Sonnenfeld, 1984). It is now absent because of dissolution either by contemporaneous rainwash or flash floods or, more likely, by abutting against an oscillating interface to surface currents of lower salinity. These surface currents were circling the basin in response to the Coriolis effect and were delivering nutrients to the reef tops.

The sylvinites are encased in halites toward the center of the basin, but abut against the overlying "A-1" carbonate along the northwestern basin margin (Tremper, 1973). This corroborates the idea that they, together with associated halites, were subject to redissolution near the shores. Indeed, an unconformity between "A-1" anhydrites and "A-1" carbonates and redissolution of "A-1" halite has been noted in some marginal areas (Tremper, 1973).

Near Bay City, i.e., away from the main sylvite occurrence, formation waters of specific gravity 1.458 contain as much as 642,798 ppm of solids, mainly $CaCl_2$, negligible NaCl, but significant amounts of $K_2Cl_2$ and $MgCl_2$ in nearly equal molar proportions (Case, 1945). The precipitate in an open brine would normally be a bed of rapidly nucleating carnallite. However, only some scattered crystals, as well as pitted and rounded grains of carnallite, have been identified in halite crystals (Lucas, 1954). All potash seams are composed of a mixture of sylvite and halite (sylvinite). This suggests a deposition in brine covered by an influx that continued to nourish reef tops. The cited water analysis reports the presence of nitrogen hydrides, which are derivatives of proteins. These easily prevent $MgCl_2$ from precipitating and thus foster slower sylvite crystallization. A substantial enrichment in ammonia has also been reported in waters of the overlying Sylvania aquifer (Egleson and Querio, 1969) that may have leaked from underlying beds.

Similarly, the substantial amount of iron in the ferrous state, reported in these formation waters, probably occurred originally in the form of bivalent Fe-organic complexes, which are mostly derivatives of chlorophyll or hemoglobin, because the pH value decreased when the brine was heated, and organic compounds hydrolized. No ferric oxide as hematite platelets or needles in red sylvite has been found. Both the nitrogen hydrides and the soluble iron are unstable in a brine exposed to the atmosphere; their presence thus suggests a brine covered by less saline surface waters (Sonnenfeld, 1984).

While some pyrite has been found in associated halite on the basin margins, toward the central basin the halite contains hematite (Dellwig, 1955), and so does the underlying carbonate (Mesolella and others, 1974). The pyrite is likely due to a bacterial destruction of goethite (Mesolella, 1972), while the hematite is a dehydration product after goethite. Inasmuch as oxygen solubility in a NaCl-saturated brine is negligible, this hematite-bearing halite cannot be a primary deposit. Halite containing iron oxides is not common; it acquires iron oxides either by the brine leaching nearby clays or as a residue after the leaching of hematite-bearing carnallitite. On the basis of available rock samples, it cannot be established whether the hematite-bearing halite is such a reprecipitated leachate after red carnallitite (carnallite in a halite matrix). No Br/Cl studies have been undertaken to determine whether the hematite-bearing halite has shed some of its original bromine content.

Within the "A-1" evaporite, smaller sylvite grains are often almost spherical, larger ones are amoeboid, and halite-sylvite contacts are concave (Nurmi and Friedman, 1977). This suggests some synsedimentary etching of sylvite cubes. Most sylvite grains are milky white, but some are reddish white. Studies are needed to determine whether the red-stained sylvites contain hematite needles inside the crystals or merely are hematite smears along already formed crystal faces, an important clue to their origin (Sonnenfeld, 1990a; Sonnenfeld and Perthuisot, 1989).

With a water surface progressively enlarged in subsequent evaporite cycles, the water deficit must have increased. None of these cycles, however, leads to potash deposition, suggesting that the inflow strait of the basin had widened sufficiently to allow only concentration to halite saturation. At the same time, rates of evaporation may also have gradually declined, as evidenced by the increasing frequency of intercalations of terrigenous material in the upper Salina sequence.

### Site of entrance strait

At this time it is possible to speculate about the location of the entrance to the Silurian Michigan Basin (Fig. 4) and with it the provenance of seawater inflow. A narrow connecting channel across the Frontenac arch, called the Georgian or Sudbury sag (cf., Briggs and others, 1980), analogous to the Chatham sag across the Algonquin arch, was detected in a carbonate lithofacies study (Briggs, 1958). It showed a finger-shaped maximum of the carbonate fraction in the evaporite sequence and, consequently, an influx from the northeast through the present Georgian Bay. Since other evaporite basins show basin growth by subsidence of distal shelves, it would have been suggestive to look for the influx into the basin along the north rim and consider the Ohio Basin as the distal part. And, indeed, Briggs and others (1980) suggested one source of Michigan Basin brines to come through the Sudbury sag.

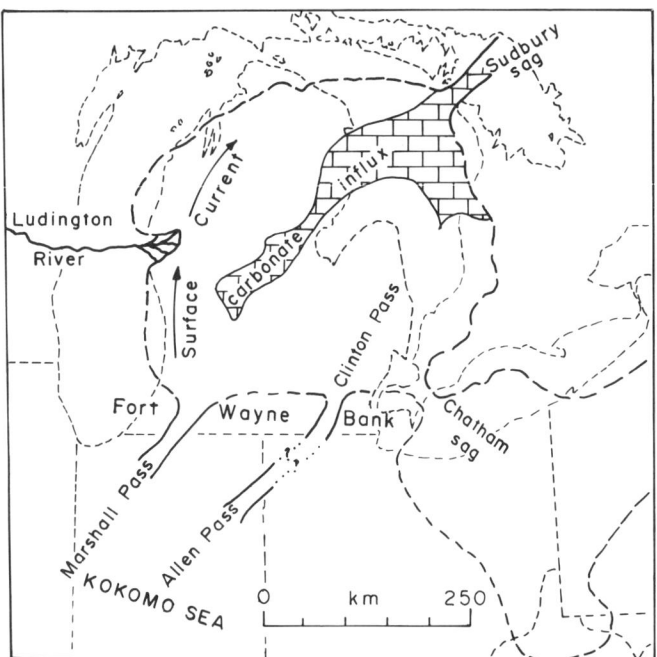

Figure 4. Silurian paleogeography of the Michigan Basin (after Briggs, 1958; Briggs and others, 1980; Droste and Shaver, 1983, 1987).

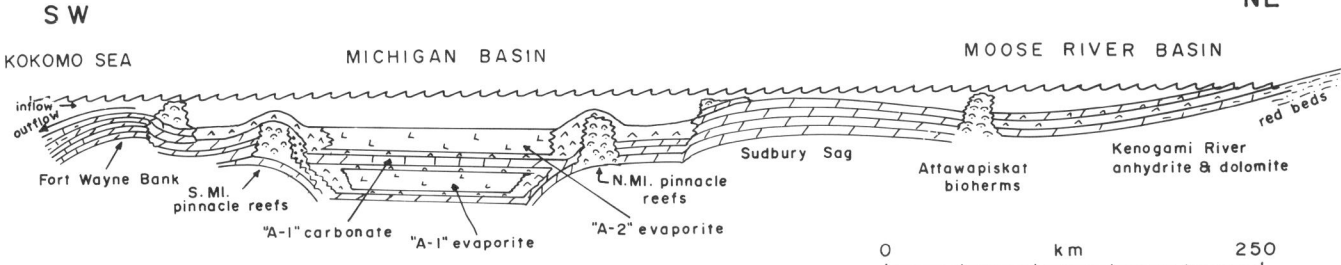

Figure 5. Schematic cross section through Michigan and Moose River Basins.

To the northeast of the Michigan Basin lay the Moose River Basin (Fig. 5), itself connected to the Hudson's Bay Basin, which was land until inundated by a mid-Silurian transgression (Sanford and others, 1968; Norris and Sanford, 1969; Dingwall, 1986). Both depressions contained, at their southwestern margins, algal and stromatoporoidal bioherms of the Attawapiskat Formation, which are equivalent to the Niagaran reefs encircling the Michigan Basin. This formation rests on limestones containing some dolomite and anhydrite.

The bioherms, in turn, are overlain by thick-bedded, finely crystalline, brown dolomites of the lower Kenogami River Formation that contain white anhydrite beds. The middle part of this formation is represented by evaporitic red beds (mudstones, siltstones, sandstones, and dolomites) of Pridolian age (Dingwall, 1986). These red beds would have been in the same latitudinal position as the Late Silurian portion of the Old Red Sandstone of the British Isles (Anderton and others, 1979).

The upper part of the Kenogami Formation contains an oolitic dolomite akin to oolitic bands in the Bass Islands Formation of the Michigan Basin and in outcrop some breccia, likely the residue in outcrop of gypsiferous interbeds. This represents the lithology of a nearshore environment, not of a connection to the open sea. The Moose River Basin thus could not have been a major supplier of marine brines.

Toward the end of the Silurian, the Moose River Basin became land, indicated by an erosional surface present below overlying Lower Devonian carbonates. There is no corresponding widespread unconformity in the Michigan Basin. Brines in the Moose River Basin did not take evaporite precipitation beyond the anhydrite facies, suggesting that hypersaline brines did not accumulate at depth.

The Moose River Basin could have in time supplied some of the clastics dominating the upper Salina Group in the northern part of the Michigan Basin. Any river draining into the Moose River Basin would have carried enough calcium and bicarbonate ions to account for the overflow of carbonate precipitation into the Michigan Basin detected by Briggs (1958); in wetter years it could also account for a major contribution to the shaliness of the upper Salina units in northern Michigan (Fig. 5). An inflow of calcium and bicarbonate ions into a marine brine ought to produce gypsum and magnesite (Braitsch, 1962). That it did not do so in this case, proves that there was a lack of sulfate ions, probably due to extensive bacterial scavenging. The sulfur isotope curves contain spikes indicative of redox recycling due to bacterial activity (Fritz and others, 1988).

An entrance from the Kokomo Sea of Indiana is a more likely source of brines in the Michigan Basin. Such an entrance was postulated near Clinton (west of Toledo, Ohio; Fig. 4; Briggs, 1958, 1962; Briggs and others, 1980). However, because of the prevalence of open-marine limestones to the south of the Kankakee arch, Droste and Shaver (1987) suggested that the Michigan Basin was not fed through a single strait (Fig. 6). A series of narrow passes across this arch and the Fort Wayne bank in front of it in northern Indiana and southwestern Michigan (also known as the Battle Creek trough, the continuation of the Logansport sag across the Wabash prong of the Kankakee arch) were thought to be entrance straits. The maximum water depths in the passes were deemed to be less than 15 m (Droste and Shaver, 1987). A

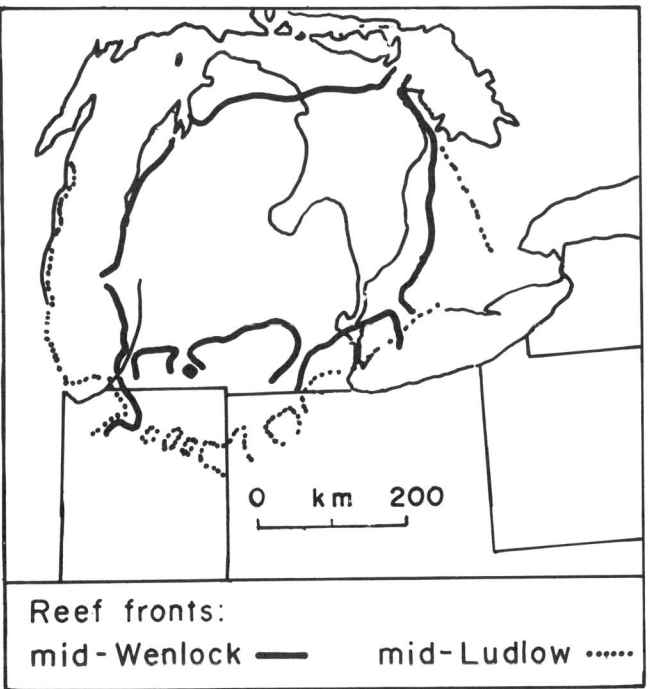

Figure 6. Location of reef fronts around the Michigan Basin in mid-Wenlockian (early Salina) and mid-Ludlovian (middle Salina) time (after Droste and Shaver, 1987).

gradual transgression of the individual Salina units displaced the front of the Fort Wayne bank southward (Fig. 6). One of these passes, the Allen Pass, lines up directly with the Clinton Pass and may well have been the same passage. Melhorn (1958) thought that this connection had become inoperative by the end of Niagaran reef development; nonetheless, all the passes could have been active at high water levels during carbonate phases of the individual cycles.

Inasmuch as the cross-sectional area of the inflow channel dictates the degree of brine concentration in the basin (Briggs, 1958), it is possible that not all the indicated passes from the Kokomo Sea were active at the same time, or that they were even shallower than anticipated following a slight drop in sea level (of the order of 3 to 10 m, or even 12 m, as suggested by Droste and Shaver, 1987). Lucia (1972) related the cross-sectional area of the entrance strait to the water surface of an evaporite basin and found that their ratio must exceed about $10^8$ for halite saturation to occur, and $10^9$ for potash saturation. Sonnenfeld (1984) provided the equations for outflow/inflow ratios, which are less than 0.1154 for halite and 0.0757 for potash saturation, the outflow being mandated, since there always prevails a deficiency of salts compared to those of next lower solubility. Based on such considerations, the total cross-sectional area of the passages to the Michigan Basin could not have exceeded about 10,000 m$^2$ (assuming 0.8 m/yr of evaporation; Sonnenfeld, 1985), or later 60,000 m$^2$ (assuming 1.5 m/yr of evaporation; Briggs, 1958) to supply the combined Michigan, Ohio, and Allegheny Basins. Either way it becomes unlikely that more than one pass has operated at any one time. Changing the assumed typical water deficits even to maximum values observed today would not significantly alter the cross-sectional size of the required entrance strait.

At present one cannot discount the possibility that a very precise balance existed in inflow sources. One passage delivered seawater at high sea-level positions and diluted the brine to carbonate saturation, but was shut out when sea level dropped slightly and the remaining passage alone was not sufficient to prevent brine concentration. Similarly, there also was a balance in the outflow, either over sills or as subsurface seepage. In no evaporite unit are the individual components represented in an isochemical ratio; in all of them there is a substantial deficiency in rocks of successively higher solubility.

## Brine level in the basin

Gill (1977) proposed an evaporative drawdown of more than 130 m. Ghyben-Herzberg lenses of fresh water get thinner with increasing difference in density between brine and freshwater, and thus there would have been an extremely thin zone of vadose alterations inside the reefs during halite or sylvite precipitation. A drawdown of even much less than 130 m would have cut off any surface intake into the basin.

However, Dombkowski (1978) found no evidence for such reef exposure, only very slight fluctuations in sea level. The pinnacle reefs attained a height of as much as 180 m (Mantek, 1973; Gill, 1979) or more, with an evaporite unit draping over them. This evaporite unit attained a thickness of only a few tens of meters above any reef and as much as 145 m in the basin center. Taking into consideration later thickness changes due to compaction, this still does not allow for very deep submersion in the halite precipitating basin. Halite precipitation levels out inhomogeneities in a sea floor, and anhydrite capping the reefs on the margin is the lateral equivalent of the halite. Before compaction of the evaporitic rocks, the reef height was at best only marginally less than that of the uncompacted halite section. Moreover, if each of the halite units did represent a major drawdown of sea level, it disagrees with the evidence (Tremper, 1973; Mesolella, 1978) that at the same time they each overstepped preceding depositional edges (Fig. 1).

Subsurface seepage of marine waters through the reefs could only occur if there is a significant difference between sea and basin levels. Beneath the brine level in the basin, an encroachment of basin brines onto less dense interstitial brines prevents any such seepage.

Seepage volume has to equal the volume of evaporation loss to prevent a rapid drying out of the basin. The seepage area (width of the basin times height of sea level above brine level in the basin, assuming no tight beds in the barrier) is bound to be smaller than the area (width times length) of the basin; it can thus be calculated how much seawater can be delivered through porous beds separating the basin from the open sea. The rate of seepage is 10 to 5 m/s through porous reefs (Maiklem, 1971; Cercone, 1988), but is 10 to 8 m/s or less (Maiklem, 1971; Freeze and Cherry, 1979) through consolidated dolomudstones, platform carbonates of the basin margin, and offreef areas (Sarg, 1982). The lower figure is of the same order of magnitude as the evaporative deficit generated by an evaporation rate of 0.8 to 2 m/yr, typical for semiarid regions. In the case of the Silurian Michigan Basin, a throughput even as high as 10 to 5 m/s through a barrier as much as several hundred meters high would have generated a shortfall of one to two orders of magnitude in compensatory flow, resulting in drying out within 100 to 200 yr.

The Dead Sea is separated from the Red Sea by a land bridge about 177 km long, from the Mediterranean Sea by only 78 km, but is not fed from either source despite a hydraulic gradient of more than 412 m. There is thus no reason to believe that waters from the Kokomo Sea in southern Indiana would have been able to seep through the Kankakee arch (also referred to as Wabash platform) over a distance of some 140 km. Moreover, a significant drawdown would have exposed off-reef carbonates and evaporites, and there is no evidence to be found of such a subaerial exposure (Dombkowski, 1978; Sarg, 1982). No rubble produced by exposure during a drawdown could be found, only multiple levels of brief subaerial exposures within the reef and supratidal, intertidal, and eventually subtidal sediments on top of the reef, indicating at best only modest fluctuations in sea level (Dombkowski, 1978).

Carbonate muds flanking the basin (Dellwig, 1955) were not very permeable and quickly lost most of their initial porosity.

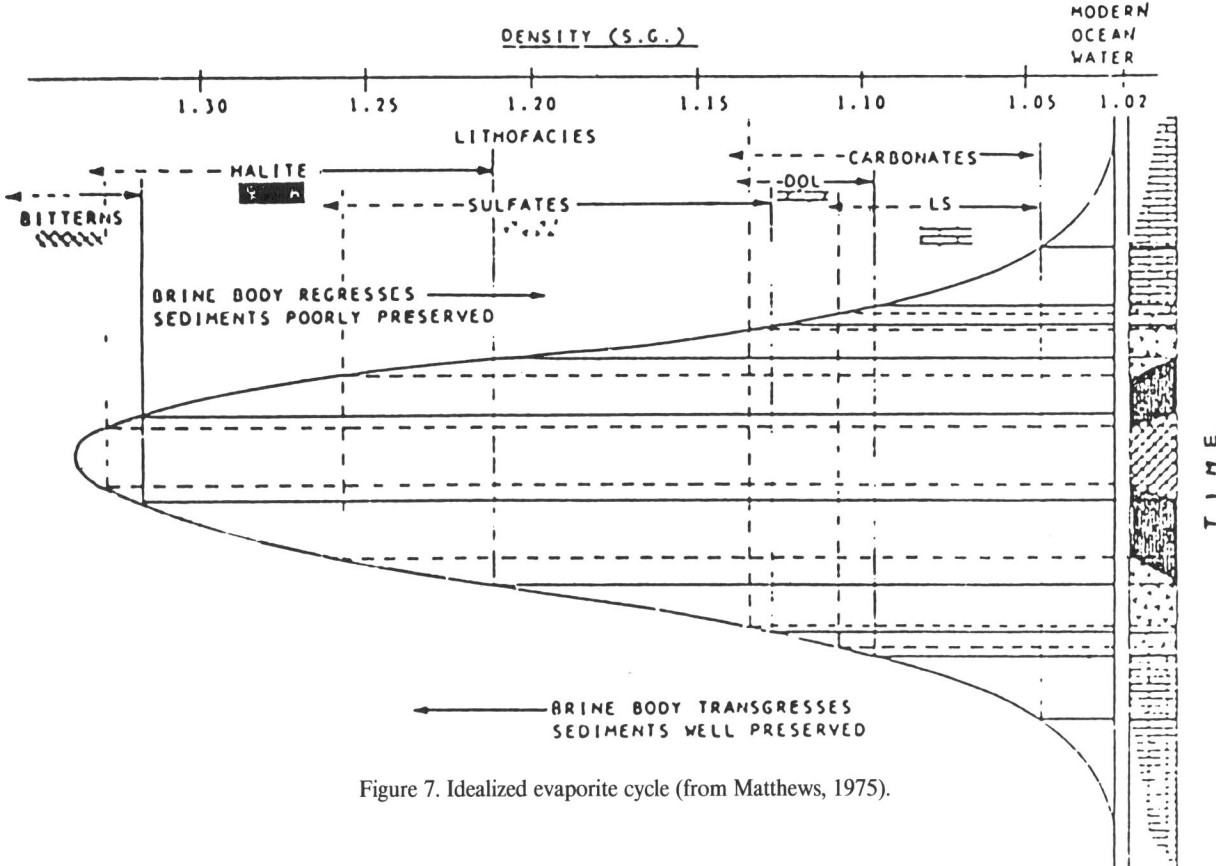

Figure 7. Idealized evaporite cycle (from Matthews, 1975).

Compaction reduced their volume by about 35 percent; this compaction commenced at about 100 m of overburden and was virtually completed at a little more than 200 m of overburden, as there is no draping above the "B" halite (Jodry, 1969). With the overlying calcium sulfates relatively impermeable, compaction may have driven out connate waters through the protruding, more porous reefs; some of the anhydritic caprocks of reefs may, indeed, have received their calcium ions in this manner. In the same manner, some of the anhydrite knobs in the vicinity of reefs may have grown from this oozing out of calcium-saturated brines.

Without major quantities of sustained inflow, the whole basin, exposed to 1 to 2 m/yr of evaporation, would dry out in short order. There is no basin-center evidence to suggest that this happened at any time during Salina deposition. Instead, a delicate balance was maintained between synsedimentary subsidence, evaporation, and water supply so that at no time did evaporation approach completion (Dellwig, 1955). This is not to say that the shoreline and sea level did not oscillate by small amounts. Indeed, multiple evidence exists for occasional subaerial exposure of the basin margin (Nurmi and Friedman, 1977).

Furthermore, a seepage of seawater would deliver all dissolved compounds isochemically, in proportions nearly equal to those extant in seawater. That means that for every 1 m$^3$ of anhydrite there should have been 31 m$^3$ of halite present, and for every 10 m$^3$ of halite there should have been more than 7 m$^3$ of potassium and magnesium chlorides precipitated. Seepage at the mentioned travel rate could neither deliver enough seawater to prevent a rapid desiccation of the basin nor produce the progressively greater deficiency in solutes of higher solubility that can be observed.

An extensive lowering of sea level would have exposed shelf deposits and would have initiated the erosion of emergent gypsum flats. Once meteoric waters reached the halite edge, gypsum solubility would be drastically increased, and oxygenated waters saturated with calcium and sulfate ions could have entered the then still inadequately compacted basin sediments. They would have altered the sylvinite deposit into an array of bedded K-Mg-sulfates, as they have done in Permian and Neogene evaporites (Sonnenfeld, 1989a). It did not happen in the Silurian Michigan Basin. This is not to say that locally such an exposure to sulfatization by meteoric waters did not occur, as evidenced by the scant polyhalitization of some gypsum crystals (Dellwig, 1955).

*Brine depth in the basin*

Halite precipitation occurs during a transgression of the sea (Matthews, 1975; Jauzein and Hubert, 1984), because the water surface area has to increase in order to achieve the required water losses (Fig. 7). Straw (1985) suggested that evaporite cycles in the Michigan Basin marks such sea-level rises. The overstepping of the "A-1" halite by the "A-2" halite, and in turn by the "B" halite, indicates that a net transgression did occur during the deposition of the Salina Group (Tremper, 1973). The "A-2" halite is even transgressive compared to the extent of the "A-2" anhydrite as it is at times overlain by an anhydrite bed and at other times occurs sandwiched between anhydrites (Johnson, 1971).

Tremper's maps indicate that the northern shelf has had an original slope of about 0°48′ (14 m/km) when the "A-2" evaporite transgressed over the "A-1" unit, but the "A-2" carbonate and "B" evaporite then transgressed onto a shore that was dipping only 0°06′ (1.75 m/km) into the basin. In both cases, the rise in sea level required to produce such a transgression is of the order of several tens of meters. However, because of the gentle slope, even a minor oscillation of sea level would either bare broad shelf areas and expose them to erosion, or submerge a wide coastal flat.

On the basis of bromide curves, Holser (1966) calculated brine depth to have been about 50 m near the center of the basin where the "A-1" halite was precipitated. The maximum depth of the Allegheny Basin during "B" salt deposition was estimated at less than 30 m (L. V. Rickard in Dellwig and Evans, 1969). These estimates are of the same order of magnitude as Holser (1966), and more recently Tucker and Cann (1986) or Sonnenfeld (1990b) have calculated for other halite-precipitating basins. If these calculated brine depths are of the right order, to fill the basin with halite required thus a synsedimentary subsidence of 125 m for the "A-1" halite, and 260 m for the "B" halite.

Halite laminites or varvites are preferentially found toward the center of the basin (Dellwig, 1955), but intercalations of anhydrite and dolomite are thicker toward the margin. In the "B" salt in the Retsof mine in New York State, the absence of laminae indicates a shallow-water regime, where turbulence has disrupted the bedding (Dellwig and Evans, 1969).

The extensive occurrence of such halite varvites (Kaufman and Slawson, 1950; Briggs and others, 1980) permits a depth estimate independent of the bromide curve. Assuming that the varvites are a record of seasonal salinity oscillations ("Jahresringe" of German evaporite literature) and were precipitated on the basin floor, we need only consider each couplet as a record of a brine freshening dropping out of the halite saturation field, followed by a phase of brine concentration back into the halite saturation field, all happening within one set of seasons.

To saturate seawater for halite requires the evaporative removal of about 91 percent of the water. Reduction of the salinity of a column of brine saturated for halite by 1 percent requires 0.067 m of seawater of normal salinity entering per meter of brine depth. This water has to then evaporate during the same set of seasons to start the precipitation of the next couplet. The evaporative water deficit (evaporation losses less rains, dew, and runoff) over evaporite basins is rarely as high as 2 m/yr. A 1-percent reduction in salinity would only allow the total brine column to freshen if it is no more than 30 m deep. For a 2-percent reduction in salinity, the brine depth drops to 14.25 m. It is unlikely that the total evaporative water deficit over the late Silurian Michigan Basin was higher than 2 m/yr. It should be noted, however, if the annual water deficit is overestimated, the resulting brine depth calculations will turn out to be too high.

### Basin subsidence

Considering an average synsedimentary subsidence of the order of only a modest 0.5 mm/yr, the total Salina sequence (including breaks in deposition) can be accommodated within the Ludlovian stage. Present subsidence rates in the Chesapeake Bay and elsewhere may exceed 8 mm/yr, and Avedik and Hieke (1981) have established an average subsidence rate of 0.5 mm/yr for the Ionian Sea over the past 5 m.y. Therefore, an average value of 0.5 mm/yr for the Silurian Michigan Basin is reasonable.

Rates of subsidence need not be uniform. The accumulation of several salt cycles, separated by carbonate banks of reasonably uniform thickness, suggest an oscillating rate of subsidence. Reef growth can match moderate rates of subsidence, of the order of 1 mm/yr; reefs only drown in spurts of sea-level rises that can be as much as one order of magnitude more rapid (Schlager, 1981). Conversely, a slight lowering of sea level exposes the reefs to atmospheric precipitation and weathering, establishes a vadose zone connected with the Ghyben-Herzberg lens (Dombkowski, 1978; Bay, 1983), and thus temporarily arrests vertical reef growth. However, the existence of such an extensive lens of meteoric waters could not be confirmed in the majority of pinnacle reefs, and thus, reef exposure could at best have been only very brief (Cercone and Lohmann, 1985).

The occurrence of calcium borate nodules in the sylvinite facies (Nurmi and Friedman, 1977; Straw, 1985) has been taken to constitute evidence for desiccation or exposure of the basin. However, their presence does not confirm a drying out, as such minerals have been found in the Zechstein and other marine evaporite basins and are merely related to early and late phases of concentration (Mueller and Fabricius, 1978).

### Evaporite thicknesses

The maximum thicknesses of evaporite units in the Michigan Basin are not random values. As random values they should yield different thicknesses for each salt. Instead, each of the "A-1," "A-2," and "B" evaporites attains a maximum thickness of 145 m, the "F" evaporite exactly twice that figure. This implies that a common limiting factor must have existed that affected thickness. It is here suggested that this consistent thickness is an indicator of how long climatic parameters remained constant and favorable for halite precipitation.

Though speculative, if we take the net accumulation rates of Salina halites, determined from micrometeoritics by Barnett and Straw (1983) not to exceed 4 mm/yr, and apply them to the "A-1," "A-2," and "B" evaporites, we arrive at a duration of about 36,000 yr of evaporite deposition for the units, double that for the "F" unit. Taking into account the uncertainty in estimates of halite preservation rates, there are at this time not enough corroborating data available to consider this to be an expression of Milankovitch intervals (to invoke the cyclicity of the earth's obliquity of 41,000 yr would imply an average halite preservation rate of 3.54 mm/yr).

## OTHER ROCKS ASSOCIATED WITH EVAPORITES IN THE SALINA GROUP

### Reef growth

It is not known whether Niagaran reefs in the Michigan Basin were entirely contemporaneous or grew alternating with evaporite cycles, or entirely preceded evaporite deposition (Huh and others, 1977). Reefs grow primarily at the top, i.e., in inflow waters, and can have their dead trunk standing in more concentrated brines. However, to assume a wave base in concentrated seawater to be as deep as some 100 to 125 m below sea level (Sears and Lucia, 1980) is not in keeping with the increased viscosity of such brines. In analogy to modern reefs that are exposed at times above sea level, the Niagaran reefs likewise display a vadose zone suggesting at least ephemeral emergence (Meloy, 1974; Gill, 1977). Calichification of the reefs before the deposition of the "A-1" halite has been noted (Bay, 1983).

The reefs increase in height from the shelf edge basinward (Gill, 1979). They may have been located on shelf-slope breaks, in analogy to coeval reefs in the Illinois Basin to the south (Thomas and others, 1989). It may also be true that all reefs located on the inclined sea floor do not have the same history; rather, those situated basinward had to grow higher or drown while those closer to shore continued to flourish. Although they developed sufficient relief to allow continued growth in a stratified basin, one can recognize at least two hiatal breaks recording periodic onlap of hypersaline basin water over the reefs, causing cessation of reef growth (Sarg, 1982).

Reef growth was controlled by eustatic sea-level changes (Sarg, 1983). It continued on surrounding platforms to very near the end of Silurian time; up to six generations of Silurian reef growth can be distinguished (Droste and Shaver, 1985), as the Michigan Basin retained a climatically, hydrographically, and nutritionally favorable environment for reef growth in surface inflow. The various reef generations on the fringes of the Michigan Basin did not necessarily abort their further growth at precisely the same time. The upper portions of the reefs, texturally differing from their substrate, appear to be equivalents of the Salina Group, specifically the "A-1" carbonate, rather than Niagaran in age, and have been rejuvenated by algal secretion during Ludlow-Pridolian time (Hadley, 1970; Mesolella, 1972; Leibold and Cercone, 1989). Some reefs thus seem to have continued to grow during the precipitation of "A-1" evaporites; the "A-2" carbonate unit thickens over salt-filled reefs, but does not do so over reefs not filled with halite (Johnson, 1971).

Many of the reefs in the southeastern Michigan Basin display thickened knobs of anhydrite on their southeastern flank (Johnson, 1971), facing away from the deeper basin toward nearby shores. This would have been not only the upslope side but also the shadow of the reefs, away from optimum sunlight for photosynthesis. The amount of anhydrite present is, however, too small to suggest a delivery of calcium bicarbonates from nearby shores through the substrate.

Many reefs have salt- or anhydrite-filled vugs in the upper part, above vugs filled with oil and gas (Johnson, 1971). Since the "A-1" halite contains nearly twice as much bromine as the "A-2" halite, it could be determined that most of the halite vug-filling, even that in the upper part of reefs, has come from the "A-1" unit, but that halite-filling spaces in the "A-1" carbonate are derived from the "A-2" halite (McCollough, 1975). Interaction of brines from the "A-2" evaporite was minimal in down-slope reefs that probably did not protrude over the uncompacted precipitate; mid-slope and, even more so, upper-slope reefs were affected by organically modified bicarbonate from brines refluxing from the "A-1" carbonate (Cercone and Lohmann, 1985).

Where vugs are filled with halite, wall rock is often separated from halite by a paper-thin layer of anhydrite, white to dirty brown in color, with included micritic dolomitic limestone. In each case, evaporite precipitation seems to postdate hydrocarbon infiltration (Konyen, 1982) and is probably caused by postsedimentary seepage of saturated brines from overlying "A-2" evaporites. Hydrocarbon migration into the reefs appears to have occurred after dolomitization and prior to salt infill; some of the hydrocarbons were later transformed into bitumen, reducing the porosity in some reservoirs (Bay, 1983). That is not to preclude further hydrocarbon migrations at a later date.

### Dolomites

Dolomites associated with the Salina evaporites grade southwestward away from the Michigan Basin into limestones (Droste and others, 1975) and thus into an environment of normal marine salinity. Toward the deeper part of the basin, the "A-1" carbonate is limestone, but it is dolomite around the basin margins. The transition zone to dolomitization is nearly coincident with a carbonate ridge that thins both basin- and shoreward. This ridge follows the older reef trend, and it may be due to earlier removal of halite and sylvinite by dissolution (Tremper, 1973). Sears and Lucia (1980) detected a difference in grain fabric between dolomitization of marginal reefs due to brine refluxing and that due to a mixing of brine with fresh water from the vadose zone of an ephemeral Ghyben-Herzberg lens.

Some of the dolomite bands in each halite bed, or between halite cycles, are not associated with any significant amounts of anhydrite. There is no unconformity recognized here, but the transition from halite to dolomite appears to be abrupt. Because a reduction in salinity below halite saturation fosters the precipitation of gypsum, a scarcity of such laminations in contact with dolomite is noteworthy. Anaerobic bacteria will digest gypsum in the presence of organic matter to produce a carbonate bank (Perthuisot, 1975; Sonnenfeld and Perthuisot, 1989).

There was no lack of organic matter for anaerobic bacteria

to digest. Dolomite-anhydrite couplets in the Michigan Basin often have a petroliferous film at the dolomite base (Briggs, 1957). In the basal part of a series of halite-anhydrite couplets, magnesite is found (Briggs and Lucas, 1954), which is a clear derivative of bacterial consumption of anhydrites in the presence of organic matter (Perthuisot, 1975). O'Shea and others (1988) noted significantly higher amounts of organic matter preserved in the "A-1" and "A-2" carbonates than in the remaining units of the Salina Group. Similarly, caprocks of Gulf Coast salt domes often contain calcite and sulfur, both derived from anhydrite in the presence of organic matter. In cores cut into a gypsum precipitate on the island of Gran Roque, Venezuela (Sonnenfeld and others, 1977), at a depth of 0.3 to 1.0 m, the Recent gypsum gave way to a fine- to medium-grained loose calcite sand rich in organic matter, which upon dolomitization would turn into a very finely crystalline, possibly even bituminous dolomite. However, washing the walls of the Detroit salt mine, Kaufman and Slawson (1950) discovered a series of paper-thin anhydrite laminae in the halite between the more widely spaced dolomite banding, suggesting that destruction of gypsum and the resulting carbonate production was an episodic event.

Existing models of dolomitization do not do justice to the observation that dolomitization in the Michigan Basin is not affecting all underlying carbonates equally. Dolomitization appears to be restricted to a rim on either side of the reef chains circling the basin, but has not affected the deeper interior of the basin, where the concentrated brine resided.

Dolomitization of either calcite or aragonite requires not only the delivery of magnesium but also carbon dioxide in solution. The solubility of $CO_2$ decreases with rising brine concentration and rapidly becomes minute. Even in early stages of a shallow basin there would have been an interface between inflow and resident brine, with the resident brine covering and protecting underlying limestones. During synsedimentary subsidence, this interface gradually rises. Possibly, most dolomitization by basin brines may then have occurred in rocks exposed to this interface, as bicarbonate solubility would have been much higher in the low-salinity surface waters than in the resident brine. This concept represents another form of brine mixing and requires further investigation.

The degree of dolomitization decreases mainly outward away from the evaporite basin, upward into younger beds, and also to some degree inward. This concurs with Walther's Law that lateral facies changes correspond to vertical ones. Dolomitization, therefore, may be directly related to the effects of brines seeping out of the evaporite pan and carrying with them magnesium-bearing dolomitization fluids. It cannot be related to an inflow, by seepage or otherwise, of seawater from the open-marine environment, as has been suggested (Cercone, 1988). Klement (1895), Van Tuyl (1916), and Linck (1937) have repeatedly shown that limestones are not dolomitized by exposure to seawater or to magnesium chloride solutions over long periods of time, but that the magnesium bond to the anion has first to be broken either by biota or by other means. On the other hand, unless later illitized, all clay intercalations in the chloride facies of marine evaporite basins contain only mixed-layer varieties of the Mg-chlorite family with brucite pillars. There is no reason to believe that all available magnesium hydroxides or Mg-organic compounds (derived from the bacterial destruction of marine magnesium sulfates) were soaked up by clays.

## Clay influx

After the deposition of the "B" salt, influx of suspended clay particles dominated the depositional regime and effectively terminated most life on the carbonate banks. At least some of these clays came largely from a delta produced by a river discharging near present-day Ludington, Michigan (Briggs, 1958; Alling and Briggs, 1961), where a low area developed throughout Salina deposition (Tremper, 1973). An alternating sequence of evaporites (halite and anhydrite) and shales was the result; carbonates only resumed dominant precipitation in the latest Silurian, as documented by the Bass Islands Group.

Clastics delivered by a river are deflected by longshore currents, moving along the coast in counterclockwise fashion in the Southern Hemisphere. The fine fraction of terrigenous material could thereby float out in suspension at the interface between river waters and resident brine far into the basin. Overall, the ratio of clastic to nonclastic rocks ranges from less than 1:9 to more than 1:3 along the northern margin of the Michigan Basin, but rises rapidly in an easterly direction in the Ohio Basin and thence in the Allegheny Basin, where at the far eastern end it reaches 1:1 (Alling and Briggs, 1961).

The shales do not display a distribution of continental or even normal marine clays, but have been considerably altered through hyperhalmyrolysis to a suite typical of marine evaporite intercalations, i.e., either members of the chlorite family that are low in bivalent iron and high in magnesium or members of the illite family, low in magnesium and trivalent iron, but with potassium instead (Bodine and Standaert, 1977).

It is unlikely that all the clastics in the Michigan Basin were supplied by one river delta coming from the Wisconsin dome or from the meeting of Kankakee and Transcontinental arches. On the occasion of subtropical flash floods, some clays were probably swept in also through the Sudbury sag from the red beds of the Moose River Basin (Figs. 1 and 5). The increase of clastics in the upper parts of the Salina Group (units "E" and "G") suggests that periods of reduced water deficit become progressively longer toward the end of the interval. This entails higher river discharges, greater amounts of rainfall in the hinterland, and lower evaporation losses.

The appearance of shales as such cannot be interpreted as direct evidence for a deepening basin; it merely means that the basin is no longer starved of terrigenous material and its circulation has possibly improved. Nevertheless, since the rate of terrigenous clay input is substantially smaller per unit of time than the rate of precipitation of any of the evaporite minerals, a cosntant rate of subsidence during shale deposition would have resulted in a progressive deepening of the basin.

## POST-SALINA EVAPORITES

The semiarid environment continued into the Early Devonian. The first major break is an unconformity above the various salt cycles, cut and then filled with an exceptionally clean sand (the Sylvania Sandstone, today a major aquifer). It suggests a sudden influx of large quantities of water and, thus, a sudden reduction in salinity. However, evaporitic conditions resumed shortly thereafter. The final evaporite sequence within finely crystalline dolomites is the Eifelian Lucas Formation (Gardner, 1974; Lilienthan, 1978). All the halite occurs in the center of the basin within the thickest portion of the formation. Anhydrite occurs above the halite, and interstratified with it and beneath it. There were 23 (Briggs, 1959) to 35 evaporite cycles (Matthews, 1975), of which the earlier ones only produced anhydrite. Later ones also produced halite, but no sylvinite or carnallitite. Four (Ehman, 1964) or eight (Briggs, 1959) or 25 salts (Matthews, 1975) have so far been distinguished; the difference depends on the methods of counting evaporite cycles and subcycles. In the same manner, the six "B" or two "D" salts could have been considered independent formations of the Salina Group.

Conditions favoring evaporite precipitation ended abruptly. Overlying beds of the Dundee and younger formations are no longer the product of an evaporitic environment, but consist of open-marine limestones and shales. In Early Pennsylvanian time the Michigan Basin became emergent.

Groundwater has then been able to enter, especially from the north, gradually dissolve salts at the margin of the basin, causing overlying strata to collapse into a breccia, the Mackinac Formation, and produce a depression occupied today by Lakes Michigan and Huron. It has not been established whether minor amounts of evaporites in younger Devonian beds are due to brief resumptions of semiarid conditions or are recycled Salina evaporites (in analogy to late Paleozoic and Mesozoic evaporites in North Dakota that represent recycled Middle Devonian Elk Point evaporites; Sonnenfeld, 1984).

## CONCLUSIONS

In the Michigan Basin, Upper Ordovician rocks contain some anhydrite; the Silurian units "A-1," "A-2," "B," "D," and "F" contain anhydrite/halite cycles, as does the Devonian Detroit River Group. However, in all cases, the units can be divided into subunits of alternating evaporite and carbonate deposition, producing a complex cyclicity. Some cycles contain anhydrites, and some contain anhydrite only, enclosing halite (in the center of the Michigan Basin). Each cycle is capped by carbonates deposited in a temporarily less concentrated brine. Each cycle can by divided into subcycles of various orders by major and minor lithologic intercalations representing conditions of less solubility. The "A-1" potash unit, a single bed in the northwestern interreef region, separates into numerous laminae within the basin. Each sylvite/halite couplet indicates a concentration and freshening phase.

The main supply of seawater came from the Kokomo Sea in Indiana. Continental waters could also have entered at times from the Moose River Basin, Ontario. Michigan Basin brines eventually entered Ohio and Pennsylvania through the Chatham sag. A progressive overstepping of one evaporite unit over the preceding one indicates either a basinal subsidence affecting wider areas or rising sea level, the latter suggested by the inundation of northern Ohio beginning with the "B" unit of salt deposition. At this time, only an outline of events during this evaporitic phase in the history of the Michigan Basin can be given. A considerable amount of study needs yet to be done to unravel a detailed history of these events.

## ACKNOWLEDGMENT

I. A. wishes to acknowledge partial support by N.S.E.R.C. and the University of Windsor.

## REFERENCES CITED

Alling, H. L., and Briggs, L. I., 1961, Stratigraphy of Upper Silurian Cayugan evaporites: American Association of Petroleum Geologists Bulletin, v. 45, p. 515–547.

Anderson, R. J., and Egleson, G. C., 1970, Discovery of potash in the A-1 Salina salt in Michigan, in Kneller, W. A., ed., Sixth Forum on Geology of Industrial Minerals: Michigan Geological Survey Miscellaneous Paper 1, p. 15–19.

Anderton, R., Bridges, P. H., Leeder, M. R., and Sellwood, B. W., 1979, A dynamic stratigraphy of the British Isles; A study in crustal evolution: London, George Allen and Unwin, 301 p.

Avedik, F., and Hieke, W., 1981, Reflection seismic profiles from the Central Ionian Sea (Mediterranean) and their geodynamic interpretation: "Meteor'-'-Forschungsergebnisse, Ser. C, no. 34, p. 49–54.

Barnett, J. M., and Straw, W. T., 1983, Sedimentation rate of salt determined by micrometeorite analysis: Geological Society of America Abstracts with Programs, v. 15, p. 521.

Bay, T. A., 1983, The Silurian of the northern Michigan Basin, in Harris, P. M., ed., Carbonate buildups; A core workshop: Society of Economic Paleontologists and Mineralogists Core Workshop 4, p. 53–72.

Bodine, M. W., Jr., and Standaert, R. R., 1977, Chlorite and illite composition from upper Silurian rock salts, Retsof, New York: Clays and Clay Mineralogy, v. 25, p. 57–71.

Bratisch, O., 1962, Die Entstehung der Schichtung in rhythmisch geschichtetan Evaporiten: Geologische Rundschau, v. 52, p. 405–417.

Briggs, L. I., 1957, Quantitative aspects of evaporite deposition: Michigan Academy of Science, Arts, Letters, Papers, v. 42, p. 115–123.

—— , 1958, Evaporite facies: Journal of Sedimentary Petrology, v. 28, p. 46–56.

—— , 1959, Physical stratigraphy of Lower Middle Devonian rocks in the Michigan Basin: Michigan Basin Geological Society Annual Field Excursion Guidebook, p. 39–58.

—— , 1962, Niagaran Cayugan sedimentation in the Michigan Basin: Michigan Basin Geological Society Annual Field Excursion Guidebook, p. 58–60.

Briggs, L. I., and Lucas, P. T., 1954, Mechanism of Salina salt deposition in the Michigan Basin: Geological Society of America Bulletin, v. 65, p. 1233.

Briggs, L. I., Gill, D., Briggs, D. Z., and Elmore, R. D., 1980, Transition from open marine to evaporite deposition in the Silurian Michigan Basin, in Nissenbaum, A., ed., Hypersaline brines and evaporitic environments: Amsterdam, Netherlands, Elsevier, p. 253–270.

Budros, R., 1974, The stratigraphy and petrogenesis of the Ruff Formation, Salina Group, in southeastern Michigan [M.Sc. thesis]: Ann Arbor, University of Michigan, 178 p.

Budros, R., and Briggs, L. I., 1977, Depositional environment of Ruff Formation (Upper Silurian) in southeastern Michigan, in Fisher, J. H., ed., Reefs and evaporites; Concepts and depositional models: American Association of Petroleum Geologists Studies in Geology 5, p. 53–71.

Burns, J. W., 1962, Regional study of the Upper Silurian Salina evaporites in the Michigan Basin [M.Sc. thesis]: East Lansing, Michigan State University, 90 p.

Case, L. C., 1945, Exceptional Silurian brine near Bay City, Michigan: American Association of Petroleum Geologists Bulletin, v. 29, p. 567–570.

Catacosinos, P. A., 1973, Cambrian lithostratigraphy of Michigan Basin: American Association of Petroleum Geologists Bulletin, v. 57, p. 2404–2418.

Cercone, K. R., 1988, Evaporative sea-level drawdown in the Silurian Michigan Basin: Geology, v. 16, p. 387–390.

Cercone, K. R., and Lohmann, K. C., 1985, Early diagenesis of Middle Silurian pinnacle reefs, northern Michigan, in Cercone, K. R., and Budai, J. M., eds., Ordovician and Silurian rocks of the Michigan Basin and its margins: Michigan Basin Geological Society Special Paper 4, p. 109–117.

Cohee, G. V., and Landes, K. K., 1958, Oil in the Michigan Basin in Weeks, L. G., ed., Habitat of oil: American Association of Petroleum Geologists, p. 473–493.

Dellwig, L. F., 1955, Origin of the Salina salt of Michigan: Journal of Sedimentary Petrology, v. 25, p. 83–110.

Dellwig, L. F., and Evans, R., 1969, Depositional processes in the Salina salt of Michigan, Ohio and New York: American Association of Petroleum Geologists Bulletin, v. 53, p. 949–956.

Dingwall, R. G., 1986, The exploration of Hudson's Bay: Ontario Petroleum Institute 25th Annual Conference, Paper 4, 41 p.

Dombkowski, F. S., 1978, The Silurian depositional environments and importance of the Maumee Quarry (Ohio) and Capac Gas Field (Michigan) carbonates [M.Sc. thesis]: Detroit, Michigan, Wayne State University, 65 p.

Droste, J. B., and Shaver, R. H., 1983, Atlas of early and middle Paleozoic paleogeography of the southern Great Lakes area: Indiana Geological Survey Special Report 32, 32 p.

——, 1985, Comparative stratigraphic framework for Silurian reefs; Michigan Basin to surrounding platforms, in Cercone, K. R., and Budai, J. M., eds., Ordovician and Silurian rocks of the Michigan Basin and its margins: Michigan Basin Geological Society Special Paper 4, p. 73–93.

——, 1987, Paleoceanography of Silurian seaways in the Midwestern basins and arches region: Paleoceanography, v. 2, p. 213–227.

Droste, J. B., Shaver, R. H., and Lazor, J. D., 1975, Middle Devonian paleogeography of the Wabash platform, Indiana, Illinois, and Ohio: Geology, v. 3, p. 269–272.

Egleson, G. C., and Querio, C. W., 1969, Variations in the composition of brine from the Sylvania Formation near Midland, Michigan: Environmental Science and Technology, v. 3, p. 367–371.

Ehman, D. A., 1964, Stratigraphic analysis of the Detroit River Group in the Michigan Basin [M.Sc. thesis]: Ann Arbor, University of Michigan, 63 p.

Ekblaw, G. E., 1938, Kankakee arch in Illinois: Geological Society of America Bulletin, v. 49, p. 1425–1430.

Evans, C. S., 1950, Underground hunting in the Silurian of southwestern Ontario: Geological Association of Canada Proceedings, v. 3, p. 55–85.

Fincham, W. J., 1975, The Salina Group of the southeastern part of the Michigan Basin [M.Sc. thesis]: East Lansing, Michigan State University, 195 p.

Fisher, J. H., Barratt, M. W., Droste, B., and Shaver, R. H., 1988, Michigan Basin, in Sloss, L. L., ed., Sedimentary cover–North American craton: Boulder, Colorado, Geological Society of America, The Geology of North America, v. D-2, p. 361–382.

Freeze, R. A., and Cherry, J. A., 1979, Groundwater: Englewood Cliffs, New Jersey, Prentice-Hall Inc., 604 p.

Fritz, P., and 5 others, 1988, Stable isotopes in sulphate minerals from the Salina Formation of southwestern Ontario: Canadian Journal of Earth Sciences, v. 25, no. 2, p. 195–205.

Gardner, W. C., 1974, Middle Devonian stratigraphy and depositional environments in the Michigan Basin: Michigan Basin Geological Society Special Paper 1, 133 p.

Gill, D., 1973, Stratigraphy, facies, evolution, and diagenesis of productive Niagaran Guelph reefs and Cayugan sabkha deposits, the Bell River Mills gas field, Michigan [Ph.D. thesis]: Ann Arbor, University of Michigan, 284 p.

——, 1977, Salina A-1 sabhkha cycles in the late Silurian paleogeography of the Michigan Basin: Journal of Sedimentary Petrology, v. 47, p. 979–1017.

——, 1979, Differential entrapment of oil and gas in Niagara pinnacle-reef belt of northern Michigan: American Association of Petroleum Geologists Bulletin, v. 63, p. 608–620.

Hadley, C. J., 1970, Subsurface studies of reef debris will help reveal the presence and quality of pinnacle reefs in southwestern Ontario: Ontario Petroleum Institute 9th Annual Conference, Paper 8, 15 p.

Holser, W. T., 1966, Bromide geochemistry of salt rocks, in Rau, J. L., ed., Second Symposium on Salt: Cleveland, Northern Ohio Geological Society, v. 1, p. 248–275.

Huh, J. M., Briggs, L. I., and Gill, D., 1977, Depositional environments of pinnacle reefs, Niagara and Salina Groups, northern shelf, Michigan Basin, in Fisher, J. H., ed., Reefs and evaporites; Concepts and depositional models: American Association of Petroleum Geologists Studies in Geology 5, p. 1–21.

Jacoby, C. H., 1969, Correlation, faulting, and metamorphism of Michigan and Appalachian Basin salt: American Association of Petroleum Geologists Bulletin, v. 53, p. 136–154.

Janssens, A., 1977, Silurian rocks in the subsurface of northwestern Ohio: Ohio Geological Survey Report of Investigations 100, 96 p.

Jauzein, A., and Hubert, P., 1984, Les bassins oscillants; Un modèle de genese des series salines: Science Geologique, Bulletin, v. 37, p. 267–282.

Jodry, R. L., 1969, Growth and dolomitization of Silurian reefs, St. Clair County, Michigan: American Association of Petroleum Geologists Bulletin, v. 53, p. 957–981.

Johnson, K., 1971, The interrelationship of the lower Salina Group and Niagaran reefs (Silurian) in St. Clair and Macomb Counties, Michigan [M.Sc. thesis]: East Lansing, Michigan State University, 46 p.

Johnson, K. S., and Gonzales, S., 1978, Salt deposits in the United States and regional geological characteristics important for storage of radioactive waste: Athens, Georgia, Earth Resources Association Report Y/OWI/SUB-7414/1, 180 p.

Kaufman, D. W., and Slawson, C. B., 1950, Ripple marks in rock salt of the Salina Group: Journal of Geology, v. 58, p. 24–29.

Klement, C., 1895, Ueber die Bildung des Dolomits: Tschermak's Mineralogisch Petrographische Mitteilungen, N.F., v. 14, p. 526–544.

Konyen, A., 1982, The nature and origin of reefal vug linings in the Salina Formation, southwestern Ontario [B.Sc. thesis]: Windsor, Ontario, Canada, University of Windsor, 62 p.

Landes, K. K., 1945, The Salina and Bass Island rocks in the Michigan Basin: U.S. Geological Survey Oil and Gas Investigation Map 40, scale 1:1,457,280.

Leibold, A. W., and Cercone, K. R., 1989, The Niagara/Salina contact in Michigan pinnacle reefs; A new interpretation and implications for ocean chemistry: Geological Association of Canada Program with Abstracts, v. 14, p. A64.

Lilienthal, R. T., 1978, Stratigraphic cross sections of the Michigan Basin: Michigan Geological Survey Report on Investigations 19, 125 p.

Linck, G., 1937, Bildung des Dolomits und Dolomitisierung: Chemie der Erde, v. 11, p. 278–286.

Lucas, P. T., 1954, Environment of Salina salt deposition [M.Sc. thesis]: University of Michigan, quoted in Dellwig, 1955.

Lucia, F. J., 1972, Recognition of evaporite-carbonate shoreline sedimentation, in

Rigby, J. K., and Hamblin, W. K., eds., Recognition of ancient sedimentary environments: Society of Economic Paleontologists and Mineralogists Special Publication 16, p. 160–191.

Maiklem, W. R., 1971, Evaporative drawdown; A mechanism for water-level lowering and diagenesis in the Elk Point Basin: Bulletin of Canadian Petroleum Geology, v. 15, p. 434–467.

Mantek, W., 1973, Niagaran pinnacle reefs in Michigan: Michigan Basin Geological Society Annual Field Excursion Guidebook, p. 35–46.

Matthews, R. D., 1970, The distribution of Silurian potash in the Michigan Basin, in Kneller, W. A., ed., Sixth Forum on Geology of Industrial Minerals: Michigan Geological Survey Miscellaneous Paper 1, p. 20–33.

——, 1975, Evaporite cycles in the Devonian of Michigan: Ontario Petroleum Institute, 14th Annual Conference, Paper 11, 13 p.

Matthews, R. D., and Egleson, G. C., 1974, Origin and implications of a midbasin potash facies in the Salina salt of Michigan, in Coogan, A. H., ed., Fourth Symposium on Salt: Cleveland, Northern Ohio Geological Society, v. 1, p. 15–34.

McCollough, C. N., Jr., 1975, Origin of pore filling salt in the Niagaran reefs of northern Michigan [abs.]: American Association of Petroleum Geologists Bulletin, v. 59, p. 1737.

Melhorn, W. N., 1958, Stratigraphic analysis of Silurian rocks in the Michigan Basin: American Association of Petroleum Geologists Bulletin, v. 42, p. 816–838.

Meloy, D. U., 1974, Depositional history of the northern carbonate bank in the Michigan Basin [M.Sc. thesis]: Ann Arbor, University of Michigan, 78 p.

Mesolella, K. J., 1972, Reciprocal deposition within Niagara and early Cayugan (Silurian) carbonates and evaporites, northern Michigan Basin: Ontario Petroleum Institute 11th Annual Conference, Paper 8, 34 p.

——, 1978, Paleogeography of some Silurian and Devonian reef trends; Central Appalachian Basin: American Association of Petroleum Geologists Bulletin, v. 62, no. 9, p. 1607–1644.

Mesolella, K. J., Robinson, J. D., McCormick, L. M., and Ormiston, A. R., 1974, Deposition of Silurian carbonates and evaporites in Michigan Basin: American Association of Petroleum Geologists Bulletin, v. 58, p. 34–62.

Michigan Department of Conservation, 1964, Stratigraphic succession in Michigan: East Lansing, Michigan, Department of Natural Resources, 1 p.

Miles, M. C., O'Shea, K., Lapcevic, P. A., Fritz, P., and Frape, S. K., 1985, Isotope study of the Salina Formation of southern Ontario: Geological Association of Canada Program with Abstracts, v. 10, p. A40.

Mueller, J., and Fabricius, F. H., 1978, Luneburgite [$Mg_3(PO_4)_2 \cdot B_2O(OH)_4 \cdot 6H_2O$] in Upper Miocene sediments of the eastern Mediterranean Sea, in Hsu, K. J., and others, eds., Initial reports of the Deep Sea Drilling Project: Washington, D.C., U.S. Government Printing Office, v. 42A, p. 661–664.

Norris, A. W., and Sanford, B. V., 1969, Paleozoic and Mesozoic geology of the Hudson's Bay Lowlands (Operation Winisk): Geological Survey of Canada Paper 68-53, p. 169–205.

Nurmi, R. D., and Friedman, G. M., 1977, Sedimentology and depositional environments of basin-center evaporites, Lower Salina Group (Upper Silurian), Michigan Basin, in Nissenbaum, A., ed., Hypersaline brines and evaporitic environments: Amsterdam, Netherlands, Elsevier, p. 23–52.

O'Shea, K. J., Miles, M. C., Fritz, P., Frape, S. K., and Lawson, D. E., 1988, Oxygen-18 and carbon-13 in the carbonates of the Salina Formation of southwestern Ontario: Canadian Journal of Earth Sciences, v. 25, no. 2, p. 182–194.

Perthuisot, J. P., 1975, La sebkha el Melah de Zarzis; Genese et evolution d'un bassin salin paralique: Ecole Normale Superieure, Laboratoire de Geologie, Paris, France, Travaux, v. 9, 252 p.

Pirtle, G. W., 1932, Michigan structural basin and its relationship to surrounding areas: American Association of Petroleum Geologists Bulletin, v. 16, p. 145–152.

Prouty, C. E., 1986, Tectonic development of the Michigan Basin [abs.]: American Association of Petroleum Geologists Bulletin, v. 70, p. 1069.

Sanford, B. V., 1969, Silurian of southwestern Ontario: Ontario Petroleum Institute 8th Annual Conference, Paper 5, 44 p.

Sanford, B. V., Norris, A. W., and Bostock, H. H., 1968, Geology of the Hudson's Bay Lowlands (Operation Winisk 1967): Geological Survey of Canada Paper 67-60, 45 p.

Sanford, B. V., Thompson, F. J., and McFall, G. H., 1984, Phanerozoic and recent tectonic movements in the Canadian Shield and their significance to the nuclear fuel waste management program, in Heinrich, W. F., ed., Proceedings, Workshop on Transitional Processes: Atomic Energy of Canada Ltd., Whiteshell Nuclear Research Establishment Report AECL-7822, p. 73–96.

Sarg, J. F., 1982, Off-reef Salina deposition (Silurian), southern Michigan Basin; Implications for reef genesis, in Handford, C. R., Loucks, R. G., and Davies, G. R., eds., Depositional and diagenetic spectra of evaporites; A core workshop: Society of Economic Paleontologists and Mineralogists Core Workshop 3, p. 354–372.

——, 1983, Eustatic control of cyclic carbonate-evaporite deposition (Silurian), southern Michigan Basin: Geological Society of America Abstracts with Programs, v. 15, p. 678.

Schlager, W., 1981, The paradox of drowned reefs and carbonate platforms: Geological Society of America Bulletin, v. 92, p. 197–211.

Sears, S. O., and Lucia, F. J., 1980, Dolomitization of northern Michigan Niagara reef by brine refluxion and freshwater/seawater mixing, in Zenger, D. H., and others, eds., Concepts and models of dolomitization: Society of Economic Paleontologists and Mineralogists Special Publication 28, p. 215–235.

Sonnenfeld, P., 1984, Brines and evaporites: Orlando, Florida, Academic Press, Inc., 613 p.

——, 1985, Flow regimes in the Silurian of the Michigan Basin in Cercone, R., and Budai, J., eds., Ordovician and Silurian rocks of the Michigan Basin and its margins: Michigan Basin Geological Society Special Paper 4, p. 157–171.

——, 1990a, Marine evaporite facies: Geologisches Jahrbuch (in press).

——, 1990b, On depths of ancient marine evaporite basins: Geologisches Jahrbuch (in press).

Sonnenfeld, P., and Perthuisot, J. P., 1989, Brines and evaporites, in 28th International Geological Congress Short Course Notes 3: Washington, D.C., American Geophysical Union, 128 p.

Sonnenfeld, P., Hudec, P. P., Turek, A., and Boon, J. A., 1977, Base-metal concentration in a density-stratified evaporite pan in Fisher, J. H., ed., Reefs and evaporites; Concepts and depositional models: American Association of Petroleum Geologists Studies in Geology 5, p. 181–187.

Straw, W. T., 1985, Facies and depositional setting of the A-1 carbonate (Silurian) in the Michigan Basin in Cercone, R., and Budai, J., eds., Ordovician and Silurian rocks of the Michigan Basin and its margins: Michigan Basin Geological Society Special Paper 4, p. 143–156.

Thomas, G. E., Sonnenberg, F. P., and Hulse, W., 1989, A paleotopographic approach to the Silurian reefs of the Illinois Basin: Oil and Gas Journal, v. 87, no. 17, p. 78–84.

Treesh, M. I., 1973, Depositional environments of the Salina Group (Upper Silurian) in New York State [Ph.D. thesis]: Troy, New York, Rensselaer Polytechnical Institute, 127 p.

Tremper, L. R., 1973, Lithofacies and stratigraphic analysis of the Salina Group of the "North Slope" of the Michigan Basin [M.Sc. thesis]: Ann Arbor, University of Michigan, 58 p.

Tucker, R. M., and Cann, J. R., 1986, A model to estimate the depositional brine depths of ancient halite rocks; Implications for ancient subaqueous depositional environments: Sedimentology, v. 33, p. 401–412.

Van der Voo, R., 1988, Paleozoic paleogeography of North America, Gondwana, and intervening displaced terranes; Comparison of paleomagnetism with paleoclimatology and biogeographic patterns: Geological Society of America Bulletin, v. 100, p. 311–324.

Van Tuyl, F. M., 1916, The origin of dolomite: Iowa Geological Survey Reports, v. 25, p. 251–422.

MANUSCRIPT ACCEPTED BY THE SOCIETY JUNE 1, 1990

Printed in U.S.A.

Geological Society of America
Special Paper 256
1991

# Upper Devonian biostratigraphy of Michigan Basin

**Raymond C. Gutschick**
*Volunteer, U.S. Geological Survey, and Professor Emeritus, University of Notre Dame, 2901 Leonard Street, Medford, Oregon 97504*

**Charles A. Sandberg**
*U.S. Geological Survey, MS 940, Box 25046, Denver Federal Center, Denver, Colorado 80225*

## ABSTRACT

The Late Devonian Michigan Basin was floored by the Middle and Upper Devonian Squaw Bay Limestone, which was deposited during the downwarping that produced the basin within a former Middle Devonian carbonate platform. The Squaw Bay comprises three beds, each having a different conodont fauna. The two upper beds, deposited during the *transitans* Zone, have different conodont biofacies that reflect this deepening. The basin was largely filled by the deep-water, anaerobic to dysaerobic, organic-rich, black Antrim Shale, which has a facies relationship with the prodeltaic, greenish gray Ellsworth Shale that prograded into the basin from the west. The Upper Devonian (Frasnian to Famennian) Antrim Shale is divided into four members, from base to top: the Norwood, Paxton, Lachine, and upper members. These members are more or less precisely dated by conodonts. The Norwood was deposited during the *transitans* Zone to *Ancyrognathus triangularis* Zone, and the Paxton was deposited from that zone probably through the *linguiformis* Zone at the end of the Frasnian. The overlying Lachine was deposited during the early Famennian and has yielded faunas of the Upper *crepida* and Lower *rhomboidea* Zones. Only the lower part of the upper member is exposed, and near Norwood, Michigan, it yielded conodonts of the Lower *marginifera* Zone. The widespread Famennian floating plant *Protosalvinia (Foerstia)* has not yet been found in outcrops of the Antrim, and should not be expected to occur except in the upper member or highest part of the Lachine Member. Its range in terms of conodont zones is from the Upper *trachytera* Zone through the Lower *expansa* Zone and possibly into the Middle *expansa* Zone. One known subsurface occurrence might be datable as *rhomboidea* or Lower *marginifera* Zone, depending on gamma ray correlations to outcrops. Black shale deposition ended when the Late Devonian mud delta of the Bedford Shale prograded across the Michigan Basin from the east and then retreated as the regressive Berea Sandstone was being deposited during the major eustatic sea-level fall that ended the Devonian. The Bedford was deposited during the Upper *expansa* to Lower *praesulcata* Zones, and the Berea was deposited during the Middle to Upper *praesulcata* Zones. Both formations contain the spore *Retispora lepidophyta*, which is a global indicator of latest Devonian age.

## INTRODUCTION

The Late Devonian depositional Michigan Basin now occupies all of lower Michigan, and parts of western Ontario, northwestern Ohio, northern Indiana, northeastern Illinois, and eastern Wisconsin (Fig. 1). Continuous concurrent marine sedimentation outside the map area occurred in the Moose River basin to the north, Appalachian basin to the southeast, and Illinois basin to the southwest. The Wisconsin lowlands flanked the basin to the west. The treated Devonian stratigraphic sequence in the Michigan Basin (Fig. 2) comprises, in ascending order: the Middle Devonian Thunder Bay Limestone (the uppermost formation of

Gutschick, R. C., and Sandberg, C. A., 1991, Upper Devonian biostratigraphy of Michigan Basin, *in* Catacosinos, P. A., and Daniels, P. A., Jr., eds., Early sedimentary evolution of the Michigan Basin: Geological Society of America Special Paper 256.

the Traverse Group); the Middle and Upper Devonian Squaw Bay Limestone; and the Upper Devonian Antrim, Ellsworth, and Bedford Shales and Berea Sandstone. Sources of primary geologic information are bedrock exposures along the outcrop belt marginal to the basin and subsurface well data from within the basin. The entire basin is veneered by glacial drift, so that outcrops of nonresistant shales are scarce. Formations dip basinward and are concealed beneath younger beds or below Lakes Michigan and Huron.

## EVIDENCE FROM THE FOSSIL RECORD

Conodonts are the most important fossils found in the Upper Devonian of the Michigan Basin. This is because the global standard Late Devonian conodont zonation contains 29 zones, all but two of which have been recognized in North America (Ziegler, 1962, 1971; Sandberg and Ziegler, 1973; Sandberg, 1979; Ziegler and Sandberg, 1984; Sandberg and others, 1988; Sandberg and others, 1989a; Sandberg and others, 1989b).* The large number of zones, each averaging only 0.5 m.y. in length, permits very precise correlation of events not only within the Michigan Basin, but also with events in other parts of the world. Conodonts from limestone beds and calcareous concretions in outcrops and well cores in Michigan, Ohio, and Indiana that have been determined and zonally assigned by Sandberg are here reported and listed for the first time. Not only do these conodont collections date the members of the Antrim Shale proposed herein, but they also change the ages of many well-known stratigraphic units and some long-standing correlations with other regions. Conodont biofacies in the Antrim Shale and equivalents are assigned according to Late Devonian biofacies models (Sandberg, 1976; Sandberg and Dreesen, 1984). They aid in interpreting the changing paleogeography, paleotectonics, and paleoecology of depositional environments (Sandberg, 1988) and the changes in stratification of the water column during deposition (Sandberg and others, 1989a).

Another important fossil for regional correlations is *Protosalvinia (Foerstia)*, which is generally agreed to be a fragment of thallose alga with a pelagic, sargassoid habit (Arnold, 1954; Schopf and Schwietering, 1970; Cross, 1983). Opinions differ, however, as to whether this tropical, floating aquatic plant was marine or terrestrial (Schopf, 1978; Gray and Boucot, 1979). We interpret that *Protosalvinia* grew in marine swamps, mainly on deltas bordering the Michigan Basin, and that its abundant occur-

Figure 1. Index map of the Michigan Basin, encircled by outcrop belt of Upper Devonian (Du) rocks, shown by lined pattern. Restored limits of Late Devonian basin and straits connecting it to adjacent basins are shown by heavy dashed lines. Only quarries, outcrop areas, and key wells discussed in text are located. P = *Protosalvinia* occurrences. Shaded rectangle shows area of Figure 3.

rences in black shales resulted from drowning of its habitat and seaward dispersal during major Famennian rises of sea level, such as those depicted on the Devonian sea-level curve (Johnson and Sandberg, 1989). We further interpret that these sea-level changes were glacio-eustatic in origin and were associated with interglacial stages of Southern Hemisphere glaciation (Sandberg and others, 1988). *Protosalvinia* has wide distribution in the Antrim, Chattanooga, New Albany, and Ohio Shales in the eastern United States, in the Kettle Point Formation of Ontario, in the Bakken Formation of the Williston basin in North Dakota (Thrasher, 1987), and even in the Amazon Basin of Brazil (Gray and Boucot, 1979).

*Protosalvinia* has been recently considered to represent a single floral zone, equated to only the Lower *expansa* conodont Zone, and has been used as a virtual time line for regional correlations (Conkin, 1985; Ettensohn and others, 1988). However, we now document that *Protosalvinia* is longer ranging, at least from the middle Famennian Upper *trachytera* Zone to the late Famennian Lower *expansa* Zone (Fig. 2). Moreover, we believe that circumstantial evidence exists for both earlier and later Famennian range extensions. We have found *Protosalvinia* in association with conodont faunas of the Upper *trachytera* Zone in two

---

*The Late Devonian standard conodont zonation, which is the basis for all dating in this paper, was revised recently by Ziegler and Sandberg (1990). Zonal-name modifiers such as Lower and Upper have been changed to Early and Late, respectively, and some zonal names have been totally changed. For example, the former Upper (*M.*) *asymmetrica* Zone, as used herein (e.g., Figs. 2 and 5), has been replaced by Early *hassi* Zone; the former *Ancyrognathus triangularis* Zone has been divided into the Late *hassi* Zone and *jamieae* Zone, in ascending order; and the former Lower *gigas* Zone has been changed to Early *rhenana* Zone. For a complete discussion and listing of the new standard zones, the reader should refer to Ziegler and Sandberg (1990).

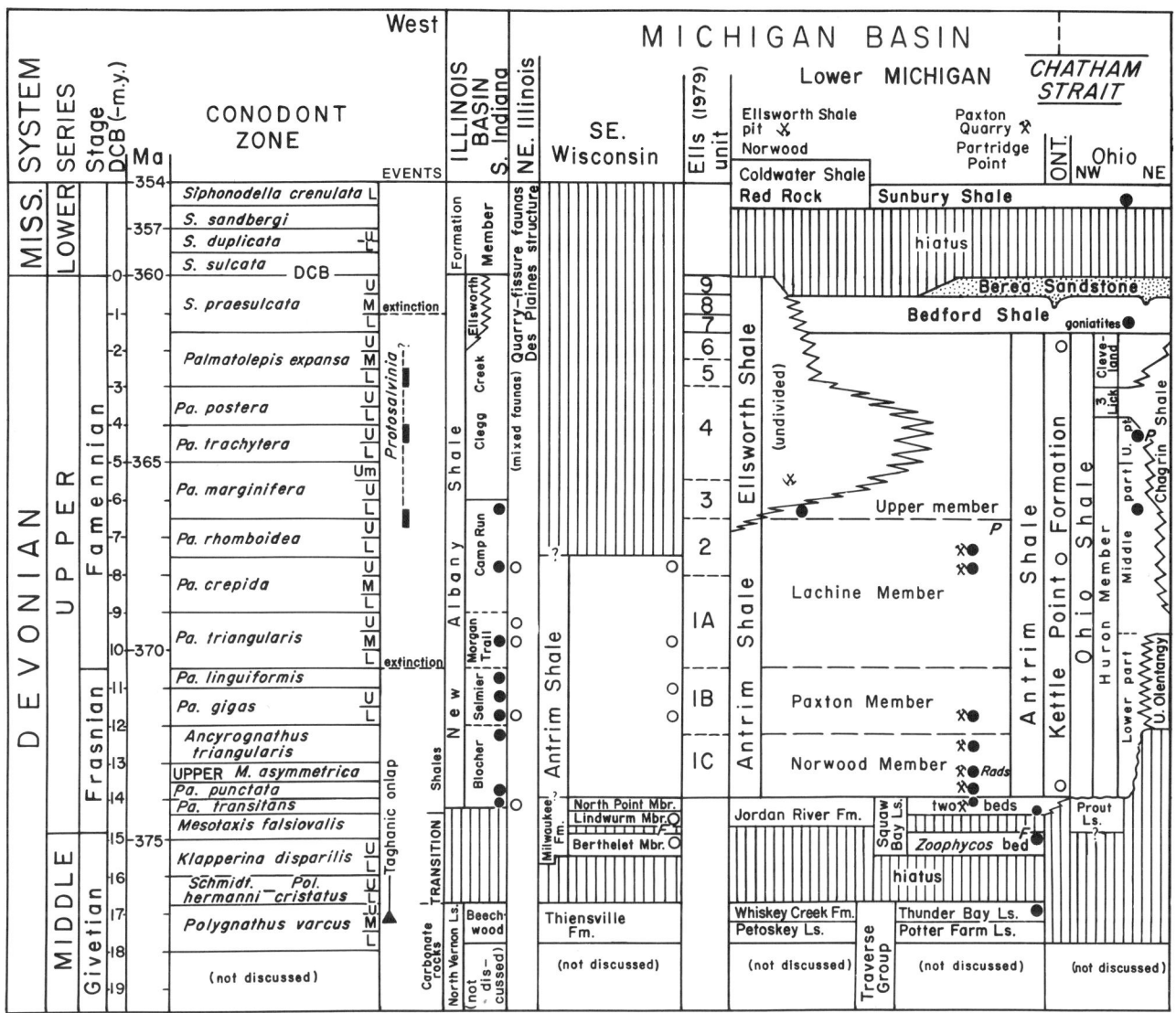

Figure 2. Correlation chart showing Upper Devonian and adjacent strata in Michigan Basin, gamma ray units of Ells (1979), and correlations with Illinois basin and northern Ohio. Rocks are dated by standard conodont zones; radiometric time scale (Palmer, 1983); and biostratigraphic time scale (Sandberg and others, 1989b), giving date in –m.y. before starting point of 0 m.y. at Devonian–Carboniferous boundary (DCB). Conodont collections identified by Sandberg are shown by closed black circles in stratigraphic position opposite assigned zone; zonal assignments of other authors are shown by open circles. Also shows positions of late Frasnian and latest Famennian mass extinctions; range and occurrences of floating aquatic plant *Protosalvinia (P)*; and occurrences of agglutinated foraminiferans *(F)*.

concretions within the upper part of the Huron Member of the Ohio Shale at the type locality of the Huron Member, as described by Broadhead and others (1980, p. 10, locality 1, unit 50). These conodont faunas, which unequivocally represent the Upper *trachytera* Zone, contain *Palmatolepis glabra distorta, Pa. perlobata helmsi, Pa. perlobata sigmoidea, Pa. gracilis sigmoidalis, Scaphignathus velifer, Mehlina strigosa, Bispathodus stabilis,* and *Polygnathus brevilaminus*. The joint occurrence of *Pa. glabra distorta* and *Pa. gracilis sigmoidalis* is particularly significant and mandates an assignment to the Upper, rather than Lower, *trachytera* Zone. Ziegler and Sandberg (1984) reported the first occurrence of *Pa. gracilis sigmoidalis* to be in the Upper *trachytera* Zone. They also reported the last occurrence of *Pa. glabra distorta* to be in the Lower *trachytera* Zone, but this subspecies was subsequently found to range higher, into the Upper *trachytera* Zone, at Borghausen, West Germany. Matthews (1983) has reported *Protosalvinia* from a core of Antrim Shale in the Dow-DOE #110 well (Fig. 1) on the eastern side of the Michigan Basin in a position that he equated to the upper part of subsurface gamma ray unit 2 of Ells (1979). Assuming that Matthews' positioning is correct, his discovery would date *Protosalvinia* as occurring in the early Famennian *rhomboidea* or Lower *marginifera*

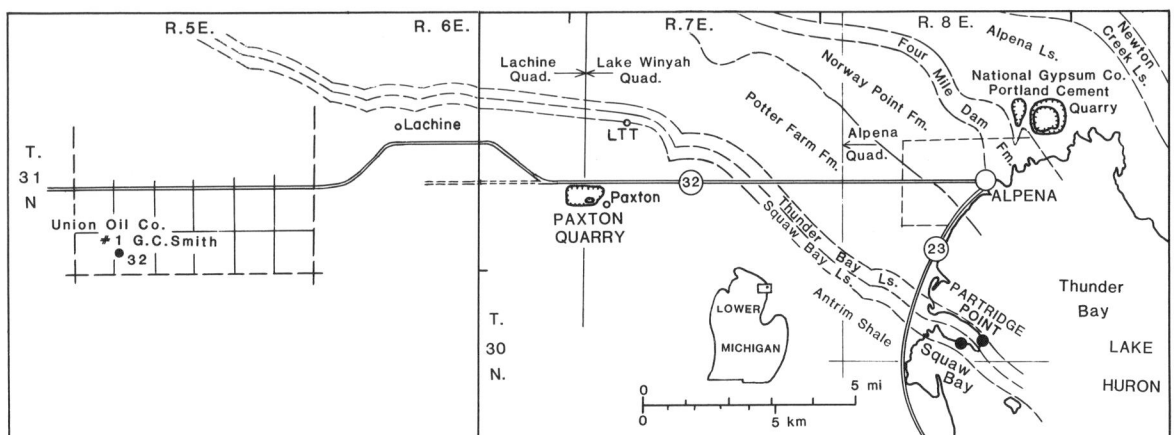

Figure 3. Geologic map (adapted from Ehlers and Kesling, 1970) of part of Alpena County, Michigan, showing locations of Paxton Quarry, Union Oil Co. #1 Smith well, Partridge Point area, and Lancaster Truck Trail locality (LTT). Gamma-ray log of well is correlated with exposures in the Paxton Quarry in Figure 5. Black dots show type localities.

Zone. This age would confirm his correlation of the subsurface occurrence with the middle part of the Huron (Matthews, 1983, Fig. 3), below the *Protosalvinia* that we date as Upper *trachytera* Zone in the upper part of the Huron. We suspect that the upper range of *Protosalvinia,* though not occurring as abundantly as in the bloom in the Lower *expansa* Zone, may continue until the glacially induced extinction (Sandberg and others, 1988) at the start of the Middle *praesulcata* Zone. We speculate that in the Michigan Basin, there may have been as many as three Famennian blooms of *Protosalvinia,* each associated with progressively greater glacio-eustatic rises, corresponding roughly to these sea-level rises in the western United States: event 12, Lower *marginifera* Zone; event 14, Lower *trachytera* Zone; and event 16, a major eustatic rise, Lower *expansa* Zone (Sandberg and others, 1989b).

A third important group of fossils are spores and palynomorphs (Cross and Bordner, 1980), especially the ubiquitous, biostratigraphically significant spore *Retispora lepidophyta* (Kedo) Playford. The extinction of this species is now recognized to occur globally at the end of the Devonian, and its disappearance supports the internationally accepted Devonian-Carboniferous boundary placed between the *Siphonodella praesulcata* and *S. sulcata* Zones (e.g., Sandberg and others, 1972; Eames, 1974; Streel and Traverse, 1978; Sandberg, 1981; Richardson and McGregor, 1986; Streel, 1986; Clayton and others, 1989). *Retispora lepidophyta* occurs widely in the Berea Sandstone in the Michigan and Appalachian basins (Sanford, 1967; de Witt, 1970). Its presence supports our reassignment to the Upper Devonian of the Bedford Shale and Berea Sandstone, which is also based on other fossil groups such as conodonts, ammonoids, and brachiopods, as well as a precise fit with the Devonian sea-level curve (Johnson and others, 1985, 1986; Johnson and Sandberg, 1989). Higher occurrences of *R. lepidophyta* in the Lower Mississippian Sunbury Shale (Lower *crenulata* Zone) are now interpreted to be reworked (Streel, 1986).

Other fossil groups that were useful in making our biostratigraphic interpretations are: foraminiferans (Roberts, 1972; Gutschick, 1987), radiolarians (Foreman, 1963; Holdsworth, 1977), and tasmanitids (Gutschick and Wuellner, 1983). An important source of paleontologic references for black shales in Upper Devonian rocks in the Michigan Basin is by Barron and Ettensohn (1980).

## BIOSTRATIGRAPHY OF THE NORTHERN CENTRAL BASIN

Some of the best surface exposures of Devonian rocks in lower Michigan are found in the northern part of the Michigan Basin where classic sections of the Middle Devonian Traverse Group occur in natural outcrops and quarries. The stratigraphy of the Traverse Group and equivalent units has been discussed by Kesling and others (1974) and Ehlers and Kesling (1970). Our reconstruction of the sequence starts in the uppermost Traverse carbonate rocks on Partridge Point along the shore of Lake Huron, south of Alpena, Michigan (Fig. 3), and continues up section in the Paxton Quarry, where the upper part of the Squaw Bay Limestone is in contact with the lower part of the Antrim Shale. The highest exposure of the lower Antrim Shale is along Lake Michigan south of Norwood, Michigan, and the section continues with the lower Ellsworth Shale in the pit at Ellsworth, Michigan. Although surface exposures are few along this northern outcrop belt, they contain the lower two-thirds of the total Antrim sequence. Higher Upper Devonian strata and the upper one-third of the Antrim are known only from the subsurface.

### Important localities

The most useful exposures are at Partridge Point, type area of the Thunder Bay and Squaw Bay Limestones, and Paxton Quarry, where the most nearly complete section of the Antrim

Shale occurs (Fig. 1; Gutschick, 1987). Farther west the type Antrim Shale crops out along the east shore of Lake Michigan, south of Norwood, Michigan, and the type Ellsworth Shale (part) is well exposed in an abandoned pit at Ellsworth, Michigan. Devonian rocks also occur in the Milwaukee area, southeastern Wisconsin, and Devonian fossils are found in fissure fillings preserved in stratigraphic leaks into Silurian reef crevices at three quarries in northeastern Illinois. The type Kettle Point Shale at Kettle Point along the east shore of Lake Huron in Ontario rounds out the outcrop belt.

More information on the extent of the basin can be obtained from pre-Antrim exposures and from the subsurface Antrim Shale in the Milwaukee, Wisconsin, area. Fissure fillings with conodonts occur in three quarries in northeastern Illinois, and fillings in some quarries contain *Protosalvinia* fragments. The type locality of the Kettle Point Shale in Ontario contains conodonts and many concretions, the age of which is equivalent to those in the lower part of the Antrim in the Paxton Quarry.

*Partridge Point.* Partridge Point is a small peninsula, subparallel to the strike of the strata, that projects southeastward into Lake Huron (Fig. 3). A large-scale geologic sketch map of the tip of the peninsula was shown by Gutschick (1987, Fig. 6). The type sections of the Thunder Bay and Squaw Bay Limestones (Fig. 4) are exposed along the shoreline in that area, the former in the W½SE¼ Sec. 11, and the latter in the SW¼SW¼ Sec. 11, both in T.30N.,R.8E., in the Alpena 7.5-minute Quadrangle, Alpena County, Michigan. The Upper Devonian upper part of the Squaw Bay Limestone and the Antrim Shale are concealed under Squaw Bay. However, Partridge Point provides biostratigraphic information for the Middle Devonian units that directly underlie these two Upper Devonian units in the nearby Paxton Quarry (Fig. 3). Previous stratigraphic and paleontologic studies on Partridge Point include Warthin and Cooper (1935, 1943), Ehlers and Kesling (1970), Müller and Clark (1967), Bultynck (1976), and Gutschick (1987).

*Paxton Quarry.* The Paxton Quarry west of Alpena, Michigan (Figs. 1 and 3), contains an exposure of upper Squaw Bay Limestone in contact with the lower 41 m of Antrim Shale. It is the best exposure of the Antrim Shale in the outcrop belt of the Michigan Basin. The sequence is described in more detail in the Appendix because three new members of the lower Antrim Shale are proposed from this locality. Sources of information useful to our study are: Ver Wiebe (1927), Morse (1938), Ehlers and Kesling (1970), Kesling and others (1976), Ells (1979), and Gutschick (1987).

The biostratigraphic section in the Paxton Quarry is shown graphically, and the lithologic units are correlated with the gamma ray units of Ells (1979) in Figure 5. The gamma ray log used for this correlation is adapted from the Union Oil Co. of California #1 Smith well in Sec.32,T.31N.,R.5E., 18 km west of the quarry (Fig. 3). Quarry operations have penetrated the base of the Antrim Shale and the upper 1.5 m of Squaw Bay Limestone, which is accessible in the sump. The upper Squaw Bay consists of greenish gray, pyritic, calcareous shale and argillaceous limestone, which contains a few conodonts of the *transitans* Zone. The contact with the overlying Antrim Shale is sharp and overlain by fissile, pyritic, black shale.

*Norwood area.* Along the shore of Lake Michigan north and south of Norwood, Michigan, an important exposure includes the Middle Devonian Traverse Group and overlying Upper Devonian Jordan River Formation, the informal "Norwood Shale" (of Ehlers, 1938), and the type Antrim Shale (Lane, 1901, 1902; Kesling and others, 1974). The Whiskey Creek Formation at the top of the Traverse Group there is correlated with the Thunder Bay Limestone of the Alpena area (Fig. 2). The Whiskey Creek is gray to brown limestone, containing some dolostone and chert; biostromal layers are stromatolitic and contain benthic colonial corals, stromatoporoids, and shelly fossils.

The Jordan River Formation, lying between Middle Devonian carbonate rocks and Upper Devonian black shales, corre-

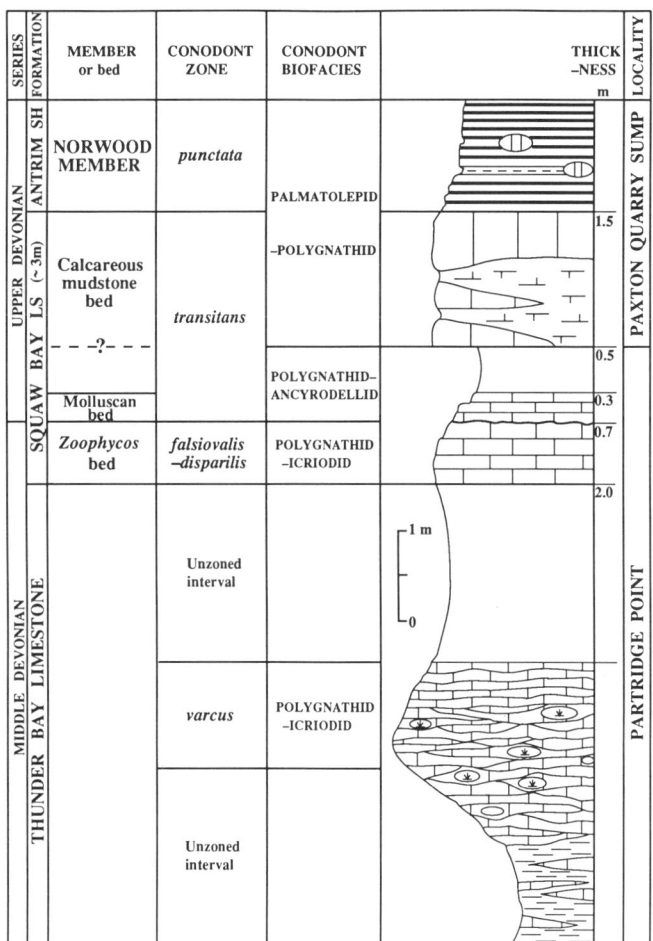

Figure 4. Composite columnar section, showing conodont zones and biofacies, for type localities of Thunder Bay and Squaw Bay Limestones along shoreline of Lake Huron, Alpena County, Michigan (Fig. 3). Section is interrupted by glacial drift or lacustrine sediment cover and underwater concealed intervals across formational contacts. Section is offset to Paxton Quarry, where it continues with a third, upper bed of Squaw Bay Limestone in contact with basal part of Antrim Shale.

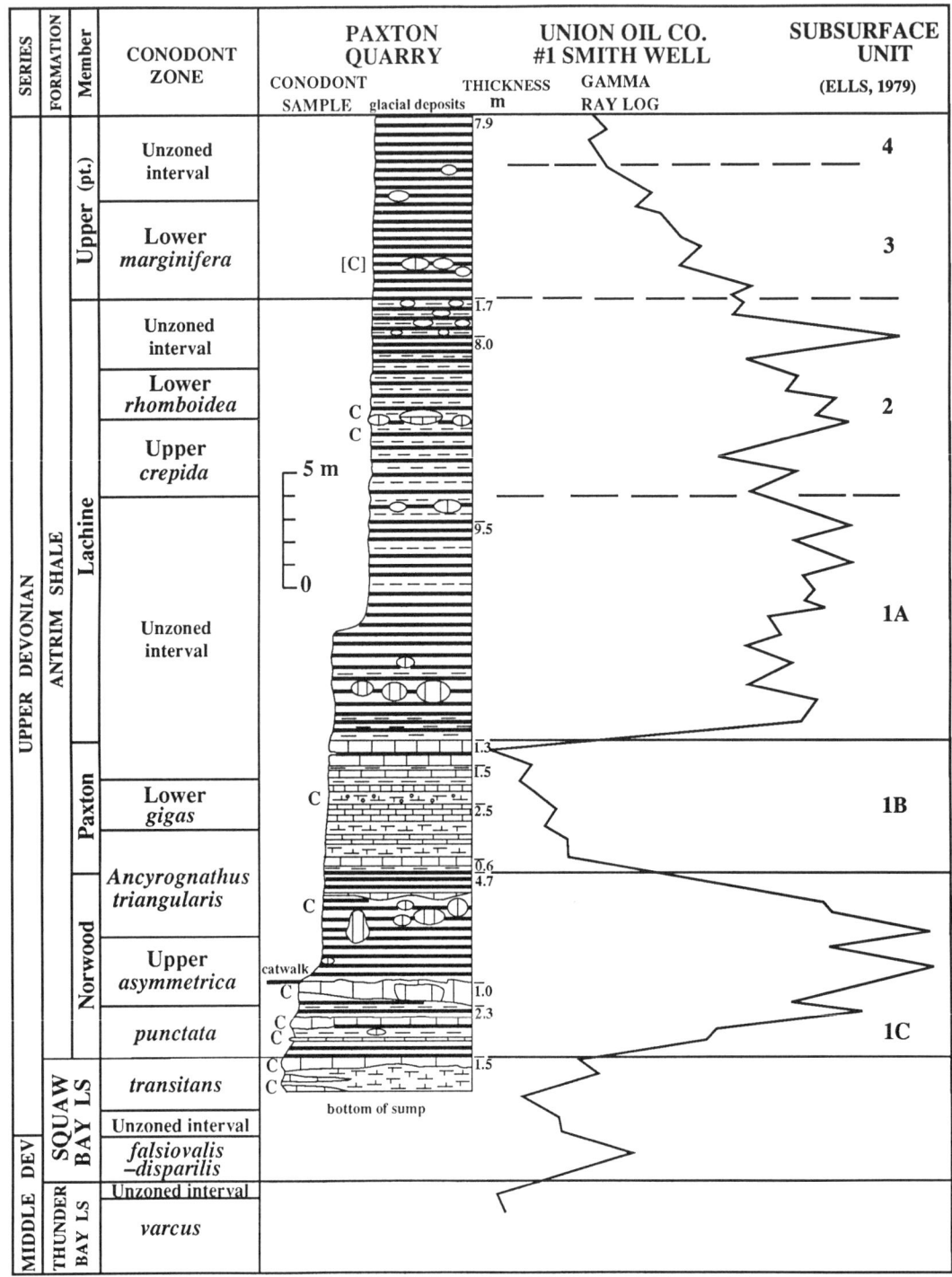

Figure 5. Detailed columnar section and conodont zonation of upper part of Squaw Bay Limestone and Antrim Shale in Paxton Quarry and correlation with gamma-ray log, showing subsurface units of Ells (1979), of Union Oil Co. #1 Smith well, about 18 to 20 km to west (Fig. 3). C = conodont collection determined and dated by C. A. Sandberg. [C] = collection projected into quarry section from concretion on shoreline near Norwood, Michigan.

lates with the Squaw Bay Limestone to the east and Milwaukee Formation to the west. The type locality and only known exposure of the Jordan River is at the Lake Michigan shoreline. The formation is composed of very thin limestones and nonresistant shales, about 1 to 2 m thick. Bottom and upper contacts are not exposed, and the formation is poorly defined. Two shelly faunas in the Jordan River were both assigned to the Late Devonian (Cooper and others, 1942), but their age is now uncertain according to recent redefinition (Sandberg and others, 1989a) of the Middle–Upper Devonian boundary (Fig. 2).

The type Antrim Shale crops out 1.6 km south of Norwood (Lane, 1901, 1902; Kesling and others, 1974), along Lake Michigan, where it forms a 4- to 5-m cliff along the beach strandline. The sequence is primarily compact, concretionary, black pyritic shale. Thin alternating layers of black and bioturbated, greenish gray shales with large *Zoophycos* traces occur below lake level but are occasionally exposed during low water. Large calcareous concretions occur sparingly in the shale and have fallen loose on the beach. One such concretion, about 1 m in diameter, contains an abundant conodont fauna of the Lower *marginifera* Zone. The gamma ray log of National Associates Petroleum Co. #1 Wolgamott well in the NW¼NW¼SE¼Sec.19,T.32N.,R.7W., about midway between the type localities of the Antrim and Ellsworth Shales, was used to correlate the Antrim type section. Below the Ellsworth, the well log indicates the presence, in descending order, of the Lachine, Paxton, and Norwood Members of the Antrim (units 3, 2, 1A, 1B, and 1C of Ells, 1979).

*Ellsworth area.* The type Ellsworth Shale in the Petoskey Portland Cement Company's quarry pit was visited by Ehlers, Case, and Ulrich in 1926; they recorded 18.6 m of greenish gray, plastic shale with a few sandstone interbeds (Kesling and others, 1974). A bore hole drilled below the floor of the pit encountered 12 m of greenish gray shale before entering black shale, thus indicating a minimum thickness of 30.6 m for the Ellsworth at its type locality.

Re-examination of the abandoned pit, in October 1982, revealed a well-exposed shale sequence measuring 24.4 m from floor of the pit to glacial drift cover. The type Ellsworth Shale is entirely medium greenish gray, sparsely fossiliferous, noncalcareous clay shale or mudstone, which is plastic when wet. Shale in the upper 3 m is medium gray to light olive gray. Three thin, light olive-gray siltstone beds, on a terrace about 15 m above the base of the exposure, exhibit crossbedding, sole marks, and trace fossils. Comparison of the type section with gamma ray log A-22 (Ells, 1979) indicates that the Ellsworth in the pit correlates with his upper unit 3 and/or lower unit 4, and lies above the type Antrim Shale at Norwood.

### Underlying Middle Devonian unit

The Thunder Bay Limestone, at the top of the Middle Devonian Traverse Group, is well exposed along the northeast-facing shoreline of Partridge Point (Fig. 3). It consists of light-colored fossiliferous limestone and shaly beds with coralline and shelly faunas (Fig. 4). Solitary and colonial corals, stromatoporoids, bryozoans, crinoids, and blastoids are most abundant in this sessile-benthic biostromal community (Ehlers and Kesling, 1970). The top 0.6 m consists of medium gray, fine-grained, unfossiliferous limestone, darker and less fossiliferous than lower beds. We obtained the conodont *Icriodus latericrescens latericrescens* from this bed. A low-diversity conodont fauna representing the polygnathid-icriodid biofacies of the Upper *varcus* Subzone was extracted from the Thunder Bay Limestone by Bultynck (1976).

The contact between the Thunder Bay Limestone and overlying Squaw Bay Limestone is concealed by low, swampy terrain. The strike and low dip of the rocks southwest into the basin suggest that only about 2 m of section is concealed. A well-exposed rock pavement with *Zoophycos* swirls on the upper bedding surface forms the first outcrop above the Thunder Bay.

### Outcropping Middle and Upper Devonian units

*Squaw Bay Limestone (Middle and Upper Devonian).* The Squaw Bay Limestone, which spans the Middle–Upper Devonian boundary from the Middle Devonian *disparilis* or lower part of the *falsiovalis* Zone to the Late Devonian *transitans* Zone (Fig. 2), was named for a thin exposure, 1 m thick, of brown, fossiliferous limestone along the south shore of Partridge Point between the Thunder Bay Limestone below and the Antrim Shale above (Warthin and Cooper, 1935, 1943). Concealed intervals bound the exposure, so formational contacts are not exposed in the type section, estimated to be about 3 m thick (Ehlers and Kesling, 1970). The type Squaw Bay (Fig. 4) comprises two beds of different lithology, the *Zoophycos* bed and the overlying molluscan bed, which are separated by a hardground. A third bed, the calcareous mudstone bed, which is concealed at the type locality, is recognized to occur directly below the Antrim Shale at the bottom of the sump in Paxton Quarry. All three beds contain conodont faunas of different zones or different biofacies.

*Zoophycos bed (Middle Devonian).* The *Zoophycos* bed is light olive-gray to brownish gray, fine-grained, silty, hard, pyritic to limonitic, dolomitic limestone that is rusty weathering, glauconitic, and burrowed. Only the upper 0.5 m is exposed as a rock pavement, and the lower 0.2 m is generally concealed below water level. The top surface displays many *Zoophycos* swirls parallel to bedding and several scattered vertical *Skolithos* tubes that evidence a hardground and a hiatus before deposition of the overlying molluscan bed. The *Zoophycos* bed has two important microfossil faunas, conodonts and siliceous agglutinated foraminiferans.

The *Zoophycos* bed, referred to as gray compact dolomitic limestone, was sampled for conodonts by Bultynck (1976, p. 140, locality 8, sample 1). He reported a conodont fauna including *Icriodus symmetricus, Polygnathus cristatus, Pol.* aff. *Pol. dengleri, Pol. dubius, Pol. ordinatus,* and *Pol. pennatus.* Bultynck (1976) concluded that the fauna was from either the Upper *hermanni-cristatus* Zone or the Lowermost *asymmetricus* Zone.

Later, Uyeno (*in* Uyeno and others, 1982) assigned the fauna to the *dengleri* Zone, which is equivalent to either the Middle Devonian *disparilis* Zone or the Middle Devonian part of the *falsiovalis* Zone of current usage (Fig. 2). Our collection from this bed is dominated by *Polygnathus pennatus, Pol. dubius,* and *Icriodus expansus,* but also contains *Schmidtognathus wittekindti, Pol. ordinatus, I.* cf. *I. symmetricus,* and *Elsonella rhenana. Schmidtognathus wittekindti* ranges into the *falsiovalis* Zone, and *I.* cf. *I. symmetricus* occurs at the same time. Hence, we are in agreement with the time span of Uyeno's assignment and show the *Zoophycos* bed as spanning the boundary between the *disparilis* and *falsiovalis* Zones (Fig. 2).

The *Zoophycos* bed yielded a well-preserved, low-diversity fauna of agglutinated foraminiferans. The fauna includes *Psammosphaera, Webbinelloidea, Tolypammina,* and single and multilocular *Oxinoxis* (Plate 1) and is assigned to the saccamminid biofacies (Sandberg and Gutschick, 1984). *Hyperammina* is questionably present. Many spherical globular forms are composed entirely of well-rounded glauconite grains, commonly cemented by pyrite. The fauna from the *Zoophycos* bed is similar, if not identical, to that described by Roberts (1972) from the Berthelet and Lindwurm Members of the Milwaukee Formation in the Milwaukee area, Wisconsin (Figs. 1 and 2). Multilocular *Oxinoxis* was found only in the upper 0.3 m of the Berthelet and lower 0.3 m of the Lindwurm (Roberts, 1972).

Other fossils in the *Zoophycos* bed include scolecodonts, and assorted water-worn phosphatic and glauconitic steinkerns of styliolinids, ostracodes, bryozoans, tiny brachiopods and snails, and pelmatozoans. These fossils were abraded by traction transport on the sea floor. The *Zoophycos* bed is also present at road level at the Lancaster Truck Trail locality (Fig. 3, LTT).

*Molluscan bed (Upper Devonian).* The few limestone layers in the upper part of the Squaw Bay Limestone are characterized by an abundant molluscan fauna. This molluscan bed crops out as an irregular patchy pavement along the beach or lies beneath the lake, depending on its fluctuating level. The limestone is brown and medium crystalline with thin, irregular bedding and coquinoid fossil remains. Linear conical tapering shells of styliolinids and orthocone cephalopods are present with their long axis parallel to current transport direction. Major taxa of this low-diversity fauna include the cephalopods *Bactrites warthini, Koenenites cooperi, Tornoceras (T.) uniangulare;* the bivalves *Buchiola* sp., *Praecardium* sp. (*Paneka* sp.); the styliolinid *Styliolina fissurella;* the gastropod *Diaphorostoma pugnus;* and poorly preserved fossil-wood fragments (Ehlers and Kesling, 1970). The goniatite *Tornoceras (T.) arcuatum* is also listed by Kirchgasser and House (1981). This fauna appears to be a pelagic, nektonic, and epiplanktonic assemblage. The only known sessile-benthic organism is a 3-mm-long, agglutinated tube (Plate 2, fig. 1) formed by coarse quartz grains, possibly of polychaete or oligochaete worm origin (Cuomo and Rhoads, 1985, and written communication, 1988).

The type Squaw Bay Limestone also yields excellent, well-preserved conodonts, which were studied by Müller and Clark (1967) and Bultynck (1976). The most recent study by Sandberg and others (1989a) reported a large (550 conodonts/kg) conodont fauna, representing the polygnathid-ancyrodellid biofacies of the Late Devonian *transitans* Zone, and including *Ancyrodella alata, A. rugosa, A. rotundiloba, A. soluta, Mesotaxis? dengleri, Polygnathus dubius,* and *Pol. pennatus.* They also reported that this fauna was unique among 11 studied faunas of the same zone from Euramerica because of the absence of otherwise ubiquitous *Icriodus symmetricus* (the ancestor of which, *I.* cf. *I. symmetricus,* does occur sparsely in the underlying *Zoophycos* bed) and that this absence had important implications concerning depth and oxygenation, namely that either the basin was shallow or, as we believe more likely, that the oxygen-minimum layer occupied a large lower part of the stratified water column. The molluscan bed is also present, though not well exposed, at the Lancaster Truck Trail locality (Fig. 3, LTT).

*Calcareous mudstone bed (Upper Devonian).* The only known exposure of the top of Squaw Bay Limestone and its contact with the overlying Antrim Shale is in the lower part of the sump at Paxton Quarry (Fig. 5; Appendix). It comprises greenish gray, calcareous mudstone at least 1.5 m thick, with thin lenses and interbeds of argillaceous limestone. The calcareous mudstone bed contains conodonts of the palmatolepid-polygnathid biofacies of the Late Devonian *transitans* Zone. The lowest limestone lens, at the bottom of the sump, yielded a sparse conodont fauna including *Ancyrognathus* cf. *A. ancyrognathoideus* (Sandberg and others, 1989a, p. 201), *Palmatolepis transitans, Mesotaxis asymmetrica,* and *Polygnathus dubius.* A thin limestone interbed at the top yielded an abundant fauna dominated by *Pa. transitans,* but including several specimens of *Pol. dubius* and a single specimen of *Icriodus* cf. *I. symmetricus.* Thus, *I. symmetricus* s.s. was not recovered from any bed of the Squaw Bay.

**Antrim Shale (Upper Devonian).** The type locality of the Antrim Shale is 1.6 km south of Norwood (Fig. 1; Kesling and others, 1974). The Paxton Quarry is here proposed as the principal reference section. The Antrim is present throughout the Michigan Basin and attains a maximum thickness of 200 m near the basin center. Its distribution and thickness are shown by Gutschick and Sandberg (this volume). A rich assortment of calcareous concretions in the Antrim was studied in detail at Paxton Quarry, and some yielded important biostratigraphic and paleoecologic data. Conodont faunas from concretions in the Antrim permit precise dating of beds and correlations with surrounding areas. The most common types of concretions in the quarry are the radially recrystallized Paxton type and the stratified Huron type (Plate 2, figs. 4A, B, C, D; Gutschick, 1987, Fig. 5). Wardlaw and Long (1982) concluded that the Paxton-type concretions grew below the seawater/sediment interface, before major compaction, and within the top 10 m of sediment in a system open to seawater.

Large calcareous concretions and small phosphatic concretions that occur in seemingly unfossiliferous deep-water shales, such as the Antrim, Huron, and other Upper Devonian and Mississippian shales, commonly yield important clues to interpreting age, sedimentology, biotic content, paleoecology, and geologic history (e.g., Foreman, 1959, 1963; Gutschick and Wuellner, 1983; Sandberg and Gutschick, 1984, p. 147, Fig. 13; Dix and Mullins, 1987; Gutschick, 1987, p. 299, Fig. 5; and Schwimmer and others, 1987). Distribution of the many surface sections from which concretions have been recovered from the Antrim and equivalent shales in the Michigan Basin and surrounding areas are shown on a regional map by Gutschick and Sandberg (this volume, Fig. 4).

*Norwood Member (new).* The Norwood Member (new) of the Antrim Shale is completely exposed in the Paxton Quarry (Fig. 5; Appendix). This section is here proposed as the principal reference section to formalize the Norwood as the basal member of the Antrim and to identify the poorly defined Norwood Shale of Ehlers (1938), north of Norwood, Michigan. That unit was not formally proposed but was mentioned in a mimeographed field-trip handout. A photo of the Norwood Shale in the Paxton Quarry was shown by Kesling and others (1976, Fig. 47). The Norwood Shale of Ehlers (1938) is invalid as a formational name, but because of long-standing usage, it is retained and used as the basal member of the Antrim Shale. Ehlers (1938) did not define the Norwood Shale, give its geographic extent and stratigraphy, nor differentiate it from the Antrim. However, he did list the Norwood Shale as the formation directly overlying the Squaw Bay Limestone in the Alpena area and the Petoskey Limestone in the Norwood area. The presumed type locality of the Norwood is 1.6 km north of the village of Norwood, whereas the type locality of the Antrim is 1.6 km south of the village. Thus, because of the southward (basinward) dip, the Norwood must lie at the base of the Antrim.

The Norwood Member, which is early Frasnian in age, is here defined as the lowest 8 m of black Antrim Shale between the top of the conformably underlying Squaw Bay Limestone and the base of the conformably overlying light-colored Paxton Member in the Paxton Quarry (Fig. 5; Appendix). The lower 3.3 m of the Norwood, present in the quarry sump, consists of, in ascending order: black fissile pyritic shale with thin fossiliferous limestone beds; several calcareous concretions; a sequence of 13 alternations of greenish gray, bioturbated shale and black fissile shale, which represent bioturbated rhythmites (Plate 2, fig. 2); and a laminated concretionary limestone layer, which rims the sump. The Norwood occupies all but the southwestern part of the Michigan Basin and attains a maximum thickness of 13 m near its center. The member is well defined lithologically in outcrop and is also recognizable in the subsurface by its highly radioactive response on gamma ray logs, contrasting sharply with less radioactive underlying and overlying units (Fig. 5). The Norwood was called subsurface unit 1C by Ells (1979).

The lower part of the Norwood Member in Paxton Quarry is richly fossiliferous, containing *Styliolina fissurella,* lingulid brachiopods, *Tasmanites,* and conodonts of the palmatolepid-polygnathid biofacies of the *punctata* Zone.

Our two conodont collections are dominated by *Palmatolepis transitans, Pa. punctata,* and *Mesotaxis asymmetrica.* A complete analysis of one of the two samples (PX-1 from 0.75 m above the base of the member) produced interesting results. The yield is 10,000+ conodonts/kg, of which 80 percent are ramiform (non-platform elements). Of the 5,463 counted platform elements, all but 10 specimens are *Palmatolepis transitans* and *Pa. punctata* (58 percent), *Mesotaxis asymmetrica* (22 percent, totalled as *Palmatolepis* in making the biofacies analysis), and *Polygnathus dubius* (20 percent). This produces a biofacies count of 80 percent *Palmatolepis* and 20 percent *Polygnathus.* The remaining 10 specimens are: *Ancyrodella gigas* (5), *Ancyrognathus ancyrognathoideus* (3), and *A.* cf. *A. ancyrognathoideus* (2). No specimens of *Icriodus* or of icriodontid simple cones were found. This low-diversity fauna is the archetype of deep-water, pelagic conodont biofacies (Sandberg, 1976; Sandberg and Dreesen, 1984).

The Norwood is also rich in other pelagic faunas. In the middle part of the member, a concretion yielded sparse conodonts, including the lowest occurrence of *Palmatolepis hassi,* and a large radiolarian fauna including *Ceratoikiscum spinosiarcuatum, C. planistellare, Entactinosphaera fredericki, E. eostrongyla, Palaeoscenidium cladophorum,* and other sphaerellarians (Plate 3). This fauna, representing the Upper *asymmetrica* conodont Zone, is similar to one in the Canol Shale of the North West Territories, Canada (Holdsworth, 1977). The upper part of the Norwood, which forms the lower wall of the second level of the quarry, is also highly radioactive black shale. It contains calcareous concretions, 1 m in maximum diameter, that are spherical, oblate to prolate spheroid, or acorn shaped. Conodonts from one of these concretions date the upper part of the Norwood as *Ancyrognathus triangularis* Zone.

*Paxton Member (new).* The informal "light-colored calcareous unit" of Gutschick (1987), which is composed of interbedded, light gray argillaceous limestone and greenish gray calcareous shale, is here designated the Paxton Member (new) of the Antrim Shale, with its type locality in the Paxton Quarry. The name is derived from the quarry name. A graphic section of the 5.9-m-thick member is shown in Figure 5, and a detailed description is given in the Appendix. The Paxton Member, which is bounded at bottom and top by mottled limestone beds, makes a conspicuous light-colored band across the south quarry wall (Plate 2, fig. 3). The upper and lower contacts are readily recognized by the changes to black shale of the Norwood Member below and to black shale of the Lachine Member above. The Paxton Member is present in most of the Michigan Basin but is very thin to absent in the southwestern part. It has two depocenters and attains a thickness of 15 m in the southern one and a maximum thickness of 23 m in the northern one (Gutschick and Sandberg, this volume). The Paxton, called subsurface unit 1B by Ells (1979), is easily identified on gamma ray logs by its low radioactive response, which contrasts sharply with the highly radioactive black shales below and above (Fig. 5).

The middle of the member is precisely dated as Lower *gigas* Zone by a conodont fauna from gray limestone nodules. The fauna includes *Palmatolepis foliacea, Pa. gigas, Pa. hassi, Pa.*

nasuta, Pa. unicornis, Ancryodella lobata, and *Polygnathus decorosus.* Overall, the member is late Frasnian in age, ranging from the *Ancyrognathus triangularis* Zone probably through the *linguiformis* Zone.

The entomozoan ("fingerprint") ostracode, *Entomoze prolifica,* occurs pyritized in thin, fissile, black shale layers in the basal 1 m of the Paxton Member. Entomozoans have a worldwide distribution within hemipelagic strata and have not been reported from nearshore or shallow-water deposits (Bless and others, 1986; Duffield and Warshauer, 1981). A layer of pelmatozoan columnals, many articulated but without calyxes, occurs in greenish gray mudstone closely associated with fossil burrows. Large double-valved leiorhynchoid brachiopods are common on nodular bedding surfaces of mottled limestones in contact with greenish gray shale. Thayer (1981) indicated that some articulate brachiopods may survive quite well in poorly oxygenated or temporarily anoxic environments, and Ettensohn (1985) suggested that *Leiorhynchus* probably represents a widespread fauna adapted for epifaunal life in dysaerobic environments.

*Lachine Member (new).* The name Lachine Member of the Antrim Shale is applied to rocks exposed in the upper walls of the Paxton Quarry (Plate 2, fig. 3). The name is derived from the small town of Lachine, after which the topographic quadrangle that includes the west half of Paxton Quarry is named (Fig. 3). The member is mostly radioactive, black shale, but it contains many thin layers of greenish gray shale and several zones of concretions. Its type section, 19.2 m thick in Paxton Quarry, is shown graphically in Figure 5 and is described more completely in the Appendix. The Lachine Member is early Famennian in age. Its lower contact with the underlying Paxton Member is based on a lithologic and color change from predominantly greenish gray shale and limestone below to predominantly black shale of the Lachine above. The bottom 1.5 m of the Lachine contains bioturbated rhythmites of black and greenish gray shale similar to those in the Norwood Member (Plate 2, fig. 2). A well-dated zone of concretions occurs in the middle of the Lachine. The top unit of the Lachine, 1.7 m thick, consists of alternating greenish gray concretionary shale and black shale. The contact with the overlying upper member is placed where alternations of greenish gray shale stop and the sequence changes to entirely black shale. The Lachine Member is present throughout the Michigan Basin and attains a maximum thickness of 27 to 30 m in several areas (Gutschick and Sandberg, this volume). The member was called subsurface gamma ray units 1A and 2 by Ells (1979).

Several large concretions, which occur 4.7 to 5.7 m below the top of the member in the quarry wall and loose on the floor of the quarry, yield abundant, biostratigraphically significant conodont faunas dominated by *Palmatolepis subperlobata, Pa.* cf. *Pa. regularis, Pa. glabra prima, Pa. glabra lepta* (early form), *Pa. minuta minuta, Pa. minuta loba,* and *Polygnathus glaber.* In addition, a few concretions contain common *Pa. quadrantinodosalobata;* these are assigned to the Upper *crepida* Zone. One concretion, however, lacks this species, but does contain a descendant, *Pa. poolei;* hence, it is assigned to the Lower *rhomboidea* Zone. Thus, this zone of concretions apparently straddles the Upper *crepida*–Lower *rhomboidea* zonal boundary (Fig. 5).

One of the concretions contains a heretofore poorly known, early Famennian conodont species, *Ancyrognathus symmetricus.* We have also found this species in the Lower *marginifera* Zone in the upper member of the Antrim Shale at Norwood and in the Lower or Upper *marginifera* Zone in the middle of the Huron Member of the Ohio Shale at Milan, Ohio (Fig. 1). Thus, the range of this species, which may become highly useful in dating other black shales, is now dated by these occurrences to be from the Upper *crepida*–Lower *rhomboidea* zonal boundary to the Lower–Upper *marginifera* zonal boundary.

*Protosalvinia* has not been reported from outcrops of the Antrim Shale in Michigan. However, it has been reported in a well core on the eastern side of the Michigan Basin by Matthews (1983) from a position that can be equated by gamma-ray log correlation to the upper part of the Lachine Member. Although this interval is unzoned by conodonts (Fig. 5), it is bracketed by dated conodont faunas, and so must represent either the Upper *rhomboidea* or Lower *marginifera* Zone (Fig. 2).

Large (7.5 cm in compressed diameter by 1.4 m in length), primary xylem-stem segments of the calamitean *Asterocalamites* occur with abundant *Tasmanites* cysts along bedding planes in black fissile shale, 3 to 4.5 m above the base of the member. The stems are composed of coalified vitrain. Coal, which may be calamitean vitrain, also occurs in the Antrim Shale of the Milwaukee area (Raasch, 1935). This suggests that these terrestrial arborescent sphenopsids were washed into the basin during deposition of the Lachine Member.

*Upper member.* Above the Lachine Member and below glacial cover over the Paxton Quarry is a uniform black shale unit, about 8 m thick, containing a few large calcareous concretions (Fig. 5; Appendix). This unit, which correlates with subsurface gamma ray log unit 3 and possibly the lower part of unit 4 of Ells (1979), is here treated as an unnamed, upper member of the Antrim Shale. It could be formally named in the future, if a complete section, exposing both contacts, were found. The upper member is early to late Famennian in age. Its upper contact, in the subsurface, is at the unconformity with the overlying Upper Devonian Bedford Shale (Fig. 2). A large concretion found loose on the beach at Norwood, Michigan (Fig. 1), is believed to come from this member. The concretion contained a conodont fauna of the Lower *marginifera* Zone, including *Ancyrognathus symmetricus.* This collection is correlated to the only observed zone of concretions in the member at Paxton Quarry and is located there in brackets in the columnar section (Fig. 5).

**Ellsworth Shale.** The Ellsworth Shale was introduced by Newcombe (1932, 1933) with its type section in the Petoskey Portland Cement Co. quarry pit, 2.4 km south of Ellsworth, in the NE¼NE¼Sec.26,T.32N.,R.8W., Antrim County, Michigan; and a complete subsurface section in a well in Muskegon County. The Ellsworth is greenish gray, bluish gray, light gray, and dark gray shale with some sandy and oolitic limestone interbeds in the subsurface. As defined, the base of the Ellsworth is the change from brown or black Antrim Shale to greenish gray shale; and the top is an unconformity overlain by Mississippian rocks, either the black Sunbury Shale or, in the southwestern part of the basin, a red limestone unit at the base of the Coldwater Shale. The Ellsworth is a western facies of all but the lowest part of the Antrim in the Michigan Basin. This facies relation is an intertonguing or alternation of greenish gray deltaic Ellsworth Shale with basinal, brown or black Antrim Shale. Above the Antrim, the Ellsworth also intertongues with similarly colored, greenish gray Bedford Shale.

## Subsurface Upper Devonian units

For the past decade informed micropaleontologists and biostratigraphers worldwide have known and widely published the fact that the age of the Bedford Shale and Berea Sandstone is Devonian because of the realignment of the Devonian-Mississippian boundary in the United States with the accepted Devonian-Carboniferous boundary in Europe (e.g., Streel and Traverse, 1978; Richardson and McGregor, 1986; Streel, 1986). This alignment was enabled because of the worldwide study of the Devonian-Carboniferous Boundary Working Group (DCBWG) since 1975, under the auspices of the International Union of Geological Sciences (IUGS). This study by experts on many fossil groups enabled greater precision in the ranges of conodonts, spores, and brachiopods on both sides of the boundary. Although the Devonian-Mississippian boundary has already been correctly realigned in the upper Mississippi Valley and in the western United States, the same new fossil ranges have not been applied heretofore to correct long-standing misconceptions about the supposedly Mississippian age of the Bedford Shale and Berea Sandstone in the Michigan and Appalachian basins (e.g., de Witt, 1970; Cohee, 1979; Matthews, 1983; Gutschick, 1987). Herein, we present a chronologic account of how these important biostratigraphic changes took place in the United States and how they occurred so gradually that there was an unfortunate breakdown in communication between conodont and spore biostratigraphers of the DCBWG and many regional lithostratigraphers working on the Michigan and Appalachian basins. Significantly, at least one worker kept abreast of the changes; Conkin (1985) placed both the Bedford and Berea in the Devonian and correctly correlated them with the Saverton Shale and Louisiana Limestone in Illinois and Missouri.

The harbinger of major changes that would have to be made in the position of the Devonian-Mississippian boundary occurred nearly 30 years ago, when Scott and Collinson (1961) and Collinson and others (1962), on the basis of conodonts, reassigned the Louisiana Limestone, the macrofauna of which had long been considered Mississippian (e.g., Williams, 1943), to the Late Devonian. This change was made because these authors had advance knowledge of the comprehensive Late Devonian conodont zonation that was to be published in Germany (Ziegler, 1962); they realized that the type Mississippian in the United States had not been properly aligned with the base of the European Carboniferous. The characteristic brachipod fauna of the Louisiana Limestone, most importantly *Syringothyris hannibalensis* and *Rhipidomella missouriensis,* which also occur widely in the western United States in the Sappington Member of the Three Forks Formation and in the Leatham Formation, was reported by Sandberg (*in* Klapper and others, 1971) to span all three parts of the *Spathognathodus costatus* conodont Zone; in terms of the current conodont zonation, this age range spans four conodont zones from the Upper *expansa* through the Upper *praesulcata* Zones (Fig. 2).

In the meantime, important advances were being made in the knowledge of spore ranges in Europe, beginning with Streel (1966) and continuing through many other papers by Streel and other authors, as listed by Sandberg and others (1972). These changes showed that *Hymenozonotriletes lepidophytus* (now *Retispora lepidophyta*), once considered to span the Devonian-Carboniferous boundary, was restricted to the Devonian. In the meantime, based on determinations by D. C. McGregor, Sanford (1967) had reported this spore from the Michigan Basin in Ontario and Michigan, not only from the Berea Sandstone, but also from the Sunbury Shale. We now know that this higher range must be in error and that the spores in the Sunbury must be reworked, because conodonts show that the base of the Sunbury is assignable to the late Kinderhookian *Siphonodella crenulata* Zone. This dating is five conodont zones later than the start of the Carboniferous, and nowhere else in the world does *R. lepidophyta* extend into even the oldest conodont zone of the Carboniferous, the *Siphonodella sulcata* Zone.

The next discussion of the age of the Bedford Shale and Berea Sandstone in the Michigan Basin was by de Witt (1970). As a regional lithostratigrapher, not as a paleontologist, he accurately summarized the then-known spore and conodont evidence for what he believed was the Mississippian age of the Berea and most of the Bedford in the Appalachian basin of Ohio, Pennsylvania, and West Virginia, and he projected this age into the Michigan Basin. He was unaware of the sweeping changes in the knowledge of spore ranges that were taking place in Europe, and his discussion of ranges was confined to the findings of Winslow (1962), Streel (1966), and Sanford (1967). Soon thereafter, these changes were incorporated into a paper by Sandberg and others (1972) that compared the conodont zonation and spore assemblages at the Devonian-Carboniferous boundary in the western and central United States and in Europe. These authors reported a new conodont fauna, the *Siphonodella praesulcata* Fauna, at the top of the Devonian and recommended that the base of the Carboniferous be placed above this at the incoming of *Siphonodella sulcata,* a position that was later adopted unanimously by the DCBWG, meeting in Washington, D.C., in May 1979. These authors also reported the finding of a *Hymenozonotriletes lepidophytus* (now *R. lepidophyta*) spore assemblage in Montana in the middle part (unit 3) of the Sappington Member, now dated as Middle *praesulcata* Zone (Fig. 2; Sandberg and others, 1989b) and a slightly older spore asemblage in Illinois in the Saverton Shale, now dated as Lower *praesulcata* Zone. They summarized and reinterpreted the findings of Winslow (1962) as indicating the presence of *R. lepidophyta* (identified by her as *Endosporites lacunosus*) in the Berea Sandstone in Ohio, its absence in the overlying Mississippian Sunbury Shale, and its younger, undoubtedly reworked occurrences above the Sunbury, higher in the Cuyahoga Group.

During the middle to late 1970s, evidence for the Late Devonian age of the Berea Sandstone, based on spore assemblages that were not corroded and hence were distinguishable from those that might be reworked, began to accumulate. In an important, but little known, unpublished study, Eames (1974) reported the widespread presence of *R. lepidophyta* in the Berea Sandstone in Ohio and concluded that it, as well as the underlying Bedford Shale and Cleveland Member of the Ohio Shale, was Late Devonian in age. Streel and Traverse (1978) accepted Eames' findings for Ohio; to the east, on the basis of new spore collections, they placed the Devonian-Carboniferous boundary even higher, at least 65 m above the middle sandstone and shale member of the Pocono Formation at the Horseshoe Curve section in Pennsylvania! In an account of his report to the DCBWG summarizing the Devonian-Carboniferous boundary in North America, Sandberg (1981), on the basis of spore evidence, placed this boundary between the Berea Sandstone and unconformably overlying Sunbury Shale. In summaries of Late Devonian miospore stratigraphy and spore zones around the Old Red continent, Streel (1986) and Richardson and McGregor (1986) independently demonstrated conclusively that globally there are no known unreworked occurrences of *Retispora lepidophyta* above the highest Devonian Upper *praesulcata* conodont Zone. This conclusion is now universally accepted by palynologists, the DCBWG, and the IUGS Subcommissions on Devonian and Carboniferous Stratigraphy (e.g., Clayton and others, 1989).

Although not as complete as the studies of spores, recent work on other fossil groups, such as conodonts, ammonoids, and brachiopods, has not found any evidence for Mississippian fossils in the Bedford Shale or Berea Sandstone. Sandberg (*in* Streel and Traverse, 1978) reported a conodont fauna from the basal Bedford Shale at Granger roadcut, Cleveland, Ohio. The faunal list, updated to incorporate later taxonomic changes, includes *Branmehla fissilis, Branmehla culminidirectus, Bispathodus aculeatus anteposicornis,* and possibly a broken fragment of *Siphonodella praesulcata*. This fauna, which is comparable to those in the upper part of the Saverton Shale or lower part of the Louisiana Limestone in southern Illinois, would now be dated as Upper *expansa* Zone or if *S. praesulcata* is present, as Lower *praesulcata* Zone (Fig. 2), which would allow ample time still within the Devonian for deposition of the upper part of the Bedford Shale and all of the Berea Sandstone. Devonian ammonoids are also reported from the basal Bedford Shale in Ohio by Kirchgasser and House (1981). Although there are no recent reports on brachiopods of the Berea Sandstone, the fauna is not unlike those of the Louisiana Limestone of Illinois and Missouri, the Sappington Member of the Three Forks Formation of Montana, and Leatham Formation of Utah, all of which, as previously discussed, have been reassigned to the Late Devonian.

Equally as convincing as the fossil evidence is the fit of the Bedford Shale and Berea Sandstone with the Devonian sea-level curve (Johnson and others, 1985, 1986; Johnson and Sandberg, 1989). This curve shows a eustatic rise in the Upper *expansa* Zone and eustatic fall, coinciding with a mass extinction (Sandberg and others, 1988), in the Middle *praesulcata* Zone. Thus the sea-level curve would date the transgressive Bedford Shale as Upper *expansa* to Lower *praesulcata* Zone and the regressive Berea Sandstone as Middle to Upper *praesulcata* Zone. This eustatic dating closely matches the conodont, brachiopod, and spore evidence. Thus, lithologically and paleontologically, the Bedford is comparable to units 2, 3, and 4, and the Berea is identical to unit 5 of the Sappington Member in Montana (as these units are defined and described by Sandberg and others, 1972). Likewise, the Bedford is lithologically and faunally comparable to the upper part of the Saverton Shale, which intertongues with the Louisiana Limestone in Illinois and Missouri, and the Berea is faunally comparable to most of the overlying shallow-water Louisiana Limestone.

***Bedford Shale.*** The Bedford Shale was named by Newberry (1870) for exposures in the Cleveland area, north-central Ohio. There, the Bedford is a sequence of bluish gray shale and siltstone, 26 m thick. The regional distribution, depositional history, and paleogeography of the Bedford were described in a classic study by Pepper and others (1954). In the Michigan Basin, the formation is covered by Pleistocene glacial drift and is known only from the subsurface (Ells, 1979). There, it is a gray shale that overlies black shale of the upper member of the Antrim Shale (Fig. 2). Its upper part is silty and sandy and grades upward into the overlying Berea Sandstone. The Bedford is more than 65 m thick along the eastern side of the basin and thins westward near the basin center before merging with deltaic shale in the upper part of the Ellsworth Shale along the western side. It thus represents the maximum westward progradation of the Catskill deltaic facies.

***Berea Sandstone.*** The Berea Sandstone (originally called Berea grit) was also named by Newberry (1870) for exposures in the Cleveland area, Ohio. There, the Berea consists of a lower unit of medium- to fine-grained, massive, crossbedded channel sandstone as much as 61 m thick, and an upper unit of thin-bedded blanket sandstone 6 to 12 m thick. The upper unit has steeply dipping foreset beds in the lower part and oscillation ripple marks in the upper part. Like the Bedford, the Berea of the Michigan Basin is known only from the subsurface (Ells, 1979), where it consists of light gray, fine- to medium-grained sandstone. The Berea attains a maximum thickness of more than 35 m in the eastern half of the basin but thins westward. It is unconformably overlain by the Mississippian black Sunbury Shale. The Berea represents the regressive, eastward-retreating phase of the Bedford-Berea deltaic complex. In direction of regression, it is the mirror image of the coeval, lithogenetically equivalent, westward-retreating, upper massive sandstone (unit 5) of the Sappington Member of the Three Forks Formation and its equivalents west of the Transcontinental arch in the western United States (Sandberg and others, 1989b).

## Overlying Lower Mississippian units

The name Sunbury Shale was originally applied in central Ohio to a black, bituminous shale (Hicks, 1878). Subsequently, it was found to be widely distributed in the Appalachian and Michigan basins. An unconformity separating the Sunbury from the underlying Berea Sandstone or Bedford and Ellsworth Shales in northern Ohio and lower Michigan commonly is marked by a thin pyritic bone bed. The highly radioactive black shale of the Sunbury represents the basal deposit of a Mississippian cycle related not only to Acadian tectonism and Catskill delta response, but also to a major North American, possibly eustatic, sea-level rise coinciding with the start of a major depophase in the Lower *crenulata* Zone in the western United States (Sandberg and others, 1986). The Sunbury Shale is 45 m thick along the eastern edge of the Michigan Basin but thins basinward. It is absent in southwestern Michigan, where the Ellsworth is overlain by the "Red Rock" unit, a red, shaly, fossiliferous limestone at the base of the Coldwater Shale. Neither the age nor the relation of this red unit to the Sunbury is certain. Ells (1979) stated that red limestone and shale beds overlie the Sunbury, whereas others have correlated it with the Berea. We suggest that it may be a correlative of the Lower Mississippian Rockford Limestone of Indiana.

## BIOSTRATIGRAPHY OF BASIN MARGINS

### Eastern Wisconsin

The Michigan Basin extends beneath Lake Michigan to eastern Wisconsin, where a very narrow band of Middle and Upper Devonian rocks occurs along the western margin of Lake Michigan (Fig. 1). Rock formations related to this study in the Milwaukee area, southeastern Wisconsin, are shown in Figure 2, which compares them with equivalent strata east of Lake Michigan. Of significance is the Milwaukee Formation, which has three members: in ascending order, the Berthelet, Lindwurm, and North Point Members. The overlying Upper Devonian Antrim Shale, the westernmost occurrence in the Michigan Basin, is confined to a small syncline covered by glacial drift, but it is known from a tunnel excavation and from wells. Published and unpublished reports found useful in this study are by Cleland (1911), Raasch (1935), Schumacher (1971), Roberts (1972), Mikulic and Kluessendorf (1988), and Kluessendorf and others (1989).

The Berthelet Member consists of gray, medium- to thick-bedded, dolomitic shales and dolostone. At its base are reworked pebbles and cobbles of Thiensville origin (Raasch, 1935), bone fragments, glauconite, and foraminiferans (Roberts, 1972), which evidence a disconformity. A massive, dense, hard, buff-weathering dolostone, 1.8 m thick, marks its top. This dolostone is strikingly similar to the *Zoophycos* bed at Partridge Point, Michigan. Both units have in common a polygnathid-icriodid biofacies conodont fauna assignable to the interval between the *varcus* and *transitans* Zones, agglutinated foraminiferan faunas with multilocular *Oxinoxis* described by Roberts (1972) from the Milwaukee Formation, and *Zoophycos* traces on the top-surface hardground. Both units may be the continuation of the persistent key brown dolostone marker unit, which Tarbell (1941) traced throughout the western and central Michigan Basin at this same stratigraphic position.

The Lindwurm Member comprises fossiliferous dolomitic shales and limestones, and contrasts sharply with the underlying dolostone. Twinned pyrite, marcasite, sphalerite, and millerite crystals are similar to those in the Squaw Bay Limestone. A bone bed at the base contains fish bones, phosphatic nodules, glauconite grains, and agglutinated foraminiferans, and marks a hiatus (Roberts, 1972; Kluessendorf and others, 1989). Nevertheless, the conodont and foraminiferan faunas are quite similar on either side of the contact. Megafossils, however, change from abundant cephalopods and pelecypods at the top of the Berthelet to pelmatozoans, bryozoans, brachiopods, corals, mollusks, trilobites, *Tentaculites,* and ostracodes in the Lindwurm.

The North Point Member, consisting of gray shales, shaly limestones, and dolomitic siltstones or silty mudstone, is known only from the subsurface. Fossils are abundant in some layers, and *Chonetes, Tentaculites,* and *Paleoneilo* are most common (Raasch, 1935). From the Berthelet and lower part of the Lindwurm, Schumacher (1971) described conodonts that suggest a latest Middle Devonian age. Lacking conodonts from the upper part of the Lindwurm and from the North Point Member, Schumacher speculated that these are early Late Devonian in age. Fossil zones in the North Point probably represent storm deposits in shallow-subtidal, normal-marine environments (Kluessendorf and others, 1989).

Information on the Antrim Shale of eastern Wisconsin is sketchy because there are no known outcrops. Maximum thickness of the formation is 16.8 m (Raasch, 1935). Two bore holes encountered a maximum of 4 m of Antrim, but conventional biostratigraphic studies of these sections are lacking. Available information comes from Edwards and Raasch (1921), Raasch (1935), Schumacher (1971), Roberts (1972), Kluessendorf and others (1989), and Mikulic and Kluessendorf (1988). The last reference is a comprehensive report on the Devonian of Milwaukee County. The Wisconsin Geological Survey has advised that no gamma-ray logs are available for wells drilled into the Antrim in the Milwaukee area.

The Antrim Shale in Wisconsin is a finely laminated, bituminous, pyritic, fissile black mudstone with interbeds of nonlaminated greenish gray, brownish gray, and bluish gray mudstone. The basal contact of the Antrim on the Milwaukee Formation in one well is sharp, irregular, and apparently unconformable. The basal Antrim in two other wells is gray silty mudstone. Antrim fossils recorded from wells are lingulid brachiopods, conodonts, fish bones, pelecypods, gastropods, *Tasmanites,* and, in greenish gray and bluish gray shales, agglutinated foraminiferans.

Shale chips and larger pieces of rock sampled from the Linwood Avenue Intake tunnel excavation and stored in the Milwaukee Public Museum were studied by Schumacher (1971). From these he obtained a random collection of conodonts with integrity only for each piece, but without stratigraphic control. The taxa were identified and plotted in a chronostratigraphic range scheme from which four zones were recognized (Fig. 2): from oldest to youngest, they are the Lower *gigas*, Upper *gigas*, Middle *triangularis*, and Upper *crepida* Zones. The two older zones came from samples that had scour-and-fill structures with concentration of the conodonts in lenticular stringers in the shale, which contained phosphatic pellets, pyrite, fish teeth, and bones characteristic of reworked bone beds. Examination of two cores for megafossils revealed only lingulid brachiopods (Mikulic and Kluessendorf, 1988).

### Northeastern Illinois

In the Elmhurst quarry (Fig. 1), Weller (1899) observed joint fillings composed of black and bluish gray fossiliferous clay, extending down into Silurian bedrock. He concluded from the fossils they contained that the fillings were of Late Devonian age. Two other examples of exotic fissure and joint fillings have been found nearby in northeastern Illinois at the Lyons and Thornton quarries (Fig. 1). All three occurrences are fillings in Silurian reefs. Open fissures apparently trapped Devonian marine muds and fossil fragments when the seas washed over the dead reefs, and some fillings were preserved after the covering Devonian sediments were stripped away by subsequent erosion. Conodonts from fissure fillings in the Elmhurst quarry were sampled and studied by Eicher (1939), who also reported the presence of *Protosalvinia* in the fillings. The recovered conodonts represent a totally mixed assemblage. Eicher's identified conodonts were tabulated chronologically, according to published ranges, by Kluessendorf and others (1989) to assess what zones they might represent. The listed conodonts suggest to us the former presence of sediments ranging in age from the early Frasnian *punctata* Zone to the early Famennian Upper *crepida* Zone. The faunas compare reasonably well with those from the Antrim Shale at the Paxton Quarry (Fig. 5) and from the lower Antrim at Milwaukee (Schumacher, 1971). The anomalous subsurface Des Plaines structural disturbance (Emrich and Bergstrom, 1962; Kluessendorf and others, 1989) is located in northeastern Illinois near the three quarries (Fig. 1) and has an important bearing on interpretations of the fissure fillings. Downfaulted blocks in the Des Plaines disturbance contain a sequence of stratified rocks that are younger than those in the surrounding area. Some of the sequence is possibly equivalent to the Ellsworth Formation in Michigan (Willman, 1971), and as much as 150 m of strata resembling the Antrim Shale is present in this probable meteor impact structure (McHone and others, 1986). As in the fissure fillings in the three quarries, *Protosalvinia* is common in cores of these strata (Kluessendorf and others, 1989). Age of the structural disturbance is within the post-Llandoverian to pre-Frasnian time span (Kluessendorf and others, 1989).

Considered together, the conodonts in the fissure fillings and the strata preserved in a probable impact crater answer some important questions, but also pose others. They strongly suggest that Late Devonian deposition occurred in this area, which must have been part of a strait connecting the Michigan and Illinois basins (Fig. 1). They pose the question of whether or not the fissures in the quarries were related to the impact, or whether they were merely the normal crevices that develop in eroding, dead reefs. They also pose the question as to whether the crevice fillings might occur on structurally high areas, on which only thin lags were deposited by successive transgressions through the strait. If so, all the conodonts in the fillings could have been derived from a multiple, reworked lag, and been introduced at the same time as the early to late Famennian transgression that introduced *Protosalvinia* to the fissure fillings at Elmhurst quarry. In that event, the formation of fissures and even of the impact structure could have occurred as late as the end of the Frasnian, because the overlying Antrim at Des Plaines is as yet undated by conodonts.

### Southwestern Ontario

The Kettle Point Formation in southwestern Ontario is the lithologic and time equivalent of the Antrim Shale (Fig. 2). The formation consists of dark brown to black bituminous shale with interbeds of greenish gray, silty shale. The only important outcrop is the type section at Kettle Point (Fig. 1), which was named after the kettle-like concretions. There, 3.6 m of organic, fissile, concretionary black shale, containing mostly Paxton-type concretions, is exposed (Uyeno and others, 1982). Conodont faunas from the type section were studied by Winder (1966, conodonts identified by G. Klapper and zoned by W. Ziegler), who reported the Upper *crepida* to *rhomboidea* Zones. Conodonts were also recovered from a concretion at the type section by Uyeno and others (1982) and one of the two listed species, *Palmatolepis quadrantinodosa*, suggests the Lower *marginifera* Zone. Overall, conodonts in the Kettle Point Formation range from the *punctata* Zone to possibly the *expansa* Zone, paralleling the ages of conodonts in the Antrim Shale. Other fossils reported from the formation include *Protosalvinia* in unit 2 (Russell, 1985), radiolarians, *Tasmanites, Lingula,* trace fossils (burrows), sponge spicules, scolecodonts, fish fragments, and fossil wood (Winder, 1966, 1967).

## CONCLUSIONS

1. The thin Squaw Bay Limestone, deposited during initial subsidence of the Michigan Basin and apparently including the Middle–Upper Devonian boundary, comprises three beds, each yielding a different conodont fauna. The middle and upper beds both represent the *transitans* Zone, but the middle bed contains the moderately deep-water polygnathid-ancyrodellid biofacies, whereas the upper bed contains the deep-water palmatolepid-polygnathid biofacies.

2. The *Zoophycos* bed, lowest bed of the Squaw Bay Limestone, contains an agglutinated foraminiferan fauna that is similar, if not identical, to that in the Berthelet and Lindwurm Members of the Milwaukee Formation in Wisconsin.

3. The Upper Devonian Antrim Shale comprises four members, which are more or less precisely dated depending on how many levels of concretions or limestone lenses yielded conodonts. The lowermost Norwood Member is early Frasnian in age and contains conodont faunas of the palmatolepid-polygnathid biofacies, representing the *punctata,* Upper *asymmetrica,* and *Ancyrognathus triangularis* Zones. The next higher Paxton Member is late Frasnian in age. It contains, near the middle, the palmatolepid-polygnathid biofacies of the Lower *gigas* Zone. The Paxton probably ranges from the *A. triangularis* Zone to the Upper *gigas* or *linguiformis* Zone, and its top may mark the late Frasnian extinction event. The overlying Lachine Member is early Famennian in age; its lower part is undated, but its upper part contains the palmatolepid-polygnathid biofacies of the Upper *crepida* and Lower *rhomboidea* Zones. The upper member is early to late Famennian in age; its basal part contains the palmatolepid-polygnathid biofacies of the Lower *marginifera* Zone, but its upper part, which is known only from the subsurface, has not been sampled for conodonts.

4. The middle part of the Norwood Member, dated as Upper *asymmetrica* Zone, contains a radiolarian fauna similar to that in the Canol Shale, North West Territories, Canada.

5. The tropical floating plant *Protosalvinia* may range through much of the Famennian. Jointly occurring conodonts unequivocally date it as middle to late Famennian from the Upper *trachytera* through the Lower *expansa* Zone, although an extension into the next younger, Middle *expansa* Zone is possible. A single subsurface occurrence on the east side of the Michigan Basin can be dated, in the absence of conodonts, only by gamma-ray correlation to outcropping beds that were deposited during either the early Famennian *rhomboidea* or Lower *marginifera* Zone.

6. *Protosalvinia* probably occurs in three, progressively greater blooms or spikes, associated with correspondingly greater, probable glacio-eustatic sea-level rises related to interglacial stages of Southern Hemisphere glaciation.

7. The Bedford Shale and Berea Sandstone are dated as Late Devonian (late Famennian Upper *expansa* through Upper *praesulcata* Zones) by occurrences of the globally significant spore *Retispora lepidophyta*; by sparse occurrences of conodonts, ammonoids, and a few other fossils; and by close fit to the Devonian sea-level curve. The Bedford was deposited as a prograding mud delta, whereas the regressive Berea was deposited, first as distributary channel fills and then as bar or sheet sandstones, during the major eustatic fall, probably resulting from Southern Hemisphere glaciation, that produced a mass extinction near the end of the Devonian.

## ACKNOWLEDGMENTS

Thanks are extended to Richard Schiemke, quarry foreman, who made the senior author feel at home in the Paxton Quarry, and also to the National Gypsum Company, Alpena, for permitting geologic study of this magnificent exposure. We also extend our appreciation to the following persons for the indicated assistance: Elisabeth M. Brouwers, U.S. Geological Survey, Denver, for assisting Gutschick with the SEM photomicrography; Jamie L. Butler, U.S. Geological Survey, for redesigning and redrafting Figures 1, 4, and 5; Jane Gray, University of Oregon, and Arthur Boucot, Oregon State University, for sharing their perspectives on *Protosalvinia* and its environmental significance; Gregg F. Gunnell, University of Michigan, for providing a photocopy of Ehler's unpublished 1938 field trip guide; Curtis Klug, University of Iowa, for sending a copy of Roberts' (1972) thesis long before its usefulness was realized; Carl B. Rexroad, Indiana Geological Survey, for loaning Sandberg conodont collections picked by Nancy R. Hasenmueller from cores of the New Albany Shale in southern Indiana and for permission to use the important zonal determinations; John P. Szabo and James W. Teeter, University of Akron, for collecting important concretions from the Ohio Shale; and Dirck Wuellner and Tom Hendrick, former students at Notre Dame, for helping in the field and laboratory. We are grateful to Ronald R. Charpentier, Mitchell E. Henry, and Marjorie E. MacLachlan, U.S. Geological Survey, Denver, who reviewed the manuscript and offered helpful suggestions.

**Plate 1**

Agglutinated foraminiferans from *Zoophycos* bed of type Squaw Bay Limestone, Partridge Point, Michigan. Fauna is similar to that described from Milwaukee Formation in Wisconsin by Roberts (1972). Scale 100 μm, except as noted.

**Figs. 1–5.** *Psammospaera* sp. test constructed from round glauconite grains and some pyrite cement; **2, 3,** broken tests showing wall construction and interiors.

**Figs. 6, 8.** *Sorosphaerella cooperi;* **6,** scale 300 μm.

**Fig. 7.** *Webbinelloidea similis,* single-chambered specimen, view of underside; scale 300 μm.

**Fig. 9.** Polygonal epimorphic foraminiferan in gastropod(?) mold or *Spirorbis* tube(?).

**Figs. 10–13.** *Oxinoxis* spp., single and multilocular forms; **11, 12,** scale 300 μm.

**Plate 2**

**Fig. 1.** Scanning electron photomicrograph (stereo pair) of curved polychaete or oligochaete worm tube, composed of agglutinated quartz grains. Scale 1 mm. This is the only sessile-benthic organism found in molluscan bed of type Squaw Bay Limestone.

**Fig. 2.** Sump of Paxton Quarry, showing bioturbated rhythmite sequence (ribbed bedding) composed of 13 couplets within 0.8-m interval of Norwood Member (photographed Aug. 2, 1982). Even-bedded, well-jointed layers that project in relief are fissile, black Antrim-type shales, whereas burrowed, bioturbated interbeds are irregular, less resistant, greenish gray, Ellsworth-type shales. For scale, Jacob staff with Brunton compass is 1.5 m long and black handle is 0.6 m long.

**Fig. 3.** South wall of Paxton Quarry, showing, from base to top, upper part of Norwood Member, Paxton Member (light band), and Lachine Member. Lower part of Norwood is exposed in sump at left (east) edge of photograph. For scale, Paxton Member is 5.9 m thick.

**Fig. 4.** Types of calcareous concretions in Antrim Shale at Paxton Quarry: **4A.** Acorn-type, with small top and large bottom, scale same as 4C; **4B.** Paxton-type, with characteristic internal radial crystallization and concentric color banding accentuated by weathering, scale is 15 cm long; **4C.** Huron-type, showing well-banded horizontal laminations (lighter bands are commonly fossil debris and/or concentrations of quartz grains); scale, all divisions on right are centimeters; **4D.** Paxton-type, with outer pyritic shell (crust) exfoliated to reveal growth of large crystals and pattern that is surface manifestation of internal radial crystallization shown in 4B. Angularity of prism intersections, concentric growth patterns, and penetration intergrowth are recognizable; card scale shows inch divisions on left and centimeter divisions on right.

**Plate 3**

Radiolarians and *Tasmanites* from limestone, 7.0 to 7.15 m above base of Norwood Member of Antrim Shale, at Paxton Quarry, Michigan. Fauna is similar to that described from Canol Shale, North West Territories, Canada, by Holdsworth (1977). All stereo pairs except Fig. 2. Scale 100 μm.
**Fig. 1.** *Ceratoikiscum spinosiarcuatum* Foreman.
**Fig. 2.** *Ceratoikiscum planistellare* Foreman.
**Figs. 3, 4.** Unidentified sphaerellarians.
**Fig. 5.** *Tasmanites* cyst.
**Fig. 6.** *Entactinosphaera fredericki?* Foreman. Adhering behind and below this large, spherical form is specimen of smaller *Palaeoscenidium cladophorum,* four spines of which are visible.
**Fig. 7.** *Entactinosphaera eostrongyla* Foreman.

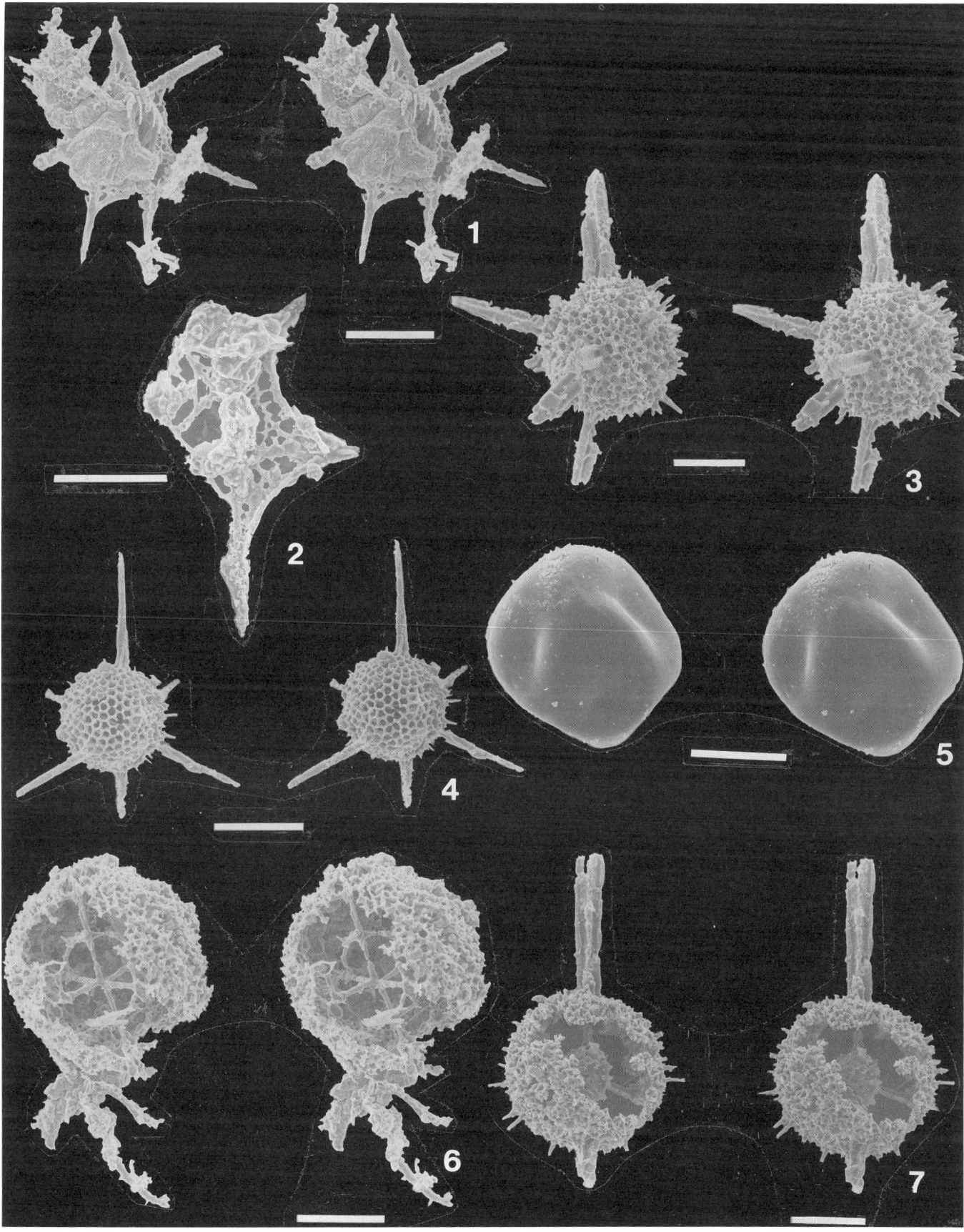

# APPENDIX: DESCRIPTION OF ANTRIM SHALE IN PAXTON QUARRY

The following measured section in the Paxton Quarry serves as the principal reference section for the Antrim Shale and its Norwood Member and as the type section for its Paxton and Lachine Members. The quarry is located in the N½Sec.30,T.31N.,R.7E., Lachine and Lake Winyah 7.5-minute Quadrangles, Alpena County, along the south side of Michigan Highway 32, about 14 km west of Alpena, Michigan (Fig. 3; Gutschick, 1987). The section starts in a sump, continues up the south quarry wall, and offsets to the west wall. Equivalent subsurface gamma ray units of Ells (1979) are shown in a columnar section (Fig. 5). All thicknesses are in meters.

Bedrock surface, top of quarry (shallow syncline on west wall).
**Antrim Shale (Upper Devonian):**
Upper member (part), (units 3 and 4 of Ells, 1979):
15. Shale, black, with a few large calcareous concretions and a few scattered, thin, greenish gray shale layers .................... 7.9
    **Lachine Member (units 2 and 1A of Ells, 1979):**
14. Shale, concretionary, alternating black and greenish gray, nonbioturbated; composed of rhythmites; each of 5 greenish gray shale interbeds, 10 cm thick, spawns calcareous, silty, unfossiliferous concretions, as large as 0.5 by 0.75 m ................................ 1.7
13. Shale, black, with many, thin, greenish gray shale interbeds; unsampled concretions near base; concretions 3 to 4 m above base contain conodonts of Upper *crepida* and Lower *rhomboidea* Zones; *Zoophycos* and *Tasmanites* occur above concretions. Pancake-shaped concretions, 0.6 by 2.0 m, have horizontal mid-separation with mottled bioturbated limestone below and brown fibrous calcite above .............. 8.0
Offset to south wall.
12. Shale, black, partly concretionary, with thin, greenish gray shale interbeds; bioturbated rhythmites in lower 1.5 m; coalified calamitean stems and *Tasmanites* 2.6 to 4.6 m above base; large calcareous concretions 1.5 to 3 m above base ............................... 9.5
Total Lachine Member ...................................... 19.2
    **Paxton Member of Antrim Shale (unit 1B of Ells, 1979):**
11. Tripartite marker beds: lower mottled limestone bed, middle black shale bed, upper acicular limestone bed with cone-in cone structures ................................................ 1.3
10. Shale, calcareous, greenish gray, with many fossiliferous limestone nodules and two thin limestone interbeds; nodules contain conodonts of Lower *gigas* Zone ........................................ 1.5
9. Limestone, light gray, fine-grained, medium-bedded, with greenish gray calcareous shale interbeds and minor black shale .......... 2.5
8. Limestone and shale; lower half shaly, nodular, with thin, black, fissile shale interbeds; contains pyritized *Entomoze prolifica*; upper half, mottled limestone marker bed .............................. 0.6
Total Paxton Member .................................... 5.9

**Norwood Member (unit 1C of Ells, 1979):**
7. Shale, concretionary, black, with a few thin, greenish gray shale layers; burrows. Large calcareous concretions, 1.8 to 3.35 m above base, about 1 m maximum diameter, are oblate to prolate spheroidal to acorn shaped and contain conodonts of *Ancyrognathus triangularis* Zone. (Floor of second quarry level is undulatory surface atop concretions) Boudinage-like limestone bed, 0.05 to 0.45 m thick, about 4 m above base ........................................................ 4.7
Catwalk level.
6. Shale, black, fissile, with limestone layers, and mammillary, concretion-like limestone bed, with lens of pelmatozoan columnals at base. Laminated, pyritic, bituminous limestone in lower 15 cm contains radiolarian fauna similar to that described from Canol Shale of Canada by Holdsworth (1977) and sparse conodonts, including *Palmatolepis hassi*, assignable to Upper *asymmetrica* Zone. Concretionary limestone bed, 46 to 51 cm thick, rims top of sump ..................... 1.0
5. Shale, black and greenish gray, composed of repeated couplets (bioturbated rhythmites). Lower part of couplet is laminated, fissile, black shale burrowed by vertical to horizontal compacted tubes, backfilled with greenish gray shale of upper part. Top of greenish gray shale in contact with overlying black shale is planar and undisturbed ...... 0.8
4. Limestone, light greenish gray, argillaceous, with mottled layer in middle; contains abundant *Styliolina fissurella*, lingulid brachiopod fragments, ostracode molds, fish fragments, framboidal and crystalline pyrite, and conodonts of *punctata* Zone. Underlying black shale burrowed and backfilled with greenish gray shale ................ 0.12
3. Shale, black, with some interbedded, thin, greenish gray shale layers. Three, small, 15-cm-diameter, oblate, ovoid, calcareous concretions with horizontal bedding cut by vertical and oblique tubular burrows in middle of unit ................................................ 0.55
2. Limestone, greenish gray, impure, fine-grained, containing abundant light brown compacted *Tasmanites*, styliolinids, abundant lingulid brachiopods, pyrite cubes, sole burrow casts, and conodonts of *punctata* Zone ...................................................... 0.07
1. Shale, grayish black, fissile, pyritic; a few, thin, greenish gray shale laminae ................................................ 0.75
Total Norwood Member ...................................... 8.0
Total measured Antrim Shale ............................. 41.0
Sharp, planar contact.
**Squaw Bay Limestone (upper, calcareous mudstone bed; Upper Devonian):**
Mudstone, greenish gray, calcareous, with argillaceous limestone interbeds; limestone, 0.25 m thick, at top contains styliolinids; conodonts of *transitans* Zone at base and top; tiny gastropod molds, and pyrite .................................................. 1.5
Bottom of sump.

# REFERENCES CITED

Arnold, C. A., 1954, Fossil sporocarps of the genus *Protosalvinia* Dawson, with special reference to *P. furcata* (Dawson) comb. nov.: Svensk Botanisk Tidskrift, v. 48, p. 292–300.

Barron, L. S., and Ettensohn, F. R., 1980, A bibliography of the paleontology and paleoecology of the Devonian–Mississippian black-shale sequence in North America: U.S. Department of Energy, Morgantown Energy Technology Center Report DOE/METC/5202-13, 86 p.

Bless, M.J.M., Crasquin, S., Groos-Uffenorde, H., and Lethiers, F., 1986, Late Devonian to Dinantian ostracodes (comments on taxonomy, stratigraphy, and paleontology): Société Géologique de Belgique Annales, v. 109, pt. 1, Special volume "Aachen 1986", p. 1–8.

Broadhead, R. F., Kepferle, R. C., and Potter, P. E., 1980, Lithologic description of cores and exposures of Devonian shale and associated strata in Ohio along Lake Erie: U.S. Geological Survey Open-File Report 80-719, 96 p.

Bultynck, P. L., 1976, Comparative study of Middle Devonian conodonts from north Michigan (U.S.A.) and the Ardennes (Belgium–France): Geological Association of Canada Special Paper 15, p. 119–141.

Clayton, G., Loboziak, S., Streel, M., Turnau, E., and Utting, J., 1989, Palynological events in the Mississippian of Europe, North America, and North Africa, *in* Brenckle, P. L., chairman, Lower Carboniferous: International Union of Geological Sciences, Subcommission on Carboniferous Stratigraphy, Provo Meeting, 1989, 2 p.

Cleland, H. F., 1911, The fossils and stratigraphy of the Middle Devonic of Wisconsin: Wisconsin Geological and Natural History Survey Bulletin, v. 21, p. 1–217.

Cohee, G. V., 1979, Michigan basin region, *in* Paleotectonic investigations of the Mississippian System in the United States: U.S. Geological Survey Professional Paper 1010, p. 49–57.

Collinson, C., Scott, A. J., and Rexroad, C. B., 1962, Six charts showing biostratigraphic zones, and correlations based on conodonts from the Devonian and Mississippian rocks of the Upper Mississippi Valley: Illinois State Geological Survey Circular 328, 32 p.

Conkin, J. E., 1985, Late Devonian New Albany–Ohio–Chattanooga Shales and their interbasinal correlation in Indiana, Ohio, Kentucky, and Tennessee: Kentucky Energy Cabinet, 1985, Eastern Oil Shale Symposium, p. 217–259.

Cooper, G. A., and others, 1942, Correlation of the Devonian sedimentary formations of North America: Geological Society of America Bulletin, v. 53, p. 1729–1794.

Cross, A. T., 1983, Plants of Devonian–Mississippian black shales, Eastern Interior, U.S.A. [abs.]: American Association of Petroleum Geologists Bulletin, v. 67, p. 444–445.

Cross, A. T., and Bordner, M. A., 1980, Palynology and environmental interpretations of the Antrim Shale of central Michigan: U.S. Department of Energy DOE Topical Report FE-2346-1, 41 p.

Cuomo, M. C., and Rhoads, D. C., 1985, Biogenic sedimentary fabrics associated with pioneering polychaete assemblages: Geological Society of America Abstracts with Programs, v. 17, p. 557.

de Witt, W., Jr., 1970, Age of the Bedford Shale, Berea Sandstone, and Sunbury Shale in the Appalachian and Michigan basins: U.S. Geological Survey Bulletin 1294-G, 11 p.

Dix, G. R., and Mullins, H. T., 1987, Shallow, subsurface growth and burial alteration of Middle Devonian calcite concretions: Journal of Sedimentary Petrology, v. 57, p. 140–157.

Duffield, S. L., and Warshauer, S. M., 1981, Upper Devonian (Frasnian) conodonts and ostracodes from the subsurface of western West Virginia: Journal of Paleontology, v. 55, p. 72–83.

Eames, L. E., 1974, Palynology of the Berea Sandstone and Cuyahoga Group of northeastern Ohio [Ph.D. thesis]: East Lansing, Michigan State University.

Edwards, I., and Raasch, G. O., 1921, Notes on the occurrence of Upper Devonian strata in Wisconsin: Milwaukee Public Museum Yearbook for 1921, v. 1, p. 88–93.

Ehlers, G. M., 1938, Eighth annual field excursion, May 28–29, 1938: Michigan Academy of Sciences, Arts and Letters, Section of Geology and Mineralogy, mimeographed handout.

Ehlers, G. M., and Kesling, R. V., 1970, Devonian strata of Alpena and Presque Isle Counties, Michigan: North-Central Section, Geological Society of America and Michigan Geological Society Guidebook, p. 1–130.

Eicher, D. B., 1939, Conodonts from the Elmhurst quarry fissure fillings [M.S. thesis]: Chicago, Illinois, University of Chicago.

Ells, G. D., 1979, Stratigraphic cross sections extending from Devonian Antrim Shale to Mississippian Sunbury Shale in the Michigan Basin: Michigan Department of Natural Resources, Geological Survey Division Report of Investigation 22, 186 p.

Emrich, G. H., and Bergstrom, R. E., 1962, Des Plaines disturbance, northeastern Illinois: Geologial Society of America Bulletin, v. 73, p. 959–968.

Ettensohn, F. R., 1985, Controls on development of Catskill Delta complex basin-facies, *in* Woodrow, D. L., and Sevon, W. D., eds., The Catskill Delta: Geological Society of America Special Paper 201, p. 65–77.

Ettensohn, F. R., Goodman, P. T., Norby, R. D., and Shaw, T. H., 1988, Stratigraphy and biostratigraphy of the Devonian–Mississippian black shales in west-central Kentucky and adjacent parts of Indiana and Tennessee: University of Kentucky, Institute of Mining and Minerals Research, 1988 Eastern Oil Shale Symposium, p. 237–245.

Foreman, H. P., 1959, A new occurrence of Devonian Radiolaria in calcareous concretions from the Huron Member of the Ohio Shale: Journal of Paleontology, v. 33, p. 77–80.

—— , 1963, Upper Devonian Radiolaria from the Huron Member of the Ohio Shale: Micropaleontology, v. 9, p. 267–304.

Gray, J., and Boucot, A., 1979, The Devonian land plant *Protosalvinia*: Lethaia, v. 12, p. 57–63.

Gutschick, R. C., 1987, Devonian shelf-basin, Michigan Basin, Alpena, Michigan, *in* Biggs, D. L., ed., North-Central section of the Geological Society of America: Boulder, Colorado, Geological Society of America, Centennial Field Guide, v. 3, p. 297–302.

Gutschick, R. C., and Wuellner, D., 1983, An unusual benthic agglutinated foraminiferan from Late Devonian anoxic basinal black shales of Ohio: Journal of Paleontology, v. 57, p. 308–320.

Hicks, L. E., 1878, The Waverly group of central Ohio: American Journal of Science, ser. 3, v. 16, p. 216–224.

Holdsworth, B. K., 1977, Paleozoic Radiolaria; Stratigraphic distribution in Atlantic borderlands, *in* Swain, F. M., ed., Stratigraphic micropalentology of Atlantic Basin and Borderlands: Amsterdam, Elsevier Publishing Co., p. 167–184.

Johnson, J. G., and Sandberg, C. A., 1989, Devonian eustatic events in the western United States and their biostratigraphic responses, *in* McMillan, N. J., Embry, A. F., and Glass, D. J., eds., Devonian of the World: Calgary, Canadian Society of Petroleum Geologists Memoir 14, v. 3, p. 171–178.

Johnson, J. G., Klapper, G., and Sandberg, C. A., 1985, Devonian eustatic fluctuations in Euramerica: Geological Society of America Bulletin, v. 96, p. 567–587.

—— , 1986, Late Devonian eustatic cycles around margin of Old Red Continent: Société Géologique de Belgique Annales, v. 109, pt. 1, Special volume "Aachen 1986", p. 141–147.

Kesling, R. V., Segall, R. T., and Sorensen, H. O., 1974, Devonian strata of Emmet and Charlevoix Counties, Michigan: Ann Arbor, University of Michigan Museum of Paleontology Papers on Paleontology no. 7, 187 p.

Kesling, R. V., Johnson, A. M., and Sorensen, H. O., 1976, Devonian strata of the Afton-Onaway area, Michigan: Ann Arbor, University of Michigan Museum of Palentology Papers on Paleontology no. 17, 148 p.

Kirchgasser, W. T., and House, M. R., 1981, Upper Devonian goniatite biostratigraphy, *in* Oliver, W. A., Jr., and Klapper, G., eds., Devonian biostratigraphy of New York; Part 1, Text: International Union of Geological Sciences Subcommission on Devonian Stratigraphy, p. 39–55.

Klapper, G., and others, 1971, North American Devonian conodont biostratigraphy, *in* Sweet, W. C., and Bergström, S. M., eds., Symposium on conodont biostratigraphy: Geological Society of America Memoir 127, p. 285–316.

Kluessendorf, J., Mikulic, D. G., and Carmen, M. R., 1989, Distribution and depositional environments of the westernmost Devonian rocks in the Michigan Basin, *in* McMillan, N. J., Embry, A. F., and Glass, D. J., eds., Devonian of the World: Calgary, Canadian Society of Petroleum Geologists Memoir 14, v. 1, Regional Syntheses, p. 251–263.

Lane, A. C., 1901, Suggested changes in nomenclature of Michigan formations: Michigan Miner, v. 3, no. 1, p. 9.

—— , 1902, Report of the State Board of the Geological Survey of Michigan for the Year 1901: Michigan Geological Survey, v. 8, pt. 2, p. 252–253.

Matthews, R. D., 1983, *Foerstia* from the Antrim Shale (Devonian) of Michigan: Geology, v. 11, p. 327–330.

McHone, J. F., Sargent, M. L., and Nelson, W. J., 1986, Shatter cones and other shock effects at Des Plaines, Illinois; Evidence for meteoroid impact: Geological Society of America Abstracts with Programs, v. 18, p. 689.

Mikulic, D. G., and Kluessendorf, J., 1988, Subsurface stratigraphic relationships of the Upper Silurian and Devonian rocks of Milwaukee County, Wisconsin: Geoscience Wisconsin 13.

Morse, M. L., 1938, Conodonts from the (Devonian) Norwood and Antrim shales of Michigan [Ph.D. thesis]: Ann Arbor, University of Michigan.

Müller, K. J., and Clark, D. L., 1967, Early Late Devonian conodonts from the Squaw Bay Limestone in Michigan: Journal of Paleontology, v. 41, p. 902–919.

Newberry, J. S., 1870, Report on the progress of the Geological Survey of Ohio in 1869: Ohio Geological Survey (Report of Progress for 1869), pt. 1, p. 3–53.

Newcombe, R. B., 1932, Geology of Muskegon oil field, Muskegon, Michigan: American Association of Petroleum Geologists Bulletin, v. 16, p. 153–168.

—— , 1933, Oil and gas fields of Michigan: Michigan Geological Survey Publication 38, Geology Series 32, 293 p.

Palmer, A. R., 1983, The Decade of North American Geology 1983 geologic time scale: Geology, v. 11, p. 503–504.

Pepper, J. F., de Witt, W., Jr., and Demarest, D. F., 1954, Geology of the Bedford shale and Berea sandstone in the Appalachian basin: U.S. Geological Survey Professional Paper 259, 111 p.

Raasch, G. O., 1935, Devonian of Wisconsin: Kansas Geological Society 9th Annual Field Conference Guidebook, p. 261–267.

Richardson, J. B., and McGregor, D., 1986, Silurian and Devonian spore zones of the Old Red Sandstone continent and adjacent areas: Geological Survey of Canada Bulletin 364, 79 p.

Roberts, J. E., 1972, Agglutinated foraminifera from the Devonian of Wisconsin [M.S. thesis]: University of Wisconsin–Milwaukee, 125 p.

Russell, D. J., 1985, Depositional analysis of black shale by using gamma-ray stratigraphy; The Upper Devonian Kettle Point Formation of Ontario: Canadian Petroleum Geology Bulletin, v. 33, p. 236–253.

Sandberg, C. A., 1976, Conodont biofacies of Late Devonian *Polygnathus styriacus* Zone in western United States: Geological Association of Canada Special Paper 15, p. 171–186.

—— , 1979, Devonian and Lower Mississippian conodont zonation of the Great Basin and Rocky Mountains: Provo, Utah, Brigham Young University Geology Studies, v. 26, pt. 3, p. 87–106.

—— , 1981, The Devonian–Carboniferous boundary in North America: Geological Society of America Abstracts with Programs, v. 13, p. 315.

—— , 1988, Role of conodont biofacies in Late Devonian and Early Mississippian paleobiogeographic reconstructions of western United States: Geological Society of America Abstracts with Programs, v. 20, p. 227.

Sandberg, C. A., and Dreesen, R., 1984, Late Devonian icriodontid biofacies models and alternate shallow-water conodont zonation, *in* Clark, D. L., ed., Conodont biofacies and provincialism: Geological Society of America Special Paper 196, p. 143–178.

Sandberg, C. A., and Gutschick, R. C., 1984, Distribution, microfauna, and source-rock potential of Mississippian Delle Phosphatic Member of the Woodman Formation and equivalents, Utah and adjacent States, *in* Woodward, J., Meissner, F. F., and Clayton, J. L., eds. Hydrocarbon source rocks of the Greater Rocky Mountains region: Denver, Colorado, Rocky Mountain Association of Geologists, p. 135–178.

Sandberg, C. A., and Ziegler, W., 1973, Refinement of standard Upper Devonian conodont zonation based on sections in Nevada and West Germany: Geologica et Palaeontologica, v. 7, p. 97–122.

Sandberg, C. A., Steel, M., and Scott, R. A., 1972, Comparison between conodont zonation and spore assemblages at the Devonian–Carboniferous boundary in the western and central United States and in Europe: Congrès International de Stratigraphie et de Géologie du Carbonifère, 7th, Compte rendu, v. 1, p. 179–203.

Sandberg, C. A., Gutschick, R. C., Johnson, J. G., Poole, F. G., and Sando, W. J., 1986, Middle Devonian to Late Mississippian geologic history of the Overthrust belt region, western United States: Société Géologique de Belgique Annales, v. 109, Part 1, Special volume "Aachen 1986", p. 205–207.

Sandberg, C. A., Ziegler, W., Dreesen, R., and Butler, J. L., 1988, Late Frasnian mass extinction; Conodont event stratigraphy, global changes, and possible causes: Courier Forschungsinstitut Senckenberg, v. 102, p. 263–307.

Sandberg, C. A., Ziegler, W., and Bultynck, P., 1989a, New standard conodont zones and early *Ancyrodella* phylogeny across Middle–Upper Devonian boundary: Courier Forschungsinstitut Senckenberg, v. 110, p. 195–230.

Sandberg, C. A., Poole, F. G., and Johnson, J. G., 1989b, Upper Devonian of western United States, *in* McMillan, N. J., Embry, A. F., and Glass, D. J., eds., Devonian of the World: Calgary, Canadian Society of Petroleum Geologists, Memoir 14, v. 1, Regional Syntheses, p. 183–220.

Sanford, B. V., 1967, Devonian of Ontario and Michigan, *in* Oswald, D. H., ed., Proceedings, International Symposium on the Devonian System: Calgary, Alberta Society of Petroleum Geologists, v. 1, p. 973–999.

Schopf, J. M., 1978, *Foerstia* and recent interpretations of early vascular land plants: Lethaia, v. 2, p. 139–143.

Schopf, J. M., and Schwietering, J. F., 1970, The *Foerstia* Zone of the Ohio and Chattanooga Shales: U.S. Geological Survey Bulletin 1294-H, 20 p.

Schumacher, D., 1971, Conodonts from the Middle Devonian Lake Church and Milwaukee Formations; and Conodonts and biostratigraphy of the "Kenwood" Shale, *in* Clark, D. L., ed., Conodonts and biostratigraphy of the Wisconsin Paleozoic: Wisconsin Geological and Natural History Survey Information Circular 19, p. 55–77.

Scott, A. J., and Collinson, C., 1961, Conodont faunas from the Louisiana and McCraney Formations of Illinois, Iowa, and Missouri: Kansas Geological Society 26th Annual Field Conference Guidebook, p. 100–141.

Schwimmer, B. A., Hannibal, J. T., Feldmann, R. M., and Stukel, D. J., III, 1987, The paleontology and depositional environment of the Chagrin Shale (Famennian) in northeastern Ohio: Calgary, Second International Symposium on the Devonian System, Aug. 17-20, 1987, Program and Abstracts, p. 204.

Streel, M., 1966, Critères palynologiques pour une stratigraphie détaillée du Tnla dans les bassins Ardenno–Rhénans: Société Géologique de Belgique Annales, v. 89, Bull. 3, p. 65–96.

—— , 1986, Miospore contribution to the Upper Famennian–Strunian event stratigraphy: Société Géologique de Belgique Annales, v. 109, Part 1, Special volume "Aachen 1986", p. 75–92.

Streel, M., and Traverse, A., 1978, Spores from the Devonian/Mississippian transition near the Horseshoe Curve setion, Altoona, Pennsylvania, U.S.A.: Review of Paleobotany and Palynology, v. 26, p. 21–39.

Tarbell, E., 1941, Antrim-Ellsworth-Coldwater Shale formations in Michigan: American Association of Petroleum Geologists Bulletin, v. 25, p. 724–733.

Thayer, C. W., 1981, Ecology of living brachiopods, *in* Broadhead, T. W., ed., Lophophorates, notes for short course: Knoxville, University of Tennessee Department of Geological Sciences Studies in Geology 5, p. 110–126.

Thrasher, L. C., 1987, Macrofossils and stratigraphic subdivisions of the Bakken Formation (Devonian–Mississippian), Williston Basin, North Dakota, *in* Carlson, C. G., and Christopher, J. E., eds., Fifth International Williston Basin Symposium: Regina, Saskatchewan Geological Society Special Publication 9, p. 53–67.

Uyeno, T. T., Telford, P. G., and Sanford, B. V., 1982, Devonian conodonts and stratigraphy of southwestern Ontario: Geological Survey of Canada Bulletin 332, 45 p.

Ver Wiebe, W. A., 1927, The stratigraphy of Alpena County, Michigan: Papers of Michigan Academy of Sciences, Arts and Letters, v. 7, p. 181–192.

Wardlaw, M. M., and Long, D. T., 1982, Mineralogy, chemistry, and physical setting of carbonate concretions in the Antrim Shale (Devonian, Michigan Basin); Clues to origin: Geological Society of America Abstracts with Programs, v. 14, p. 291.

Warthin, A. S., Jr., and Cooper, G. A., 1935, New formation names in the Michigan Devonian: Journal of Washington Academy of Sciences, v. 38, p. 105–113.

——, 1943, Traverse rocks of Thunder Bay region, Michigan: American Association of Petroleum Geologists Bulletin, v. 27, p. 571–595.

Weller, S., 1899, A peculiar Devonian deposit in northeastern Illinois: Journal of Geology, v. 7, p. 483–488.

Williams, J. S., 1943, Stratigraphy and fauna of the Louisiana limestone of Missouri: U.S. Geological Survey Professional Paper 203, 133 p.

Willman, H. B., 1971, Summary of geology of the Chicago area: Illinois Geological Survey Circular 460, 77 p.

Winder, C. G., 1966, Conodont zones and stratigraphic variability in Upper Devonian rocks, Ontario: Journal of Paleontology, v. 40, p. 1275–1293.

——, 1967, Micropaleontology of the Devonian in Ontario, *in* Oswald, D. H., ed., Proceedings, International Symposium on the Devonian System: Calgary, Alberta Society of Petroleum Geologists, v. 2, p. 711–719.

Winslow, M. R., 1962, Plant spores and other microfossils from Upper Devonian and Lower Mississippian rocks of Ohio: U.S. Geological Survey Professional Paper 364, 93 p.

Ziegler, W., 1962, Taxionomie und Phylogenie Oberdevonischer Conodonten und ihre stratigraphische Bedeutung: Hessisches Landesamt Bodenforschung Abhandlungen, v. 38, 166 p.

——, 1971, Conodont stratigraphy of the European Devonian, *in* Sweet, W. C., and Bergström, S. M., eds., Symposium on conodont biostratigraphy: Geological Society of America Memoir 127, p. 227–284.

Ziegler, W., and Sandberg, C. A., 1984, *Palmatolepis*-based revision of upper part of standard Late Devonian conodont zonation, *in* Clark, D. L., ed., Conodont biofacies and provincialism: Geological Society of America Special Paper 196, p. 179–194.

——, 1990, The Late Devonian standard conodont zonation: Courier Forschungsinstitut Senckenberg, v. 121, 115 p.

Manuscript Accepted by the Society June 1, 1990

Printed in U.S.A.

# Late Devonian history of Michigan Basin

**Raymond C. Gutschick**
*Volunteer, U.S. Geological Survey, and Professor Emeritus, University of Notre Dame, 2901 Leonard Street, Medford, Oregon 97504*
**Charles A. Sandberg**
*U.S. Geological Survey, MS 940, Box 25046, Denver Federal Center, Denver, Colorado 80225*

## ABSTRACT

The Upper Devonian sequence in the Michigan Basin is a westward extension of coeval cyclical facies of the Catskill deltaic complex in the Appalachian basin. Both basins and the intervening Findlay arch express the tectonic and sedimentational effects of foreland compression and isostatic compensation produced by the Acadian orogeny. The Late Devonian Michigan Basin formed as one of several local deeps within the long Eastern Interior seaway that separated the North American craton, backboned by the Transcontinental arch, on the west from the Old Red continent, Avalon terrane (microplate), and possibly northwest Africa on the east. Basin development began in the late Middle Devonian (late Givetian *varcus* Zone) with subsidence of a shallow-water carbonate platform formed by rocks of the Traverse Group. Subsidence was contemporaneous with Taghanic onlap of the North American craton. During subsidence, a thin transitional sequence of increasingly deeper water limestones separated by hardgrounds was deposited in the incipient Michigan Basin during the latest Givetian to earliest Frasnian *disparilis* to *falsiovalis* Zones. Deposition of this sequence culminated during the early Frasnian *transitans* Zone with a calcareous mudstone bed at the top of the Squaw Bay Limestone. Subsidence was followed by a 12-m.y.-long Late Devonian episode of slow, hemipelagic, basinal sedimentation of organic black muds that formed the Antrim Shale, interrupted basinwide only by deposition of its prodeltaic Paxton Member. Westward, the basinal Antrim black muds intertongued with greenish gray, deltaic and prodeltaic muds of an eastward-prograding delta platform formed by the Ellsworth Shale. Basinal black shale deposition ceased in latest Devonian (late Famennian Lower *praesulcata* Zone) time, when the Bedford deltaic complex prograded westward, completely filling the Antrim Basin and even covering part of the older Ellsworth deltaic complex on the west. As sea level was lowered eustatically near the end of the Devonian, the regressive Berea Sandstone terminated deltaic deposition. After an Early Mississippian erosional episode, widespread deposition of the unconformably overlying Lower Mississippian Sunbury Shale began during the next transgression, associated with a major eustatic rise in the Lower *crenulata* Zone.

## INTRODUCTION

The sedimentary basin fill has been described as the recorder of geologic history. Analysis of basin fills using concepts of genetic stratigraphic sequence analysis and recognizing the complex interplay of tectonics, eustatic, and internal as well as external sedimentary controls may resolve major problems in earth science. [Galloway, 1989]

The Michigan Basin contains more than 4,000 m of Paleozoic rocks deposited during approximately 280 m.y., based on the

Gutschick, R. C., and Sandberg, C. A., 1991, Late Devonian history of Michigan Basin, *in* Catacosinos, P. A., and Daniels, P. A., Jr., eds., Early sedimentary evolution of the Michigan Basin: Geological Society of America Special Paper 256.

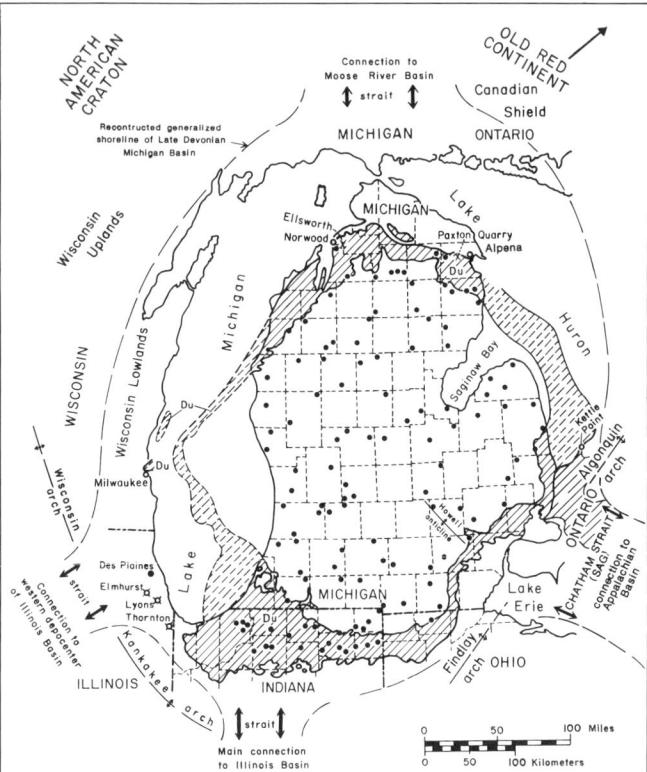

Figure 1. Geologic map of the Michigan Basin encircled by Upper Devonian (Du) outcrop belt (shaded area). Locations of wells (black dots) that provide subsurface lithologic and thickness data are from Ells (1979) for Michigan, and Hasenmueller and Bassett (1980) for Indiana. In northeastern Illinois, localities of older rocks that contain crevice fillings with Devonian conodonts are shown by barbed, open circles. Information for Milwaukee area is from Mikulic and Kluessendorf (1988), for northeastern Illinois from Kluessendorf and others (1989), under Lake Michigan from Wold and others (1981), and in Ontario and under Lake Huron from the Geologic Map of Canada (Geological Survey of Canada, 1969).

Geological Society of America radiometric time scale (Palmer, 1983). Our report focuses on a small segment of basin history from its reactivation at the close of the Middle Devonian (Givetian) through the end of its first major depophase at the close of the Late Devonian (Famennian). This interval involves a maximum of 260 m of rocks deposited during ~17 m.y. at an average rate of ~15 m/m.y. The Late Devonian depositional Michigan Basin now occupies all of Lower Michigan and parts of western Ontario, northwestern Ohio, northern Indiana, northeastern Illinois, and eastern Wisconsin (Fig. 1).

The treated Devonian stratigraphic sequence in the Michigan Basin comprises, in ascending order: the Middle Devonian Thunder Bay Limestone, uppermost formation of the Traverse Group; the Middle and Upper Devonian Squaw Bay Limestone; and the Upper Devonian Antrim, Ellsworth, and Bedford Shales, and Berea Sandstone. This sequence is essentially a westward extension of coeval cyclical facies of the Catskill deltaic complex in the Appalachian basin. A chart (Fig. 2) summarizes the correlations of this sequence with stratigraphic units in adjacent areas, and especially with the Appalachian basin, in terms of the chronostratigraphic, biostratigraphic, and radiometric time scales, the global standard Late Devonian conodont zonation (referenced by Gutschick and Sandberg, this volume), subsurface gamma ray units of Ells (1979), numbered Acadian orogenic tectophases (Ettensohn, 1985a), and numbered Catskill deltaic cycles (Ettensohn, 1985b).

The present-day Michigan Basin has the appearance of being an oval, yoked basin with a northeast-southwest axis (Fig. 1). The basin is surrounded by arches, which separate it from other sedimentary basins, except on its north side, where depositional connections have been eroded from the Precambrian Canadian shield. Clockwise around the Michigan Basin: to the east, the Algonquin and Findlay arches, breached by the Chatham sag (a Late Devonian paleogeographic strait), separate it from the Appalachian basin; to the south and southwest, the Findlay, Kankakee, and Wisconsin arches, breached by two unnamed sags (Late Devonian paleogeographic straits), separate it from the Illinois basin; and to the west, the stable Wisconsin uplands (Fig. 1) and Transcontinental arch separate it from the Williston basin.

We interpret the Late Devonian Michigan Basin to have formed as a small deep within the long Eastern Interior seaway, which separated the North American craton, backboned by the Transcontinental arch, from the Old Red continent (Fig. 3). The southern part of this seaway occupied the same area as the mid-Mississippian Eastern Interior trough depicted by Gutschick and Sandberg (1983). The Late Devonian basin was underlain at Precambrian basement level by a divergent arm of the Midcontinent rift system and the Grenville orogen margin (Hinze and others, 1975; Fisher and others, 1988). Reactivation of the basinal area, over an aulacogen that may have been as young as Late Ordovician, started with subsidence of a shallow-water carbonate platform formed by the Traverse Group during the Taghanic onlap. This onlap affected the entire craton (Johnson, 1970; Johnson and others, 1985, 1986) in the late Middle Devonian (Givetian Middle *varcus* Subzone). Initial subsidence was followed by a 12-m.y.-long episode, entirely during the Late Devonian, of slow, hemipelagic, starved basinal sedimentation of organic-rich, black muds that became the Antrim Shale. In the western part of the basin, the Antrim basinal facies intertongued continuously with greenish gray muds of the Ellsworth deltaic facies that prograded from the west, whereas in the eastern part, the Antrim Shale was deposited as an almost continuous, anoxic black shale sequence. However, Antrim deposition was twice interrupted nearly basinwide by the influx of oxygenated sediments: (1) during the Frasnian *Ancyrognathus triangularis* to *linguiformis* Zones, when prodeltaic mud of the Paxton Member was deposited across most of the basin, and (2) during the early to late Famennian *marginifera, trachytera,* and *postera* Zones, when Ellsworth prodeltaic sediments prograded eastward across more than half of the basin, at about the same time that prodeltaic muds of the Chagrin Shale prograded westward in Ohio. Black shale deposition ceased at the start of the very late Famennian Lower *praesulcata* Zone, when the Bedford gray mud delta pro-

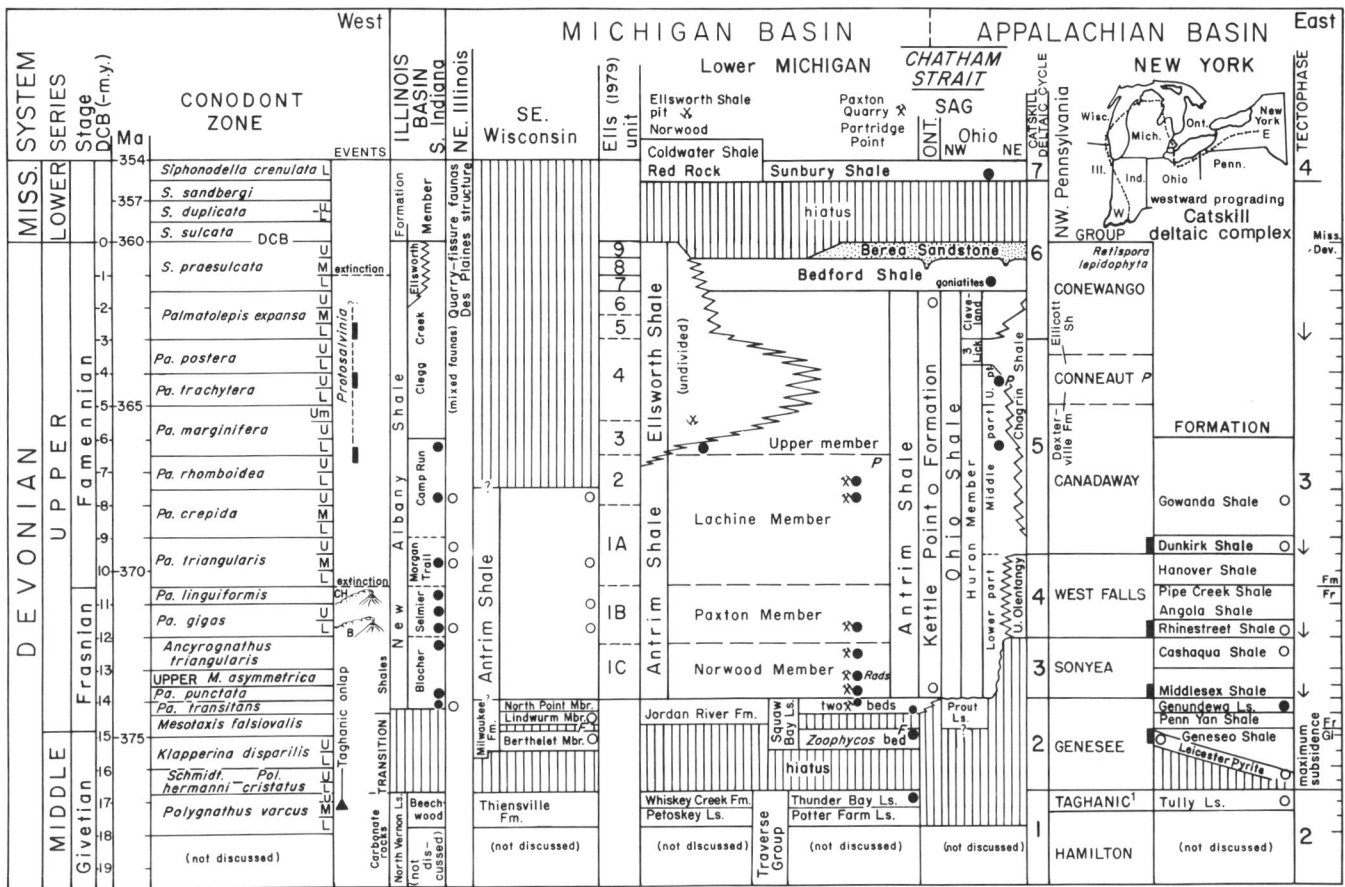

Figure 2. Correlation chart showing Upper Devonian and adjacent strata in Michigan Basin, subsurface gamma-ray units of Ells (1979), and correlations with Illinois basin, northern Ohio, and Appalachian basin. Appalachian basin column from Rickard (1975); bars indicate black shales; [1], Taghanic is New York Stage. Rocks are dated by standard conodont zones; radiometric time scale (Palmer, 1983); and biostratigraphic time scale (Sandberg and others, 1989a), giving date in –m.y. before starting point of 0 m.y. at Devonian–Carboniferous boundary (DCB). Conodont collections identified by Sandberg are shown by closed black circles in stratigraphic position opposite assigned zone; zonal assignments of other authors are shown by open circles. Also shown are times of mass extinctions and other events; range and occurrences of aquatic plant *Protosalvinia* (P); and occurrences of agglutinated foraminiferans (F). Timing of volcanoes that produced bentonites in Pennsylvania and Ohio is after Conkin (1985); B, Belpre; CH, Center Hill. Tectophase refers to pulse of Acadian orogeny (after Ettensohn, 1985a); Catskill deltaic cycle after Ettensohn (1985b); arrows show times of subsidence.

graded from the east across the basin to merge with the Ellsworth greenish gray mud delta. The regressive Berea Sandstone was deposited on the eastern side of the basin as sea level fell eustatically and the delta retreated. This fall terminated a major Michigan Basin depophase that was coeval with the one in the western United States described by Johnson and Sandberg (1989). After an earliest Mississippian erosional interval, the Lower Mississippian black Sunbury Shale was deposited at the start of a second major depophase that is coincident with a eustatic rise in the Northern Hemisphere (Sandberg and others, 1986).

Primary sources of biostratigraphic information are bedrock exposures along the outcrop belt at the north end and along the margins of the basin and subsurface well data from within the central basin (Gutschick and Sandberg, this volume). Most of the basin is veneered by glacial drift, so outcrops of nonresistant shales are scarce. Formations dip basinward and are concealed beneath younger beds or below Lakes Michigan and Huron. The most useful exposures, described by Gutschick and Sandberg (this volume), are in lower Michigan at Alpena (Partridge Point), Paxton Quarry, Norwood, and Ellsworth (Fig. 1). Devonian rocks also occur in the Milwaukee area in southeastern Wisconsin, and at Kettle Point along the east shore of Lake Huron in Ontario. Their former presence is recorded by Devonian fossils found in fissure fillings preserved in stratigraphic leaks into Silurian reef crevices at three quarries in northeastern Illinois (Fig. 1).

Interpretation of the sedimentational history of the Late Devonian Michigan Basin draws heavily on the models of Gutschick and Sandberg (1983) and Gutschick (1987). For subsur-

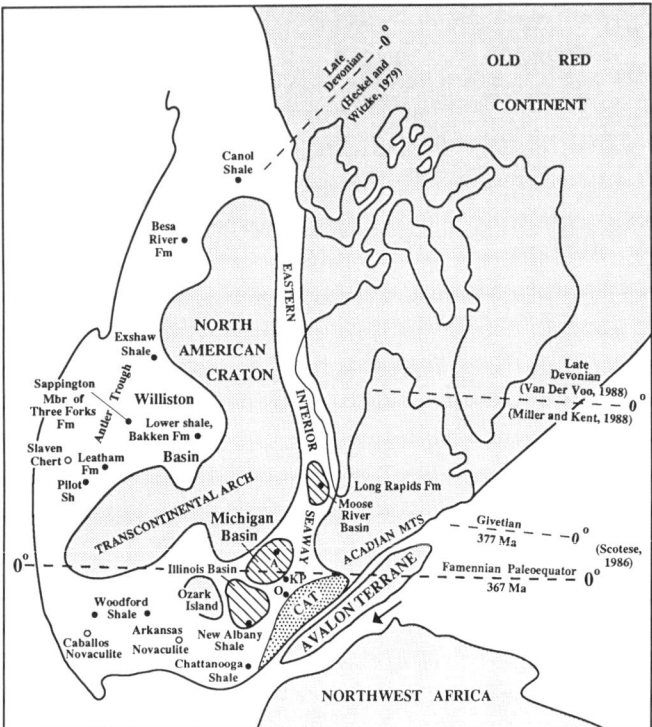

Figure 3. Late Devonian paleogeographic reconstruction (adapted from Ziegler and others, 1979) of North America and adjacent continents, showing relation of Antrim Shale (A) in Michigan Basin (diagonally lined area) to coeval black shales (black dots) in other basins (also diagonally lined) of Eastern Interior seaway and in other areas around North American craton. Coeval siliceous rocks shown by open circles. Land areas are screened. CAT, Catskill deltaic complex; O, Ohio Shale; KP, Kettle Point Formation. Positions of various Devonian paleoequators by different authors are shown, but 367 Ma paleoequator of Scotese (1986), which is used in interpreting Michigan Basin history, is emphasized by heavier line.

face correlations it relies on the comprehensive gamma-ray log analysis and data base provided by Ells (1979). In his report, gamma-ray logs and records of 99 wells distributed throughout the Michigan Basin (Fig. 1) were used to construct six regional cross sections. Gamma-ray log formational signatures were recognized for the Antrim, Bedford, and Sunbury Shales and for the Berea Sandstone; and the Antrim and Bedford were subdivided into ten gamma-ray log units (Fig. 2). The shortcoming of Ells's study was its inability, because of the low radioactive response, to subdivide the Ellsworth Shale into Antrim-equivalent units.

Information on regional thicknesses and distribution patterns of Upper Devonian rocks in the subsurface of the Michigan Basin was obtained from the following sources: isopach and structure maps compiled by Cohee and others (1951) and Fisher (1980); gamma-ray logs from wells in northern Indiana furnished by the Indiana Geological Survey (Hasenmueller and Bassett, 1980); tunnel and well-log information in the Milwaukee area (Mikulic and Kluessendorf, 1988); and the anomalous Des Plaines structural deformation, in northeastern Illinois (Emrich and Bergstrom, 1962; McHone and others, 1986; Kluessendorf and others, 1989). Additional useful subsurface information is contained in these studies: Tarbell (1941), Cohee and Underwood (1944), Fisher (1969), Mounds Facility Laboratory Report (1980), Hasenmueller and Woodward (1981), Hydrogeologic Atlas of Michigan (1981), Matthews (1983), Russell (1985), Hasenmueller and Leininger (1987), and Fisher and others (1988).

Graphic illustration of the Late Devonian history of the Michigan Basin is provided herein by 13 sequential isopach maps and two paleogeologic maps, all compiled from log tops picked by Ells (1979) prior to making interpretations. These maps improve Ells's data base through biostratigraphic constraints provided by our companion study of the outcropping Antrim and Ellsworth Shales in Michigan (Gutschick and Sandberg, this volume) and extend it southward into northern Indiana by gamma-ray log correlations (Hasenmueller and Bassett, 1980).

## RELATIONS TO APPALACHIAN BASIN

The Michigan Basin resembles the Appalachian basin in geographic, tectonic, climatic, and sedimentologic factors. It differs in distance from the ancestral Acadian Mountains, an important source of sediments and cause of rain-shadow effects on precipitation. References that were most useful in making interbasinal comparisons and correlations are: Rickard (1975, 1981), Ettensohn and Barron (1981), Oliver and Klapper (1981), Potter and others (1981), Ettensohn (1985a, b), Pashin and Ettensohn (1987), and Woodrow and others (1989).

The general pattern of sedimentation in the Michigan Basin and of the Catskill deltaic complex in the Appalachian basin started with tectonism accompanied by rising sea level, transgression and subsidence, and development of a pycnocline. Both basins were characterized by hemipelagic black shale deposition in the deep basin with input of pelagic faunas and floras and extrabasinal floras, and by stagnant starved-basin conditions. They show no evidence of a benthos, upwelling, spiculitic bedded cherts (lydites), or bedded phosphorites. Nevertheless, the black basinal muds were rich in exotic organic matter, and the resulting organic-rich shales were highly radioactive. These conditions were repeated cyclically. The close synchronization of the sedimentational pattern between the two basins is emphasized in discussions of the sequential maps of the Michigan Basin.

Formation of the Late Devonian Michigan Basin coincided with maximum Acadian tectonism and Appalachian basin subsidence from late Middle Devonian (Givetian Upper *varcus* Subzone) through early Late Devonian (Frasnian *transitans* Zone) time, which corresponds to Genesee cycle 2 (black Geneseo Shale) of the Catskill deltaic complex. The initial sea floor of the Michigan Basin was dysaerobic, not anaerobic or euxinic. The earliest deposition of the anoxic, black Antrim Shale was during the Frasnian *punctata* Zone (= black Middlesex Shale, cycle 3 in New York State; and Blocher Member of New Albany Shale in Illinois basin). Antrim deposition continued through cycle 4 (=

black Rhinestreet Shale), cycle 5 (= black Dunkirk Shale), and cycle 6 (= black Cleveland Member of Ohio Shale).

## LATE DEVONIAN PALEOGEOGRAPHY OF MICHIGAN BASIN

Understanding the Late Devonian North American continental framework and its paleogeography is the first step toward interpreting the Late Devonian paleogeography of the Michigan Basin. Review of the paleogeography of other areas follows the paths of Pepper and others (1954), Conant and Swanson (1961), Johnson (1970), Heckel and Witzke (1979), Ziegler and others (1979), Ettensohn and Barron (1981), Droste and Shaver (1983), Gutschick and Sandberg (1983), Ettensohn (1985a, b), Scotese and others (1985), Scotese (1986), Miller and Kent (1988), Sandberg (1988), Sandberg and others (1988), Van der Voo (1988), Sandberg and others (1989a), and Telford (1989).

Late Devonian paleotectonic patterns are dominated by the Catskill deltaic complex of the Appalachian basin and the widespread belt of organic-rich, radioactive black shales around the southeastern side of the North American craton (Fig. 3). Sedimentation of these shales was the result of Acadian orogenic cycles. Similar black shales rim the northwestern margin of the craton as a result of Antler orogenic influence (Fig. 3). The Old Red continent, an ancient landmass, was attached to the northeastern part of Late Devonian North America. The paleolatitudinal framework must be compatible with the paleomagnetism, paleoclimatology, and biogeography. However, three recent papers present widely divergent, conflicting paleoequatorial positions. A paper on the late Frasnian mass extinction (Sandberg and others, 1988) favors the reconstruction of Scotese (1986) over that of Van der Voo (1988) on the basis of intercontinental sedimentologic and conodont biostratigraphic evidence.

The Michigan Basin was in the tropics, straddling the paleoequator in Late Devonian time, south of it at the start of the Frasnian, and north of it at the end of the Famennian, assuming the reconstruction of Scotese (1986) is correct. Sea level was generally high and marine transgression was widespread during episodes of black shale deposition. The Eastern Interior seaway was extensive, and basins such as the Illinois, Michigan, and Moose River formed as deeps within it (Fig. 4). In the Moose River basin, the Long Rapids Formation, as described by Telford (1989), is a deep, starved-basin sequence nearly identical to the Antrim Shale, although he misinterpreted it as a shallow-water deposit on the basis of brachiopod faunas in its lower part and intertonguing prodeltaic sediments in its upper part. Throughout the region, equatorial temperatures and rainfall amounts were high. Delta marsh wetlands bordering the eastern and western sides of the Michigan Basin probably were dominated by abundant floating *Protosalvinia* (*Foerstia*) algae, and floodplains of rivers contained forests in which *Callixylon, Asterocalamites,* and other plants grew. Much terrestrial organic material was added to the pelagic plankton and nekton of the basin to be deposited and preserved in the anoxic, reducing Antrim muds.

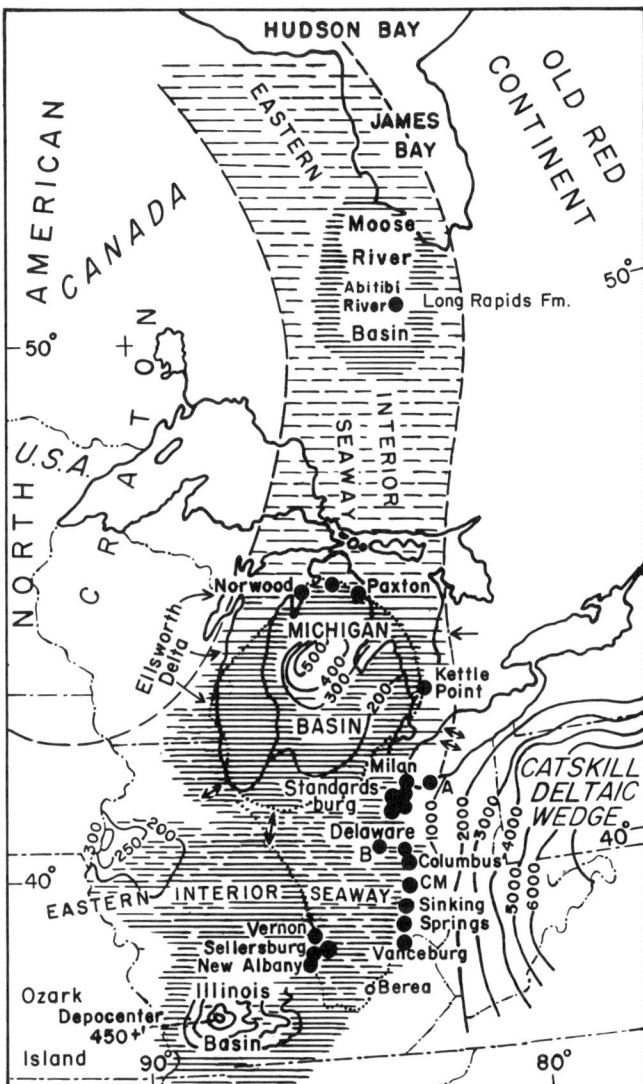

Figure 4. Map showing relation of the Michigan Basin to adjacent depositional areas within Eastern Interior seaway. Black shale shown by solid lines; greenish gray shale shown by thinner, dashed lines. Isopachs show thickness in feet. Single-tipped arrows indicate sediment transport directions; double-tipped arrows indicate location of straits. Black dots show distribution of known Upper Devonian calcareous concretions from named localities of Antrim, Kettle Point, Long Rapids, Ohio, and New Albany formations in Michigan, Ontario, Ohio, Indiana, and Kentucky. A, subsurface recognition; B, Bellefontaine outlier; and CM, Copperas Mountain.

The stratified, stagnant Michigan Basin and related Late Devonian interior basins are distinguished from open-marine seaways, such as the Cretaceous Western Interior seaway described by Parrish and Gautier (1988), by these characteristics: (1) lack of significant phosphate production due to absence of equatorial upwelling and oxygenated bottom waters, (2) little or no benthic biologic productivity, (3) low sediment-accumulation rates but significant preservation of terrestrial organic matter, and (4) an anoxic facies that is confined to basin deeps. Bedded

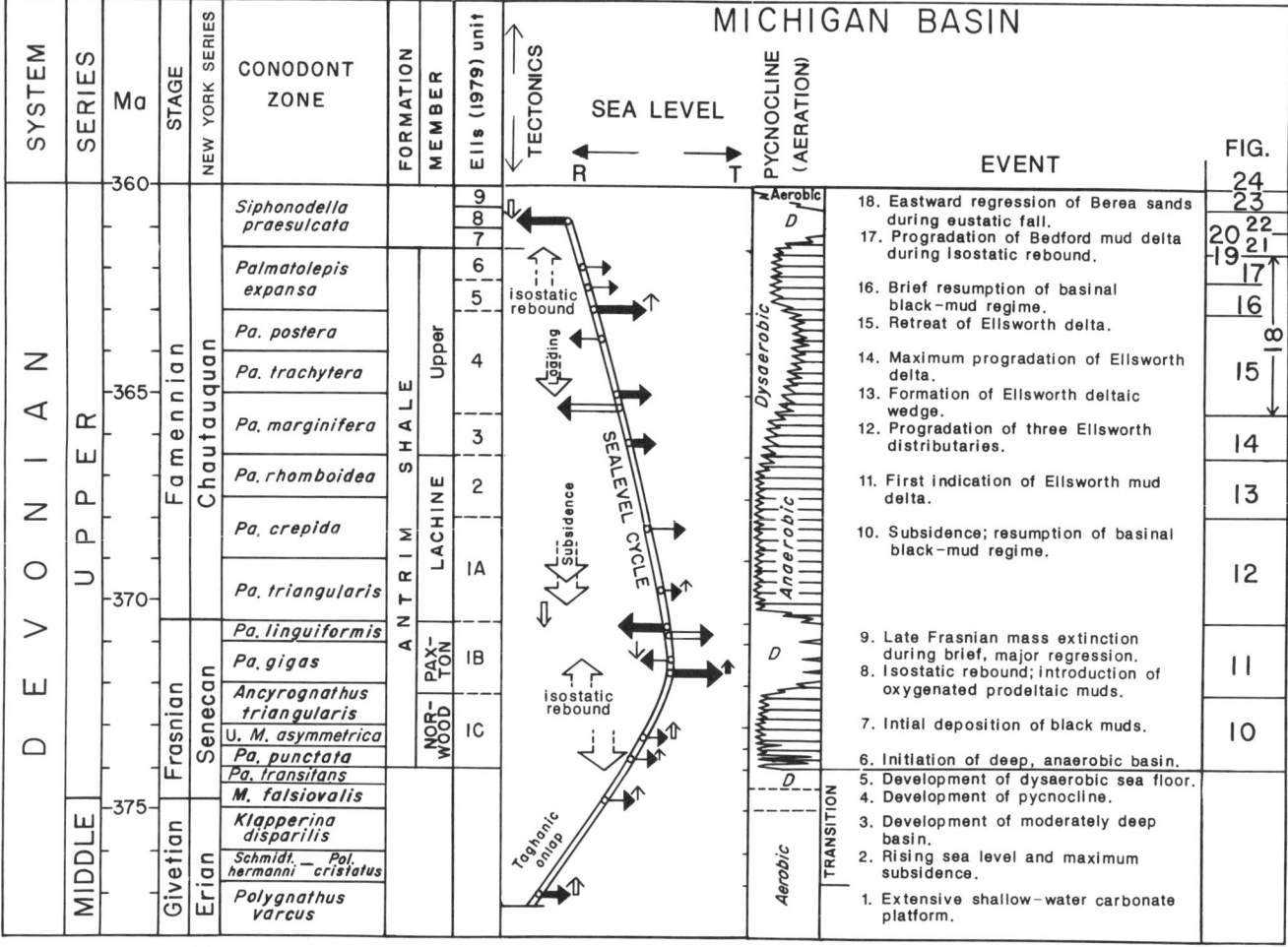

Figure 5. Event stratigraphy and sea-level and pycnocline curves of latest Middle to Late Devonian Michigan Basin. Chart serves as guide for sequential isopach maps (Figs. 7 through 18, 20 through 23), and paleogeologic maps (Figs. 19 and 24). Compare this chart with correlation chart (Fig. 2). Starts of transgressions are after Sandberg and others (1988) and Johnson and Sandberg (1989). Horizontal arrows show transgressions (T) to right and regressions (R) to left. Labelled vertical arrows inside sea-level curve show subsidence downward and rebound upward; unlabelled arrows on both sides of curve show sea-level rise upward and fall downward. Weights of all arrows show increasing magnitutde of events by single, double, and heavy lines. Within pycnocline curve, D stands for dysaerobic.

phosphorites and thin-bedded spicular cherts were not observed in outcrop sections of the black Antrim Shale, suggesting a low rate of production of benthic biogenic silica in contrast to the Lower Devonian of the central Appalachians (Newton and Thompson, 1987).

Other North American Upper Devonian black shales, such as the Woodford Shale and Leatham Formation (Fig. 3), contain bedded, spicular, radiolarian chert; the Woodford Shale (Cheng, 1986) and the greenish gray Chagrin Shale in Ohio (Schwimmer and others, 1987) contain large numbers of phosphate concretions with varied, abundant faunas. Extensive siliceous deposits such as the Arkansas Novaculite (Arkansas and Oklahoma), Caballos Novaculite (Texas), and Slaven Chert (Nevada) lie adjacent to and are interbedded with black shale, but these formations were deposited farther seaward, in deeper parts of the Ouachita and Antler foreland troughs.

## INTERPRETATIONS

The late Middle to Late Devonian event stratigraphy of the Michigan Basin is charted in Figure 5. The global standard Late Devonian conodont zonation* is subjectively related to the radiometric time scale, following Sandberg and Poole (1977) and

---

*The Late Devonian standard conodont zonation, which is the basis for all dating in this paper, was revised recently by Ziegler and Sandberg (1990). Zonal-name modifiers such as Lower and Upper have been changed to Early and Late, respectively, and some zonal names have been totally changed. For example, the former Upper *Mesotaxis asymmetrica* Zone, as used herein (e.g., Figs. 2 and 5), has been replaced by Early *hassi* Zone; the former *Ancyrognathus triangularis* Zone has been divided into the Late *hassi* Zone and *jamieae* Zone, in ascending order; and the former Lower and Upper *gigas* Zones have been changed to Early and Late *rhenana* Zones. For a complete discussion and listing of the new standard zones, the reader should refer to Ziegler and Sandberg (1990).

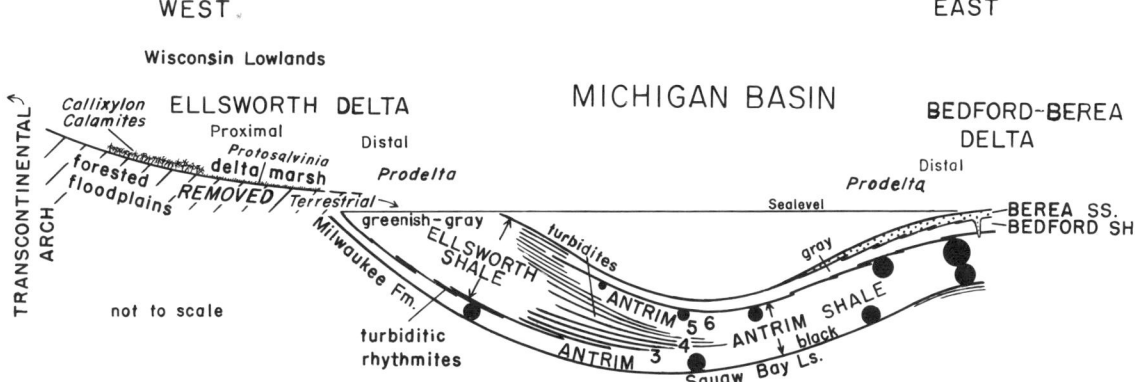

Figure 6. Schematic cross section, not to scale, showing distribution, relations, and sedimentational features of uppermost Middle and Upper Devonian stratigraphic units in Michigan Basin. Also shows relation of Ellsworth delta (preserved distal part) to Antrim Shale and illustrates erosional truncation of proximal part of delta in Wisconsin. Bedford-Berea delta represents maximum westward progradation and spillover of Catskill deltaic complex from Appalachian basin into Michigan Basin. Black dots (size proportional to amount of radioactivity) represent radioactive zones in Antrim; upper zone suggests dilution by nonradioactive Ellsworth deltaic muds transported eastward into basin.

Sandberg and others (1989a), to provide a biostratigraphic time scale for our interpretations. In addition to being highly important for dating sequences and events, conodonts are used for making paleoenvironmental and paleotectonic interpretations following the Late Devonian conodont biofacies models of Sandberg (1976) and Sandberg and Dreesen (1984) and some pertinent observations on the Michigan Basin by Sandberg and others (1989b). The lower three members of the Antrim Shale, the Norwood, Paxton, and Lachine Members, are completely exposed in the Paxton Quarry. Excellent conodont faunas there permit precise dating by the standard conodont zonation, and from there, members can be correlated to a nearby well (Fig. 5 of Gutschick and Sandberg, this volume) and thence throughout the basin, using gamma-ray logs. Consequently, gamma-ray unit tops picked by Ells (1979) permit us to map the distribution and thickness of these members as well as the four subunits of the upper member of the Antrim.

Tectonism, the Antler orogeny in the west (Poole and Sandberg, 1977) and the Acadian orogeny in the east (Ettensohn, 1985a), is of prime importance in interpreting Upper Devonian stratigraphic sequences in orogenic troughs flanking the North American craton. Cyclic Catskill deltaic facies (Ettensohn, 1985b), generated by tectonic pulses, are reflected by corresponding cycles in the uppermost Middle and Upper Devonian sequence of the Michigan Basin (Figs. 2 and 5).

Upper Devonian sedimentational relations in the basin are shown in a west-to-east cross section (Fig. 6), which serves, together with Figure 5, as a guide for interpreting the isopach maps of our study. The chronologic succession of events 1 through 18 and the facies relations of the major stratigraphic units are interpreted in relation to sea-level and pycnocline curves for the Michigan Basin (Fig. 5). Sediments were of two main types: (1) slowly deposited, starved-basin black muds of the Antrim Shale and (2) more rapidly accumulated, deltaic and prodeltaic, extrabasinal, oxygenated muds of the Ellsworth and Bedford Shales. Radioactivity is greater in organic-rich black muds but decreases by dilution in oxygenated, greenish gray and gray muds.

The first three maps (Figs. 7 through 9) provide perspective on the amount, distribution, and major types of Late Devonian sedimentation in the basin. The first map (Fig. 7) shows the total thickness and distribution of Upper Devonian rocks (Antrim, Ellsworth, Bedford, and Berea Formations). This entire sequence was deposited during Acadian tectophase 3 (Fig. 2; Ettensohn, 1985a). The map is a simple one with a major depocenter, containing rocks more than 244 m thick in the northwestern part of the Michigan Basin. A secondary depositional site is evident in the northeastern part of the basin. Thicknesses taper from these two depocenters to less than 61 m at its southern margin. The isopach pattern indicates a sediment source from the northwest and a lesser source from the northeast. The second map (Fig. 8) shows the thickness and distribution of the Antrim Shale, and the third map (Fig. 9) shows the thickness and distribution of the Ellsworth Shale. From a comparison of the Antrim and Ellsworth map patterns, it is apparent that they are complementary and that the two formations must represent partly contemporaneous facies. The Antrim was black, slowly deposited, starved-basin mud, whereas the Ellsworth was greenish gray, deltaic and prodeltaic mud deposited about three times faster than the Antrim. The irregular, narrow band of closely spaced isopachs common to both maps is the line of facies separation, which is shown on isopach maps of individual units. Increased thickness of the Antrim directly east of this line is due to greater thicknesses of Ellsworth sediments tonguing into the Antrim.

### Carbonate platform to basin transition

The time interval from the late Middle Devonian, Givetian Middle *varcus* Subzone to the earliest Late Devonian, Frasnian

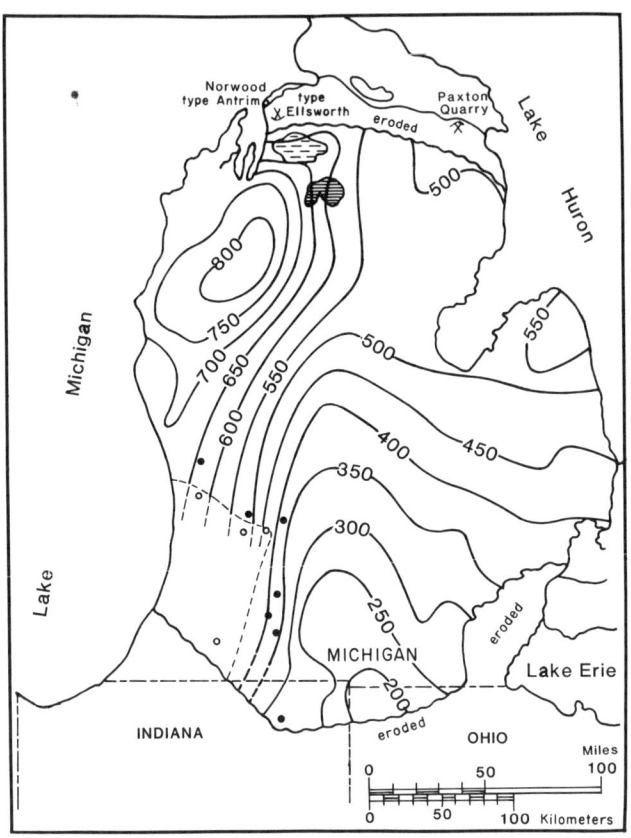

Figure 7 (left). Map showing distribution and thickness of Upper Devonian sequence (total Antrim, Ellsworth, Bedford, and Berea Formations) in Michigan Basin. Isopach interval, 50 ft. Overlying Mississippian Sunbury Shale is not recognized in southwestern Michigan (area with control wells indicated by open circles), so Upper Devonian thicknesses could not be determined there. Black dots show adjacent complete sections. Depocenters are shown for black Antrim Shale (solid horizontal lines) and greenish gray Ellsworth Shale (dashed lines).

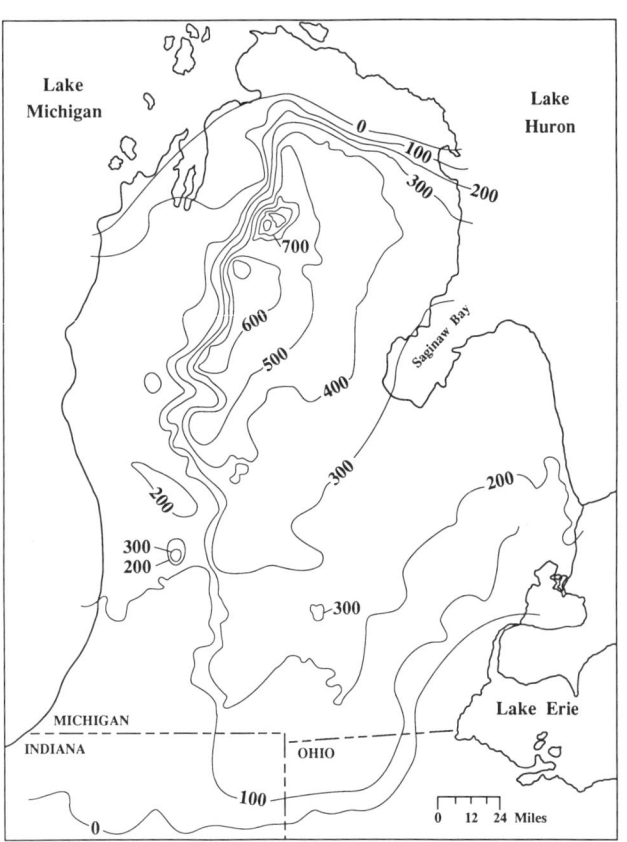

Figure 8 (lower left). Map showing distribution and thickness of Antrim Shale in Michigan Basin (after Fisher, 1980). Isopach interval, 100 ft. Note narrow band of abrupt thickness change and compare with complementary Figure 9. Thickening of basinal Antrim directly east of this band is due to intertongues of prodeltaic Ellsworth Shale.

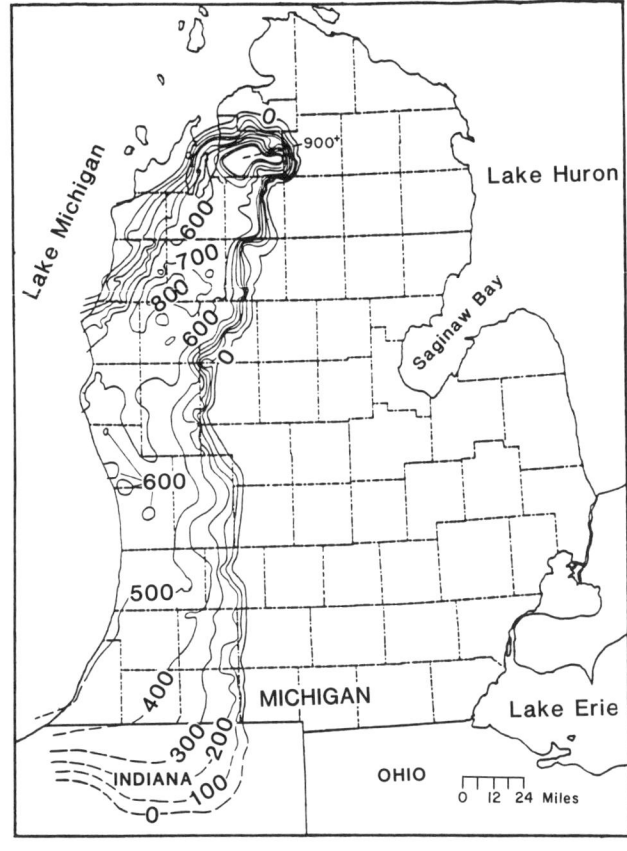

Figure 9 (lower right). Map showing distribution and thickness of Ellsworth Shale in Michigan Basin (after Fisher, 1980). Isopach interval, 100 ft. Note that Ellsworth is restricted to western side of basin and is interpreted as a mud-rich delta-platform. Compare to complementary Figure 8.

*transitans* Zone (Figs. 2 and 5) marks several important events in development of the Late Devonian Michigan Basin. The start of this transition from carbonate platform to deep basin coincides with the Taghanic onlap (Johnson, 1970). Proceeding from a shallow-water carbonate platform (Fig. 5, event 1) with rich coralline-shelly faunas (Thunder Bay Limestone), the transition is characterized by: (a) progressively rising sea level and major subsidence within the Eastern Interior seaway (event 2); (b) development of a moderately deep, incipient Michigan Basin with a stratified water column (events 3 and 4); (c) slowing of sedimentation resulting in hiatuses, such as a hardground at the top of the lower, *Zoophycos* bed of the Squaw Bay Limestone; and (d) further deepening to produce a dysaerobic sea floor (middle, molluscan and upper, calcareous mudstone beds of Squaw Bay; event 5) for the deep Late Devonian Michigan Basin (Antrim and Ellsworth Shales).

The transition from carbonate platform to deep basin was a time of dynamic change produced by the Acadian orogeny and by the influence of the resulting Catskill deltaic complex in the Appalachian basin on sedimentation in the Michigan Basin. During Acadian tectophase 2 (Fig. 2; Ettensohn, 1985a), the entire region between the Acadian Mountains and Transcontinental arch, including the Eastern Interior seaway, Michigan Basin, Appalachian basin, and intervening Findlay arch (Figs. 1 and 3), underwent compression and was downwarped. The transitional interval corresponds to the close of Acadian tectophase 2 (carbonate-platform rocks of Traverse Group) and the start of tectophase 3. This time of tectonic activity in the Acadian orogen produced major uplift of the Acadian Mountains and maximum isostatic subsidence of the Appalachian basin to initiate cycle 2 (Genesee Group) of the Catskill deltaic facies (Fig. 2; Ettensohn, 1985b). Deposition of the resulting black Geneseo Shale was largely confined to the Appalachian basin and did not spread into the Michigan Basin, as did black shales of later deltaic cycles. This lack of westward spread was probably because the intervening Findlay and Algonquin arches had not yet been sufficiently downwarped to permit sediment transport through the Chatham sag (strait).

Resolution of the stratigraphy and sedimentology of the transitional rock sequence is difficult because of sparse biostratigraphic information. Characteristics of the zone and factors that heretofore caused problems in interpretation are: (1) Available sections are few, thin, incomplete, generally poorly exposed, areally discontinuous, and difficult to correlate. (2) Uncertainty exists as to whether the diverse lithofacies and biofacies are sequential or lateral time equivalents. (3) Strata exhibit different stages of subsidence, from shallow to deep water, from aerobic to dysaerobic environments, and from a state of oxidation to one of reduction, as suggested by changes in rock colors (Hosterman and Whitlow, 1981, 1983; Gutschick and Sandberg, 1983). (4) Strata contain lag deposits; bone beds with water-worn phosphatic fish-bone fragments, phosphatic nodules, conodonts, glauconite, and sulfides; and hardgrounds with trace fossils and burrowed and bored surfaces that evidence reworking of sediments.

Rates of deposition for the 2.5-m.y. timespan of the transition period were apparently slow and indicate sediment starvation: Squaw Bay Limestone, 2 m/m.y.; Jordan River Formation, <1 m/m.y.; and more nearshore Milwaukee Formation, 7.5 m/m.y.

Conodont faunas from outcrops were used to determine the age of the transitional rocks, and these faunas and an agglutinated foraminiferan fauna were used to interpret their depositional environments (Gutschick and Sandberg, this volume). The *Zoophycos* Bed and Milwaukee Formation (Fig. 2) contain a polygnathid-icriodid conodont biofacies and a saccamminid foraminiferan biofacies, which indicate moderately deep water on the outer platform or upper foreslope. The Jordan River Formation has a shelly fauna, but conodonts have not been reported.

### Initial deposition of Late Devonian black muds

Deep subsidence toward the end of the transition period had completely changed the seawater and sediment regimes in the newly formed Late Devonian Michigan Basin (Fig. 5, event 6) and produced a stagnant, black-mud sea floor devoid of oxygen (event 7). Floating and swimming organisms thrived in the aerobic zone of the pelagic realm, and their organic remains were preserved in the reducing bottom muds. The isopach map (Fig. 10) of the basal, Norwood Member of the Antrim Shale depicts the first accumulation of Upper Devonian black shale in the deep Michigan Basin (Fig. 5, event 6). Sedimentation of the Norwood began during Catskill deltaic cycle 3, coincident with deposition of the black Middlesex Shale (Fig. 2), as sea level continued to rise and Taghanic onlap continued (Fig. 5). Circulation stagnated, and the sea floor became anoxic and starved. Sediments accumulated at a maximum rate of only about 6 m/m.y. throughout the member. Thin limestone interbeds in the lower Norwood contain conodont faunas of the *punctata* Zone, shale in the middle part contains a radiolarian fauna assignable to the Upper *asymmetrica* conodont Zone, and concretionary black shale in the upper Norwood contains a conodont fauna of the *Ancyrognathus triangularis* Zone.

Isopachs of the Norwood Member (Fig. 10) have an amoeboid pattern but trend generally northeast subparallel to the Findlay arch (Fig. 1) and Transcontinental arch (Fig. 3). Pelagic sedimentation apparently proceeded slowly on an undulating sea floor, first filling the lows. The single small area of thick accumulation in the center of the basin may be anomalous, but it is based on several wells. The Norwood thins to the southwest and apparently disappears under Lake Michigan. A Bouguer gravity high near the southwest corner of Michigan (Hinze and others, 1971) is about the same size and shape as the area in which the Norwood is apparently absent because of nondeposition (Fig. 10).

Correlation of the Norwood Member with the Blocher Member of the New Albany Shale was cited by Collinson (1967, p. 946, Fig. 7A-B), Lineback (1970, Fig. 3), and Hasenmueller and Leininger (1987, Fig. 3) as evidence that the Michigan and Illinois basins were initiated at the same time. Here we confirm that, based on conodont faunas, black shale deposition started in both basins at precisely the same time, during the *punctata* Zone (Fig. 2). Moreover, there is also a lithologic similarity of units at this stratigraphic position. The lowest limestone bed, 0.75 m above the base of the Norwood in the Paxton Quarry, contains a coquina of largely hindeodellid conodonts, tiny inarticulate brachiopods, and pyritic molds of styliolinids. The same combination, abundance, and preservation of fossils occur in bed 4 of the Blocher at section 4 (Lineback, 1970, p. 60; Conkin, 1985, p. 227; Hasenmueller and Leininger, 1987, Fig. 5, p. 12), south of Vernon, Indiana. Blocher bed 4 is 0.7 m above the top of the Middle Devonian North Vernon Limestone (= top of the Traverse Group). This is a remarkable similarity, considering that the distance between Paxton Quarry and the Vernon, Indiana, outcrop is 725 km.

Conodont faunas of the same age as those in the Norwood Member have not been identified from Antrim Shale samples in Milwaukee, Wisconsin, nor from fissure fillings in Silurian reefs in northeastern Illinois. These reefs are located on the northeastern flank of the Kankakee arch (Fig. 1) and may have been exposed during this time interval.

### Isostatic rebound and deposition of prodeltaic muds

Deposition in the Michigan Basin changed abruptly from anaerobic, organic-rich, black muds to dysaerobic, lighter colored muds and lime muds at the base of the Paxton Member (Fig. 5, event 8). The Paxton Member was deposited at the same time as the Rhinestreet Shale at the start of Catskill deltaic cycle 4. The time of Paxton deposition has special significance in the history of the Michigan Basin for several reasons: (1) The color change from black in the Norwood Member to predominantly greenish gray and gray in the Paxton Member resulted from development of a more oxygenated water column over the basin throughout Paxton deposition as prodeltaic sediments were introduced into the basin. (2) Although fossil diversity is low and sessile-benthic faunas are absent in the Paxton, nodular calcareous shales just above the middle of the member contain a conodont fauna of the Lower *gigas* Zone. The middle part of this zone, which must be represented in the member, is the time of a major eustatic rise throughout Euramerica (Sandberg, 1988). Because of this rise, which produced the maximum transgression on the Late Devonian sea-level curve (Johnson and Sandberg, 1989), black shale deposition would be expected to have continued unabated. The fact that it does not means that relative sea level must have been lowered in the Michigan Basin during Paxton deposition. In the absence of significantly increased sediment fill and load, the only logical explanation for this lowering is an episode of isostatic rebound resulting from lessened Acadian orogenic compression. (3) The end of Paxton deposition is also highly significant and is believed to mark the end of the Frasnian, and thus to coincide with the time of mass extinction (Fig. 5, event 9) during a severe sea-level fall at the end of the *linguiformis* Zone in Euramerica and North Africa (Sandberg and others, 1988).

The depositional pattern displayed by the isopach map of the Paxton Member (Fig. 11) shows two distinct areas of thicker accumulation, which we interpret to be prodeltaic distributaries that entered the basin from the southeast, as a distal extension of the upper Olentangy prodeltaic facies across the Findlay arch in Ohio (Figs. 1 and 2), and from an unknown source to the northeast. The pattern resembles, in a general way, the sub-basins of the Norwood Member, except that Paxton sediments were deposited slightly more rapidly: about 14 m/m.y in the northern depocenter, and about 9 m/m.y. in the southern depocenter. Paxton-equivalent strata are thin to absent in the southwestern third of the basin, possibly because of erosion in that area. Their former presence is suggested by reworked conodonts of the Lower *gigas* Zone found in younger Devonian lag deposits at Milwaukee (Schumacher, 1971). Conodonts of possibly the same age were also found in fissure fillings in northeastern Illinois (Kluessendorf and others, 1989).

### Resumption of black-mud deposition

Deposition of the lower part of the Lachine Member of the Antrim Shale resulted from reestablishment of the basinal black-mud regime (Fig. 5, event 10) that had prevailed throughout the Norwood Member. The lower Lachine corresponds to the upper part of cycle 4 and the lower part of cycle 5 (black Dunkirk Shale) in the Catskill deltaic facies. The thickness map (Fig. 12) of the lower part of the Lachine (subsurface unit 1A of Ells, 1979) shows that it was deposited across most of the Michigan Basin. The dominant black-mud lithology, thickness pattern, and slow rate of sedimentation clearly indicate the existence of a starved, anoxic basin at this time. A preferred northeast lineation of the axes of thicker sedimentation in Figure 12 parallels the axes of the Findlay arch (Fig. 1) and Transcontinental arch (Fig. 3).

Deposition in the southwestern part of the basin continued to be less around an "island" in southwest Michigan, but data for interpretation of its history are scarce. Conodonts of the *triangularis* Zone were found under the Milwaukee area, where mixed faunas in tunnel samples contained conodonts of both the *P. linguiformis* and/or *P. triangularis* Zones (Schumacher, 1971). Conodonts of the same ages were also found in fissure fillings in the Elmhurst quarry in Illinois (Fig. 1; Kluessendorf and others, 1989).

### First indication of Ellsworth mud delta

The upper part of the Lachine Member (unit 2 of Ells, 1979) is essentially a continuation of starved-basin black-mud deposition in the eastern two-thirds of the Michigan Basin (Fig. 13). However, abrupt thickening in the western one-third of the basin, caused by intertonguing of more oxygenated greenish gray muds, is evidence for the initial progradation by prodeltaic sediments of the Ellsworth Shale (Fig. 5, event 11). There, isopachs are closely spaced, trend generally northward, and indicate a uniform westward thickening. Deposition was at a rate of 17 m/m.y. in this western area. Three distributaries are faintly evidenced in this area of the basin (Fig. 13), but these became better differentiated during deposition of the basal part of the overlying upper member of the Antrim Shale. The appearance of *Protosalvinia* (*Foerstia*) in the Michigan Basin was during the Upper *rhomboidea* or Lower *masrginifera* Zone (Gutschick and Sandberg, this volume), coinciding with this earliest stage of the Ellsworth delta.

In the eastern two-thirds of the basin, the isopach pattern consists of alternating areas of thinner and thicker deposits, with several deeps indicated by areas of thinner deposition. In this area, isopach lineations trend northeast-southwest, parallel to the Findlay and Algonquin arches. This trend started with initial black-shale deposition and continued for about 10 m.y. Wavelengths between isopach undulations changed through time, but the elongated axes were consistently parallel to one another and to a tectonic fold pattern that is normal to orogenic compression. Deposition was hemipelagic and quite similar to that in the lower part of the Lachine Member basinwide (Fig. 12). The record of the lower part everywhere and of the upper part to the east indicates a deep-water, anoxic sea floor with little or no vertical mixing, and slow accumulation of black mud, which preserved pelagic organic remains in a reducing benthic environment.

The uppermost 1.5 m of the Lachine Member in the Paxton Quarry contains a zone of rhythmites and concretions that formed within five greenish gray shale layers (Gutschick and Sandberg, this volume). The concretions are composed of silty nonfossiliferous limestone. This zone of concretions probably represents the silt and mud deposition from the distributary tongue (Fig. 13) that is directed northeastward toward the Paxton Quarry.

The upper part of the Lachine Member was deposited during the latter part of cycle 5 (Gowanda Shale), which was not a time of black shale deposition in New York (Fig. 2).

### Changing patterns of later Famennian deposition

The depositional history of the Michigan Basin changed drastically during middle and late Famennian time (*marginifera* through *expansa* Zones). The change in sedimentary regime during deposition of the upper member of the Antrim Shale resulted from progradation of the Ellsworth delta into the western side of the basin (Figs. 2 and 6). The change had started slightly earlier, but the full impact of delta growth into the basin occurred during deposition of the lower part of the upper member (units 3 and 4 of Ells, 1979) in the upper part of cycle 5. Much of the deltaic history is reconstructed from the isopach maps, which depict only the distal-margin, prodelta progradation into and retrogradation from the basin. In the inferred fluvial, landward part of the delta, streams must have flowed eastward across the coastal Wisconsin lowlands (Fig. 1), but all traces were removed by post-Devonian erosion. Deposition on the Ellsworth deltaic platform continued at a faster rate than deposition of the uppermost part of the Antrim (units 5 and 6 of Ells, 1979), which correlates with the lower part of cycle 6 (black Cleveland Member of Ohio Shale).

The age of the upper member of the Antrim Shale, which is known mainly from the subsurface, is constrained by conodonts of the Lower *marginifera* Zone found in a concretion near the base of the member in outcrop (Gutschick and Sandberg, this volume) and by the Late Devonian (Lower *praeslucata* Zone) age of the overlying basal Bedford Shale, based on conodonts and miospores (Gutschick and Sandberg, this volume) and on goniatites (House and others, 1986). The ages of the four parts of the upper member are not well documented.

The depositional history of the upper member is interpreted through a series of four sequential isopach maps (Figs. 14, 15, 16, and 17) depicting its four parts, which are based on subsurface gamma-ray units 3, 4, 5, and 6 of Ells (1979). A fifth map (Fig. 18), which shows the combined thickness of units 4, 5, and 6, is essentially a form map of the Ellsworth delta. Low radioactivity throughout the Ellsworth Shale precludes reliable correlations with the Antrim Shale based on gamma-ray units of Ells (1979). Consequently, this fifth map was used to extrapolate the thicknesses of Antrim-equivalent units into sediments of the Ellsworth deltaic complex in constructing the four sequential isopach maps. This extrapolation was performed subjectively on the basis of estimated proportional rates of sedimentation.

Antrim black-mud deposition in the eastern half of the Michigan Basin continued largely unabated and with only minor distal intertongues of greenish gray muds and lime muds from the west through the entire upper member (units 3 through 6 of Ells, 1979). However, during the major growth of the Ellsworth delta, black-mud deposition in the lower part of the upper member (units 3 and 4) of the Antrim Shale was overwhelmed by deposition of greenish gray mud in the western part of the basin (Figs. 14 and 15). The sharp facies boundary between the Antrim and Ellsworth Shales that was determined by comparison of Figures 8 and 9 is shown by a dashed line on sequential maps of the upper member of the Antrim (Figs. 14 through 17).

### Development and progradation of Ellsworth delta

Progradation of the Ellsworth delta into the Michigan Basin (Fig. 5, event 12) is shown by the isopach map of the basal part of the upper member of the Antrim Shale (unit 3 of Ells, 1979) and equivalents (Fig. 14). Because some residual deposition of black mud continued in the area of the delta on the west, the radioactive signature of unit 3 within the Antrim extends into coeval, greenish gray strata of the Ellsworth Shale. Thus, thicknesses of unit 3 can be mapped basinwide, whereas this is not possible for higher units 4 through 6, in which deltaic, greenish gray muds completely overwhelmed black muds on the west side of the basin (compare Fig. 14 to Figs. 15 and 16). Three distributary systems, which had already entered the deep basin just before the start of the Lower *marginifera* Zone (Fig. 13), continued their courses eastward well into the basin through this time interval (Lower to Upper *marginifera* Zones), as shown by arrows on Figure 14. Deposition of greenish gray mud in the areas of greatest thickness of all three lobes, along the delta front, was at a

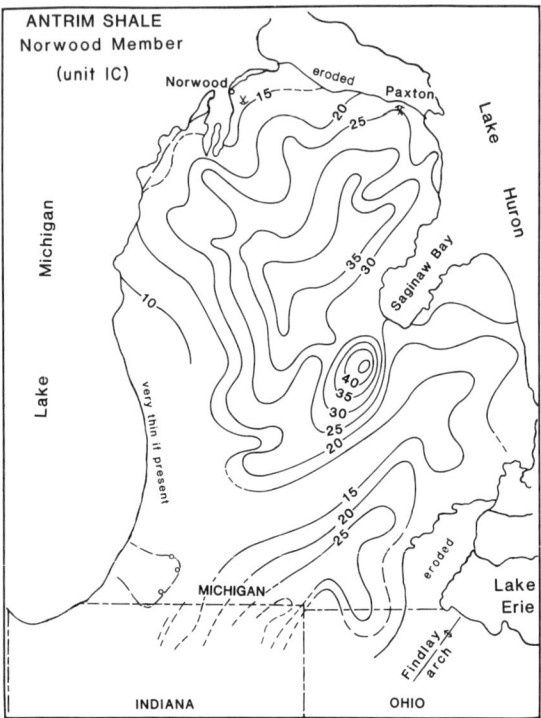

Figure 10. Map showing distribution and thickness of Norwood Member of Antrim Shale (unit 1C of Ells, 1979). Isopach interval, 5 ft. Member is mainly organic-rich black shale and is very thin to absent in area where control points are indicated by open circles.

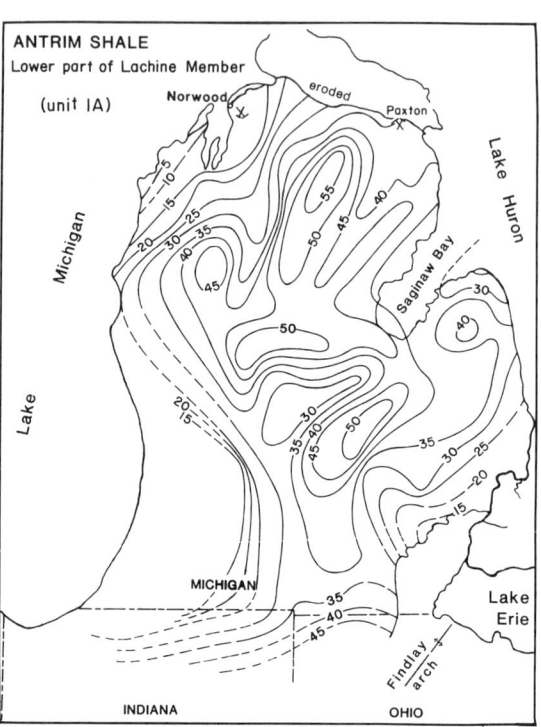

Figure 12. Map showing distribution and thickness of lower part of Lachine Member of Antrim Shale (unit 1A of Ells, 1979). Isopach interval, 5 ft.

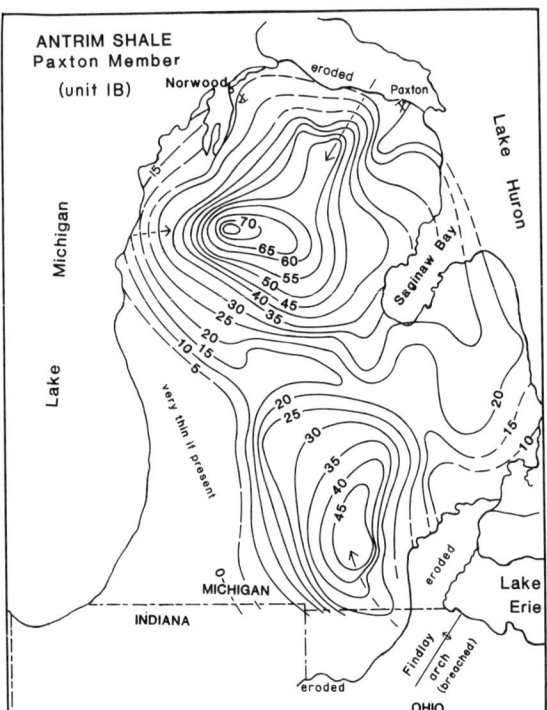

Figure 11. Map showing distribution and thickness of Paxton Member of Antrim Shale (unit 1B of Ells, 1979). Isopach interval, 5 ft. Member is mainly greenish gray limestone and calcareous shale deposited in two distinct depocenters.

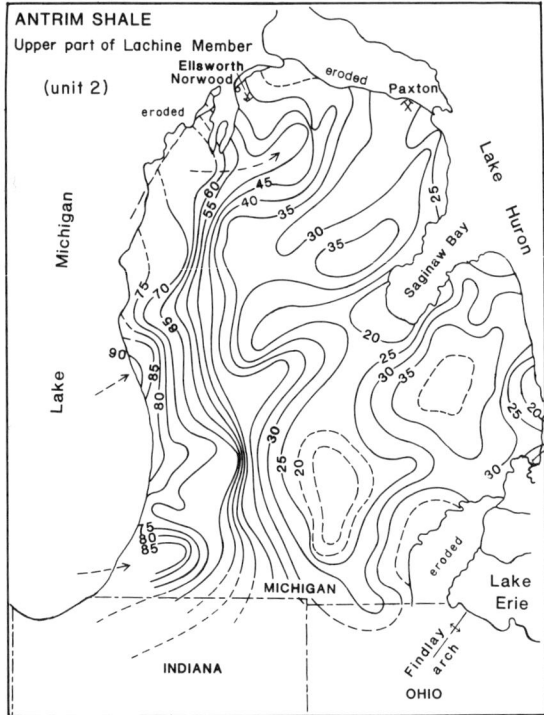

Figure 13. Map showing distribution and thickness of upper part of Lachine Member of Antrim Shale (unit 2 of Ells, 1979). Isopach interval, 5 ft. Shows initial stage of Ellsworth delta along western side of basin. Arrows indicate direction of progradation of three distributaries.

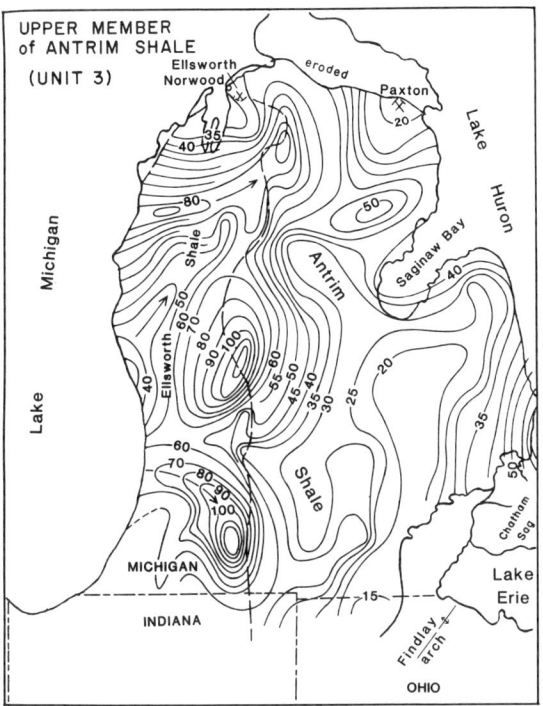

Figure 14. Map showing distribution and thickness of basal part of upper member of Antrim Shale and equivalents (unit 3 of Ells, 1979). Isopach interval, 5 and 10 ft. Note three-pronged distributary system (arrows) of Ellsworth delta entering western side of basin.

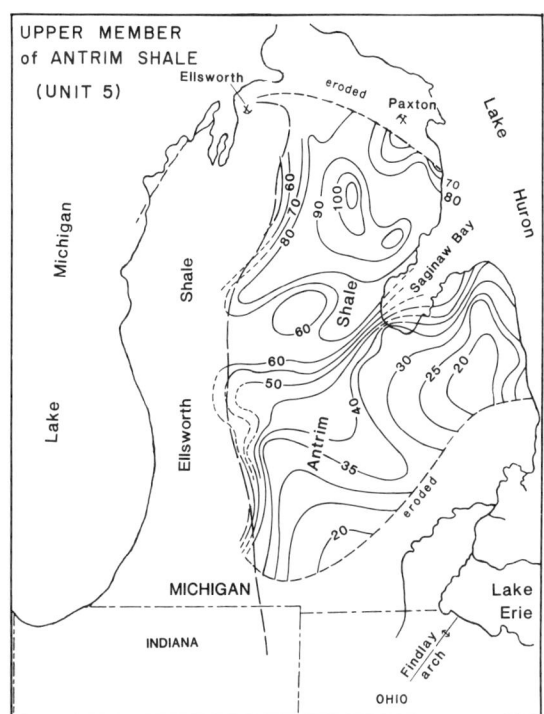

Figure 16. Map showing distribution and thickness of next to highest part of upper member of Antrim Shale and equivalents (unit 5 of Ells, 1979). Isopach interval, 5 and 10 ft.

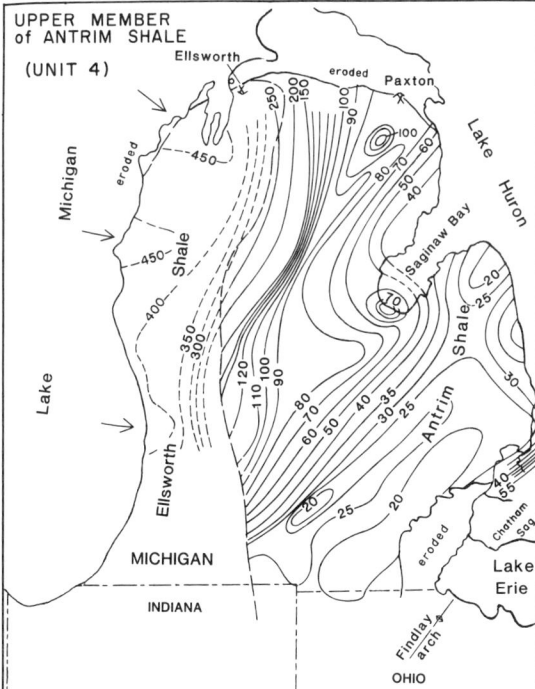

Figure 15. Map showing distribution and thickness of next to lowest part of upper member of Antrim Shale and equivalents (unit 4 of Ells, 1979). Isopach interval, 5, 10, and 50 ft. Dashed lines are extrapolated rough estimates based on Figure 18.

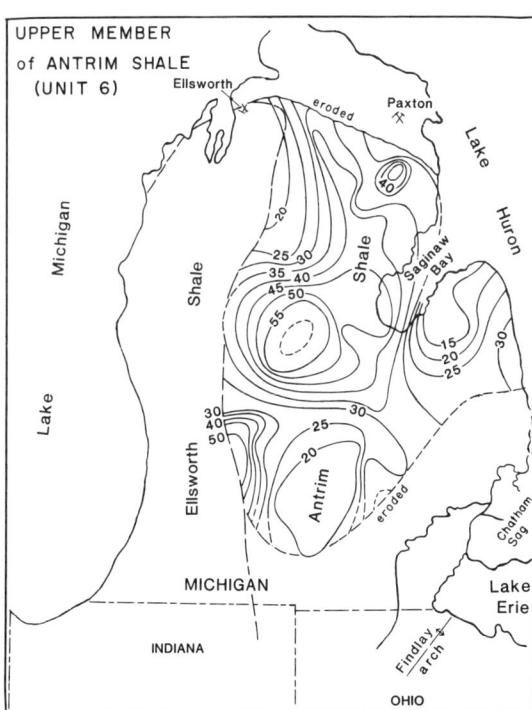

Figure 17. Map showing distribution and thickness of uppermost part of upper member of Antrim Shale and equivalents (unit 6 of Ells, 1979). Isopach interval, 5 ft.

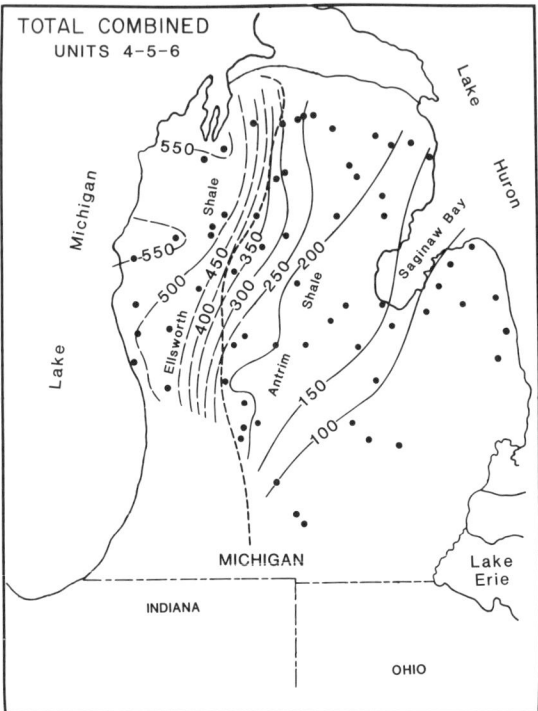

Figure 18. Map showing distribution and thickness of undivided Ellsworth Shale and equivalent three upper units of upper member of Antrim Shale (total combined units 4, 5, and 6 of Ells, 1979). Isopach interval, 50 ft. Black dots show control points. This is essentially a form map showing pattern of deposition for extrapolating Antrim-equivalent units on western side of basin (e.g., Fig. 15).

rate of about 40 m/m.y. This is about three times the rate of coeval black-mud deposition in the basin. An abrupt change in thicknesses occurs between the 35- and 50-ft isopachs (Fig. 14). This area probably was the location of a pycnocline separating dysaerobic, greenish gray muds of the Ellsworth from anaerobic, black muds of the Antrim. Streams flowing off the Transcontinental arch and Wisconsin uplands crossed the coastal lowlands into the basin with only enough energy to transport muds and fine silts. Most of the delta, especially landward in Wisconsin, has been removed by later erosion (Fig. 6). Some of the delta may be preserved beneath Lake Michigan, but the lake bottom is 100 to 150 m deep (Hough, 1958) over Upper Devonian bedrock.

The deep basin during deposition of unit 3 had an alignment subparallel to the Findlay arch, although this lineation is less apparent to the north (Fig. 14). Contemporaneous, slow, deepwater Antrim deposition and more rapid, less deep Ellsworth prodelta influx account for the transitional marginal intergradation between the two shale bodies. Contact between the muds, in the area between the unstable distal slope of the Ellsworth delta and the prodeltaic turbiditic flows into the Antrim basin, was controlled by slight tectonic perturbations, seasonal tropical rainfall variations, and sea-level fluctuations. Bioturbated rhythmites (Gutschick and Sandberg, this volume), which reflect changes in the depositional regime, were deposited in this area. Larval stages of burrowing organisms probably were transported within turbidite flows of oxygenated greenish gray muds below the pycnocline into the dysaerobic zone. The mature organisms then burrowed through the prodeltaic muds and into the underlying, organic-rich, black, hemipelagic, basinal muds, back filling and pelletizing as they fed. The bioturbated muds were eventually covered by another layer of black hemipelagic muds, and the process was repeated during successive turbidite flows, producing the rhythmites.

Progradation of the Ellsworth delta into the Michigan Basin during the Lower *marginifera* Zone is an interregionally significant event. The Ellsworth Shale can be correlated approximately with the greenish gray Chagrin Shale, which is a time-equivalent, distal-prodeltaic deposit of the westward-prograding Catskill deltaic complex, in northeastern Ohio. The Chagrin is a westward-deepening, off-shelf sequence deposited below wave base in a restricted dysaerobic environment (Schwimmer and others, 1987). The Chagrin delta has about the same facies relation to the Ohio Shale that the Ellsworth delta has to the Antrim Shale.

## Maximum progradation of Ellsworth delta

A thickness map (Fig. 15) of the next-to-lowest part of the upper member of the Antrim Shale (unit 4 of Ells, 1979) demonstrates the coalescence or shifting of distributary fans through time to form a wedge (Fig. 5, event 13). It also shows the maximum progradation of the Ellsworth delta eastward beyond the middle of the Michigan Basin (Fig. 5, event 14). Isopachs showing the thickness and shape of the Ellsworth delta, west of the north-trending Ellsworth-Antrim facies line (Fig. 15), were reconstructed using Figure 18, which shows the combined thickness of units 4, 5, and 6 of Ells (1979) basinwide. Thicknesses and trends were extrapolated from that map by proportioning the thicknesses of Ellsworth Shale according to the duration of the conodont zones during which this unit was deposited (Fig. 2).

The form and location of the delta for unit 4 (Fig. 15) are markedly changed from those of the distributary pattern for unit 3 (Fig. 14). The new form in Figure 15 is a wide wedge of deltaic and prodeltaic strata, thicker close to its northwestern source and thinner to the southeast. This map also shows by arrows the inferred directions of sediment transport. The location of the delta and its major source of sediments had shifted northward, indicating tilt of the basin due to loading. Also, the pycnocline had shifted eastward, and the basin was dominated by a dysaerobic environment.

Rate of sedimentation for greenish gray shale of the Ellsworth delta may have been as high as 55 m/m.y. during deposition of unit 4, overwhelming the coeval part of the Antrim Shale, which is restricted to the southeastern and eastern sides of the Michigan Basin (Fig. 15). The pattern of black-mud deposition there continues to show a northeast-southwest lineation parallel to the Findlay arch. Thicknesses undulate normal to this lineation, suggesting other structural controls.

The closing phase of Catskill deltaic cycle 5 (Fig. 2) had decreased tectonic activity, lower sea level, convectional equatorial precipitation, and high clastic input, resulting in rapid westward progradation of the Chagrin–Three Lick Tongue (of the Ohio Shale) from the Catskill delta, contemporaneous with eastward progradation of the Ellsworth delta.

## Retreat of Ellsworth delta

The isopach maps (Figs. 16 and 17) of the upper two parts (units 5 and 6 of Ells, 1979) of the upper member of the Antrim Shale depict first the retreat of the Ellsworth delta (Fig. 5, event 15) and then the reestablishment of the black-mud, stagnant, anoxic, starved Antrim basin (Fig. 5, event 16). East of the Ellsworth-Antrim facies line, these changes are reflected by the increasingly radioactive response of units 5 and 6, which together are referred to as the "upper radioactive zone" (Mounds Facility Laboratory Report, 1980). This increase in radioactivity does not occur west of the facies line, so extrapolation of isopachs into that area is difficult. However, because of isostatic rebound, the thickness of the Ellsworth Shale for equivalents of both units 5 and 6 on the delta platform probably was only twice as much as the greatest thicknesses east of the facies line.

The isopach map of unit 5 (Fig. 16) shows that Ellsworth greenish gray muds continued to spill into the Antrim basin at a maximum rate of about 24 m/m.y. The Antrim basin maintained a northeast-southwest trend, parallel to the axis of the Findlay arch. Deposition was greater at the north end of the basin than on the south end, where units 5 and 6 wedge out against the Findlay arch. This combination follows the pattern of northward basin tilt initiated in unit 4 (Fig. 15), and provides strong evidence that the Findlay arch was isostatically uplifted and tilted northward as the north end of the basin subsided. This uplift depressed the adjacent Chatham sag (Fig. 1), allowing depositional interconnection of the black Cleveland, Kettle Point, and Antrim muds.

The upper part of the upper member of the Antrim Shale (units 5 and 6) was correlated by Matthews (1983) with the Cleveland Member of the Ohio Shale, which corresponds to a black-shale depositional phase at the start of cycle 6 of the Catskill deltaic facies (Fig. 2; Ettensohn, 1985b). Maximum development of the Catskill delta occurred in the Appalachian basin adjacent to the Acadian orogen, and eventually the delta filled the basin and prograded westward. Deposition of the Cleveland started when deep water spread to northern Ohio and joined the Michigan Basin through the Chatham strait (Fig. 1). Thus, black-shale deposition of units 5 and 6 demonstrates the important influence of the distant Acadian orogeny on the Michigan Basin and the close correlation of basinal sedimentation to that of Catskill deltaic facies.

## Pre-Bedford Shale paleogeology

An abrupt change in gamma-ray log characteristics occurs between the highest part of the Antrim Shale (unit 6 of Ells,

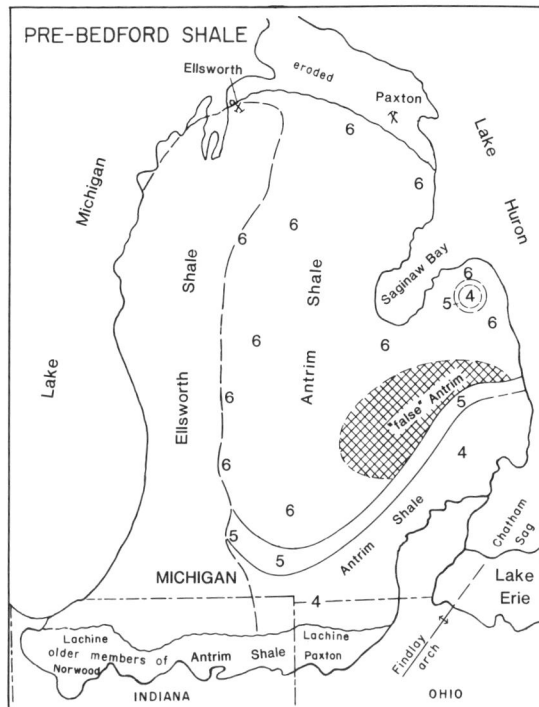

Figure 19. Post-Antrim Shale paleogeologic map, showing distribution of stratigraphic units directly underlying Bedford Shale and equivalent part of Ellsworth Shale. Numbers refer to units of Ells (1979), which are parts of upper member of Antrim Shale. In some areas of eastern Michigan, basal part of Bedford contains grayish black shale referred to as "false Antrim" (cross-hatched area), which has lower radioactive response than true Antrim.

1979) and the overlying lower part of the Bedford Shale in wells on the eastern side of the Michigan Basin. A paleogeologic map (Fig. 19) shows the distribution of members and units of the Antrim Shale directly underlying the Bedford Shale. The position of this surface within the Ellsworth could not be determined with any certainty, although Bedford equivalents are shown on the Ellsworth platform in a subsequent map. The Bedford seems to be conformable on Antrim unit 6 through most of the Antrim basin; however, in the southeastern part, truncation of the Antrim against the Findlay arch is indicated. The sequence of maps for units 4, 5, and 6 (Figs. 15, 16, and 17, respectively) suggests some offlap of strata away from the Findlay arch as it was uplifted, whereas in the Chatham strait, more Antrim was preserved (e.g., Fig. 15). The cross-hatched area of "false" Antrim (Fig. 19) refers to grayish black shale in the base of the Bedford, which resembles the Antrim but has a less radioactive gamma-ray response. This area may represent the waning stage of the anaerobic sea floor when nonradioactive, lowest Bedford muds diluted black sediments in the deepest part of the basin. Truncated margins of the Antrim on the northern and southern sides of the basin are due to post-Devonian erosion, partly by Pleistocene glaciation.

## Progradation of Bedford mud delta

The Michigan Basin underwent a pronounced change in very late Devonian (late Famennian Lower *praesulcata* Zone) time as a result of isostatic rebound (Figs. 2 and 5). After maximum progradation of the Catskill delta had occurred early in cycle 6, filling of the Appalachian basin introduced oxygenated waters into the Michigan Basin and changed the sedimentary regime there. Deposition of the black Antrim Shale ceased as gray prodeltaic muds of the Bedford Shale poured into the basin from the east (Fig. 5, event 17) and spread across the Michigan Basin to merge with greenish gray prodeltaic muds of the Ellsworth Shale on the west. The Bedford mud delta is evidenced only in the subsurface of the Michigan Basin, and as in the case of the Ellsworth delta, only its prodeltaic part is preserved. Poor preservation of fossil-plant debris found in these rocks indicates that sediment source was distant from the basin (Bordner, 1983).

A series of paleogeographic maps of the Bedford Shale and Berea Sandstone (Pepper and others, 1954) depicted depositional history in land-seascape panoramas, concentrating on their type area near Cleveland in north-central Ohio, and its relation to the western margin of the Catskill deltaic complex. They showed an epicontinental sea, including Michigan Bay, but gave no details of the area west of the Findlay arch (Cincinnati land peninsula).

The Bedford-Berea deltaic depositional pattern in the Michigan Basin is similar to one described by Pashin and Ettensohn (1987) for these same formations on the west-central margin of the Appalachian basin in northeastern Kentucky and south-central Ohio. In contrast to our present interpretation, these authors considered the Bedford-Berea to be a clastic wedge in which the Bedford Shale and Berea Sandstone are entirely time-equivalent lithofacies and not superposed. With the introduction of oxygenated Bedford muds into the anoxic basin, the offshore, dysaerobic, greenish gray foredelta muds deposited on the basinal black muds mark the pycnocline (Pashin and Ettensohn, 1987). The change from gray to black shales was also observed to mark the position of a paleopycnocline (Ettensohn and Elam, 1985; Ettensohn and others, 1989).

The isopach map (Fig. 20) of the Bedford Shale and equivalent part of the Ellsworth Shale can be compared with the isopach map (Fig. 17) of the highest part of the Antrim Shale (unit 6) to observe the abrupt change in sedimentary regime and the shift to a northeast and east source of extrabasinal sediments from the Appalachian basin. Bedford deposition dominated the eastern side of the Michigan Basin, and a parallel, linear, northeast-southwest pattern consistent with earlier lineations was maintained. However, this lineation is not as apparent in the more detailed maps of the lower and upper parts of the Bedford. On the western side of the basin, the Bedford can be correlated with part of the Ellsworth by means of gamma-ray logs, so that isopachs can be drawn west of the former Ellsworth-Antrim facies line. Maximum rate of sedimentation of the Bedford Shale was about 60 m/m.y. The Bedford of the Michigan Basin extends under the southwestern part of Lake Huron and barely reaches

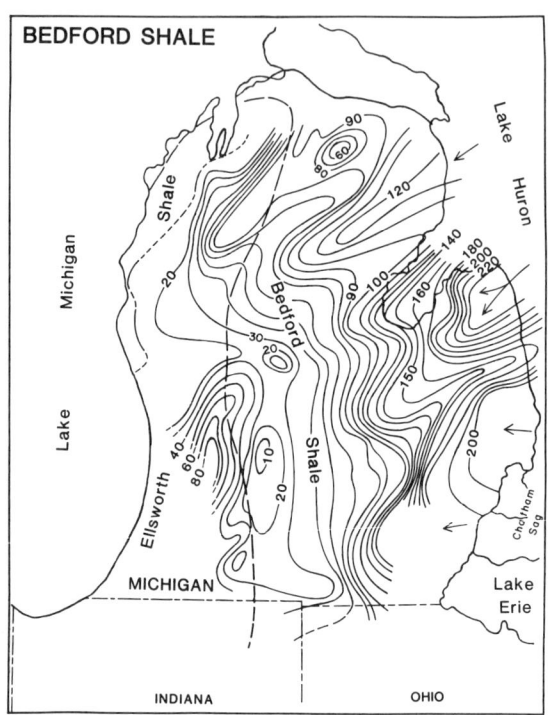

Figure 20. Map showing distribution and thickness of Bedford Shale and, in areas where gamma ray log correlation is reliable, equivalent part of Ellsworth Shale. Isopach interval, 10 ft. Arrows show direction of progradation of three distributaries.

southwestern Ontario (Sanford, 1967), but it is not connected with its type area in northeastern Ohio because of later erosion in the area of the intervening Chatham sag and Findlay arch (Fig. 1).

The distribution and thickness of the lower part of the Bedford Shale (unit 7 of Ells, 1979) east of the Ellsworth deltaic platform are shown by an isopach map (Fig. 21). The pattern is one of two deltaic distributaries entering the basin from the northeast. The main distributary dumped most of its load southeast of Saginaw Bay and continued south, forming the area of false Antrim (Fig. 19; Fig. 21, arrows). The other distributary flowed westward into the north-central part of the basin. Delta-front muds form an apron wedge, which thins westward where it coalesces with Ellsworth muds on the western platform. Sea level fell, and only the distal part of the fluvio-deltaic sediments were preserved in the Michigan Basin. From this evidence, the delta was clearly initiated west of the Algonquin arch by a stream system of low gradient that flowed from the northeast into the basin. This direction would parallel the hypothetical Ontario River (Pepper and others, 1954, pl. 13A), east of the Findlay and Algonquin arches.

The isopach map (Fig. 22) of the upper part of the Bedford Shale (unit 8 of Ells, 1979) depicts the closing phase of Bedford deposition in the eastern part of the Michigan Basin. The pattern had shifted from one of deltaic input from the northeast to one of major deltaic debouchment through the Chatham strait from the

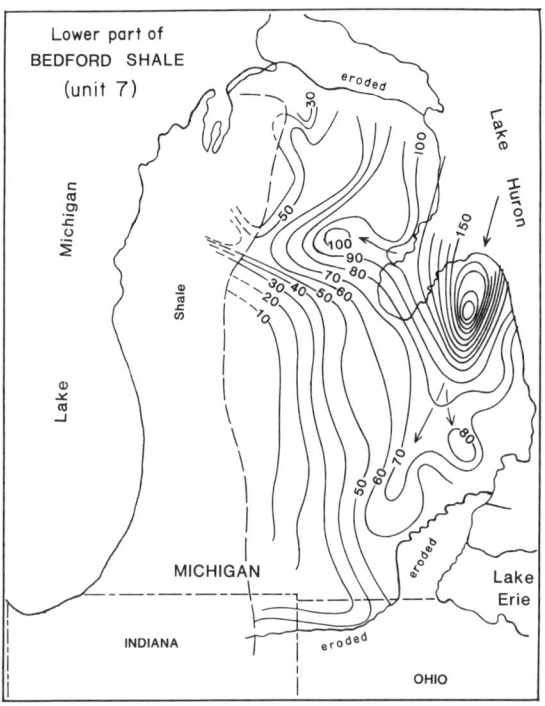

Figure 21. Map showing distribution and thickness of lower part of Bedford Shale (unit 7 of Ells, 1979). Isopach interval, 10 ft. Arrows show direction of progradation of distributaries.

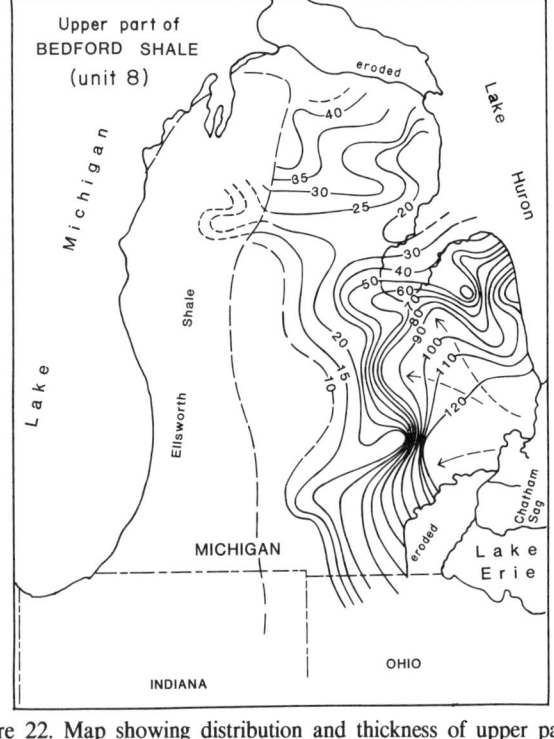

Figure 22. Map showing distribution and thickness of upper part of Bedford Shale (unit 8 of Ells, 1979). Isopach interval, 5 and 10 ft. Arrows show direction of sediment transport.

southeast, suggesting connection with the Bedford type area. The changed pattern of Bedford deposition may have resulted from rise of the Algonquin arch and downwarp of the Chatham sag (Fig. 1). The prodelta front that extended into the center of the basin was deposited at a maximum rate of 73 m/m.y. The red Bedford delta must have formed at this time in north-central Ohio, according to Pepper and others (1954), and a branch of the stream system that deposited it could have flowed westward through the then-filled Chatham sag into the Michigan Basin. However, the Bedford in the Michigan Basin is gray shale, whereas red Bedford shales are considered to represent a tidal-flat environment (Coogan and others, 1981). Ellsworth deposition on the platform during this time was subdued in comparison to deltaic input from the east.

### Eastward regression of Berea sands

The Berea Sandstone (unit 9 of Ells, 1979) in eastern Michigan is light gray, fine-grained, partly dolomitic, micaceous, pyritic sandstone with some interbeds of gray shale (Cohee and Underwood, 1944; Cohee and others, 1951). The distribution and thickness of the Berea are shown in Figure 23. The Berea is readily distinguished by its low radioactive response on gamma-ray logs from the underlying Bedford Shale (Ells, 1979).

The Berea Sandstone was deposited during a major regression (Fig. 5, event 18) that occurred in the Middle and Upper *praesulcata* Zones (Fig. 2) during a eustatic fall coinciding with a glacially induced global mass extinction (Sandberg and others, 1988). Distribution of the Berea follows the general pattern of Bedford sedimentation across the eastern margin and center of the Michigan Basin. The Berea intergraded with and then overrode only the highest part of the Bedford. It was the coarsest Late Devonian sediment to be deposited in the basin. Short prongs of Berea Sandstone, deposited by feeder channels of the delta, project into the basin from the northeast, east, and southeast. In southwestern Ontario, one such Berea shoestring sandstone, more than 60 m thick, was deposited in a channel cut through the Bedford Shale into the underlying Kettle Point Formation (Fig. 2; Sanford, 1967). The maximum rate of deposition of the Berea in the channels was the same as that for the upper Bedford Shale, about 73 m/m.y., and they were probably partly contemporary facies. Pashin and Ettensohn (1987) concluded that the Bedford and Berea represent a series of regressive fluvial-deltaic and shore-zone complexes along the margin of the Appalachian basin. A study of Berea sedimentology in central Ohio and northeastern Kentucky (Potter and others, 1983) concluded that the formation was probably deposited in storm-dominated shelf environments.

### Pre-Sunbury Shale paleogeology

The final Devonian map (Fig. 24) is a paleogeologic map of the surface on which the Lower Mississippian Sunbury Shale was

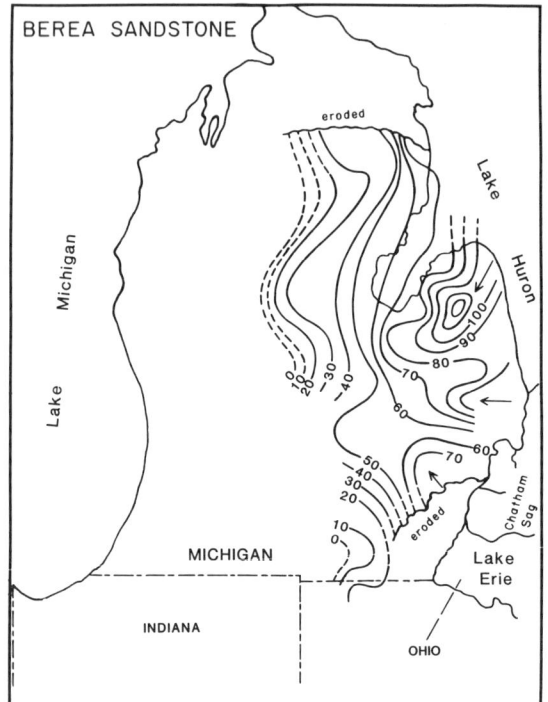

Figure 23. Map showing distribution and thickness of Berea Sandstone. Isopach interval, 10 ft. Arrows show direction of sediment transport.

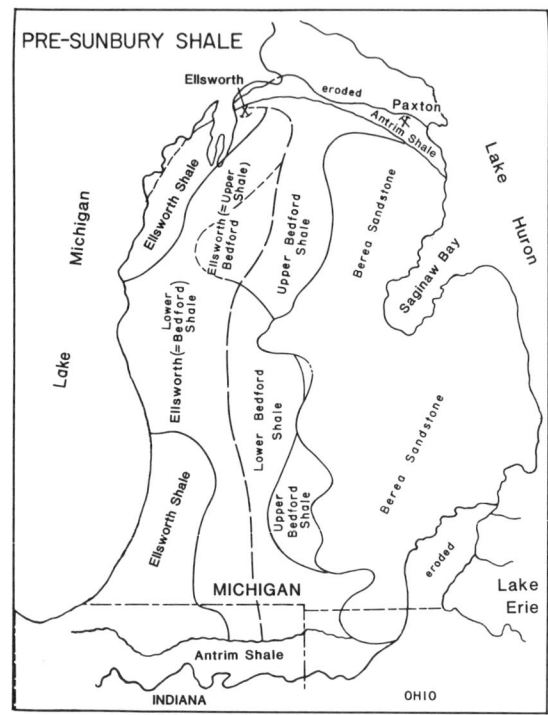

Figure 24. Post-Devonian paleogeologic map, showing distribution of Upper Devonian stratigraphic units directly underlying basal Mississippian Sunbury Shale. Contacts dashed where approximately located.

deposited after the earliest Mississippian erosional episode that followed latest Devonian deltaic deposition (Fig. 2). The map (Fig. 24) shows the extent and pattern of Bedford deltaic progradation into the Michigan Basin from the east and of the final Berea retreat. It also shows the extent of Ellsworth Shale on the western side of the basin and its relation to the lower and upper parts of the Bedford Shale and their equivalents. Only on the northern and southern sides of the basin, where the younger two Devonian formations were truncated, does the Sunbury and its equivalents rest on Antrim Shale.

### Deposition of transgressive Sunbury black muds (Mississippian)

The Sunbury Shale was deposited during a major North American rise in sea level during the Lower *crenulata* conodont Zone (Sandberg and others, 1986); this rise probably was eustatic as it is also reflected in Western European sequences. The Sunbury represents the transgressive first phase of black-mud deposition in Catskill deltaic cycle 7 (Fig. 2). Sedimentary conditions during the Early Mississippian were drastically changed from those of the latest Devonian, probably as a result of the earliest Mississippian erosional episode between the close of cycle 6 and the start of cycle 7. The changes were from regressive to transgressive seas, from light- to dark-colored sediments, from fluvial-derived clastics to pelagic starved-basin muds, from dysaerobic prodeltaic basin to deep-water anaerobic sea floor, and from rapid to slow sedimentation. Maximum Sunbury deposition occurred along the eastern margin of the Michigan Basin, suggesting rapid subsidence there due to flexure and deltaic sediment load.

## CONCLUSIONS

1. The Late Devonian Michigan Basin formed as a deep within the long Eastern Interior seaway separating the North American craton from the Old Red continent.

2. Depositional events in the Michigan Basin were closely related to Late Devonian eustatic sea-level changes as well as being synchronized with Acadian orogenic tectophases and Catskill deltaic facies in the Appalachian basin.

3. Formation of the Michigan Basin, probably over an early Paleozoic aulacogen, coincided with Taghanic onlap of the North American craton, beginning in the late Middle Devonian (Givetian Middle *varcus* Subzone), and with maximum tectonism in the Acadian Mountains, east of the Catskill deltaic complex.

4. A transition from carbonate platform to deep starved basin occurred during the time interval between the Middle *varcus* Subzone and the Late Devonian (early Frasnian) *transitans* Zone. Basin subsidence resulted from tectonic downflexing, not from sediment load.

5. Initial deposition of black, anoxic, hemipelagic muds in the deep, stagnant, starved Michigan Basin occurred during the

early Frasnian *punctata* Zone and produced the Norwood Member of the Antrim Shale.

6. An episode of isostatic rebound of the Michigan Basin, coinciding with several episodes of eustatic fall, resulted in deposition of the prodeltaic, greenish gray Paxton Member of the Antrim Shale, which interrupted black-mud deposition basinwide.

7. A stagnant starved-basin regime, in which black muds were deposited, resumed in the early Famennian *triangularis* Zone and resulted in deposition of the lower part of the Lachine Member of the Antrim Shale.

8. The first minor indication of the Ellsworth delta prograding into the Michigan Basin from the west occurred during deposition of the upper part of the Lachine Member in the early Famennian Upper *rhomboidea* or Lower *marginifera* Zone.

9. Deposition of black mud continued unabated on the eastern side of the Michigan Basin during deposition of the upper member of the Antrim Shale between the early and late Famennian Lower *marginifera* to Upper *expansa* Zones.

10. During deposition of the upper member of the Antrim Shale, a facies relation existed between the black basinal Antrim Shale on the east and the eastward-prograding, greenish gray, prodeltaic and deltaic Ellsworth Shale on the west side of the Michigan Basin.

11. Within an episode of isostatic rebound, the Bedford mud delta prograded westward at the start of the very late Famennian Lower *praesulcata* Zone from the then-filled Appalachian basin into the Michigan Basin, ending Late Devonian black-mud deposition there and merging with the Ellsworth mud delta farther west.

12. The closing phase of the Bedford-Berea deltaic complex was the deposition of the eastward-regressing Berea Sandstone, which occurred during a major eustatic fall in the latest Famennian Middle and Upper *praesulcata* Zones, coincident with a glacially induced global mass extinction.

13. Because of the fall in sea level, the margins of the Michigan Basin became largely emergent, and some Upper Devonian rocks were removed there during an erosional episode in earliest Mississippian time.

14. Our study of the Late Devonian geologic history of the Michigan Basin has an important economic application because of the high potential for natural gas production from Upper Devonian rocks there. Organic-rich, highly radioactive black shales of the Antrim Shale are probable source rocks, and more porous, intervening and overlying, greenish gray shales of the Ellsworth Shale are possible reservoir rocks. The Ellsworth delta front, with intertonguing facies of the Ellsworth and Antrim Shales, contains distributary-channel sandstones and siltstones that provide attractive targets for conventional petroleum exploration. This front coincides with the Ellsworth-Antrim facies line on our isopach maps. Migration of fluids from source rocks probably occurs along fracture systems; more porous strata of the Ellsworth Shale, which might serve as reservoirs, lie updip.

## ACKNOWLEDGMENTS

This chapter is dedicated to Garland Delos Ells, geologist for many years with the Michigan Department of Natural Resources, Geological Survey Division, Subsurface and Petroleum Geology unit of the Oil and Gas Section, who died August 21, 1987. His 1979 report on the subsurface Upper Devonian of the Michigan Basin provided the data base and gamma-ray log interpretations that combined with our outcrop and conodont studies to make this chapter possible.

The help of the following persons is gratefully acknowledged: R. L. Milstein of the Michigan Geological Survey, who furnished important literature; Robert Shaver and Nancy Hasenmueller of the Indiana Geological Survey, who furnished copies of gamma-ray logs from northern Indiana wells; Christopher Scotese, Shell Research Lab, Houston, who shared his latest paleogeographic reconstructions and concern for closer resolution of Late Devonian paleogeography; Donald Rhoads and his colleagues, Science Applications International Corporation, Woods Hole, who shared their studies of oxygen in marine environments and significance of the dysaerobic zone; Donald Mikulic, Illinois Geological Survey, who sent preprints of two important papers; and M. E. Ostrom and M. G. Mudrey, Jr., Wisconsin Geological Survey, who alerted us to their closely related current research.

Formulation of the interpretations and conclusions expressed here benefited from critiques of earlier versions by Frank Ettensohn, University of Kentucky, Lexington; George Moore and his colleagues, Chevron Oil Field Research Company; and Joaquin Rodriguez, Hunter College. We are grateful to Ronald R. Charpentier and Mitchell E. Henry, U.S. Geological Survey, Denver, who reviewed the manuscript and offered helpful suggestions, and to Jamie L. Butler, U.S. Geological Survey, Denver, who redesigned and redrafted Figures 3 and 8. Last, but not least, editors Paul Catacosinos and Paul Daniels deserve praise for their patience and understanding.

# REFERENCES CITED

Bordner, M. A., 1983, Palynology of the Antrim Shale and associated sediments from Manistee County, Michigan to Lambton County, Ontario: Geological Society of America Abstracts with Programs, v. 15, p. 256.

Cheng, Y. N., 1986, Taxonomic studies on upper Paleozoic Radiolaria: Taiwan, National Museum of Natural Science Special Publication 1, 311 p.

Cohee, G. V., and Underwood, L. B., 1944, Maps and sections of the Berea Sandstone in eastern Michigan: U.S. Geological Survey Oil and Gas Investigation Series Preliminary Map OM-17.

Cohee, G. V., Macha, C., and Halk, M., 1951, Thickness and lithology of Upper Devonian and Carboniferous rocks in Michigan: U.S. Geological Survey Oil and Gas Investigation Series Chart OC-41, 5 sheets.

Collinson, C., 1967, Devonian of the north-central region, United States, *in* Oswald, D. H., ed., International Symposium on the Devonian System: Calgary, Alberta Society of Petroleum Geologists, v. 1, p. 933–962.

Conant, L. C., and Swanson, V. E., 1961, Chattanooga Shale and related rocks of central Tennessee and nearby areas: U.S. Geological Survey Professional Paper 357, 91 p.

Conkin, J. E., 1985, Late Devonian New Albany–Ohio–Chattanooga Shales and their interbasinal correlation in Indiana, Ohio, Kentucky, and Tennessee: Commonwealth of Kentucky, Kentucky Energy Cabinet 1985 Eastern Oil Shale Symposium, p. 217–259.

Coogan, A. H., Heimlich, R. A., Malcuit, R. J., Bork, K. B., and Lewis, T. L., 1981, Early Mississippian deltaic sedimentation in central and northeastern Ohio, *in* Roberts, T. G., ed., Stratigraphy, sedimentology; Geological Society of America Cincinnati 1981 Field Trip Guidebook: American Geological Institute, p. 113–152.

Droste, J. B., and Shaver, R. H., 1983, Atlas of early and middle Paleozoic paleogeography of the southern Great Lakes area: Indiana Department of Natural Resources, Geological Survey Special Report 32, 32 p.

Ells, G. D., 1979, Stratigraphic cross sections extending from Devonian Antrim Shale to Mississippian Sunbury Shale in the Michigan Basin: Michigan Department of Natural Resources, Geological Survey Division Report of Investigation 22, 186 p.

Emrich, G. H., and Bergstrom, R. E., 1962, Des Plaines disturbance, northeastern Illinois: Geological Society of America Bulletin, v. 73, p. 959–968.

Ettensohn, F. R., 1985a, The Catskill delta complex and the Acadian orogeny; A model, *in* Woodrow, D. L., and Sevon, W. D., eds., The Catskill delta: Geological Society of America Special Paper 201, p. 39–49.

—— , 1985b, Controls on development of Catskill delta complex basin-facies, *in* Woodrow, D. L., and Sevon, W. D., eds., The Catskill Delta: Geological Society of America Special Paper 201, p. 65–77.

Ettensohn, F. R., and Barron, L. S., 1981, Depositional model for the Devonian–Mississippian black shales of North America; A paleogeographic-paleoclimatic approach, *in* Roberts, T. G., ed., Geological Society of America Guidebook, Cincinnati 1981, v. 2: Economic Geology, structure, p. 344–357.

Ettensohn, F. R., and Elam, T. D., 1985, Defining the nature and location of a Late Devonian–Early Mississippian pycnocline in eastern Kentucky: Geological Society of America Bulletin, v. 96, p. 1313–1321.

Ettensohn, F. R., and 8 others, 1989, Characterization and implications of the Devonian–Mississippian black shale sequence of eastern and central Kentucky, U.S.A.; Pycnoclines, transgression, regression, and tectonism, *in* McMillan, N. J., Embry, A. F., and Glass, D. J., eds., Devonian of the World: Calgary, Canadian Society of Petroleum Geologists Memoir 14, v. 2, p. 323–345.

Fisher, J. H., chairman, 1969, Stratigraphic cross sections, Michigan Basin: Michigan Basin Geological Society, 2 sheets.

—— , 1980, Traverse Limestone structure, and thickness of the Sunbury Shale, Berea Sandstone, Bedford Shale, Ellsworth Shale, and Antrim Shale: Dow Chemical Company, U.S. Department of Energy Report FE-2346-80, Plates 4, 8–12.

Fisher, J. H., Barratt, M. W., Droste, J. B., and Shaver, R. H., 1988, Michigan Basin, *in* Sloss, L. L., ed., Sedimentary cover, North American craton, U.S.: Boulder, Colorado, Geological Society of America, The Geology of North America, v. D-2, p. 361–382.

Galloway, W. E., 1989, Genetic stratigraphic sequences in basin analysis; 1, Architecture and genesis of flooding-surface bounded depositional units: American Association of Petroleum Geologists Bulletin, v. 73, p. 125–142.

Geological Survey of Canada, 1969, Geological map of Canada: Geological Survey of Canada, Map 1250A, scale 1:5,000,000 (compiled by R.J.W. Douglas).

Gutschick, R. C., 1987, Devonian shelf-basin, Michigan Basin, Alpena, Michigan, *in* Biggs, D. L., ed., North-Central Section of the Geological Society of America: Boulder, Colorado, Geological Society of America, Centennial Field Guide, v. 3, p. 297–302.

Gutschick, R. C., and Sandberg, C. A., 1983, Mississippian continental margins of the conterminous United States, *in* Stanley, D. J., and Moore, G. T., eds., The shelfbreak; Critical interface on continental margins: Society of Economic Paleontologists and Mineralogists Special Publication 33, p. 79–96.

Hasenmueller, N. R., and Bassett, J. L., 1980, Stratigraphic cross-section showing the Sunbury, Ellsworth, and Antrim Shale (Devonian and Mississippian) in northern Indiana: Indiana Geological Survey chart.

Hasenmueller, N. R., and Leininger, R. K., 1987, Oil-shale prospects for the New Albany Shale in Indiana: Indiana Geological Survey Special Report 40, 31 p.

Hasenmueller, N. R., and Woodward, G. S., eds., 1981, Studies of the New Albany Shale (Devonian and Mississippian) and equivalent strata in Indiana: Indiana Geological Survey, U.S. Department of Energy Report, 100 p.

Heckel, P. H., and Witzke, B. J., 1979, Devonian world paleogeography determined from distribution of carbonates and related lithic paleoclimatic indicators: Paleontological Association Special Papers in Paleontology 23, p. 99–123.

Hinze, W. J., Kellog, R. L., and Merritt, D. W., 1971, Gravity and aeromagnetic anomaly maps of the Southern Peninsula of Michigan: Michigan Geological Survey Report of Investigations 14, 15 p.

Hinze, W. J., Kellog, R. L., and O'Hara, N. W., 1975, Geophysical studies of basement geology of Southern Peninsula of Michigan: American Association of Petroleum Geologists Bulletin, v. 59, p. 1562–1584.

Hosterman, J. W., and Whitlow, S. I., 1981, Munsell color value as related to organic carbon in Devonian shales of Appalachian Basin: American Association of Petroleum Geologists Bulletin, v. 65, p. 333–335.

—— , 1983, Clay mineralogy of Devonian shales in the Appalachian Basin: U.S. Geological Survey Professional Paper 1298, 31 p.

Hough, J. L., 1958, Geology of the Great Lakes: Champaign, University of Illinois Press, 313 p.

House, M. R., Gordon, M., Jr., and Hlavin, W. J., 1986, Late Devonian ammonoids from Ohio and adjacent states: Journal of Paleontology, v. 60, p. 126–144.

Hydrogeologic Atlas of Michigan, 1981, Department of Geology, Western Michigan University, Kalamazoo, Michigan: U.S. Environmental Protection Agency Underground Injection Control Program.

Johnson, J. G., 1970, Taghanic onlap and the end of North American Devonian provinciality: Geological Society of America Bulletin, v. 81, p. 2077–2106.

Johnson, J. G., and Sandberg, C. A., 1989, Devonian eustatic events in the western United States and their biostratigraphic responses, *in* McMillan, N. J., Embry, A. F., and Glass, D. J., eds., Devonian of the World: Calgary, Canadian Society of Petroleum Geologists Memoir 14, v. 3, p. 171–178.

Johnson, J. G., Klapper, G., and Sandberg, C. A., 1985, Devonian eustatic fluctuations in Euramerica: Geological Society of America Bulletin, v. 96, p. 567–587.

—— , 1986, Late Devonian eustatic cycles around margin of Old Red Continent: Société Géologique de Belgique Annales, v. 109, pt. 1, Special volume

"Aachen 1986", p. 141–147.

Kluessendorf, J., Mikulic, D. G., and Carman, M. R., 1989, Distribution and depositional environments of the westernmost Devonian rocks in the Michigan Basin, *in* McMillan, N. J., Embry, A. F., and Glass, D. J., eds., Devonian of the World: Calgary, Canadian Society of Petroleum Geologists Memoir 14, p. 251–263.

Lineback, J. A., 1970, Stratigraphy of the New Albany Shale in Indiana: Indiana Geological Survey Bulletin 44, 73 p.

Matthews, R. D., 1983, *Foerstia* from the Antrim Shale (Devonian) of Michigan: Geology, v. 11, p. 327–330.

McHone, J. F., Sargent, M. L., and Nelson, W. J., 1986, Shatter cones and other shock effects at Des Plaines, Illinois; Evidence for meteoroid impact: Geological Society of America Abstracts with Programs, v. 18, p. 689.

Mikulic, D. G., and Kluessendorf, J., 1988, Subsurface stratigraphic relationships of the Upper Silurian and Devonian rocks of Milwaukee County, Wisconsin: Geoscience Wisconsin 13.

Miller, J. D., and Kent, D. V., 1988, Paleomagnetism of the Silurian–Devonian Andreas redbeds; Evidence for an Early Devonian supercontinent?: Geology, v. 16, p. 195–198.

Mounds Facility Laboratory Report, Tetra Tech., Inc., 1980, Evaluation of Devonian shale potential in the Michigan Basin: U.S. Department of Energy, Morgantown Energ;y Technology Center Report DOE/METC/123, 26 p.

Newton, C. R., and Thompson, J. B., 1987, High biological production and the origin of Lower Devonian spicular chert and black shale, central Appalachians: Calgary, 2nd International Symposium on the Devonian System, Aug. 17-20, 1987, Program and Abstracts, p. 177.

Oliver, W. A., and Klapper, G., eds., 1981, Devonian biostratigraphy of New York: International Union of Geological Sciences Subcommission on Devonian Stratigraphy, 2 pts.

Palmer, A. R., 1983, The Decade of North American Geology 1983 geological time scale: Geology, v. 11, p. 503–504.

Parrish, J. T., and Gautier, D. L., 1988, Upwelling in Cretaceous Western Interior seaway [abs.]: American Association of Petroleum Geologists Bulletin, v. 72, p. 232.

Pashin, J. C., and Ettensohn, F. R., 1987, An epeiric shelf-to-basin transition; Bedford–Berea sequence, northeastern Kentucky and south-central Ohio: American Journal of Science, v. 287, p. 893–926.

Pepper, J. F., de Witt, W., Jr., and Demarest, D. F., 1954, Geology of the Bedford shale and Berea sandstone in the Appalachian basin: U.S. Geological Survey Professional Paper 259, 111 p.

Poole, F. G., and Sandberg, C. A., 1977, Mississippian paleogeography and tectonics of the western United States, *in* Stewart, J. H., Stevens, C. H., and Fritsche, A. E., eds., Paleozoic paleogeography of the western United States: Pacific Section, Society of Economic Paleontologists and Mineralogists Pacific Coast Paleogeography Symposium 1, p. 67–85.

Potter, P. E., Maynard, J. B., and Pryor, W. A., 1981, Sedimentology of gas-bearing Devonian shales of the Appalachian Basin: Eastern Gas Shales Project, U.S. Department of Energy, Morgantown Energy Technology Center Report DOE/METC-114, 43 p.

Potter, P. E., deReamer, J. H., Jackson, D. S., and Maynard, J. B., 1983, Lithologic and paleoenvironmental atlas of Berea Sandstone (Mississippian) in the Appalachian Basin: Appalachian Basin Geological Society Special Publication 1, 157 p.

Rickard, L. V., 1975, Correlation of the Silurian and Devonian rocks in New York State: New York State Museum and Science Service Map and Chart Series No. 24.

——, 1981, The Devonian System of New York State, *in* Oliver, W. A., Jr., and Klapper, G., eds., Devonian biostratigraphy of New York; Part 1, Text: International Union of Geological Sciences Subcommission on Devonian Stratigraphy, p. 5–22.

Russell, D. J., 1985, Depositional analysis of black shale by using gamma-ray stratigraphy; The Upper Devonian Kettle Point Formation of Ontario: Canadian Petroleum Geology Bulletin, v. 33, p. 236–253.

Sandberg, C. A., 1976, Conodont biofacies of Late Devonian *Polygnathus styriacus* Zone in western United States, *in* Barnes, C. R., ed., Conodont paleoecology: Geological Association of Canada Special Paper 15, p. 171–186.

——, 1988, Role of conodont biofacies in Late Devonian and Early Mississippian paleobiogeographic reconstructions of Western United States: Geological Society of America Abstracts with Programs, v. 20, p. 227.

Sandberg, C. A., and Dreesen, R., 1984, Late Devonian icriodontid biofacies models and alternate shallow-water conodont zonation, *in* Clark, D. L., ed., Conodont biofacies and provincialism: Geological Society of America Special Paper 196, p. 143–178.

Sandberg, C. A., and Poole, F. G., 1977, Conodont biostratigraphy and depositional complexes of Upper Devonian cratonic-platform and continental-shelf rocks in the Western United States, *in* Murphy, M. A., Berry, W.B.N., and Sandberg, C. A., eds., Western North America; Devonian: Riverside, University of California Campus Museum Contribution 4, p. 144–182.

Sandberg, C. A., Gutschick, R. C., Johnson, J. G., Poole, F. G., and Sando, W. J., 1986, Middle Devonian to Late Mississippian geologic history of the Overthrust belt region, western United States: Société Géologique de Belgique Annales, v. 109, pt. 1, Special volume "Aachen 1986", p. 205–207.

Sandberg, C. A., Ziegler, W., Dreesen, R., and Butler, J. L., 1988, Late Frasnian mass extinction; Conodont event stratigraphy, global changes, and possible causes: Courier Forschungsinstitut Senckenberg, v. 102, p. 263–307.

Sandberg, C. A., Poole, F. G., and Johnson, J. G., 1989a, Upper Devonian of Western United States, *in* McMillan, N. J., Embry, A. F., and Glass, D. J., eds., Devonian of the World: Calgary, Canadian Society of Petroleum Geologists Memoir 14, v. 1, Regional Syntheses, p. 183–220.

Sandberg, C. A., Ziegler, W., and Bultynck, P., 1989b, New standard conodont zones and early *Ancyrodella* phylogeny across Middle–Upper Devonian boundary: Courier Forschungsinstitut Senckenberg, v. 110, p. 195–230.

Sanford, B. V., 1967, Devonian of Ontario and Michigan, *in* Oswald, D. H., ed., Proceedings, International Symposium on the Devonian System: Calgary, Alberta Society of Petroleum Geologists, v. 1, p. 973–999.

Schumacher, D., 1971, Conodonts from the Middle Devonian Lake Church and Milwaukee Formations; and Conodonts and biostratigraphy of the "Kenwood" Shale, *in* Clark, D. L., ed., Conodonts and biostratigraphy of the Wisconsin Paleozoic: Wisconsin Geological and Natural History Survey Information Circular 19, p. 55–77.

Schwimmer, B. A., Hannibal, J. T., Feldmann, R. M., and Stukel, D. J., III, 1987, The paleontology and depositional environment of the Chagrin Shale (Famennian) in northeastern Ohio: Calgary, Second International Symposium on the Devonian System, Aug. 17-20, 1987, Program and Abstracts, p. 204.

Scotese, C. R., 1986, Early Famennian (367 Ma) equatorial world view, *in* Roy, S., ed., The Devonian; A portfolio of maps 1978–1986: Anchorage, Alaska Pacific University, The Devonian Institute, Plate 12.

Scotese, C. R., Van der Voo, R., and Barrett, S. F., 1985, Silurian and Devonian base map: Royal Society of London Philosophical Transactions, v. B309, p. 57–77.

Tarbell, E., 1941, Antrim–Ellsworth–Coldwater Shale formations in Michigan: American Association of Petroleum Geologists Bulletin, v. 25, p. 724–733.

Telford, P. G., 1989, Devonian stratigraphy of the Moose River Basin, James Bay lowland, Ontario, *in* McMillan, N. J., Embry, A. F., and Glass, D. J., eds., Devonian of the World: Calgary, Canadian Society of Petroleum Geologists Memoir 14, v. 1, Regional Syntheses, p. 123–132.

Van der Voo, R., 1988, Paleozoic paleogeography of North America, Gondwana, and intervening displaced terranes; Comparisons of paleomagnetism with paleoclimatology and biogeographical patterns: Geological Society of America Bulletin, v. 100, p. 311–324.

Wold, R. J., Paull, R. A., Wolosin, C. A., and Friedel, R. J., 1981, Geology of central Lake Michigan: American Association of Petroleum Geologists Bulletin, v. 65, p. 1621–1632.

Woodrow, D. L., Dennison, J. M., Ettensohn, F. R., Sevon, W. T., and Kirchgasser, W. T., 1989, Middle and Upper Devonian stratigraphy and paleogeography of the central and southern Appalachians and eastern Midcontinent, U.S.A., *in* McMillan, N. J., Embry, A. F., and Glass, D. J., eds., Devonian of the World: Calgary, Canadian Society of Petroleum Geologists Memoir 14, v. 1, Regional Syntheses, p. 277–301.

Ziegler, A. M., Scotese, C. R., McKerrow, W. S., Johnson, M. E., and Bambach, R. K., 1979, Paleozoic paleogeography: Annual Review of Earth and Planetary Sciences, v. 7, p. 473–502.

Ziegler, W., and Sandberg, C. A., 1990, The Late Devonian standard conodont zonation: Courier Forschungsinstitut Senckenberg, v. 121, 115 p.

MANUSCRIPT ACCEPTED BY THE SOCIETY JUNE 1, 1990

Printed in U.S.A.

# Mississippian System of the Michigan Basin; Stratigraphy, sedimentology, and economic geology

**James A. Harrell and Craig B. Hatfield**
*Department of Geology, University of Toledo, Toledo, Ohio 43606*
**George R. Gunn**
*Sun Exploration and Production Company, Campbell Center 2, 8150 North Central Expressway, Dallas, Texas 75221*

## ABSTRACT

The Mississippian System has the largest subcrop area of any Phanerozoic system in the Michigan Basin, and attains a maximum thickness of 719 m (2,360 ft) northeast of the basin center. The Mississippian formations include, in ascending stratigraphic order: Antrim Shale, the laterally equivalent Bedford and Ellsworth Shales (all Upper Devonian to Kinderhookian); Berea Sandstone (Kinderhookian); Sunbury Shale (Kinderhookian); Coldwater Shale (Kinderhookian to Osagian); Marshall Sandstone (Osagian); Michigan Formation (Osagian to Meramecian); and Bayport Limestone (Meramecian). There are no Chesterian sediments in the Michigan Basin. The Mississippian sediments accumulated conformably on Devonian strata but are overlain with disconformity by Pennsylvanian and, very locally, Jurassic strata.

The Kinderhookian, Osagian, and Meramecian series record a decreasing rate of Michigan Basin subsidence through time. Subsidence ceased temporarily during the Chesterian Epoch, and some Mississippian units were eroded from local anticlines in the central basin area during this interval of nondeposition. As a result, basal Pennsylvanian strata rest directly on Meramecian rocks and locally on older Mississippian formations.

The Mississippian sediments are primarily shallow-marine deposits consisting largely of shale with subordinate amounts of sandstone, siltstone, carbonates, and evaporites. Fluvial-deltaic deposits make up a significant portion of the section only in the eastern half of the basin. Terrigenous clastics were derived mainly from a source to the northeast of the basin in the Canadian Shield and, to a lesser extent, from the northwest in the Wisconsin Highlands.

Significant quantities of oil and gas have been produced from sandstones in the Berea, Marshall, and Michigan Formations, and from carbonates in the Ellsworth Shale. Sandstones in the Coldwater and Marshall Formations were, at one time, extensively quarried for grindstones and construction flagstone, respectively. The Michigan Formation is the chief source of gypsum in Michigan, and the Bayport supplies some of the state's limestone.

## INTRODUCTION

The Mississippian System of the Michigan Basin (Fig. 1) is spread over a roughly circular area approximately 400 km (250 mi) in diameter, extending northward from northern Indiana and northwestern Ohio to nearly the northernmost counties of Michigan's Southern Peninsula, and eastward from central Lake Michigan to within Lake Huron. In the central part of this area, where Phanerozoic sedimentary rocks are thickest, Mississippian strata are covered by the Pennsylvanian and Jurassic systems, producing a surrounding doughnut-shaped Mississippian subcrop belt larger than that of any other system in the Michigan Basin (Figs. 2

Harrell, J. A., Hatfield, C. B., and Gunn, G. R., 1991, Mississippian System of the Michigan Basin; Stratigraphy, sedimentology, and economic geology, *in* Catacosinos, P. A., and Daniels, P. A., Jr., eds., Early sedimentary evolution of the Michigan Basin: Geological Society of America Special Paper 256.

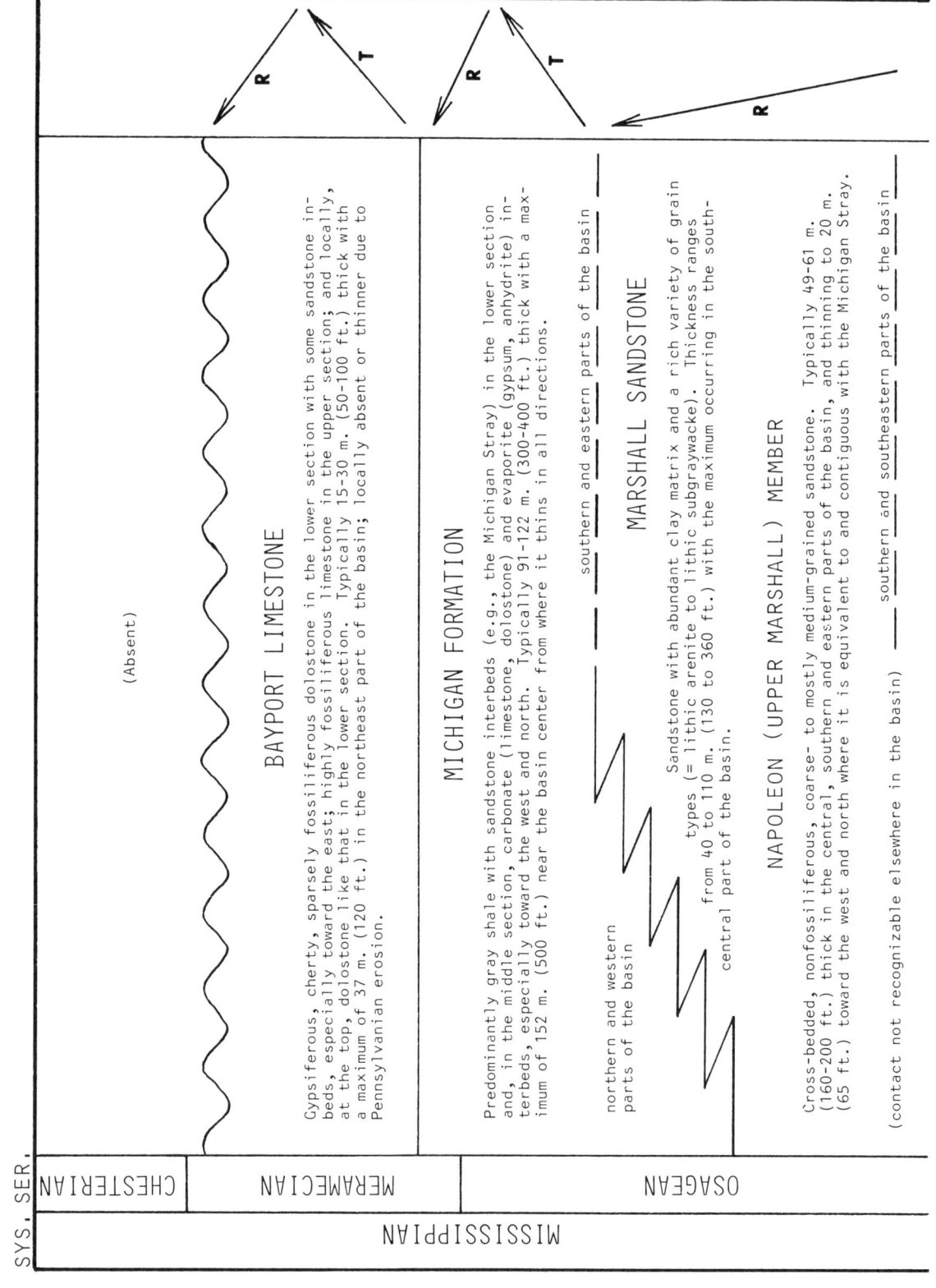

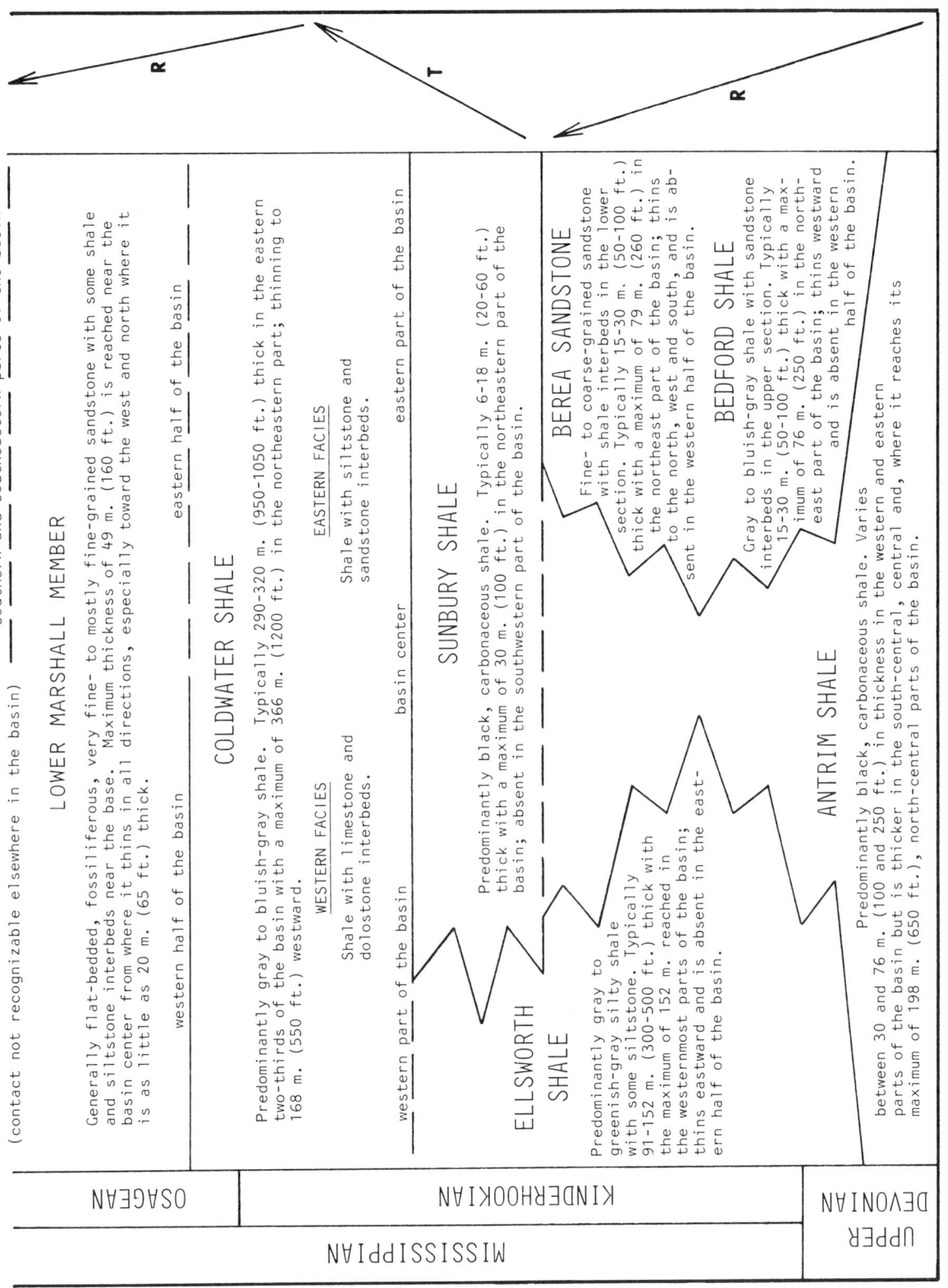

Figure 1. Stratigraphic column for the Mississippian System of the Michigan Basin. Contacts are solid where sharp and definite, dashed where gradational and uncertain, zigzagged where gradational and interfingering, and wavy where unconformable. Transgressive and regressive sequences are indicated on the right. Positions of series boundaries are taken from the AAPG COSUNA Correlation Chart (Shaver, 1985).

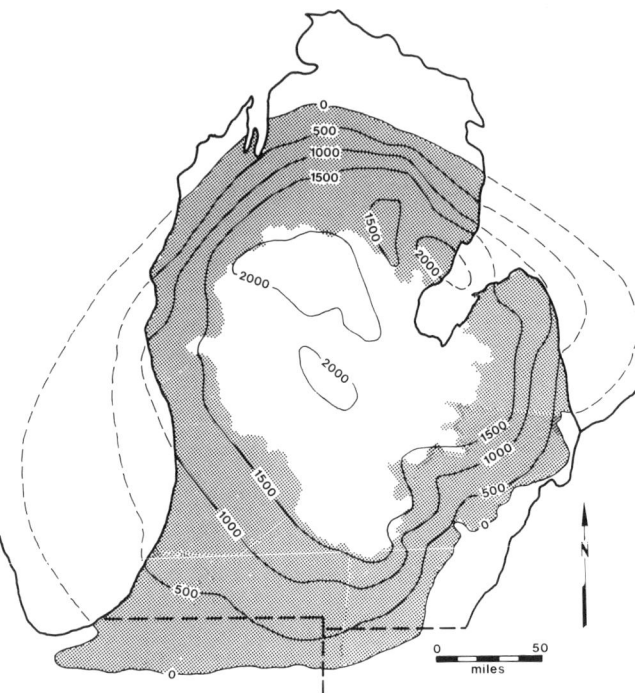

Figure 2. Map showing thickness (in feet) and subcrop area (stippling) for the Mississippian System (Antrim Shale included) of the Michigan Basin (redrawn from USGS, 1979, Plates 1 and 7 of Sheet 1).

and 3). Exposures, however, are limited to those few parts of this subcrop belt that are not blanketed by unconsolidated Pleistocene sediment. This glacial till and outwash is locally thicker than 305 m (1,000 ft) but is in most places less than 122 m (400 ft) thick (Western Michigan University, 1981, Plate 15).

Mississippian rocks are generally thickest in the central part of the basin, where they are covered by Pennsylvanian strata and range from about 457 m (1,500 ft) to 640 m (2,100 ft) thick, although the maximum thickness reported is from a well in Arenac County on the northeast side of the basin just north of Saginaw Bay. Here, the system is 719 m (2,360 ft) thick (Cohee, 1979).

Throughout its extent in the Michigan Basin, the Mississippian System apparently was deposited conformably on Upper Devonian strata. Ellsworth, Antrim, and Bedford Shales, intertonguing in that order from west to east across the basin, are in part equivalent in age, and each of these formations probably contains the Devonian-Mississippian contact. Paucity of index megafossils, however, lends a degree of uncertainty to this conclusion, and controversy concerning ages of conodonts and plant microfossils has moved the Devonian-Mississippian contact up and down through the lithostratigraphic column over the years. The paleontologically depauperate Antrim black shale in particular, which is the only one of the three shale units to occur throughout the basin, has provided essentially no information on the position of the sytemic contact within it.

In contrast to this enigmatic conformable basal contact, the top of the Mississippian System in Michigan is overlain unconformably by the Lower Pennsylvanian Saginaw Formation (Strutz, 1978) and, very locally at the boundary of Newaygo and Kent Counties in the west-central part of the basin, by Upper Jurassic "red beds," which overstep the Pennsylvanian to rest directly on Mississippian formations. The assignment of a Jurassic age to these youngest basinal deposits has been controversial, and many geologic maps place them in the Pennsylvanian. The age determination based on plant microfossils, however, is Kimmeridgian (Late Jurassic) (Shaffer, 1968). Chesterian sediments either were not deposited in Michigan or were removed by erosion prior to deposition of the Pennsylvanian, which rests directly on Meramecian and older Mississippian sequences.

The Mississippian of Michigan consists largely of shallow-marine terrigenous detritus, most of which is shale. Extensive sandstones are second in volume, and carbonates and evaporites are subordinate components. The shales and sandstones from all parts of the Mississippian section yield evidence of having been derived primarily from the Canadian Shield to the northeast of the basin, where they entered in the vicinity of the Michigan Thumb, and secondarily from the Wisconsin Highlands to the northwest of the basin (Cohee and Underwood, 1944; Potter and Pryor, 1961; Cohee, 1965; Asseez, 1969; Pryor and Sable, 1974; Craig and Varnes, 1979). Sediments derived from the northwest made significant contributions only to the Ellsworth Shale and possibly also to the Marshall Sandstone. The reduced terrigenous input in the northern and western parts of the basin allowed in these areas the widespread development of carbonates in the Coldwater Shale, and carbonates and evaporites in the Michigan Formation. The Canadian Shield and Wisconsin Highlands source areas were emergent throughout Mississippian time. This produced well-defined east-west lithologic and thickness trends in most of the Mississippian formations (Figs. 1 and 4).

The sediment supply from the northeast was more constant throughout the Mississippian Period than the absence of the Chesterian Series would suggest. During this last epoch of the period, when deposition had ceased in Michigan, voluminous mud and sand were carried for the first time southwestward across the Michigan Basin and Kankakee arch to accumulate in the Illinois Basin, where carbonate rock predominates in the earlier portions of the Mississippian System (Potter, 1963; Swann, 1963, 1964). Apparently the essentially continuous subsidence of the Michigan Basin during Kinderhookian and Valmeyeran times trapped nearly all the detritus supplied from the northeast and deprived the Illinois Basin of significant terrigenous input (Lineback, 1969). In Chesterian time, however, with temporary cessation of Michigan Basin subsidence, detritus from the northeast bypassed the Michigan Basin and accumulated as the numerous Illinois Basin Chesterian sandstone and shale units, which bear clear stratigraphic and sedimentologic evidence of a source to the northeast (Potter and others, 1958; Potter and Pryor, 1961; Potter, 1963; Swann, 1963, 1964; Pryor and Sable, 1974; USGS, 1979, Plate 12)

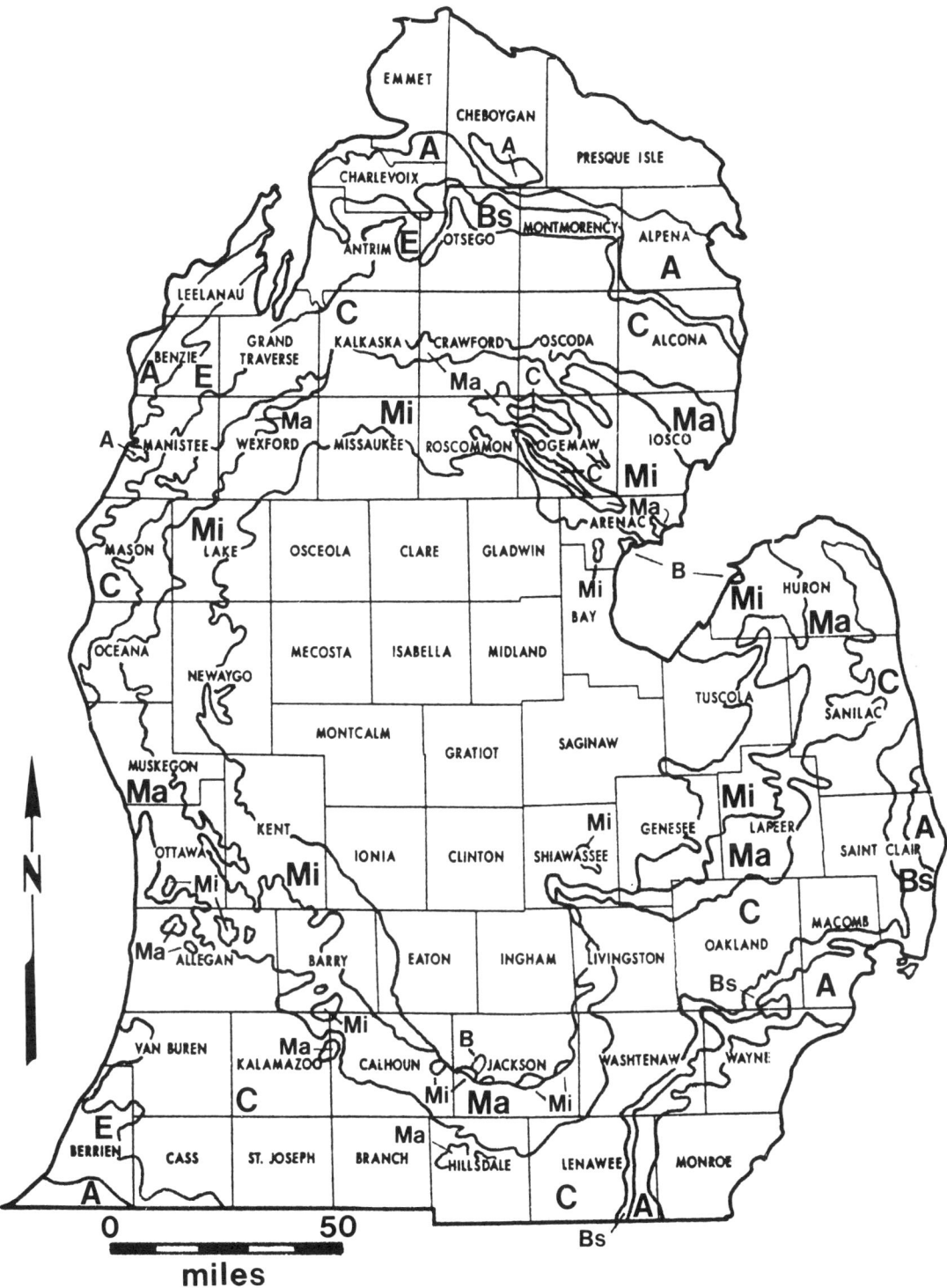

Figure 3. Bedrock geology map of the Mississippian formations in the Lower Peninsula of Michigan (redrawn from Martin and Straight, 1956, Map 11). Formation symbols: B = Bayport Limestone, Mi = Michigan Formation (includes the Bayport), Ma = Marshall Sandstone, C = Coldwater Shale, Bs = Bedford Shale–Berea Sandstone–Sudbury Shale, E = Ellsworth Shale, and A = Antrim Shale.

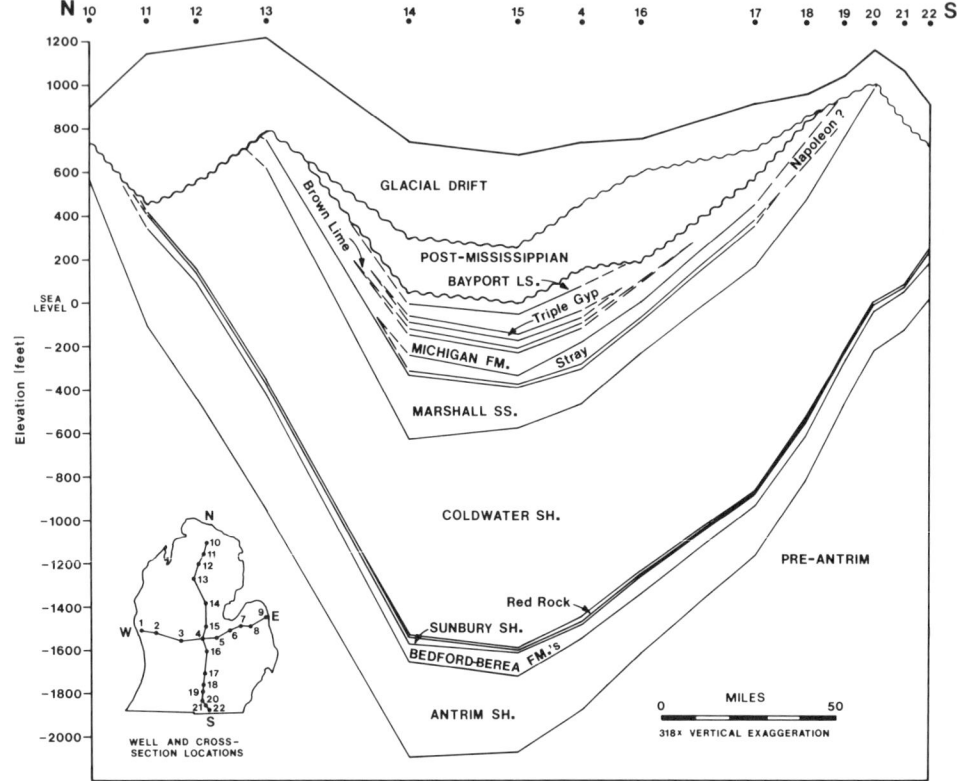
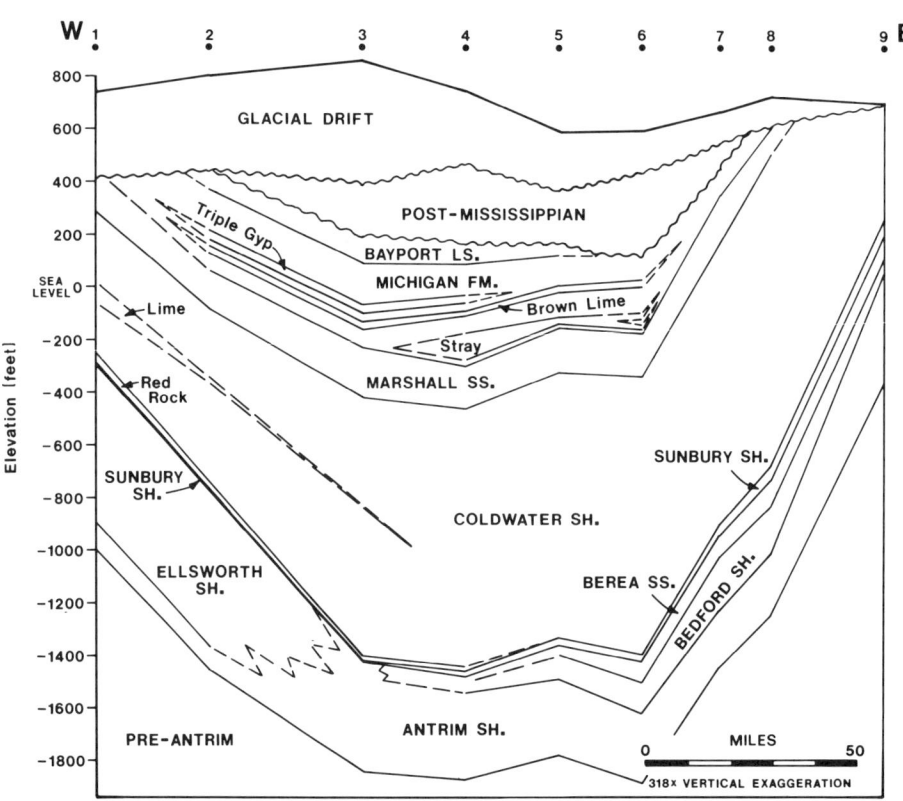

Figure 4. Stratigraphic cross-sections of the Mississippian System in the Michigan Basin (redrawn and, in part, reinterpreted from gamma-ray well logs and cross-sections in Lilienthal, 1978, Plates 9, 10, 34, 35, 56, 57, 84 and 85).

Michigan Basin subsidence was most rapid during early Mississippian (Kinderhookian) time and decelerated during the Osagean and Meramecian toward its halt in the Chesterian Epoch. The Coldwater Shale of Kinderhookian, and in small part Osagean, age accounts for approximately half—locally more than half—of the thickness of the entire system, whereas the other Osagean sediments are thinner and the Meramecian sediments are thinnest of all (Fig. 4). The Kinderhookian, however, was the briefest of the Mississippian epochs, whereas the Meramecian was the longest. Thus, the basinal subsidence rate, which had decreased during the Devonian Period from a late Silurian maximum, continued to decline during Mississippian time, and Paleozoic subsidence terminated in the middle of the Pennsylvanian Period (Craig and Varnes, 1979).

## DEVONIAN-MISSISSIPPIAN BOUNDARY

The lack of reported guide fossils in Antrim and Ellsworth Shales has resulted in varied published positions for the Devonian-Mississippian contact in central and western Michigan (McGregor, 1954; Sanford, 1967; deWitt, 1970; Lilienthal, 1974, 1978; Cohee, 1979). The Sunbury, Berea, and most of the Bedford, however, are clearly Kinderhookian in northeastern Ohio on the basis of conodonts and plant microfossils (deWitt, 1970). If these formations are assumed to be isochronous from northeastern Ohio into the Michigan Basin, then the base of the Mississippian System is probably in the lower part of the Bedford Shale in eastern Michigan (Ells, 1979). Because Bedford and Berea grade into upper Antrim Shale westward toward the basin center, the Antrim must contain the systemic contact there. Still farther westward, replacement of upper Antrim by Ellsworth in western Michigan, coupled with the absence of reported guide fossils in each of these formations, means that the systemic contact could be within either the Antrim or the overlying Ellsworth Shale.

Lilienthal (1974) suggested that the base of the Mississippian is in the upper part of the Ellsworth Shale in western Michigan, whereas McGregor (1954) and Cohee (1979) placed it near the Antrim-Ellsworth contact. Lilienthal's (1974) conclusion was based on the observation that the thickness of Antrim that intertongues with Ellsworth in western Michigan is much greater than that of the Bedford, Berea, and Sunbury sequence in eastern Michigan. Thus, assuming at least a crude relation of thickness to time represented, the systemic contact near the base of hte Bedford in eastern Michigan must be in the upper part of the Antrim and in the upper Ellsworth in central and western Michigan, respectively.

Cohee's (1979) contrasting conclusion that the Devonian-Mississippian contact is close to the base of the Ellsworth was, interestingly enough, also based on stratigraphic thickness criteria. The difference is that Cohee's thickness considerations started at the base of the Antrim instead of the top. Having observed that an average of about 46 m (150 ft) of Antrim underlie the Bedford in eastern Michigan, where basal Bedford presumably contains the systemic boundary, he assigned only the lower 46 m of Antrim to the Devonian System throughout the basin. This makes the Ellsworth almost entirely Mississippian.

Such disagreement reflects the inadequacy of using stratigraphic thicknesses as the basis for temporal correlation as well as the need for paleontologic study of Antrim and Ellsworth conodonts and microflora. Because published accounts assigning the Ellsworth Shale to the Mississippian outnumber those assigning it to the Devonian, we have deferred to custom and placed it within the Mississippian in Figure 1.

## ANTRIM SHALE

The Antrim Shale in Michigan is exposed only in the northern part of the basin in Antrim, Charlevoix, Cheboygan, and Alpena Counties. Here, in Antrim County, it was studied and named by Lane in 1901. In addition to these Michigan exposures, Antrim Shale is exposed in a few localities near the southern margin of the basin in northwestern Ohio and northern Indiana.

The Antrim, which is predominantly black carbonaceous shale, underlies all of the Mississippian System in the Michigan Basin and continues upward into the Mississippian throughout much of the basin. Its base is easily recognized everywhere except in the northern part of the basin where it is gradational with the underlying Traverse Group (Lilienthal, 1978). The Antrim is thickest in Roscommon and Gladwin Counties in the north-central basin area, where 198 m (650 ft) have been recorded (Cohee, 1979). Elsewhere in the basin, its thickness typically varies between 30 and 91 m (100 and 300 ft) (Asseez, 1969). The Antrim thins toward the basin margins, where it wedges out and so is absent over much of the surrounding structurally high areas such as the Kankakee and Findlay arches.

In southeastern and eastern Michigan, upper Antrim interfingers with and is replaced by the Kinderhookian Bedford Shale and Berea Sandstone, which overlie about 30 to 76 m (100 to 250 ft) of Antrim and are overlain by the black Sunbury Shale. Westward from this area, toward the basin center, where the Bedford and overlying Berea pinch out and are replaced by Antrim Shale, the Antrim is directly overlain by the Sunbury to form a single thick, black shale unit in which Antrim and Sunbury are virtually indistinguishable. In the western part of the basin, upper Antrim intertongues with the Ellsworth Shale (Kinderhookian?) of western Michigan and farther west, a somewhat thinner Antrim Shale (about 30 to 61 m; 100 to 200 ft) is overlain by the Ellsworth Shale. Figures 1 and 4 depict these facies relations.

Because of these east-west facies changes, the lower Kinderhookian Series (below the Coldwater Shale) is entirely black shale (Antrim and Sunbury) throughout a narrow north-south–trending belt that runs through the central part of the Michigan Basin but is largely green shale (Ellsworth) in the western part of the basin (Bishop, 1940). The Kinderhookian includes significant amounts of sandstone, siltstone, and bluish gray shale (Bedford and Berea) only in the eastern Michigan Basin (Asseez, 1969; Cohee, 1979).

The Antrim is perhaps the most enigmatic and interesting of Mississippian formations in Michigan. It cannot be adequately described or interpreted, however, by considering only the Michigan occurrence, because much of the interest in the Antrim hinges on the remarkable geographic extent of the same peculiar lithology near the Devonian-Mississippian boundary throughout much of the United States and elsewhere. Although this unit is known as Antrim Shale only in the Michigan Basin, rocks with the same salient lithologic and paleontologic characteristics and essentially the same stratigraphic position occur in Illinois, Indiana, and western Kentucky as the New Albany Shale; in Ohio and eastern Kentucky as the Ohio Shale; in Iowa, Missouri, Kansas, Oklahoma, Arkansas, Tennessee, Mississippi, and Alabama as the Chattanooga Shale; in West Virginia, Pennsylvania, and New York as the Geneseo Shale among other names; in Ontario as the Kettle Point Shale; and in other regions such as western Canada and Europe (Wilmarth, 1938; Conant and Swanson, 1961). Any proposed explanation for the unusual characteristics of this shale surely should also explain such widespread occurrence, in that some depositional environments that are otherwise feasible would not be expected to affect an area so huge.

The Antrim is dark gray or brown to largely black, highly carbonaceous, thinly laminated shale with meager fossil content except for profuse algal spores (*Tasmanites* and *Foerstia*). Some beds within the Antrim, especially in the lower part, are less carbonaceous and consequently are greenish gray rather than black. Large, dark brown, bituminous and pyritic limestone concretions occur in the lower Antrim and are typically from 0.6 to 1.5 m (2 to 5 ft) in diameter. Locally, pyrite occurs as burrow fillings and as finely disseminated grains concentrated in laminae (Lilienthal, 1978). Fragments of carbonized wood are common, and larger carbonized and silicified stems and branches as much as 15 cm (6 in) in width are abundant in Antrim exposures at Alpena and elsewhere in the northern part of the Michigan Basin (Arnold, 1931). These are remains of *Callixylon,* the same genus of which large logs (as much as 0.9 m [3 ft] in diameter) have been found in Antrim Shale equivalents in Indiana, Kentucky, Ohio, and elsewhere, as in the Devonian of the Catskill region of New York (Arnold, 1931). Specimens of the inarticulate brachiopod *Lingula* are common at some horizons, as are conodonts and scolecodonts. Fish scales and vertebrae can be found occasionally. Other kinds of fossils are virtually absent in the Antrim.

The extreme profusion of the algal spore *Tasmanites* suggests that the abnormally high organic carbon content of the Antrim (commonly as great as 25 percent and locally even greater) (Broadhead and others, 1982) may have been supplied by exceptionally dense distribution of floating marine algae. Lineback (1968, 1970) has proposed such profuse algal floatant as the primary supplier of organic matter to the Antrim and its equivalents. An alternative source for such large volumes of organic matter over such a vast marine area is not readily apparent. Some authors (Rich, 1951; Broadhead and others, 1982) have attributed the high organic content to deoxygenated bottom waters related to great water depth in the Antrim sea. In this case, the high content of organic matter would be largely attributable to lack of decomposition rather than abnormally high rate of supply, or perhaps both.

Broadhead and others (1982) concluded that the Ohio Shale (Antrim equivalent) along the southern shore of Lake Erie was deposited in water depths of at least 213 m (700 ft) because these shales are at the western end of coarser, thicker, and presumably shallower water lateral equivalents in the Catskill sequence to the east. Using the *Foerstia* zone of the Ohio Shale as a chronostratigraphic unit, Broadhead and others (1982) interpreted westward decrease in stratigraphic thickness between the base of the Ohio Shale and the *Foerstia* zone as equivalent to an increase in water depth westward. Because stratigraphic convergence of 213 m (700 ft) was observed, a minimum water depth of 700 ft was suggested for the Ohio Shale in northern Ohio. This conclusion, however, is clearly tenuous to the extent that increasing thickness eastward could reflect greater subsidence rates and, thus, more rapid sedimentation in that direction rather than variation in water depth.

Occurrences of burrows and the benthonic *Lingula* seem to argue against anoxic bottom waters, at least for those scattered horizons where these benthonic biotic components are present. *Lingula,* because it is most abundant in the modern world in shallow-water settings in and near littoral environments (Cooper, 1937; Ferguson, 1963), also raises questions concerning great water depth in the Antrim sea.

Considering the extent of this black shale in North America, one of its stratigraphic characteristics that may bear on the water depth in which it accumulated is the observation that it rests uncomfortably on formations varying in age from Devonian to Ordovician (Conant and Swanson, 1961). Thus, it is apparently a deposit of a transgressing sea and, at least in its lower part, is not likely to have accumulated in deep water.

In the case of the Antrim and Sunbury black shales of Michigan, perhaps the most compelling depth criterion is the observed lateral gradation westward of Bedford Shale and Berea Sandstone, and eastward of Ellsworth Shale into black shale, which both underlies and overlies these formations (Fig. 4). Because the Bedford-Berea and Ellsworth are, respectively, fluvial-deltaic and prodelta deposits (Pepper and others, 1954; Asseez, 1969), the Antrim and Sunbury black shales immediately above, below, and laterally adjacent to them would hardly seem to record deep-sea sedimentation in this area. Similar but more impressive evidence occurs in Ohio, where, from north to south across the central part of the state, a geographically widespread but thin Bedford-Berea complex extends into the Ohio Shale (Antrim equivalent) and suggests deltaic progradation into marine water that was very shallow indeed. This sequence, known as the Red Bedford Delta, is generally between 30 and 61 m (100 and 200 ft) thick and was deposited in seas so shallow that there was apparently almost no sloping delta front. The deposits are almost entirely delta plain beds (Pepper and others, 1954). This interpretation is further supported by sand-filled mudcracks at several horizons in the Berea, which record repeated subaerial exposure (Lewis, 1988). Potter and others (1984) also have suggested a fluvial or tidal-channel origin for the Berea Sandstone in this region.

Particularly vexing is the question of how limbs and logs of *Callixylon* were distributed so widely in this black shale throughout much of its extent in the United States and Canada. This observation is difficult to explain regardless of conclusions concerning depth and oxidation state of bottom waters.

Twenhofel (1939) and Conant and Swanson (1961) have presented the most complete summaries of contrasting ideas concerning the origin of this black shale. The Antrim Shale and its equivalents undoubtedly will provide more fuel for continued controversy and speculation about their peculiar depositional setting.

In recent years, the hope that this shale also will provide more pragmatic fuel in the form of hydrocarbons has inspired study of its natural gas-producing potential (Provo and others, 1978; Broadhead and others, 1982). The Antrim has been a very marginal producer of natural gas for more than a century, primarily within the central basin area where, unlike in the peripheral areas, its level of organic maturity is sufficient to generate gaseous hydrocarbons (Cercone, 1984).

## BEDFORD SHALE AND BEREA SANDSTONE

The type localities of the Bedford Shale and Berea Sandstone are in Cuyahoga County, northeastern Ohio. These formations are not exposed in Michigan. The Bedford Shale and Berea Sandstone of the eastern Michigan Basin record southwestward growth of a deltaic system (the Thumb Delta) into the Antrim sea during Kinderhookian time. This system was centered in the northern part of the "thumb" of eastern Michigan and was apparently a distributary of the Ontario paleoriver, which flowed southward through southwestern Ontario and into Ohio, where it was partially responsible for the much more extensive Bedford-Berea deltaic complex of that state (Pepper and others, 1954; Sawtelle, 1958; Lewis, 1988). The Ontario paleoriver and its descendant, the Michigan paleoriver, were active throughout Mississippian time and, during the Chesterian Epoch, carried detritus farther southwest to the Illinois Basin (Swann, 1964).

The Bedford Shale is a bluish to light gray, silty shale that becomes sandy in its upper part and has a gradational contact with the overlying Berea Sandstone. It is commonly 15 to 30 m (50 to 100 ft) thick, and thins and becomes finer grained to the west, where it intertongues with the Antrim (Asseez, 1969; Lilienthal, 1978; Cohee, 1979). It reaches its maximum thickness of about 76 m (250 ft) in the Michigan thumb area. The Bedford has produced small amounts of natural gas at various localities in Michigan.

The Berea Sandstone attains a thickness as great as 79 m (260 ft) in Huron County but thins northwestward, westward, and southwestward away from the thumb area and is absent in the western half of the Michigan Basin (McGregor, 1954; Fig. 4).

Throughout most of its extent, the Berea is 15 to 30 m (50 to 100 ft) thick. It consists predominantly of light gray sandstone that is fine grained in the lower and upper parts of the formation but medium to coarse grained in the middle. It is silty and pyritic, particularly in its lower part where thin beds of shale occur locally. The Berea is cemented with dolomite and quartz (Lilienthal, 1978; Cohee, 1979).

Asseez (1969) stated that the western portion of the Berea in Michigan is primarily a blanket sand with the laterally equivalent distributary and channel sands farther to the east. Several linear, channel-form thickenings in the Berea extend from northeast of the Michigan thumb area southwestward through the eastern part of the state. These linear sand bodies terminate in east-central Michigan close to longitude 84°W, the apparent delta margin. Their character suggests that the Berea Sandstone is a constructive birdfoot-type delta similar to the Mississippi Delta.

Minor amounts of oil and gas have been produced from the Berea in the Saginaw Bay area since 1927, and recent discoveries have resulted in substantial new oil production during the 1980s in Midland and Bay Counties. In the latter cases, production is from northwest-southeast–trending linear sand bodies in the more permeable middle part of the Berea. These apparently accumulated as nearshore deposits with their long axes parallel to shore and normal to the southwestward regional transport of sand from the Ontario paleoriver system. Such middle Berea sand bodies have trapped oil at their southeast (up-dip) ends, where they are replaced locally by mud-filled channels trending perpendicular to the bars. From 1980 through 1986, more than 120 wells were drilled to depths of approximately 762 m (2,500 ft) into these stratigraphic traps, with average initial production of 35 barrels per day. Approximately 1.4 million barrels of oil had been produced by late 1986 (Gunn, 1986).

## ELLSWORTH SHALE

The Ellsworth Shale, which is exposed only in Antrim and Charlevoix Counties, is restricted to the western part of the basin, where it replaces Antrim and Sunbury Shales. It averages about 152 m (500 ft) thick in the westernmost parts of the Michigan Basin, but elsewhere is typically 91 to 152 m (300 to 500 ft) thick (Asseez, 1969; Lilienthal, 1978; Cohee, 1979). On the basis of conodonts, lithology, and stratigraphic position, the Ellsworth, Antrim, and Sunbury Shales of Michigan are considered equivalent in age to the New Albany Shale of Indiana, Illinois, and Kentucky (Lineback, 1970).

The Ellsworth is predominantly silty shale of gray, greenish gray, or more commonly, green color. Siltstone and sandstone are minor components and are most abundant in the westernmost, and especially southwestern, parts of the basin. Oolitic limestone and dolostone beds are present in the upper portion of the formation, and become thicker and more numerous toward the west (Asseez, 1969; Lilienthal, 1978; Cohee, 1979). Asseez (1969) interpreted the Ellsworth as a prodelta shale proximal to a delta front to the west, where erosion has removed correlative strata.

Several workers have suggested an Ellsworth source area to the northwest in the Wisconsin Highlands (e.g., McGregor, 1954; Cohee, 1979).

In the western and southwestern parts of the basin, oil and gas are produced from the carbonate beds (Lilienthal, 1974). In the petroleum industry, these carbonate reservoir intervals in the Ellsworth are called "Berea" production but are lithologically distinct from and not continuous with the Berea Sandstone of eastern Michigan. The true Berea of eastern Michigan is laterally separated from the Ellsworth Shale by the Antrim Shale in the central part of the basin (Fig. 4).

## SUNBURY SHALE

The type section of the Sunbury Shale is in Delaware County, central Ohio. It is not exposed in Michigan. The Sunbury Shale overlies the Berea and is a black, carbonaceous shale similar to the Antrim, with which it is vertically contiguous in the central part of the Michigan Basin. It is distinguished from the Antrim by its darker color and by the scarcity of algal spores (Michigan Basin Geological Society, 1969). In the eastern half of the basin, where Bedford and Berea intervene between Antrim and Sunbury, the Sunbury is commonly about 6 to 18 m (20 to 60 ft) thick (Cohee, 1979). It reaches a maximum thickness of slightly more than 36 m (120 ft) in the northern thumb area. The Sunbury thins westward and, in the western half of the basin, is typically less than 10 m (33 ft) thick (Asseez, 1969; Lilienthal, 1978; Fig. 4). Here, it locally overlies the Ellsworth Shale, but in southwestern Michigan, the Sunbury grades into Ellsworth and so is absent (Lilienthal, 1974). The Sunbury Shale apparently records continuation of the same peculiar depositional environment that produced the Antrim beneath it.

## COLDWATER SHALE

The Coldwater Shale was named by Lane (1895) for exposures along Coldwater Creek in Branch County, Michigan (Wooten, 1951). Although the Coldwater has the largest outcrop area of any Mississippian formation, it is accessible at relatively few localities. Its exposures are limited to portions of Branch, Calhoun, and Hillsdale Counties in the southern part of the basin, and Huron and Sanilac Counties in the Michigan thumb area (Tarbell, 1941; Chung, 1973; Ells, 1979).

The Coldwater is conformable with the underlying Sunbury and Ellsworth shales and the overlying Marshall Sandstone. The lower contact is lithologically distinct, as is the upper contact, except in the eastern part of the basin where the Coldwater is gradational with the Marshall (Monnett, 1948; Chung, 1973). Fossils found in the uppermost portion of the Coldwater in the western part of the basin are Osagean in age, but the rest of the formation is Kinderhookian (Cohee and others, 1951; Miller and Garner, 1953a, b, and 1955; Chung, 1973).

The Coldwater is the thickest Mississippian formation, attaining a maximum thickness of about 366 m (1,200 ft) in Iosco and Arenac Counties just north of Saginaw Bay. In the eastern two-thirds of the basin, the Coldwater maintains a relatively constant thickness of approximately 305 m (1,000 ft) and in the western third of the basin, it thins to about 168 m (550 ft) (Hale, 1941; Tarbell, 1941; Monnett, 1948; Lilienthal, 1978; Chung, 1973; Fig. 4). Although its thickness in the west is widely agreed upon, there is still considerable uncertainty about the Coldwater's thickness in the eastern part of the basin because of its gradational contact with the Marshall. The Coldwater thins slightly over the large anticlines in the central part of the basin, suggesting that these structures were positive elements during Coldwater time (Monnett, 1948).

The Coldwater consists predominantly of gray to bluish gray shale (Tarbell, 1941; Monnett, 1948; Michigan Basin Geological Society, 1969; Chung, 1973; Cohee, 1979; Ells, 1979). Its clay minerals are chiefly illite and kaolinite with minor chlorite. The kaolinite:illite ratio increases eastward across the basin, suggesting an eastern source with settling of the smaller illite particles farther from shore (Parham, 1966; and Chung, 1973).

Other lithologies occur in the Coldwater Shale, and their distributions divide the formation into distinct eastern and western facies. In the eastern half of the basin, beds of silty and sandy shale, siltstone, and fine-grained sandstone are common, and increase in abundance and coarseness to the east and up-section (Monnett, 1948; McGregor, 1954; Chung, 1973; Cohee, 1979). In the western half of the basin, the Coldwater shales are more calcareous, and beds of glauconitic, fossiliferous limestone and dolostone occur frequently, especially in the middle and upper portions of the formation (Hale, 1941; Monnett, 1948; McGregor, 1954; Michigan Basin Geological Society, 1969; Chung, 1973; Cohee, 1979). Also in the western basin, the upper Coldwater shales become siltier and more micaceous, and fossiliferous ironstone concretions become common (Squire, 1972; Chung, 1973).

Two marker beds in the Coldwater Shale can be traced over long distances: the "Lime" and "Red Rock" beds. The Coldwater Lime is one of the carbonate beds typical of the western facies. It occurs throughout the western part of the basin, is commonly 6 to 18 m (20 to 60 ft) thick, and thickens westward to a maximum of about 24 m (80 ft) (Hale, 1941; Chung, 1973). The bed is found 61 to 110 m (200 to 360 ft) below the top of the Coldwater and drops in the section to the west (Fig. 4). If this bed is assumed to be isochronous, its changing stratigraphic position provides further evidence of decreasing age for the Coldwater westward across the basin. The Coldwater Red Rock is the more extensive of the two marker beds, as it occurs in all parts of the basin except the extreme northeast (Hale, 1941; Monnett, 1948; McGregor, 1954; Michigan Basin Geological Society, 1969; Chung, 1973; Lilienthal, 1978; Cohee, 1979; Ells, 1979; Fig. 4). The bed is typically 3 to 6 m (10 to 20 ft) thick and locally reaches 15 m (50 ft). It has a persistent reddish color but is highly variable in lithology. It is a shaly limestone or dolostone in the western and central parts of the basin and a calcareous shale in the eastern part. It is thus more appropriate to think of the Red

Rock as a stratigraphic interval rather than a specific lithologic bed. It is found at or very near (at most 9 m or 30 ft above) the contact with the Sunbury and Ellsworth Shales.

After the Marshall Sandstone, the Coldwater Shale is the most fossiliferous of the Mississippian formations. The carbonate beds and ironstone concretions in the western part of the basin have yielded coiled and straight nautiloids, ammonoids, brachiopods, pelecypods, gastropods, ostracods, horn corals, bryozoa, and crinoids (Hale, 1941; Oden, 1952; Miller and Garner, 1953a, b, 1955; Driscoll, 1965; Squire, 1972; Chung, 1973). Chondrichthyes fish remains and carbonized wood have also been reported from the sandstones in the thumb area (Monnett, 1948; Dorr and Eschman, 1970).

The lithology and paleontology of the Coldwater indicate that the formation was deposited in a shallow-marine environment bounded to the east by a fluvial-deltaic system, most likely the Thumb Delta associated with the Ontario paleoriver. Sediments in the western and central parts of the basin were probably deposited below wavebase in waters of normal salinity whereas sediments farther east were laid down in shallower, more brackish, pro-delta waters (Dorr, and Eschman, 1970; Chung, 1973; USGS, 1979, Plate 12). The terrigenous sediment in the Coldwater was derived almost entirely from the Canadian Shield to the northeast of the basin. The following east-to-west trends support this provenance interpretation: (1) the westward decrease in the age of the Coldwater-Marshall contact, (2) the westward decrease in the thickness of the formation, (3) the westward transition from coarse siliciclastics to fine siliciclastics and carbonates, (4) the westward decrease in the kaolinite:illite ratio, and (5) the westward transition to a fully marine fauna.

The Coldwater Shale has apparently been mined on a small scale in Branch and Calhoun Counties, where it was used as a source of clay for ceramics (Dorr and Eschman, 1970). Small quantities of gas have been produced from the "Weir Sand," one of the sandstone beds in the lower Coldwater in the eastern part of the basin (Lilienthal, 1978). The principal economic commodity derived from the formation, however, is the fine-grained sandstone quarried in Huron County and used for grindstones (Kirkby, 1964; Heinrich, 1979). Some authors have assigned these "grindstone" beds to the Marshall (e.g., Hard, 1938; Kirkby, 1964), but it is now widely accepted that they belong to the uppermost Coldwater (Monnett, 1948; Dorr and Eschman, 1970; Heinrich, 1979). The Coldwater grindstones were quarried from about 1840 to 1940, and were said to be among the finest available. The sandstone consists of fine, angular quartz grains set in a clayey, micaceous matrix. When used in grinding applications, the softer matrix wears away just fast enough to prevent clogging or glazing of the grindstone surface. A new crop of angular grains is thus continuously supplied to the abrasive surface.

## MARSHALL SANDSTONE

The Marshall Sandstone has been subdivided into the "Lower Marshall" and "Napoleon" (or Upper Marshall) members. The latter member is recognizable only in exposures in the southern part of the basin. The Marshall Sandstone was named by Winchell (1861) for exposures near Marshall in Calhoun County, Michigan. In adjacent Jackson County near the town of Napoleon, Houghton (1838) named similar but stratigraphically higher beds the Napoleon Sandstone. Lane (1901) merged the two units and designated them the Lower and Upper Marshall, respectively. Other exposures of Marshall Sandstone occur in Hillsdale County in the south; Ottawa County in the west; and Huron, Sanilac, and Tuscola Counties in the northeast (Monnett, 1948; Ells, 1979).

The Marshall Sandstone is conformable with the underlying Coldwater Shale and the overlying Michigan Formation. Some workers (e.g., Thomas, 1931; Kirkham, 1942) have argued that the Michigan lies with local unconformity on the Marshall. While this may be true, the consensus is that the contact is conformable in most parts of the basin (Monnett, 1948). In the eastern half of the basin, the Marshall-Coldwater contact is not well defined because of the lithologic gradation between the two formations. Although the Michigan Formation lithologies are generally quite distinct from those of the Marshall, abundant sand lenses, fingers, and tongues in the transition zone between the two formations have resulted in considerable uncertainty regarding the position of the contact in the central, western, and northern parts of the basin. The Michigan Formation has been completely removed by pre-Pennsylvanian erosion only in Livingston and Bay Counties in the eastern part of the basin, and in these areas the Lower Pennsylvanian Saginaw Formation rests with marked disconformity on the Marshall (Monnett, 1948).

The ambiguities in the positions of the lower and upper contacts of the Marshall Sandstone make thickness estimates for the formation difficult. Despite this complication, general thickness ranges and trends have been identified (Monnett, 1948; Ells, 1979). The total thickness of the Marshall appears to vary between 40 and 110 m (130 and 360 ft), and the maximum thickness occurs near the basin center in Ionia County. The formation thins toward its minimum thickness to the west and north of this area. In the southern and eastern parts of the basin the Marshall is typically 61 to 91 m (200 to 300 ft) thick.

The thickness estimates for the lower and upper members of the Marshall are even more uncertain than those for the total formation (Hard, 1938; Monnett, 1948; Cohee, 1979). Nevertheless, it is clear that the Napoleon Sandstone achieves its maximum thickness of 49 to 61 m (160 to 200 ft) in the southern part of the basin and thins rapidly to about 20 m (65 ft) to the west and north where it is apparently contiguous with the Michigan Stray sand near the base of the Michigan Formation (Monnett, 1948; Moser, 1963; Ells, 1979; Fig. 4). The Lower Marshall reaches its maximum thickness of about 49 m (160 ft) near the basin center and also thins to the west and north to as little as 11 to 21 m (35 to 70 ft) thick. The thickness trends in the Marshall and the probable subsurface correlation between the Napoleon and Michigan Stray suggest that the top of the formation becomes older westward and northward across the basin.

The lithology of the Marshall Sandstone has been described by numerous workers (e.g., Stearns and Cook, 1931; Stearns, 1933; Hard, 1938; Monnett, 1948; Potter and Pryor, 1961; Pawlowicz, 1969; Heinrich, 1979; Rorick, 1983). Generally speaking, it is a sparsely fossiliferous, sometimes cross-bedded and rippled, very fine- to coarse-grained sandstone of buff, tan, or gray color. Its most striking features are the great diversity of framework grains and the abundant interstitial clay matrix. Petrographically, the sandstone is classified as a lithic arenite or, where the clay matrix is especially abundant, as a lithic subgraywacke. The interstitial clays consist mostly of illite and kaolinite with minor chlorite. As has also been observed for the Coldwater Formation, the kaolinite:illite ratio increases toward the east (Pawlowicz, 1969). Some shale and siltstone beds are present at the base of the Marshall near its gradational contact with the underlying Coldwater Shale.

The Lower Marshall and Napoleon members are lithologically distinguishable only in the extreme southern and southeastern parts of the basin. The Napoleon is predominantly a medium-grained sandstone with some coarse sand intervals; it is nonfossiliferous and commonly cross-bedded and rippled. The Lower Marshall, in contrast, is mostly a fine-grained sandstone with some very fine sand intervals; it is fossiliferous and generally flat-bedded (Stearns, 1933; Hard, 1938; Monnett, 1948; Dorr and Kauffman, 1963; Heinrich, 1979; Cohee, 1979). The cross-bed dip directions for the Napoleon member show a preferred orientation to the southwest (Potter and Pryor, 1961; Pawlowicz, 1969). The Napoleon is also notable for another sedimentary structure, the "rippled toroid," which is a sand cast of a small depression formed by vortex currents (Dorr and Kauffman, 1963).

Although only sparsely fossiliferous, the Marshall Sandstone contains the most diverse flora and fauna of any Mississippian formation in the Michigan Basin. Apart from some spores and carbonized wood fragments of *Calamites* and *Lepidodendron* from the Upper(?) and Lower Marshall in Huron County (Lane and Cooper, 1900; Chaloner, 1954; Dorr and Eschman, 1970; See, 1980), all of the fossil material described from the formation comes from the Lower Marshall in Jackson and Calhoun Counties. These beds contain coiled and straight nautiloids, ammonoids, brachiopods, pelecypods, gastropods, ostracods, and Chondrichthyes fish, as well as *Calamites* and *Lepidodendron* (Monnett, 1948; Oden, 1952; Miller and Garner, 1953a, b, 1955; Dorr and Moser, 1964; Driscoll, 1965 and 1969; Dorr and Eschman, 1970; Squire, 1972; See, 1980).

The coarsening-upward sandstone sequence, the northward and westward thinning of the formation, the virtual absence of fossils in the upper member and diverse normal marine fauna in the lower member, and the abundant ripples, rippled toroids, and southwest-dipping cross-beds together suggest that the Marshall Sandstone was deposited in a shallow-marine, delta-front environment that was bordered on the east by a westward-prograding fluvial-delatic system associated with the Ontario paleoriver. Water depths would have been above wavebase and decreased through time. Salinities were probably at normal levels during deposition of the lower member and may have been brackish by the time the upper member was deposited. The Marshall Sandstone, and the Napoleon member in particular, represents the culmination of a marine regression that began during Coldwater deposition. Abundant terrigenous influx from the Canadian Shield was accompanied by westward progradation of the eastern shoreline.

The petrography of the Marshall Sandstone provides further evidence of the locations and character of the source areas supplying sediment to the Michigan Basin. Stearns (1933) noted that the heavy mineral assemblages are markedly different in the eastern and western halves of the basin. In the west, garnet and subordinate amounts of pink zircon and ferromagnesium minerals dominate, whereas the eastern assemblage consists mostly of tourmaline with minor leucoxene and colorless zircon (Stearns, 1933; Rorick, 1983). The heavy mineral data suggest that the Wisconsin Highlands were also contributing some sediment to the basin. Study of the light mineral fraction confirms and further refines the earlier sediment dispersal interpretations. Based on variations in framework grain composition and shape, Rorick (1983) identified four geographic anomalies in the Marshall, which he interpreted as representing shallow-marine sites located offshore from four inflowing paleorivers: the Marquette River (northwest of Manistee County) and the Wisconsin River (west of Ottawa County) carried material from the Wisconsin Highlands, while the Huron River (north of Oscoda County) and the Algonquin (or Ontario) River (east of Oakland County) supplied material from the Canadian Shield. The varieties of heavy and light mineral grains, their relative proportions, and their areal distributions indicate that a primary source area to the northeast was dominated by granite, granite-gneiss, and other related rocks, whereas a secondary source area to the west and northwest was dominated by metasediments and metavolcanics (Stearns, 1933; Rorick, 1983). These interpretations are consistent with the known paleogeology of the Canadian Shield and Wisconsin Highlands.

The Marshall Sandtone has been only a minor producer of oil, gas, and brine (Thomas, 1931; Hake, 1938; Hard, 1938; Ells and Ives, 1964; Ells, 1979). The formation is perhaps better known as a source of construction flagstone (Kirkby, 1964; Heinrich, 1979). It has been quarried in Calhoun, Hillsdale, Jackson, and Ottawa Counties, with production peaking around the turn of the century. It is currently worked only at Ray's Quarry and Jude's Quarry in Jackson County. The well-known grindstones quarried in Huron County are sometimes attributed to the Marshall Sandstone but actually come from the Coldwater Shale.

## MICHIGAN FORMATION

The Michigan Formation was named by A. Winchell (Taylor, 1839) without designation of a type locality. Exposures are found in Kent, Iosco, and Huron Counties. The formation, which is mostly shale, is Osagean and Meramecian in age and rests conformably on the Marshall Sandstone, with which it has a gradational contact. The upper beds of the Marshall interfinger with the lower part of the Michigan Formation, and an extensive tongue of Michigan sandstone (the "Michigan Stray") is apparently contiguous with the Napoleon member of the Marshall, and is laterally persistent in the lower part of the Michigan Formation in the central, western, and northern parts of the basin (Stearns and Cook, 1931; Moser, 1963; Olszewski, 1978; Ells, 1979; Fig. 4). This sandstone is petrographically indistinguishable from the Marshall except for the greater abundance of feldspar in the latter (Stearns and Cook, 1931; McGregor, 1954). The Marshall-Michigan contact becomes younger to the southeast, so that the uppermost Marshall is correlative and has an intertonguing relation with the lowermost Michigan Formation to the north and west (Cohee, 1979). The Michigan Formation is overlain conformably or with small hiatus by the Meramecian Bayport Limestone, which is the uppermost Mississippian unit in Michigan.

The Michigan Formation has its maximum thickness in Missaukee County to the north of the central basin area, where wells have penetrated approximately 152 m (500 ft) of the unit (Cohee, 1979). Thicknesses of 91 to 122 m (300 to 400 ft) are, however, more typical. Although greenish gray and dark gray shales dominate, this formation is lithologically variable and includes discontinuous beds of sandstone, limestone, dolostone, gypsum, and anhydrite (McGregor, 1954; Cohee, 1979). Each of these lithologies occurs randomly throughout most of the geographic extent of the formation, but the carbonates and evaporites are somewhat more abundant in the western and northern parts of the basin (Newcombe, 1933; Olszewski, 1978; Cohee, 1979). One of these carbonate beds, the "Brown Lime," is a 3- to 6-m-thick (10 to 20 ft) dolostone that can be traced in the subsurface across much of the basin (Michigan Basin Geological Society, 1969; Lilienthal, 1978; Fig. 4).

Aggregate maximum thickness of gypsum beds in the subsurface is about 30 m (100 ft) but commonly decreases to approximately 12 m (40 ft) in peripheral areas, where the formation is at or near the surface and where the gypsum is commercially exploited (Cohee, 1979). A 12-m-thick zone of three anhydrite beds separated by shale (the "Triple Gyp") has been traced in the subsurface through most of the central part of the basin (Briggs, 1970; Lilienthal, 1978; Fig. 4) and is present at Grand Rapids, Kent County, on the west side of the basin. At this location, the gypsum is mined at a depth of 46 m (150 ft) over an area of 1.2 km$^2$ (300 acres) (Dorr and Eschman, 1970). On the opposite side of the basin, in Huron and Iosco Counties near Saginaw Bay, the Michigan Formation is exposed, and gypsum is quarried from open pits at National City and Alabaster (Carlson, 1964; Dorr and Eschman, 1970).

Although gypsum also occurs at greater depth in the Silurian and Devonian systems in Michigan, it is commercially exploited only in the Michigan Formation, where it has accounted for a large proportion of the total value of all natural resources taken from Mississippian rocks in Michigan. Exploitation of Michigan Formation gypsum for the manufacture of wallboard, plaster, and cement has placed Michigan among the leading states in the nation in gypsum production (Dorr and Eschman, 1970).

Michigan Formation gypsum and anhydrite commonly wedge out laterally into shale and are thin or absent in the subsurface over small anticlines in the central part of the basin (McGregor, 1954). This relation suggests that these structural highs were formed either before or during deposition of the Michigan Formation and produced local shoals in the Michigan sea. McGregor (1954) suggested that these shoals restricted water circulation enough to enhance or cause evaporite accumulation in the intervening deeper-water areas. Maximum thicknesses of evaporites are indeed in the structurally low areas.

Significant quantities of natural gas and a very small amount of oil have been produced from the sandstone of the Michigan Formation (as well as from the upper part of the Marshall) in Mecosta, Isabella, Montcalm, and adjacent counties in the central basin area (Hard, 1938). The Michigan Formation was the primary producer of natural gas in Michigan during the 1940s, but production rapidly declined to insignificance thereafter (Michigan Geological Survey, 1964–1984). Until the mid-1960s, cumulative gas production from the Michigan Formation was still greater than that from all other rocks in the state combined, but since about 1960, Silurian strata have yielded more natural gas than the cumulative total of pre-1960s production in Michigan (Michigan Geological Survey, 1964–1984). Gas production from the Michigan Stray in the lower part of the formation was primarily on small northwest–southeast-trending anticlines in the central part of the basin (Ells and Ives, 1964).

## BAYPORT LIMESTONE

The Bayport Limestone was named by Lane in 1895 after exposures at Bay Port in Huron County and had earlier been called the Point Au Gres Limestone (Cohee, 1979). The Bayport is exposed in the eastern part of the Michigan Basin in Arenac County as well as at the type locality of Bay Port, where its carbonate rocks are quarried by the Wallace Stone Company. Exposures in the southern part of the basin are near Parma, Jackson County, and at Bellevue, Eaton County, where the Cheney Limestone Company quarries it. It is also exposed at Grand Rapids, Kent County, on the southwestern side of the basin. This unit, the youngest of the Mississippian formations in Michigan, contains marine invertebrate fossils taxonomically similar to those found in the St. Louis and Ste. Genevieve limestones of the Mississippi Valley region (Lasemi, 1975) and thus is Meramecian in age.

The Bayport is the least areally extensive of the Mississippian formations in Michigan, being restricted to a more central portion of the Michigan Basin than earlier units (Fig. 4). It rests conformably or with slight hiatus on the Michigan Formation and is disconformably overlain by the Pennsylvanian Saginaw Formation (Strutz, 1978; Tyler, 1980). If any Chesterian sediments were deposited in the Michigan Basin, they were subsequently removed by late Mississippian erosion. This erosion also removed the Bayport in places where the Pennsylvanian Saginaw Formation now locally rests directly on the Michigan Formation (McGregor, 1954; Lilienthal, 1974). Because of this post-Meramecian erosion, the Bayport is highly variable in thickness, and its stratigraphically higher parts are absent in exposures as well as in some places in the subsurface, where the basal Pennsylvanian lies directly on eroded structural highs in Mississippian strata (McGregor, 1954; Cohee, 1979). The maximum thickness of the Bayport is about 37 m (120 ft), but in most places it is 15 to 30 m (50 to 100 ft) thick (Cohee, 1979). In the outcrop belt, where more recent erosion also has occurred, the Bayport is generally only a few tens of feet thick (Cohee and others, 1951).

The Bayport is more lithologically variable than most older Mississippian formations because of both original lateral variation and local erosional removal of higher facies. The Bayport is unique among Michigan's Mississippian formations in being predominantly carbonate rock, both limestone and dolostone. It also contains significant amounts of sandstone, however, especially in the lower part of the formation, and there are minor quantities of shale, bedded chert, and evaporites (Bacon, 1971; Lasemi, 1975; Ciner, 1987).

Bayport rocks apparently record a continuation of the arid or semiarid climates that had produced the gypsum and anhydrite deposits of the underlying Michigan Formation. The lower part of the Bayport in most localities is a gray to tan, fine-grained, gypsiferous and cherty dolostone that is unfossiliferous except for silicified ostracodes and algal stromatolites (Bacon, 1971; Lasemi, 1975). In some places, coarse sand and dolostone gravel clasts are observed near the base (Lasemi, 1975), suggesting an unconformable contact with the underlying Michigan Formation. Bacon (1971) reported desiccation cracks on bedding surfaces in this lower part of the Bayport. Intercalated with these dolomitic carbonates are discontinuous laminae and thin beds of sandstone (Lasemi, 1975). The sandstone beds vary strikingly in thickness over short distances, and at a few places in the eastern half of the basin, the dolomitic facies is replaced entirely by an upward-fining sandstone sequence that is capped by shale (Lasemi, 1975; Cohee, 1979). These occurrences indicate that the Canadian Shield to the northeast was still supplying terrigenous detritus to the Michigan Basin.

The lower dolomitic Bayport, which is highly variable in thickness and locally absent (Lasemi, 1975), is reminiscent of modern carbonate deposits currently accumulating on supratidal and intertidal flats, several of which are in arid to semiarid climates (Alderman and Skinner, 1957; Kinsman, 1964; Deffeyes and others, 1964, 1965; Illing and Wells, 1964; Shinn and others, 1964; Hsu and Siegenthaler, 1969; Zenger, 1972). Fine-grained dolostone containing gypsum crystals, desiccation cracks, and no fossils other than ostracodes and algal stromatolites suggests a depositional setting characterized by very shallow water of abnormal and highly variable salinity and frequent subaerial exposure. The chert, like that in so many carbonate rocks, has an affinity for organic remains (stromatolites and invertebrate shells) and for sediments bearing evidence of deposition in a setting characterized by very shallow water and intermittent subaerial exposure. Bacon (1971), Lasemi (1975), Tyler (1980), and Vugrinovich (1984) all concluded that lower Bayport rocks record deposition in a tidal-flat and sabkha environment similar to those found today in various regions such as the Persian Gulf, where analogous sediments have recently accumulated.

Above the lower dolomitic sequence, the Bayport is gray dolomitic limestone containing an abundance of varied marine invertebrate fossils including corals, brachiopods, pelecypods, bryozoa, echinoderms, and foraminifera (Bacon, 1971; Lasemi, 1975). These fossiliferous limestones contain less chert, fewer algal stromatolites, and lack the desiccation cracks, gypsum, and anhydrite found in the lower dolomitic beds (Bacon, 1971; Lasemi, 1975). This apparently reflects marine transgression and more seaward deposition in somewhat deeper water of more normal marine and less variable salinity.

Lasemi (1975), in his study of Bayport lithologies in the subsurface, reports, in the uppermost Bayport, a cherty, sparsely fossiliferous dolostone similar to that found in the lower part of the formation. This upper dolomitic sequence, which seems to record a return to near-strand conditions, has been largely or entirely removed by erosion in the outcrop belt.

The Bayport Limestone records the last of the decelerating subsidence experienced by the Michigan Basin during the Mississippian Period, so that terrigenous detritus supplied from more northern sources during the Chesterian Epoch by-passed Michigan and accumulated farther south in the Illinois Basin and other areas (Potter and others, 1958; Potter, 1963; Swann, 1964; Lineback, 1969; Sleep and Snell, 1976). Michigan Basin epeirogenic downwarp ceased only temporarily, however, and, although never again as impressive as earlier in the Paleozoic, was destined to occur again during the Pennsylvanian and Jurassic Periods.

## SUMMARY

The Mississippian System of the Michigan Basin occupies a roughly circular area approximately 400 km (250 mi) in diameter and has the largest subcrop area of any Phanerozoic system in the Michigan Basin. Although Mississippian rocks are generally thickest in the central part of the basin, where they are covered by Pennsylvanian strata, the maximum thickness reported is 719 m (2,360 ft) from a well in Arenac County on the northeast side of the basin. The Mississippian formations include, in ascending stratigraphic order: Antrim Shale; the laterally equivalent Bedford and Ellsworth Shales (all Upper Devonian to Kinderhookian); Berea Sandstone (Kinderhookian); Sunbury Shale (Kinderhookian); Coldwater Shale (Kinderhookian to Osagian); Marshall Sandstone (Osagian); Michigan Formation (Osagian to Meramecian); and Bayport Limestone (Meramecian). There are no Chesterian sediments in the Michigan Basin. The Mississippian sediments accumulated conformably on Devonian strata, and paucity of index fossils in the Upper Devonian and Lower Mississippian Antrim Shale renders uncertain the position of the Devonian–Mississippian contact within it. Mississippian strata are overlain with disconformity by Pennsylvanian and, very locally, Jurassic strata.

The Kinderhookian, Osagian, and Meramecian series record a decreasing rate of Michigan Basin subsidence through time. Subsidence ceased temporarily during the Chesterian Epoch, and some Mississippian units were eroded from local anticlines in the central basin area during this interval of nondeposition. As a result, basal Pennsylvanian strata in most places rest directly on Meramecian rocks and locally on older Mississippian formations.

The Mississippian sediments are primarily shallow-marine deposits consisting largely of shale and a smaller quantity of sandstone with subordinate amounts of siltstone, carbonates, and evaporites. Fluvial-deltaic deposits comprise a significant portion of the section in the eastern half of the basin only. Terrigenous clastics were derived mainly from a source to the northeast of the basin in the Canadian Shield and entered the basin primarily in the vicinity of the Michigan thumb. To a lesser extent, terrigenous debris came from the Wisconsin Highlands to the northwest. Sediments derived from the northwest made significant contributions only to the Ellsworth Shale and possibly also to the Marshall Sandstone.

Significant quantities of oil and gas have been produced from sandstones in the Berea, Marshall, and Michigan Formations, and from carbonates in the Ellsworth Shale. Recent discoveries in the Berea have resulted in substantial new oil production during the 1980s in Midland and Bay counties. Sandstones in the Coldwater and Marshall Formations were, at one time, extensively quarried for grindstones and construction flagstones, respectively. The Michigan Formation is the chief source of gypsum in Michigan, and the Bayport supplies some of the state's limestone.

## REFERENCES CITED

Alderman, A. R., and Skinner, H.C.W., 1957, Dolomite sedimentation in the southeast of South Australia: American Journal of Science, v. 255, p. 561–567.

Arnold, C. A., 1931, On *Callixylon newberryi* (Dawson, Elkins and Wieland): Ann Arbor, University of Michigan, Contributions from the Museum of Paleontology, v. 3, no. 12, p. 207–232.

Asseez, L. O., 1969, Paleogeography of Lower Mississippian rocks of Michigan Basin: American Association of Petroleum Geologists Bulletin, v. 53, p. 127–135.

Bacon, D. J., 1971, Chert genesis in a Mississippian sabkha environment (saline plain), Bayport Limestone, Huron County, Michigan [M.S. thesis]: East Lansing, Michigan State University, 47 p.

Bishop, M. S., 1940, Isopachous studies of Ellsworth to Traverse limestone sections of southwestern Michigan: American Association of Petroleum Geologists Bulletin, v. 24, p. 2150–2162.

Briggs, L. I., 1970, Geology of gypsum in the Lower Peninsula, Michigan, *in* 6th Proceedings, Forum on geology of Industrial Minerals: Michigan Geological Survey Miscellaneous Publication 1, p. 66–76.

Broadhead, R. F., Kepferle, R. C., and Potter, P. E., 1982, Stratigraphic and sedimentologic controls of gas in shale; Example from Upper Devonian of northern Ohio: American Association of Petroleum Geologists Bulletin, v. 66, p. 10–27.

Carlson, E. T., 1964, Gypsum in Michigan, *in* Our rock riches: Michigan Geological Survey Bulletin 1, p. 77–78.

Cercone, K. R., 1984, Thermal history of Michigan Basin: American Association of Petroleum Geologists Bulletin, v. 68, p. 130–136.

Chaloner, W. G. 1954, Mississippian megaspores from Michigan and adjacent states: Ann Arbor, University of Michigan Contributions from the Museum of Paleontology, v. 12, p. 23–35.

Chung, P. K., 1973, Mississippian Coldwater Formation of the Michigan Basin [Ph.D. thesis]: East Lansing, Michigan State University, 159 p.

Ciner, A. T., 1987, Stratigraphy and depositional environment of the Bayport Limestone of the southern Michigan Basin [M.S. thesis]: Toledo, Ohio, University of Toledo, 133 p.

Cohee, G. V., 1965, Geologic history of the Michigan Basin: Washington Academy of Science Journal, v. 55, p. 211–223.

—— , 1979, Michigan Basin region, *in* Craig, L. C., and Connor, C. W., eds., Paleotectonic investigations of the Mississippian System in the United States; Part 1, Introduction and regional analyses of the Mississippian System: U.S. Geological Survey Professional Paper 1010, p. 49–57.

Cohee, G. V., and Underwood, L. B., 1944, Maps and sections of the Berea Sandstone in eastern Michigan: U.S. Geological Survey Oil and Gas Investigation Preliminary Map 17.

Cohee, G. V., Macha, C, and Holk, M., 1951, Thickness and lithology of Upper Devonian and Carboniferous rocks in Michigan: U.S. Geological Survey Oil and Gas Investigation Chart OC-41.

Conant, L. C., and Swanson, V. E., 1961, Chattanooga Shale and related rocks of central Tennessee and nearby areas: U.S. Geological Survey Professional Paper 357, 91 p.

Cooper, G. A., 1937, Brachiopod ecology and paleoecology: National Research Council, Division of Geology and Geography, Report of the Committee on Marine Ecology as Related to Paleontology, p. 26–50.

Craig, L. C., and Varnes, K. L., 1979, History of the Mississippian System; An interpretive summary, *in* Craig, L. C., and Connor, C. W., eds., Paleotectonic investigations of the Mississippian System in the United States; Part 2, Interpretive summary and special features of the Mississippian System: U.S. Geological Survey Professional Paper 1010, p. 371–406.

Deffeyes, K. S., Lucia, F. J., and Weyl, P. K., 1964, Dolomitization; Observations on the island of Bonaire, Netherlands Antilles: Science, v. 43, p. 678–679.

—— , 1965, Dolomitization of recent and Plio-Pleistocene sediments by marine evaporite waters on Bonaire, Netherlands Antilles: Society of Economic Paleontologists and Mineralogists Special Publication 13, p. 71–88.

deWitt, W., Jr., 1970, Age of the Bedford Shale, Berea Sandstone, and Sunbury Shale in the Appalachian and Michigan Basins, Pennsylvania, Ohio, and Michigan: U.S. Geological Survey Bulletin 1294-G, 11 p.

Dorr, J. A., and Eschman, D. F., 1970, Geology of Michigan: Ann Arbor, University of Michigan Press, 476 p.

Dorr, J. A., and Kauffman, E. G., 1963, Rippled toroids from the Napoleon Sandstone member (Mississippian) of southern Michigan: Journal of Sedimentary Petrology, v. 33, p. 751–758.

Dorr, J. A., and Moser, F., 1964, Ctenacanth sharks from mid-Mississippian of Michigan: Michigan Academy of Science, Arts, and Letters Papers, v. 49, p. 105–113.

Driscoll, E. G., 1965, Dimyarian pelecypods of the Mississippian Marshall Sandstone of Michigan: Palaeotographica Americana, v. 5, no. 35, p. 67–128.

—— , 1969, Animal-sediment relationships of the Coldwater and Marshall Formations of Michigan, *in* Campbell, K.S.W., ed., Stratigraphy and paleontology essays in honor of Dorothy Hill: Canberra, Australian National University Press, p. 337–352.

Ells, G. D., 1979, Michigan, *in* The Mississippian and Pennsylvanian (Carboniferous) Systems in the United States: U.S. Geological Survey Professional Paper 1110, p. J1–J17.

Ells, G. D., and Ives, R. E., 1964, Oil and gas producing formations in Michigan, *in* Our rock riches: Michigan Geological Survey Bulletin 1, p. 83–92.

Ferguson, L., 1963, The paleoecology of *Lingula squamiformis* (Phillips) during a Scottish Mississippian marine transgression: Journal of Paleontology, v. 37, p. 669–681.

Gunn, G. R., 1986, Stratigraphy and petroleum potential of Berea Sandstone in Larkin and Williams fields, Midland and Bay Counties, Michigan [abs.]: American Association of Petroleum Geologists Bulletin, v. 70, p. 1066.

Hake, B. F., 1938, Geologic occurrence of oil and gas in Michigan: American Association of Petroleum Geologists Bulletin, v. 22, p. 393–415.

Hale, L., 1941, Study of sedimentation and stratigraphy of Lower Mississippian in western Michigan: American Association of Petroleum Geologists Bulletin, v. 25, p. 713–723.

Hard, E. W., 1938, Mississippian gas sands of central Michigan area: American Association of Petroleum Geologists Bulletin, v. 22, p. 129–174.

Heinrich, E. W., 1979, Economic geology of the sand and sandstone resources of Michigan: Michigan Geological Survey Report of Investigations 21, 31 p.

Houghton, D., 1838, Report of the State Geologist: Michigan Legislature, House Document no. 14, p. 1–39.

Hsu, K. J., and Siegenthaler, C., 1969, Preliminary experiments on hydrodynamic movement induced by evaporation and their bearing on the dolomite problem: Sedimentology, v. 12, p. 11–25.

Illing, L. V., and Wells, A. J., 1964, Penecontemporary dolomite in the Persian Gulf [abs.]: American Association of Petroleum Geologists Bulletin, v. 48, p. 532–533.

Kinsman, D.J.J., 1964, Dolomitization and evaporite development, including anhydrite, in lagoonal sediments, Persian Gulf: Geological Society of America Abstracts with Programs, p. 108–109.

Kirkby, E. A., 1964, Michigan sandstones, *in* Our rock riches: Michigan Geological Survey Bulletin 1, p. 79–81.

Kirkham, V.R.D., 1942, Unconformity at the top of the Marshall Formation in Michigan [abs.]: Geologica Society of American Bulletin, v. 43, p. 137–138.

Lane, A. C., 1895, The geology of Lower Michigan with reference to deep borings: Michigan Geological Survey, v. 5, pt 2, 100 p.

——, 1901, Suggested changes in nomenclature of Michigan formations: Michigan Miner, v. 3, no. 10, p. 9.

Lane, A. C. and Cooper, W. F., 1900, Geological report on Huron County: Michigan Geological Survey, v. 7, p. 1–329.

Lasemi, Y., 1975, Subsurface geology and stratigraphic analysis of the Bayport Formation in the Michigan Basin [M.S. thesis]: East Lansing, Michigan State University, 80 p.

Lewis, T. L., 1988, Late Devonian and Early Mississippian distal basin-margin sedimentation of northern Ohio: Ohio Journal of Science, v. 88, p. 23–39.

Lilienthal, R. T., 1974, Subsurface geology of Barry County, Michigan: Michigan Geological Survey Report of Investigations 15, 36 p.

——, 1978, Stratigraphic cross-sections of the Michigan Basin: Michigan Geological Survey Report of Investigations 19, 27 p.

Lineback, J. A., 1968, Subdivisions and depositional environments of the New Albany Shale (Devonian–Mississippian) in Indiana: American Association of Petroleum Geologists Bulletin, v. 52, p. 1291–1303.

——, 1969, Illinois Basin; Sediment-starved during Mississippian: American Association of Petroleum Geologists Bulletin, v. 53, p. 112–126.

——, 1970, Stratigraphy of the New Albany Shale in Indiana: Indiana Geological Survey Bulletin 44, 73 p.

Martin, H. M., and Straight, M. T., 1956, An index of the geology of Michigan 1823–1955: Michigan Geological Survey Publication 50, 461 p.

McGregor, D. J., 1954, Stratigraphic analysis of upper Devonian and Mississippian rocks in Michigan Basin: American Association of Petroleum Geologists Bulletin, v. 38, p. 2324–2356.

Michigan Basin Geological Society, 1969, Stratigraphic cross-sections; Michigan Basin: Michigan Basin Geological Society, 22 p.

Michigan Geological Survey, 1964–84, Michigan's oil and gas fields; Annual statistical summaries: Michigan Geological Survey, nos. 2–38.

Miller, A. K., and Garner, H. F., 1953a, Lower Mississippian cephalopods of Michigan; Part 1, Orthoconic nautiloids: Ann Arbor, University of Michigan Contributions from the Museum of Paleontology, v. 10, no. 7, p. 159–192.

——, 1953b, Lower Mississippian cephalopods of Michigan; Part 2, Coiled nautiloids: Ann Arbor, University of Michigan Contributions from the Museum of Paleontology, v. 11, no. 6, p. 111–151.

——, 1955, Lower Mississippian cephalopods of Michigan; Part 3, Ammonoids and summary: Ann Arbor, University of Michigan Contributions from the Museum of Paleontology, v. 12, no. 8, p. 113–173.

Monnett, V. B., 1948, Mississippian Marshall Formation of Michigan: Association of Petroleum Geologists Bulletin, v. 32, p. 629–688.

Moser, F., 1963, The Michigan Formations; A study in the use of a computer oriented system in stratigraphic analysis [Ph.D. thesis]: Ann Arbor, University of Michigan, 96 p.

Newcombe, R. B., 1933, Oil and gas fields of Michigan: Michigan Geological Survey Publication 38, 293 p.

Oden, A. L., 1952, The occurrence of Mississippian brachiopods in Michigan [M.S. thesis]: East Lansing, Michigan State University, 52 p.

Olszewski, G. P., 1978, An interpretation of the sedimentary environment of the Michigan Formation (Mississippian) in Michigan [M.S. thesis]: Detroit, Michigan, Wayne State University, 104 p.

Parham, W. E., 1966, Lateral variations of clay mineral assemblages in modern and ancient sediments, *in* Gekker, K., and Weiss, A., eds., Proceedings of the International Clay Conference: London, Pergammon Press, v. 1, p. 135–145.

Pawlowicz, R. M., 1969, Stratigraphy of the Marshall Formation (Mississippian, Osagian) in southern Michigan [M.S. thesis]: Toledo, Ohio, University of Toledo, 43 p.

Pepper, J. F., deWitt, W., Jr., and Demarest, D. F., 1954, Geology of the Bedford Shale and Berea Sandstone in the Appalachian Basin: U.S. Geological Survey Professional Paper 259, 111 p.

Potter, P. E., 1963, Late Paleozoic sandstones of the Illinois Basin: Illinois Geological Survey Report of Investigations 217, 92 p.

Potter, P. E., and Pryor, W. A., 1961, Dispersal centers of Paleozoic and later clastics of the upper Mississippi Valley and adjacent areas: Geological Society of American Bulletin, v. 72, p. 1195–1250.

Potter, P. E., Nosow, E., Smith, N. M., and Swann, D. H., 1958, Chester crossbedding and sandstone trends in Illinois Basin: American Association of Petroleum Geologists Bulletin, v. 42, p. 1013–1046.

Potter, P. E., DeReamer, J. H., Jackson, D. S., and Maynard, J. B., 1984, Lithologic and environmental atlas of Berea Sandstone (Mississippian) in the Appalachian Basin: Appalachian Geological Society Special Publication 1, 157 p.

Provo, L. J., Kepferle, R. C., and Potter, P. E., 1978, Division of black Ohio Shale in eastern Kentucky: American Association of Petroleum Geologists Bulletin, v. 62, p. 1703–1713.

Pryor, W. A., and Sable, E. G., 1974, Carboniferous of the Eastern Interior Basin, *in* Briggs, G., ed., Carboniferous of the southeastern United States: Geological Society of America Special Paper 148, p. 281–355.

Rich, J. L., 1951, Probable fondo origin of Marcellus–Ohio–New Albany–Chattanooga bituminous shales: American Association of Petroleum Geologists Bulletin, v. 35, p. 2017–2040.

Rorick, A. H., 1983, Sediment dispersal patterns and provenance of the Marshall Formation (Mississippian) in the Michigan Basin; A petrographic analysis [M.S. thesis]: Toledo, Ohio, University of Toledo, 130 p.

Sanford, B. V., 1967, Devonian of Ontario and Michigan, *in* Proceedings, International Symposium on the Devonian System, Calgary, Alberta: Alberta Society of Petroleum Geologists, v. 1, p. 973–999.

Sawtelle, E. R., Jr., 1958, The origin of the Berea Sandstone in the Michigan Basin [M.S. thesis]: East Lansing, Michigan State University, 67 p.

See, B. E., 1980, Palynology of the Mississippian Marshall Sandstone and Coldwater Shale of Michigan [M.S. thesis]: Bowling Green, Ohio, Bowling Green State University, 66 p.

Shaffer, B. L., 1968, Palynology and geology of newly discovered Jurassic sediments in the Michigan Basin [abs.]: Geological Society of America, Special Paper 115, p. 376–377.

Shaver, R. H., ed., 1985, Midwestern basin and arches region: COSUNA Correlation Charts, American Association of Petroleum Geologists.

Shinn, E. A., Ginsburg, R. N., and Lloyd, R. M., 1964, Recent supratidal dolomitization in Florida and the Bahamas: Geological Society of America Abstracts with Programs, p. 183–184.

Sleep, N. H., and Snell, N. S., 1976, Thermal contraction and flexure of midcontinent and Atlantic marginal basins: Geophysical Journal of the Royal Astronomical Society, v. 45, p. 125–154.

Squire, G. R., 1972, A field guide to the geology of southwestern Michigan: Kalamazoo, Western Michigan University, Department of Geology Publication ES-1, 58 p.

Stearns, M. D., 1933, The petrology of the Marshall Formation of Michigan: Journal of Sedimentary Petrology, v. 3, p. 99–112.

Stearns, M. D., and Cook, C. W., 1931, A petrographic study of the Marshall Formation and its relation to the sands of the Michigan Series: Michigan Academy of Science, Arts, and Letters Papers, v. 16, p. 429–437.

Strutz, T. A., 1978, A pre-Pennsylvanian paleogeologic study of Michigan [M.S. thesis]: East Lansing, Michigan State University, 68 p.

Swann, D. H., 1963, Classification of Genevievian and Chesterian (late Mississippian) rocks of Illinois: Illinois Geological Survey Report of Investigations 216, 91 p.

——, 1964, Late Mississippian rhythmic sediments of Mississippi Valley: American Association of Petroleum Geologists Bulletin, v. 48, p. 637–658.

Tarbell, E., 1941, Antrim–Ellsworth–Coldwater Shale Formations in Michigan: American Association of Petroleum Geologists Bulletin, v. 25, p. 724–733.

Taylor, W. H., 1839, Michigan Geological Survey Report of the State Geologist: *in* Re-improvement of state salt springs; Michigan Legislature House of Representatives, Document 2, Act 1, p. 3.

Thomas, W. A., 1931, A study of the Marshall Formation in Michigan: Michigan Academy of Science, Arts, and Letters Papers, v. 14, p. 487–498.

Twenhofel, W. H., 1939, Environments of origin of black shales: American Association of Petroleum Geologists Bulletin, v. 23, p. 1178–1198.

Tyler, J. G., 1980, Subsurface geology and depositional systems of the Upper Mississippian–Lower Pennsylvanian Bayport and Saginaw Formations, central Michigan Basin [M.S. thesis]: Detroit, Michigan, Wayne State University, 71 p.

U.S. Geological Survey (USGS), 1979, Paleotectonic investigations of the Mississippian System in the United States; Part 3, Plates: Professional Paper 1010.

Vugrinovich R., 1984, Lithostratigraphy and depositional environments of the Pennsylvanian rocks and Bayport Formation of the Michigan Basin: Michigan Geological Survey Report of Investigations 17, 33 p.

Western Michigan University, 1981, Hydrogeologic atlas of Michigan: Kalamazoo, Western Michigan University Department of Geology, 35 plates.

Wilmarth, M. G., 1938, Lexicon of geologic names of the United States: U.S. Geological Survey Bulletin 896, 2396 p.

Winchell, A., 1861, First biennial report of the progress of the Geological Survey of Michigan embracing observations on the geology, zoology, and botany of the Lower Peninsula: Michigan Geological Survey, 339 p.

Wooten, M. J., 1951, The Coldwater Formation in the area of the type locality [M.S. thesis]: Detroit, Michigan, Wayne State University, 52 p.

Zenger, D. H., 1972, Dolomitization and uniformitarianism: Journal of Geological Education, v. 20, p. 107–124.

MANUSCRIPT ACCEPTED BY THE SOCIETY JUNE 1, 1990

# Geological and geophysical evaluation of the region around Saginaw Bay, Michigan (central Michigan Basin) with image processing techniques

**John D. Herman and Robert K. Vincent**
*Geospectra Corporation, 333 Parkland Plaza, P.O. Box 1387, Ann Arbor, Michigan 48106*
**Ben Drake**
*Amoco Production Company, P.O. Box 3092, Houston, Texas 77523*

## ABSTRACT

A northeast-trending graben was hypothesized to extend southwest of Saginaw Bay to the Mid-Michigan Gravity High, based on interpretation of Landsat 1 imagery, stream drainage maps, and sparse well-log data. The edges of the graben were thought to extend along and southwest of the Pinconning oil field on the northwest side, and the Quanicassee River on the southeast side. Subsequent analysis of digital terrain, magnetic, gravity, seismic, and well-log data showed that no unequivocal evidence for a discrete, simple graben within the originally defined limits could be found. However, the new data indicated that the proposed edges of the "graben" correspond to structural lineaments (monoclines and anticlines) expressed within the Paleozoic section and on the bedrock surface. These structural features correlate with basement contacts and/or fault zones inferred from interpretation of magnetic and gravity images. These possibly basement-controlled structural lineaments influenced depositional patterns intermittently during the Paleozoic, as evidenced by the presence of northeast-trending highs within the limits of the "graben" on isopach maps of Pennsylvanian, Mississippian, and Devonian stratigraphic units. Rapid thinning and facies changes in Middle and Lower Ordovician units across the southeastern edge of the "graben," coupled with its correlation with northeast-trending positive gravity and magnetic anomalies, suggest that this is a significant structural feature, possibly controlled or influenced by the Grenville Front.

## INTRODUCTION

The Michigan Basin has yielded its geological secrets grudgingly, primarily because it is blanketed by Pleistocene glacial deposits. Wisconsin-age glacial deposits have obscured most bedrock structures to both eye and seismograph until the 1970s, when Landsat satellite images made large synoptic views of the Earth's surface available for the first time on a systematic basis, and seismic processing techniques started to include more effective near-surface static corrections. The effects of synoptic views on the photointerpretation of satellite images, and static corrections required for interpretation of seismic data for effective exploration of covered basins (e.g., Michigan Basin) are similar, in that both techniques tend to separate out the characteristics of surface features caused by glacial or eolian deposition from surface features controlled by underlying bedrock topography.

Magnetic and gravity data have been an important source of information about subsurface structures in the Michigan Basin for decades because such data are relatively unaffected by glacial features. Also, advances in image processing for remote sensing in the past decade have made possible the imaging of magnetic and gravity data. This has greatly improved the amount of information that can be gained from such geophysical data. The main reason for this improvement comes from the fact that the human

Herman, J. D., Vincent, R. K., and Drake, B., 1991, Geological and geophysical evaluation of the region around Saginaw Bay, Michigan (central Michigan Basin) with image processing techniques, *in* Catacosinos, P. A., and Daniels, P. A., Jr., eds., Early sedimentary evolution of the Michigan Basin: Geological Society of America Special Paper 256.

brain is far more adept at extracting information from images than from contour maps, the traditional form in which most geophysical data has been viewed. Another improvement comes from the relative ease with which satellite image information can be mixed with gravity and magnetic images. Integration of data from such disparate sources yields far more information than any one data source can alone provide.

Geologists and geophysicists from GeoSpectra Corporation have been studying the Michigan Basin with Landsat multispectral scanner images since 1975, and with elevation, magnetic, and gravity images since 1978. This chapter describes what we learned from these image processing and data integration methods concerning the first large structural feature that our geologists ever studied for petroleum exploration: the Saginaw Bay "Graben," the on-shore part of which is located primarily in Bay, Midland, Saginaw, and Gratiot Counties, Michigan. The study area described here covers a 10,400-km$^2$ region immediately surrounding Saginaw Bay, centered just east of the structurally deepest part of the Michigan Basin (Figs. 1 and 2).

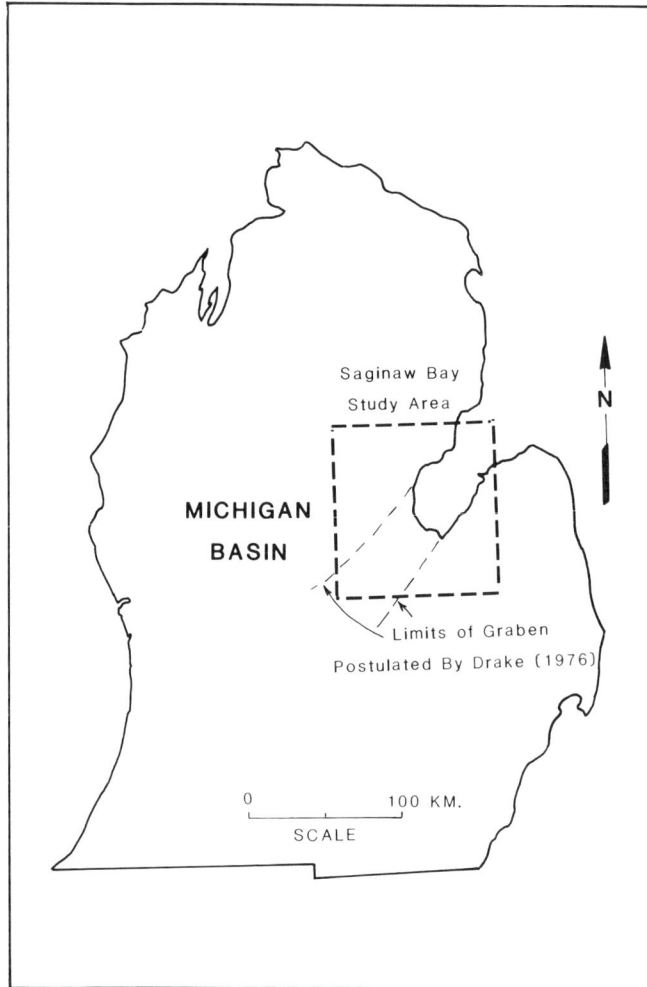

Figure 1. Locations of Saginaw Bay "Graben" and study area discussed in detail in this chapter, relative to the Michigan Basin.

## ORIGIN OF THE SAGINAW BAY GRABEN HYPOTHESIS

Using data from Landsat 1, Drake and Vincent (1975) published an article concerning the geological interpretation of Landsat 1 imagery for the greater part of the Michigan Basin. The writers found that there were numerous linear and elliptical features in the photomosaiced color composite image of the central Michigan Basin that could not be attributed to glacial origins because they extended across several glacial features, and their trends coincided with geophysical anomaly trends and known structural trends of pre-Pleistocene age

As a direct consequence of those findings, a color composite image of Landsat MSS (Multispectral Scanner) bands 4, 5, and 7, and a ratio image of bands 5 and 4 were produced for the Saginaw Bay area for the purpose of petroleum exploration research with Landsat data. Lineaments suspected of overlying deep-seated faults were mapped from these images (Drake and Vincent, 1975; Drake, 1976). Drake concluded that the most prominent of those suspected faults, a southwest-trending linear feature along the southeastern edge of Saginaw Bay, extended on shore along a probable former course of the Quanicassee River. It was further concluded that this lineament represented the southeastern edge of a down-dropped block, which Drake called the Saginaw Bay Graben.

The outline of the suspected graben, as interpreted from a Landsat image of band 7, is shown on Figures 1, 2, and 3a. Drake (1976) also concluded that the Saginaw Bay Graben extended from the Mid-Michigan Gravity Anomaly northeastward, about 110 km to the shallower inner bay of Saginaw Bay. This graben was thought to lie within the Precambrian Penokean and Keweenawan Provinces of Hinze and others (1975). Drake (1976) hypothesized that such a graben may represent the failed third arm of a triple junction that formed and collapsed in late Precambrian time, perhaps as the principal event forming the Michigan Basin. Drake theorized intermittent subsidence along the graben faults during the Phanerozoic, making the graben a local depocenter several times during Paleozoic time. Drake described the graben as being "essentially constant in width. The graben trends approximately north 50 degrees east on land, but changes trend to approximately north 30 degrees east under the inner bay. This northeast trend is parallel to one of the two dominant trends of structures in the Paleozoic rocks in the Michigan Basin."

Evidence for the existence of the Saginaw Bay Graben is as follows:

1. The previously mentioned southeast bounding fault seen in Landsat data along the old trace of the Quanicassee River.

2. The southeast bounding fault is on trend with the 1.9-m depth contour on the southeast side of the inner bay of Saginaw Bay. This trend coincides with that of the southeast shore of Saginaw Bay, suggesting that the southeast shore had been eroded back from the southeast bounding fault.

3. The Pinconning oil field in northeastern Bay County trends parallel to both the direction of the Quanicassee River

lineament and the 1.9-m depth contour on the southeast side of Saginaw Bay. Drake theorized that the Pinconning field marked the northwestern edge of the "graben."

4. Landsat image interpretations and topographic maps show that rivers flow northeast along the traces of the bounding faults of the graben, and that the drainage pattern and density of streams between the bounding faults differ from adjacent areas outside the bounding faults.

5. Elevations of tops of various formations from three nearby bore holes close to the suspected southeast-bounding fault showed that formations of Silurian and older ages occur at lower elevations within the interpreted graben. Further, the differences in depth to formation tops inside (Bay County) and outside (Tuscola County) the hypothetical graben increase stratigraphically downward, indicating that it was a local depocenter for the Salina, Niagaran, and Trenton–Black River Groups.

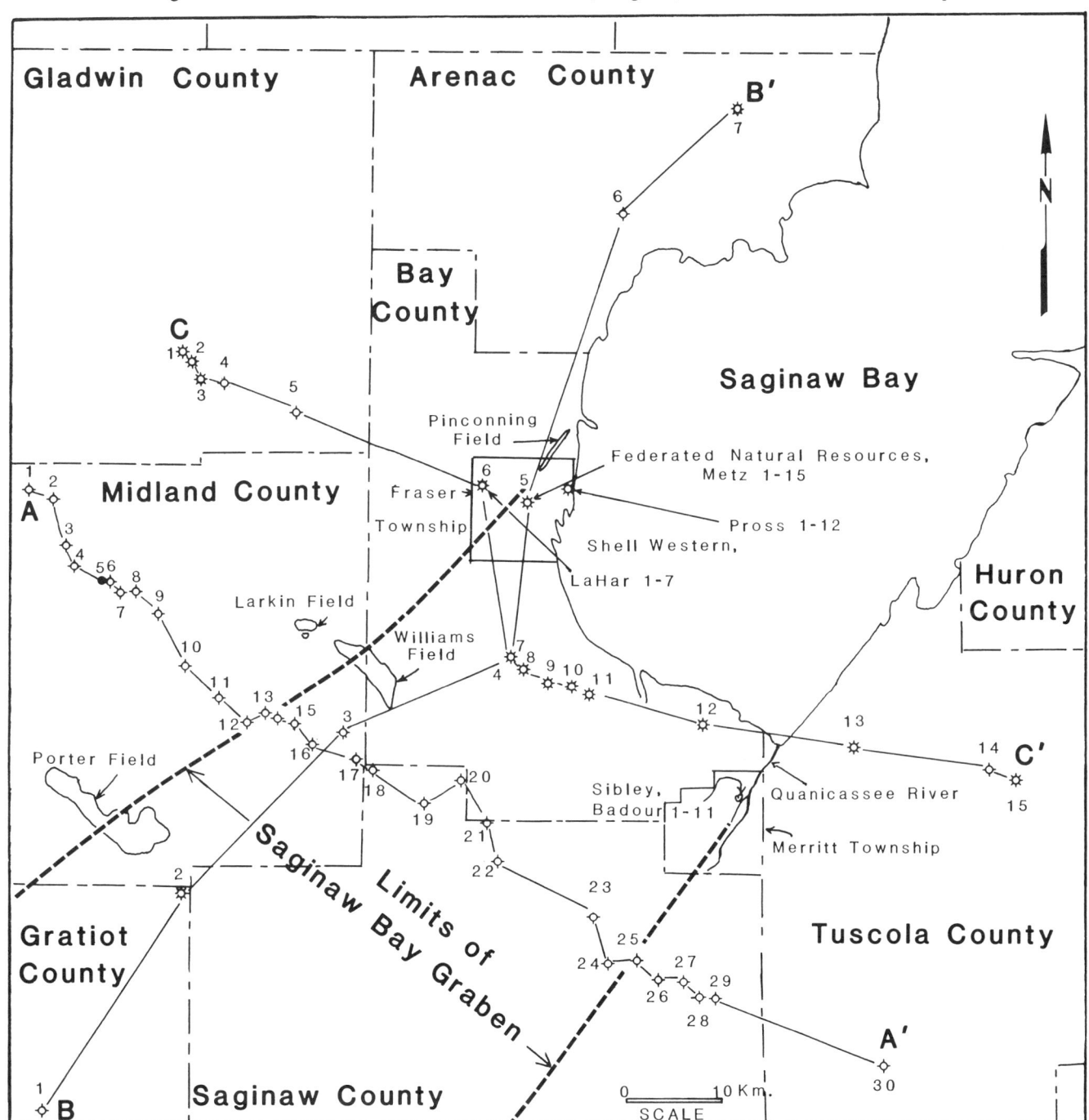

Figure 2. Enlarged view of Saginaw Bay study area, showing boundaries of the Saginaw Bay "Graben," cross-section locations, and other features discussed in the text. This area is outlined on succeeding Figures and labeled "SBSA."

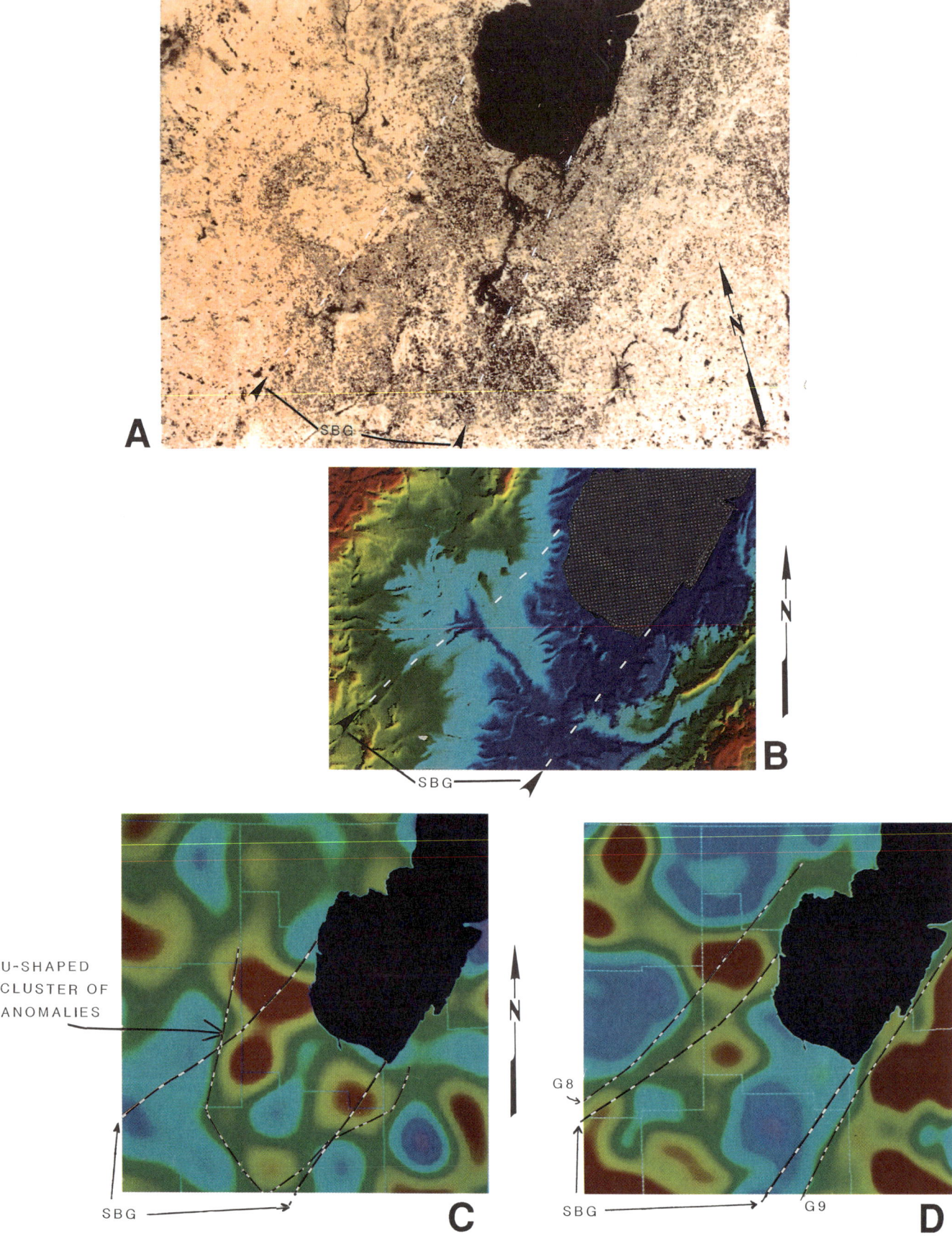

Figure 3. A. LANDSAT 1 image showing brightness values of band 7 for Saginaw Bay and vicinity. Drake (1976) thought that the northeast-trending dark zone extending southwestward from Saginaw Bay (oval black area at the top of the image) was associated with a graben. SBG = Saginaw Bay "Graben." B. 135° TOPOIMAGE of digital terrain (elevation) data. Red color represents highest elevations; dark blue represents lowest elevations. Intermediate elevations from highest to lowest are depicted by yellow, green, and light blue. Dashed white lines are boundaries of the Saginaw Bay "Graben" (SBG). Dark oval area in the upper right corner of the image is Saginaw Bay. C. Residual magnetic anomaly image of the Saginaw Bay study area. Red color represents positive anomalies, and purple and blue colors represent negative anomalies. The edges of the SBG are shown along with the outer boundaries of prominent positive anomalies possibly associated with Precambrian intrusives of mafic to intermediate composition. D. Residual gravity anomaly image of Saginaw Bay study area. Red color represents positive anomalies, and purple and blue colors represent negative anomalies. The edges of the SBG are nearly parallel to but not coincident with northeast-trending lineaments (G8 and G9), possibly associated with major boundaries within the Precambrian basement across which there are large density contrasts.

Examination of well logs from 23 Precambrian wells in the Southern Peninsula, plus any near the suspected graben that extended into the Trenton–Black River Group, led to the conclusion that some evidence existed to support an interpretation for the Niagaran Group and Trenton–Black River Group dipping toward the graben from the southeast, southwest, and possibly north. There was evidence, however, for an increase in thickness between the Niagaran and Trenton–Black River Group tops toward the northeast along a direction from southwest of the suspected graben to the inner bay, indicating a possible northeast plunge for the graben.

## SEISMIC AND DRILLING INVESTIGATIONS ALONG THE GRABEN EDGES

As a result of Drake's work, oil and gas leases were acquired by a limited partnership along the southeast and northwest edges of the "graben" in Merritt Township (T13N, R6E) (Quanicassee River prospect) and in Fraser Township (T16N, R4E) (Pinconning South prospect) both in Bay County, Michigan (Figs. 2, 4, and 5). The oil and gas exploration targets within these lease blocks were thought to be porous and permeable dolomitized zones in the Devonian Dundee Limestone and the Middle Ordovician Trenton and Black River Groups, similar to zones found in the Pinconning, North Adams, Deep River, and Albion-Scipio oil fields (Landes, 1970). It was thought that these dolomitized zones would be localized along the bounding faults of the "graben."

Additional structural traps in Paleozoic rocks related to intermittent movement along the bounding faults were also thought to exist within these lease blocks. These structural traps would be caused by truncation and sealing of reservoir beds against the faults or by drag folding of reservoir rocks along the fault zones thought to exist within the leased areas. The subsequent seismic exploration and drilling done to evaluate the commercial oil and gas possibilities of these areas was very useful in verifying the existence of the Saginaw Bay "Graben."

### Exploration along the northwest edge of the "graben," Pinconning South prospect

A 7.2-km-long, north-south–trending seismic line extending through the center of the lease block in Fraser Township was acquired to verify the existence of faulting southwest of the Pinconning field, along the postulated northwest edge of the "graben" (Fig. 4). This seismic data revealed an anticlinal structure at the Devonian Dundee and the Middle Ordovician Trenton levels. The crest of the structure at both levels appeared to be in the southwest ¼ of the northwest ¼ of section 15 in Fraser Township (T16N, R4E). The southern flank of the anticlinal structure dipped more steeply than the northern flank. In addition, the seismic data indicated that all the rock units appeared to be structurally lower on the south end of the line, i.e., within the limits of the postulated "graben," than they were at the north end outside the "graben" limits. This evidence then appeared to support the hypothesis regarding the northwest edge of the Saginaw Bay Graben.

Although it was recommended that a deep well be drilled in section 16 or 15 to test the hydrocarbon possibilities of the Ordovician Trenton–Black River Group on the anticlinal structure, the company that acquired the Pinconning South prospect lease block was only interested in testing the Devonian section. In 1979 and 1980, two wells were drilled in SE, NE, SE, section 16 of Fraser Township to test oil and gas possibilities in the Devonian Dundee Limestone (Dart, Christian 1-16, and Dart, Christian 1-16A) on the crest of the anticlinal structure identified by the seismic data. The second well, the Dart, Christian 1-16A was directionally drilled from section 16 to NW, SW, SW, section 15. A good show of gas was found in the Devonian Traverse Formation, in the Christian 1-16, but a completion attempt was not made. A completion attempt on the Christian 1-16A well was made in the Traverse Formation, but no commercial production was established, and it was subsequently abandoned with no further development work.

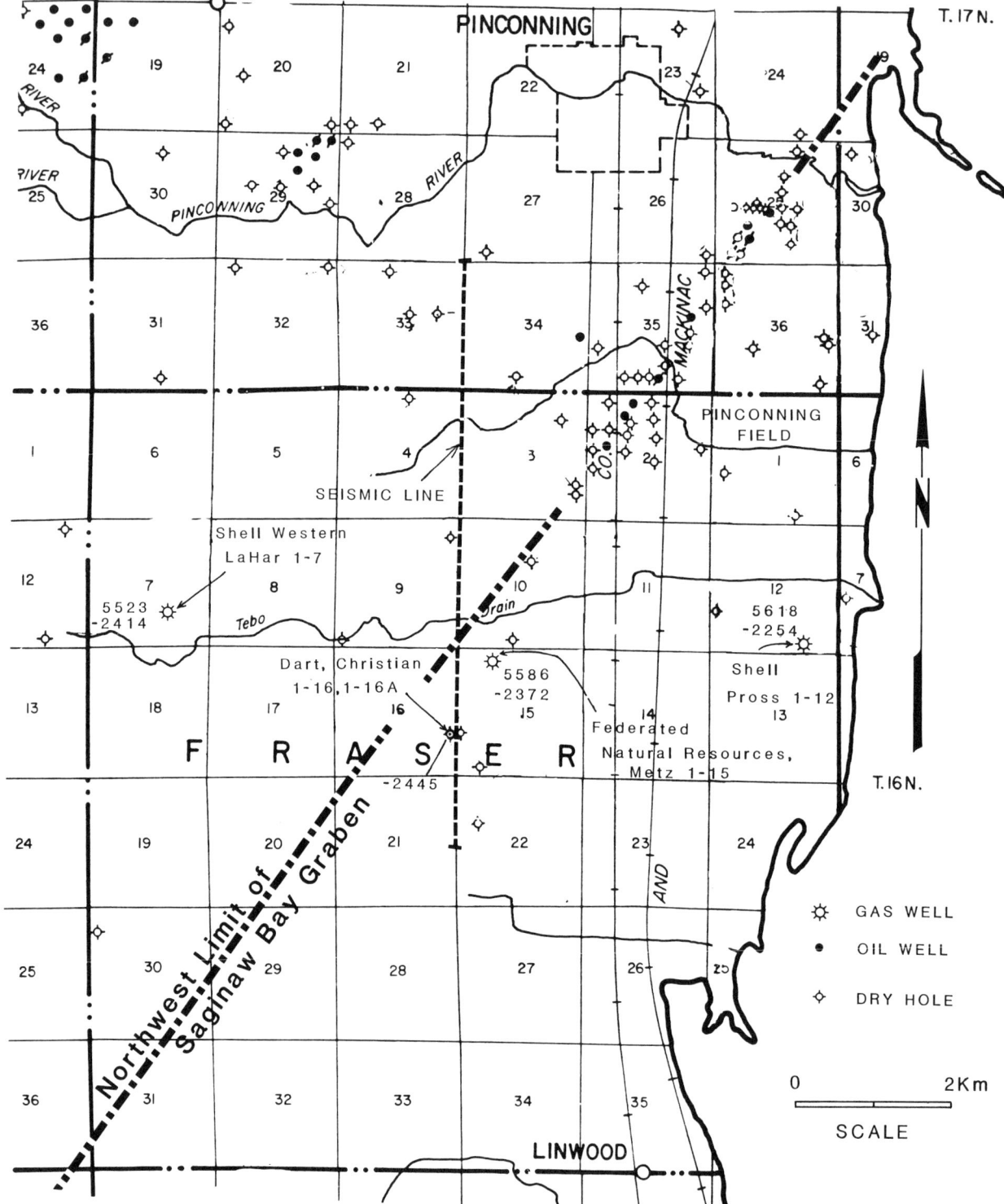

Figure 4. Detailed map of Fraser Township of Bay County, where exploration was initiated in the late 1970s to search for faults or folds that might trap oil and gas along the northwest edge of the Saginaw Bay "Graben." Two numbers are given for each labeled oil and gas well on the figure. The upper positive number is the combined thickness in feet of the entire Silurian section, the Upper Ordovician section, and the Ordovician Trenton and Black River Groups. The lower, negative numbers are the elevations in feet relative to sea level of the Dundee Limestone.

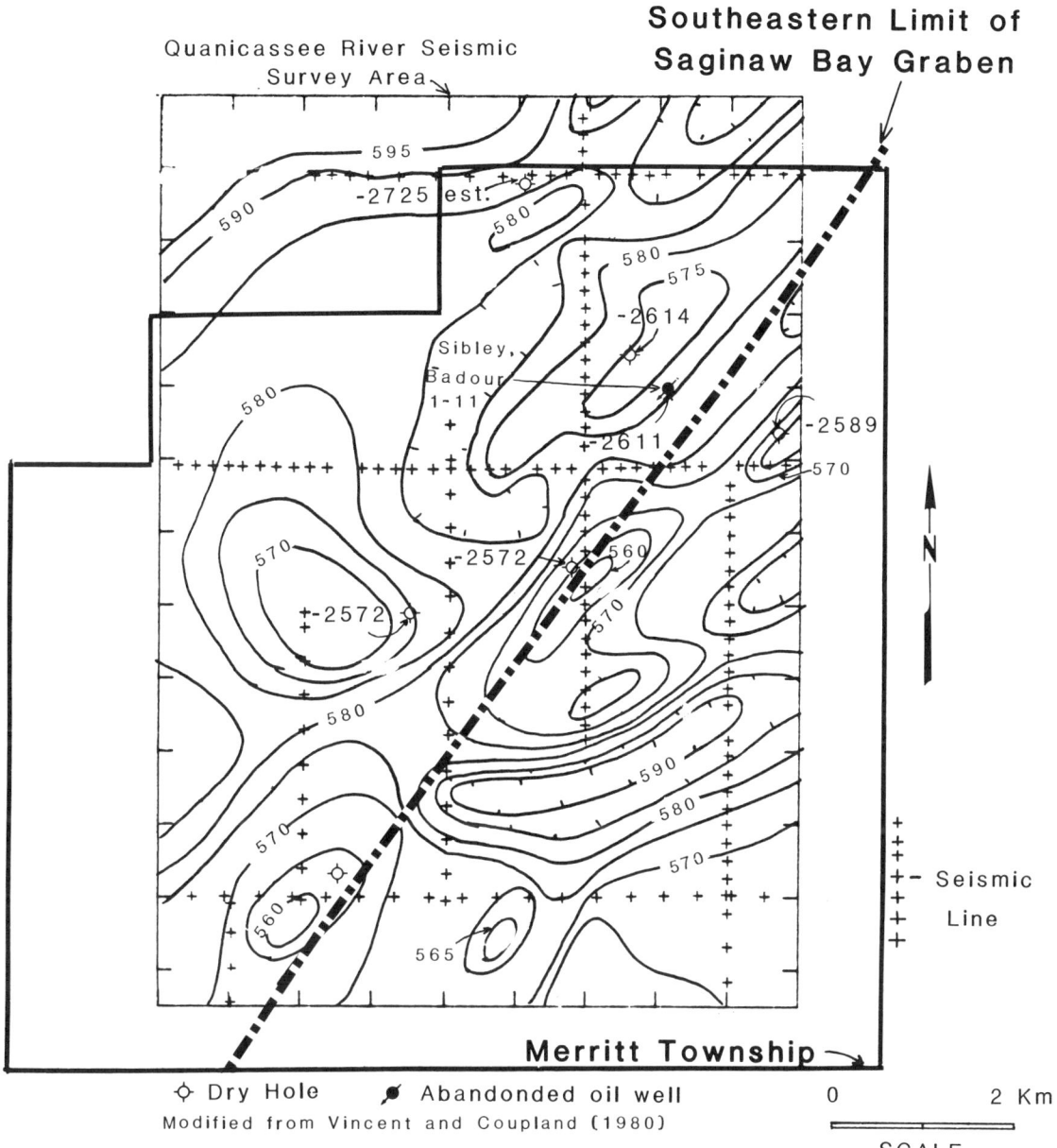

Figure 5. Map of Merritt Township in Bay County, where a detailed seismic survey was conducted to find possible faulting or folding along the southeast edge of the Saginaw Bay "Graben." The elevations in feet relative to sea level of the Dundee Limestone at several oil and gas wells are shown as the negative numbers. Contours with positive numbers from 560 to 595 depict the two-way travel time in milliseconds of a seismic reflector correlative with the Devonian Dundee Limestone. Contour interval is 5 milliseconds.

In 1987, Federated Natural Resources drilled and completed an Ordovician Prairie du Chien "deep gas" discovery well in NE, NW, NW, section 15 of Fraser Township, about 800 m northeast of the previous directional Devonian test by Dart. Shell Western Exploration and Production Incorporated subsequently drilled and completed two more "Prairie du Chien" gas wells in Fraser Township in section 7 (Shell, LaHar 1-7, NE, SW, SE, section 7) and in section 12 (Shell, Pross 1-12, NW, SE, SE, section 12) (Fig. 4) The elevations of the Dundee Limestone relative to sea level and the thickness of the Silurian, Upper Ordovician, and Middle Ordovician Trenton–Black River sections, incorporating data from these "Prairie du Chien" gas discovery wells, are shown on Fig. 4.

The Silurian and Middle Ordovician sections are thicker in the Metz 1-15, and Pross 1-12 wells than they are in the LaHar 1-7 well, thus supporting Drake's contention that the "graben" acted as a depocenter in Silurian and Ordovician time. Dundee Limestone elevation data, however, indicate structural dip to the west, showing that if the "graben" did exist, it was not an active feature during the deposition of the upper Paleozoic section.

## Exploration along the southeast edge of the "graben," Quanicassee River prospect

Approximately 65 line-km of seismic data were collected in southeast Bay County and northwest Tuscola County to locate subsurface faulting below the Quanicassee River, the hypothesized southeast edge of the Saginaw Bay Graben. Vincent and Coupland (1980) compared structural interpretations of this seismic data with the lineaments mapped on Landsat 1 images (Fig. 5). Figure 5 shows a hand-contoured plot of seismic return times of the top of the Dundee Formation (Devonian age) in the seismic study area, including the location of the Landsat lineament mapped by Vincent and Drake (1976) that follows the Quanicassee River. Figure 5 shows that the Quanicassee River Landsat lineament is nearly coincident with a northeast-trending structural low of the Dundee reflector. The seismic data also shows that the Dundee is structurally higher on the southeast side of the Quanicassee River than it is on the northwest side in support of a fault interpretation.

Figure 6 shows three-dimensional computer-generated plots of subsurface structural interpretations of the Devonian Dundee Formation and a Cambrian (or possibly Lower Ordovician) sedimentary unit, as calculated from seismic data for the study area. This figure shows evidence that the deeper, older formations dip toward north-northwest, in support of the hypothesis that the principal Landsat lineament may be the surface expression of a possibly fault-controlled fold along the southeast edge of the Saginaw Bay "Graben." Although the folding at the Dundee and lower Paleozoic levels could be associated with a deep basement fault postulated by Drake (1976), the seismic data showing this folding did not reveal such structural disturbances at the Precambrian basement level. This apparent lack of basement faulting could be due to original processing parameters being set for shal-

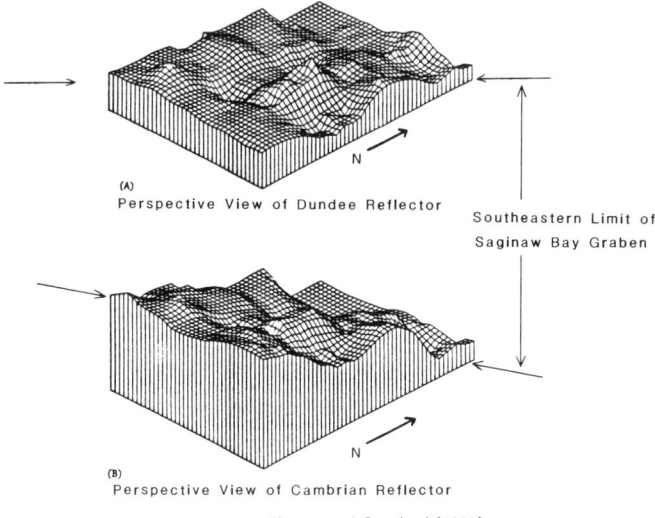

Figure 6. Perspective views of seismic reflectors correlating with the Devonian Dundee Limestone and a Cambrian unit, derived from data covering the Quanicassee River seismic study area outlined on Figure 4.

lower Paleozoic objectives within the Devonian section, limitations of the original data-input collection procedures, or the real absence of basement faulting.

Several wells were drilled to explore for oil and gas on structures identified by the seismic survey, none of which penetrated below the Devonian Richfield Formation (Fig. 5). One of these was an oil well that produced from the Devonian Detroit River Sour Zone (the Sibley, Badour 1-11 well, in T13N, R6E, section 11, recently abandoned) on the northwest side of the Quanicassee River. Elevations of the top of the Dundee Limestone derived from these oil exploration wells indicate that it is structurally lower on the northwest side of the Quanicassee River than it is on the southeast side, in concurrence with the seismic data and Drake's hypothesis (Fig. 5). It is also possible that the dip of the Dundee Limestone could be due to regional dip into the middle of the Michigan Basin.

## DATA SOURCES AND PROCESSING USED FOR THIS STUDY

### Well data

Since the initial publication of the Saginaw Bay "Graben" hypothesis, much new deep-well data has become available as a result of the so-called "deep gas play" involving exploration for hydrocarbon-bearing sandstones within the Ordovician "Prairie du Chien" Group. Drake (1976) hypothesized that the "graben" began to develop in late Precambrian time and continued to act as a depocenter intermittently during Paleozoic time. Study of data from several deep wells penetrating the Middle and Lower Ordovician section in the Saginaw Bay area allows us to test this hypothesis. Three cross sections, A-A', B-B', and C-C', were constructed from well data available from the Michigan Department of Natural Resources (Figs. 7, 8, and 9). Locations of these cross sections are shown on Fig. 2. A-A' is a northwest-southeast–trending structure section depicting the elevation of the top of the Devonian Dundee Limestone relative to sea level. Sections B-B' and C-C' are stratigraphic sections showing the thickness of the Silurian, Upper Ordovician, and Middle Ordovician stratigraphic units from the top of the Trenton Group to the top of the Glenwood Formation.

Besides the information from cross sections constructed for this study, regional structure and isopach maps of Pennsylvanian, Mississippian, Devonian, and Ordovician rock units were studied to search for evidence for the "graben" (Figs. 10a–c and 11a–c).

*Image-processing techniques applied to elevation and geophysical data*

TOPOIMAGE, a program developed by Geospectra in 1978, was used to produce shaded relief images of digital terrain, magnetic, and gravity data (Vincent and Etzler, 1982). On a shaded relief image, or TOPOIMAGE, of digital terrain or geophysical data, the highs and lows of geophysical data or the hills and valleys of digital terrain (elevation) data are artifically illuminated by a user-selected "sun" or light-source position. This enhances linear features that trend almost perpendicular to the user-selected "sun" position or the direction of illumination. In order to enhance almost all linear features in a given data set, it is necessary to make at least two TOPOIMAGES, with two "sun" azimuths that are 90° apart.

Further enhancement and sharpening of subtle gravity and magnetic anomalies can be accomplished by artificial directional illumination of shaded relief images (TOPOIMAGES) of the gravity and magnetic data. If a shaded relief image of a gravity or magnetic grid is artificially illuminated, the result is a directional second derivative image in which the curvature or the rate of change of slope of the geophysical data is depicted parallel to the direction of artificial illumination.

For example, to produce a 135° directional second-derivative gravity image, the TOPOIMAGE program would be applied to a 135° shaded relief gravity image. That is, gravity data artificially illuminated by a light source shining from the southeast toward the northwest would be treated like the original gridded gravity values, and the artificially illuminated gravity data would be artificially illuminated by a light source shining in the same direction. On a black and white second-derivative gravity image, areas of high curvature or increasing slope would appear to be bright, while zones of low curvature or decreasing slope would be dark. The zones of high curvature (bright areas) are also residual lows of the geophysical data, while the zones of low curvature (dark areas) are residual highs.

In order to make the second-derivative images more intuitively sensible, an inverse stretch is applied to them to convert the bright areas to dark areas, and the dark areas to bright ones. This inverse stretch results in a directional second-derivative image in which the bright zones and dark zones now correspond to residual highs and lows, respectively.

For ideal conditions, the boundaries between the dark and bright zones on directional second-derivative images of gravity data approximately trace contacts across which there are signifi-

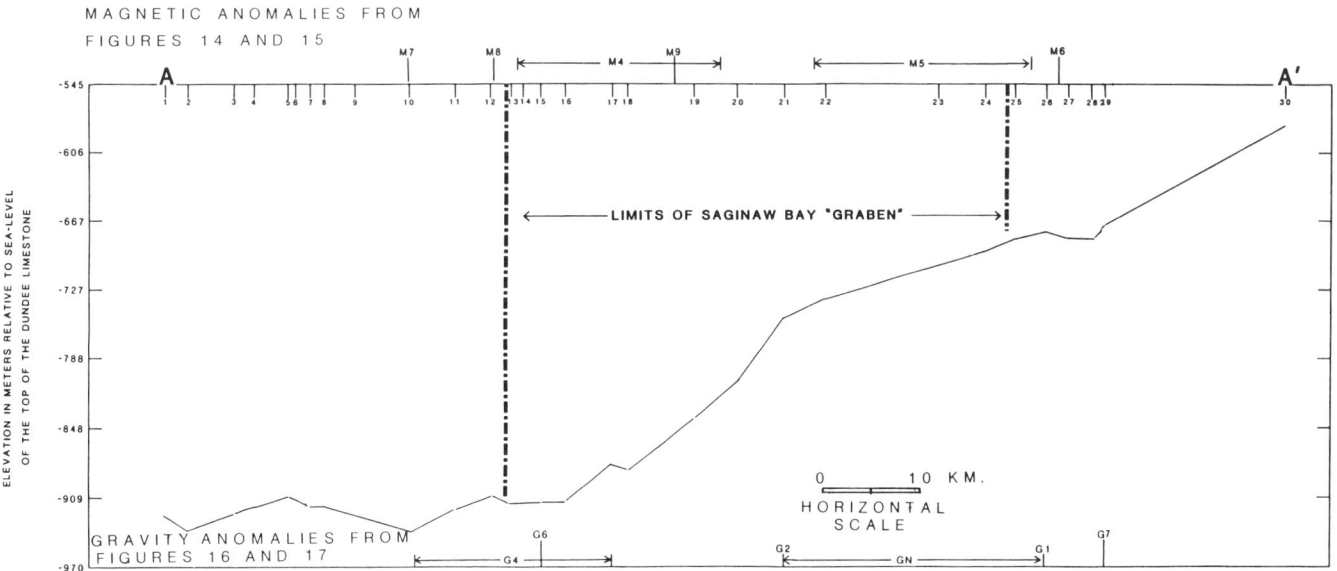

Figure 7. Structure section A-A' showing the elevation of the top of the Devonian Dundee Limestone relative to sea level. The locations where various magnetic and gravity anomalies and lineaments (mapped from images on Figs. 13, 14, 15, and 16) intersect the section are also shown. Location of the section is shown on Figure 2.

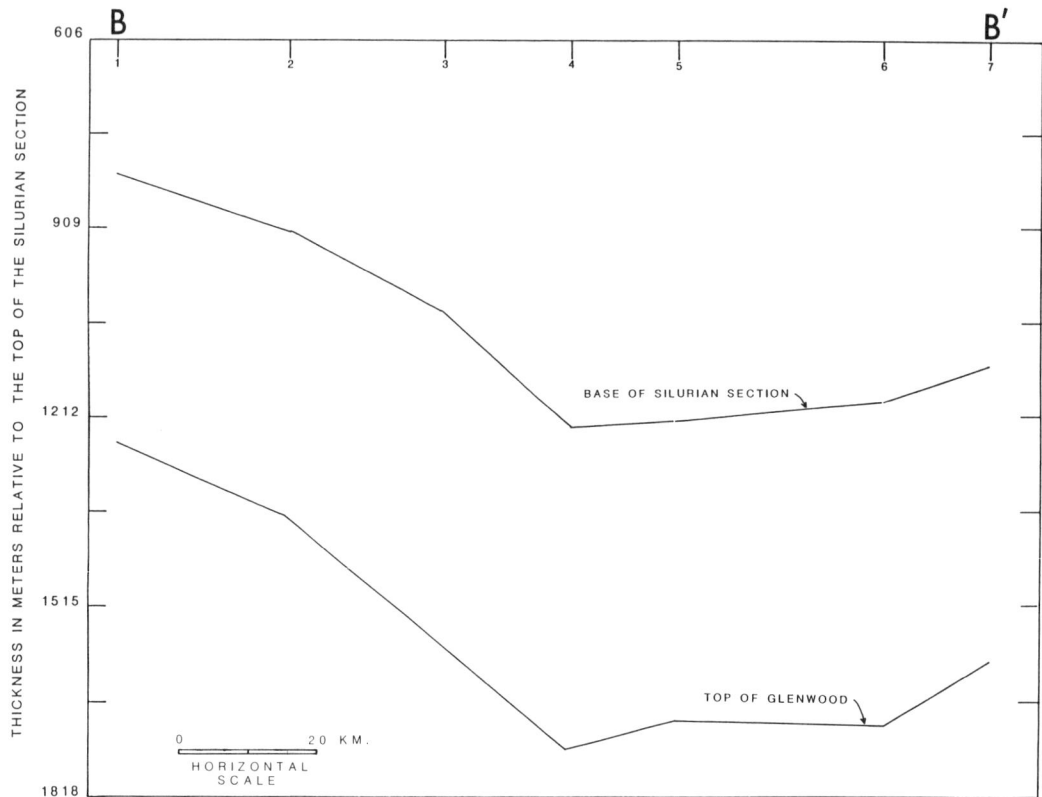

Figure 8. Thickness section B-B' showing the thickness of the Silurian section and the section from the base of the Silurian to the top of the Ordovician Glenwood Formation relative to the top of the Silurian (top of Silurian datum). Location of the section is shown on Figure 2.

cant density contrasts within the Paleozoic and Precambrian rock units. On a directional second-derivative image of magnetic data, the brightness contrast boundaries generally coincide with contacts on the surface of the Precambrian basement across which there are significant differences in magnetic susceptibility, under ideal circumstances.

### Digital terrain data sources and processing

Digital terrain (elevation) data obtained from the National Cartographic Information Center were used to produce a 135° TOPOIMAGE of the Saginaw Bay region (Fig. 3b). This data was derived from digitization of elevations from 1 × 2° USGS topographic maps. Spacing of the data points is 3 s of arc in longitude and latitude (x and y) directions. Vertical resolution of the data is dependent on the contour interval of the 1 × 2° topographic maps digitized. In the case of the Saginaw Bay area, vertical resolution or elevation resolution is approximately 15 m. The digital terrain data points were used to produce an image grid with a Transverse Mercator map projection. The map-projected data were then used to produce the TOPOIMAGE artificially illuminated by a light source shining from the southeast to the northwest (135° sun angle).

### Magnetic and gravity data sources and processing

The Bouguer gravity data used for this chapter were obtained from the National Geophysical Data Center. Average spacing of the gravity data points is 5 km. These data cover the entire Lower Peninsula of Michigan (i.e., central Michigan Basin area).

The digital aeromagnetic data covering the entire Southern Peninsula of Michigan were obtained from William Hinze of Purdue University. These data were collected along north-south flight lines spaced 5 km apart. A contour map of this data set has been published (Hinze and others, 1971).

Both data sets were map projected using a Lambert Conformal Conic projection. These map-projected data sets were converted into image grids, both with a 1,600 × 1,600 m grid cell size. The gridded data sets were then processed with the TOPOIMAGE software.

A total of four TOPOIMAGES of geophysical data were produced for the Southern Peninsula of Michigan, with each image specifically processed to enhance features in the Saginaw Bay region. Four black and white directional second-derivative TOPOIMAGES were then produced from these TOPOIMAGES, from which small windows covering the Bay County area were extracted.

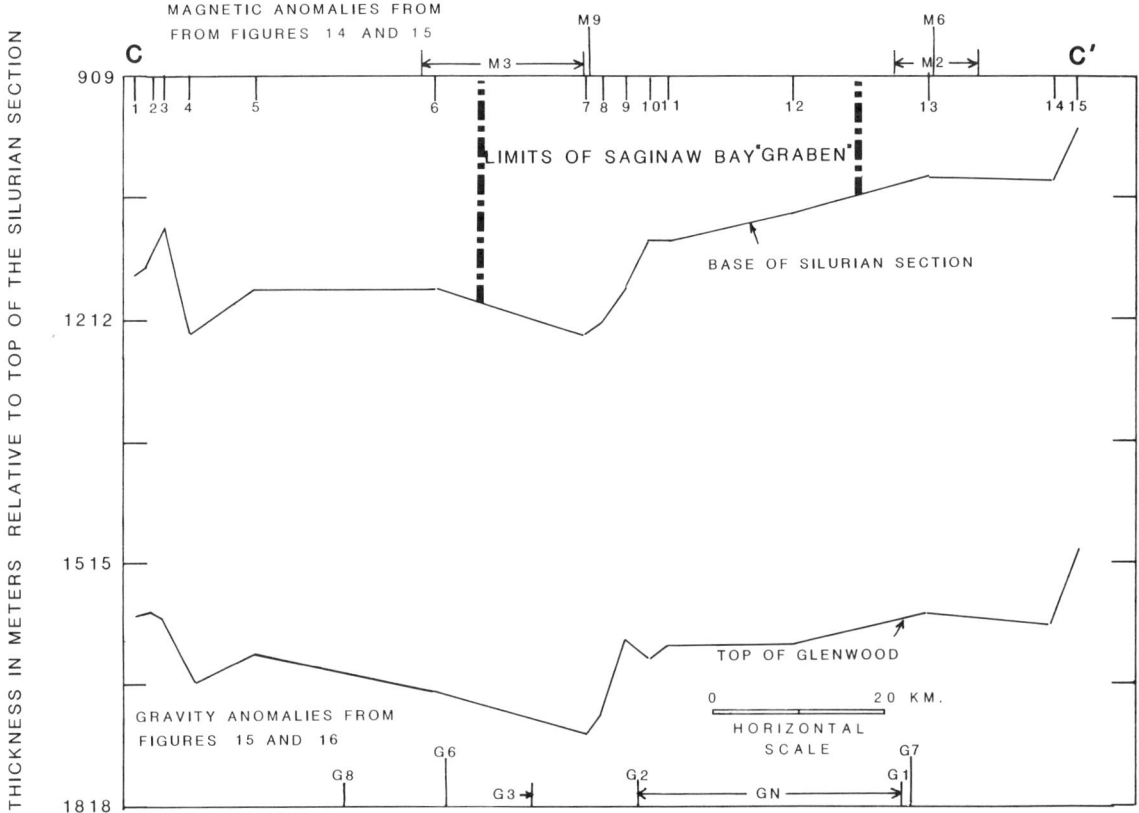

Figure 9. Thickness section C-C' showing the thickness of the same sections depicted on section B-B'. Locations where various magnetic and gravity anomalies and lineaments intersect the section are also shown. Location of the section is shown on Figure 2.

TOPOIMAGES of gravity and magnetic data, and interpretations of these data, covering the entire Southern Peninsula of Michigan are shown on Figures 12a, b, and 13a, b. Two black and white, second-derivative TOPOIMAGES with orthogonal illumination or sun angles (135° or southeast, and 45° or northeast) were produced for each data set (Figs. 14a, b, 15a, b, 16a, b, and 17a, b). On the 135° second-derivative TOPOIMAGES, the artificial light source illuminates the data from the southeast, so that northeast-trending gravity and magnetic anomalies are selectively enhanced. On the 45° second derivative TOPOIMAGES, the data are illuminated from the northeast corner of the images, and northwest-trending anomalies are selectively enhanced.

In order to obtain a simultaneous view of all the geophysical anomaly trends, the 45° and 135° directional second-derivative TOPOIMAGES of the magnetic data were added together, resulting in a residual anomaly image filtered in two directions. The same operation was performed on the second derivative gravity images. A color-level slice stretch was then applied to the resultant summed data sets (Figs. 3c and d). The most intense or prominent residual positive and negative anomalies appear in red and purple colors, respectively, on these images. Residual anomalies, intermediate between these highs and lows, appear in other colors (i.e., yellow, green, and blue, going from positive, or higher, anomalies to negative, or lower, anomalies). The color magnetic and gravity images produced by this method show much sharper definition of positive and negative anomalies, which might otherwise be difficult to detect on contour maps or color-level slices of gridded magnetic and gravity data.

The primary value of regional gravity and magnetic data is for determination of Precambrian basement lithologies and structures. Gravity data are useful for delineation of density contrasts within the basement, while magnetic data reveal magnetic susceptibility contrasts at or near the surface of the basement. Determination of basement boundaries is important for hydrocarbon exploration because these basement features may have exerted control on structures and lithologies in overlying Paleozoic rock units in at least two ways. First, if the basement boundaries are also fault contacts, they could have been reactivated during subsidence of the basin, thereby causing folding or faulting of younger strata. Second, if the basement boundaries separate lithologies of greatly contrasting erosion resistance, paleotopography developed across them during subaerial exposure could cause drape of overlying Paleozoic rock units through differential compaction during deposition and diagenesis.

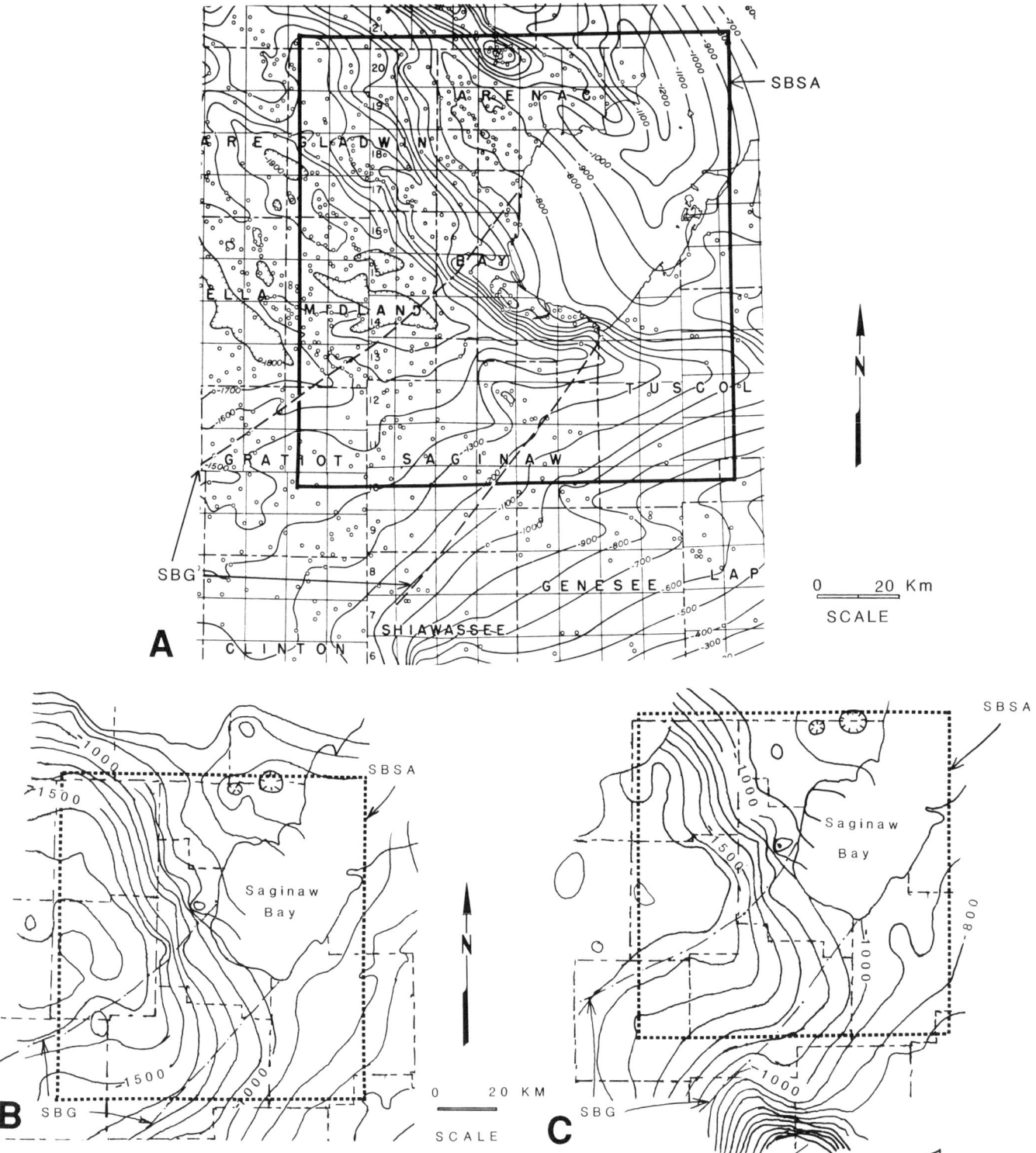

Figure 10. A. Structure map of the base of the Mississippian Coldwater Shale relative to sea-level datum. Contour interval is 30.3 m (100 ft). SBG = Saginaw Bay "Graben," SBSA = Saginaw Bay study area. (Modified from Cohee and others, 1951.) B. Structure map of the Mississippian Sunbury Shale relative to sea-level datum. Contour interval is 30.3 m (100 ft). (Modified from Fisher, 1980a.) C. Structure map of the Mississippian Berea Sandstone relative to sea-level datum. Contour interval is 30.3 m (100 ft). (Modified from Fisher, 1980b.)

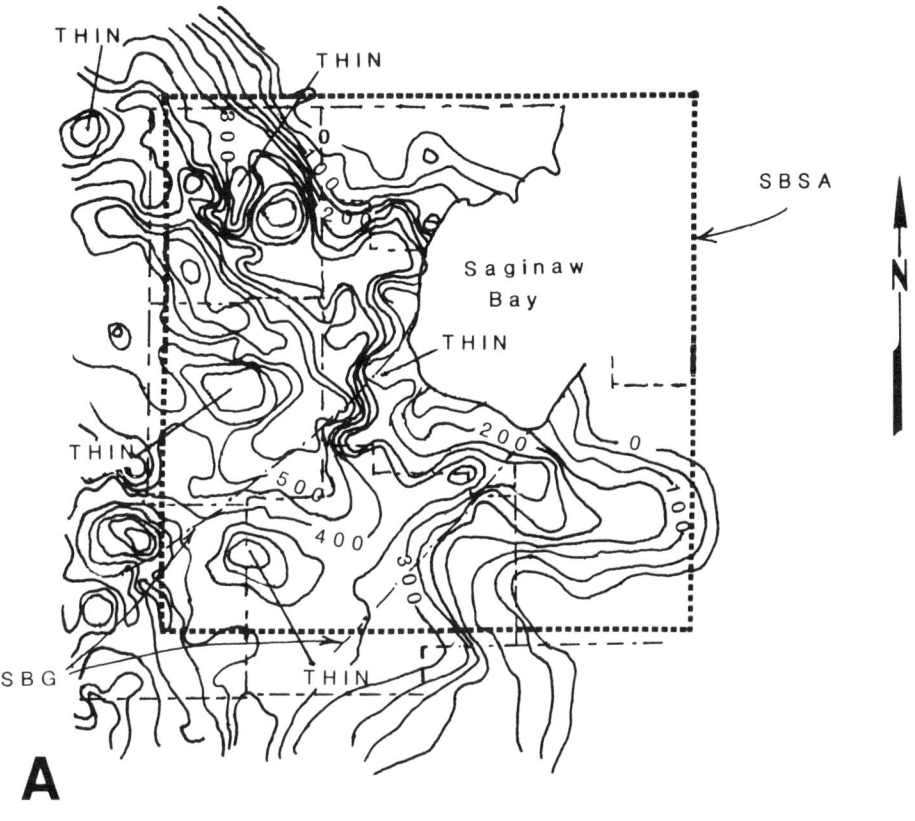
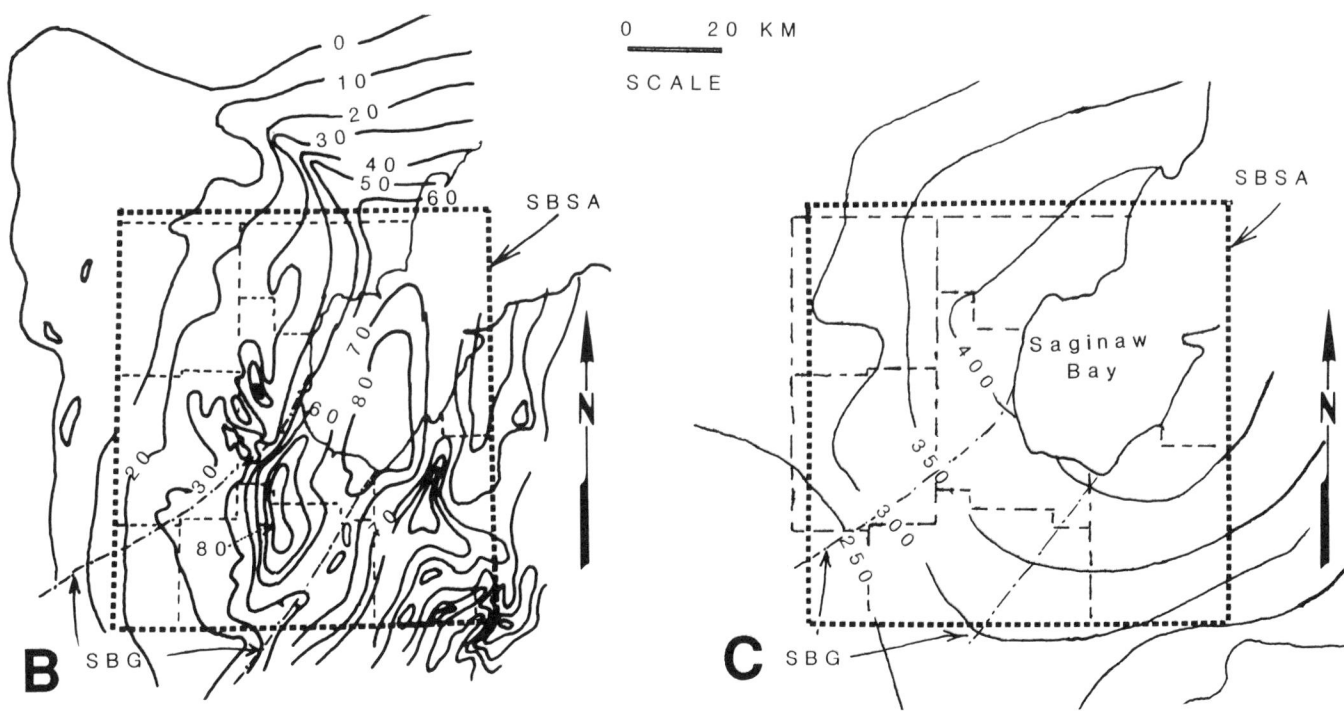

Figure 11. A. Isopach map of the Pennsylvanian section. SBG = Saginaw Bay "Graben," SBSA = Saginaw Bay study area. Isopach or thickness interval is 15.1 m (50 ft). (Modified from Strutz, 1978.) B. Isopach map of the Mississippian Berea Sandstone. Isopach or thickness interval is 3.3 m (10 ft). (Modified from Gunn, 1988.) C. Isopach map of the Devonian Dundee Limestone. Isopach or thickness interval is 15.1 m (50 ft). (Modified from Gardner, 1974.)

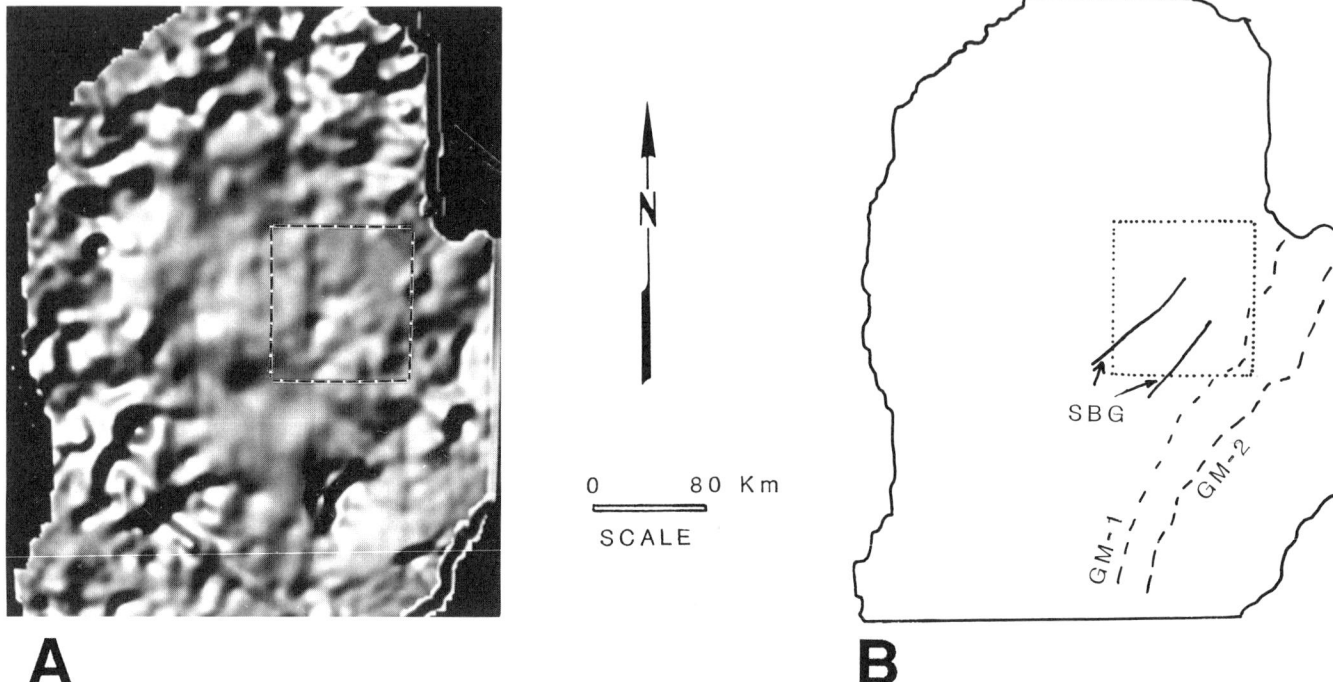

Figure 12. A. 135° TOPOIMAGE of magnetic intensity covering the entire Southern Peninsula of Michigan. The box outlined by dashed lines is the Saginaw Bay study area. B. Interpretation of Figure 12A showing location of Saginaw Bay "Graben" (SBG) relative to the study area and major positive magnetic anomalies thought to be related to the Grenville Front (GM-1 and GM-2).

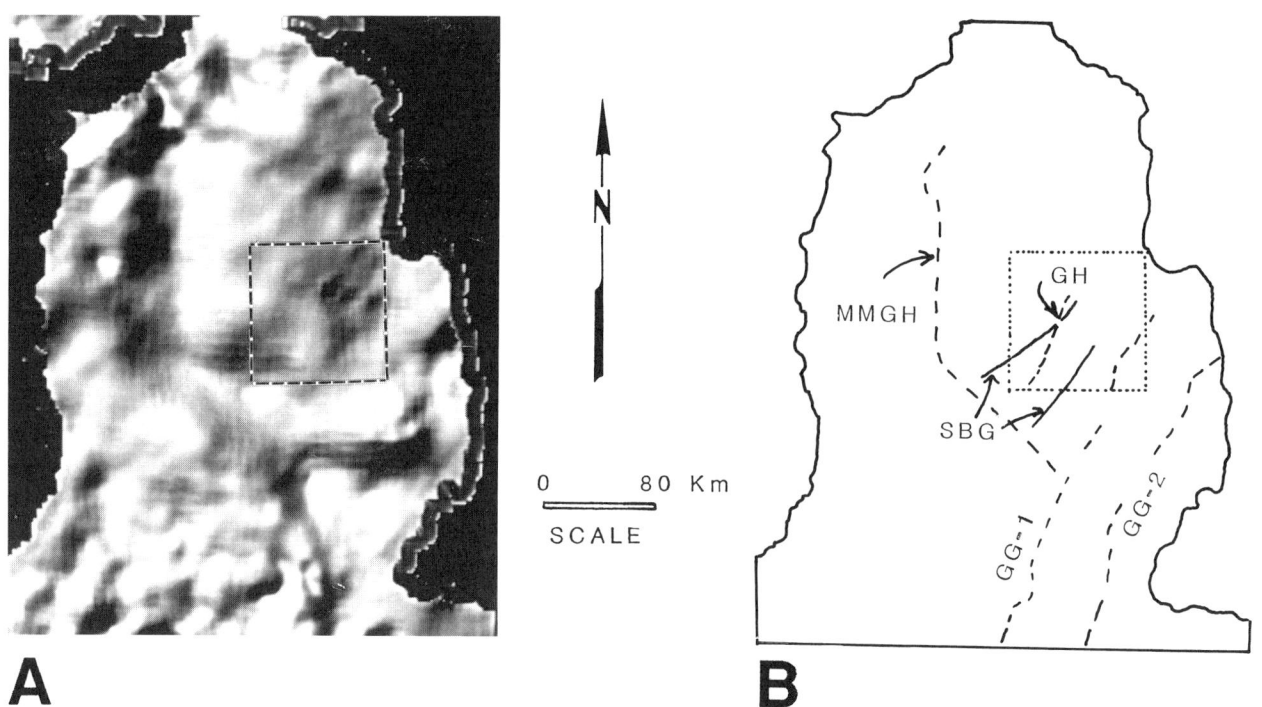

Figure 13. A. 135° TOPOIMAGE of gravity intensity covering all of the Southern Peninsula of Michigan. The box outlined by dashed lines is the Saginaw Bay study area. B. Interpretation of Figure 13A showing the location of the Saginaw Bay "Graben" relative to the study area and major positive gravity anomalies thought to be related to the Grenville Front and the possible rift system associated with the Mid-Michigan Gravity High (GG-1, GG-2, MMGH, GH-1).

## INTERPRETATION OF WELL-LOG, DIGITAL TERRAIN, AND GEOPHYSICAL DATA

### Analysis of geologic cross sections

Structure section A-A' shows a general northwestward dip, with no apparent evidence of a present-day graben-like structure at the Dundee level (Fig. 7). A-A' also shows anticlinal structures on the edges of the "graben," and apparent flattening or decrease of northwestward dip on the northwest side of the "graben." This indicates the presence of a possible structural hinge-line along the northwest edge of the "graben," as well as possible fault-controlled folding along both edges.

If the "graben" acated as a depocenter in pre-Devonian time, Silurian and older rocks should be of greater thickness within the graben limits (see thickness sections B-B' and C-C', Figs. 8, 9). Section B-B', which extends in a northeast direction within the northwest half of the graben, shows an increase in thickness of the Silurian and Ordovician rock units toward the northeast. Along this line of section, Silurian and Ordovician sections are thickest in the Shell, Whyte 1-33 well near the middle of the "graben," thus lending support to the hypothesis that the region around Saginaw Bay may have been a local depocenter in Silurian and Ordovician time. Section C-C', which extends in a northwest-southeast direction across the graben, also shows that the combined thickness of the Silurian and Ordovician sections is greatest within the "graben." However, section C-C' also indicates another thickening trend in southeast Gladwin County, Michigan, which is nearly as great as that within the hypothesized "graben." The thickness data along section C-C' then do not unequivocally support the existence of a "graben" persisting from Middle Ordovician to Silurian time, as originally proposed by Drake (1976).

### Analysis of structure and isopach maps

A map showing the elevation of the base of the Mississippian Coldwater Shale by Cohee and others (1951) shows northeast-trending, northwest-dipping, monoclinal structures along edges of the "graben" (Fig. 10a). The regional strike of the Coldwater is northeast within the limits of the "graben" except near major structural features such as the anticlines associated with the Saginaw, Porter, Kawkawlin, and Essexville oil fields. Similar associations between structures and the "graben" can be seen on regional structure maps of the Mississippian Sunbury Shale and Berea Sandstone constructed by Fisher (1980a, b); Figs. 10b, c).

Strutz (1978) mapped rapid thickness variations of the entire Pennsylvanian section along the northwest and southeast edges of the "graben" (Fig. 11a). These variations consisted of rapid thinning of the Pennsylvanian section going from northwest to southeast over portions of both edges of the "graben." However, within the limits of the "graben" the thickness of the Pennsylvanian section exhibited little variation except for the southwest limit of the "graben" near the Mid-Michigan Gravity High and the northeast onshore limit. Gunn (1988) mapped a northeast-trending zone of increased thickness of the Mississippian Berea Sandstone extending southwestward from Saginaw Bay (Fig. 11b). An isopach map of the Devonian Dundee Limestone produced by Gardner (1974) shows a northeast-trending "thick" wrapping around Saginaw Bay (Fig. 11c). Isopach and facies maps of the Middle to Lower Ordovician "Goodwell," Foster, and "Umlor" units show hints of rapid thickness variations and facies changes along a northeast-trending zone extending nearly parallel to the southeastern edge of the "graben" (Brady and DeHaas, 1988). Similar variations are not observable on the northwestern "graben" edge. This difference may be linked to closer proximity of the southeastern edge of the "graben" to the southern extension of the Grenville Front, to a lack of deep well control along the northwest edge, or the "masking effect" that regional basinal thickening would have on the northwestern edge.

These thickness variations may indicate that a structure or set of structures extending southwest of Saginaw Bay controlled sediment depositional patterns during several periods of geologic history. The structural configuration southwest of Saginaw Bay may have been a half-graben, shelf, or terrace-like structure that interrupted the regional northwest dip of the rock units during these time periods.

## DIGITAL TERRAIN IMAGE INTERPRETATION

The 135° digital terrain TOPOIMAGE shows a prominent northeast-trending low extending southwestward from the southern edge of Saginaw Bay (Fig. 3b). The axis of this low is offset southeastward from the axis of the "graben." The southeastern edge of the "graben" corresponds to a series of subtle, northwest-facing scarps. The northwestern edge of the "graben" is delineated by stream valleys in some areas, and topographic scarps in other areas. This indicates that although portions of the "graben" edges correspond to mappable topographic features, the interior section of the "graben" does not correlate with a simple topographic depression. The southeastern "graben" edge is nearly coincident with a northeast-trending lineament mapped on the bedrock surface (Rieck, 1989), and northeast-trending monoclinal structures mapped at the Mississippian Coldwater Shale, Sunbury Shale, and Berea Sandstone levels (Figs. 10a, b, c). Portions of the northwestern "graben" edge are marked by stream valleys and scarps cutting across the Porter and Williams oil fields corresponding with subsurface, northeast-trending and northwest-dipping, monoclinal structures mapped by Helmboldt (1968) and Gunn (1988) at the Devonian Dundee Limestone and Mississippian Berea Sandstone levels. These correlations suggest that topographic lineaments along the "graben" edges may be the surface expressions of bedrock topography and subsurface structures.

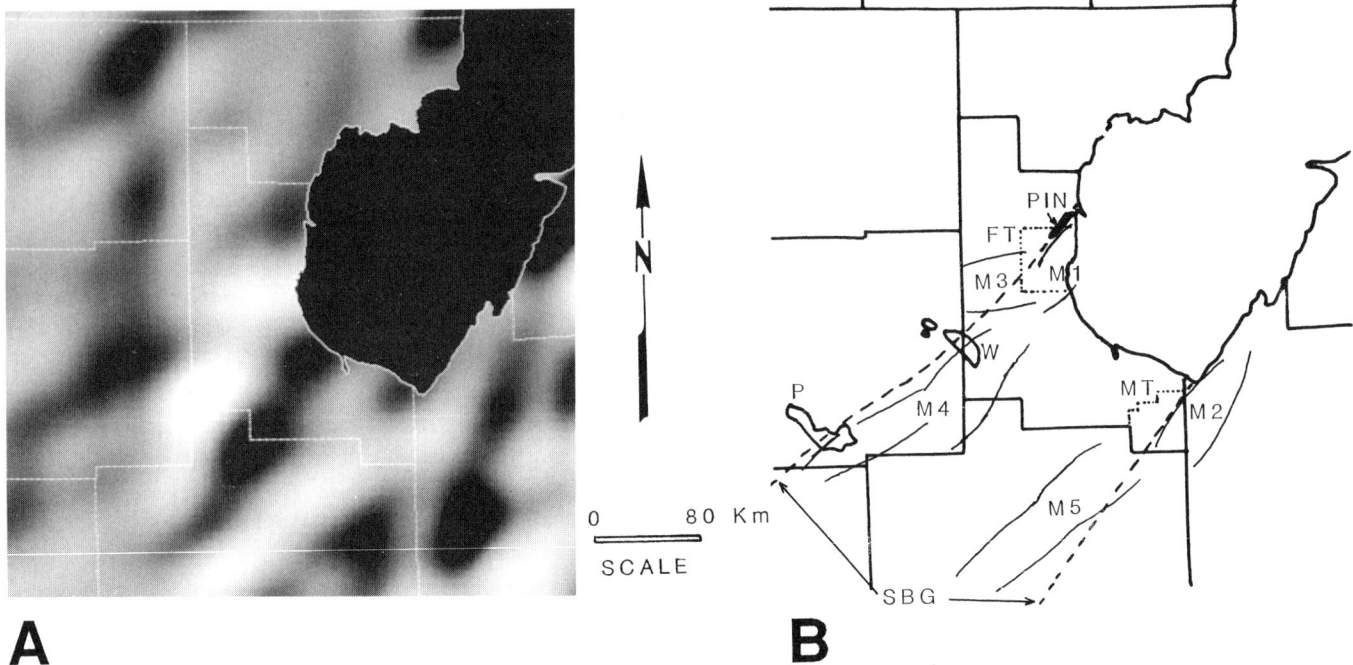

Figure 14. A. 135° directional second-derivative TOPOIMAGE of magnetic intensity covering the Saginaw Bay study area. B. Interpretation of image on Figure 14A showing major anomalies along the edges of the Saginaw Bay "Graben" (SBG). Other features include: FT = Fraser Township, MT = Merritt Township, P = Porter oil field, W = Williams oil field, Pin = Pinconning oil field.

## GRAVITY AND MAGNETIC DATA INTERPRETATION

### Regional geophysical trends

A regional geophysical view of the Saginaw Bay region and surrounding area is provided by gravity and magnetic TOPOIMAGES of the entire Southern Peninsula of Michigan (Figs. 12a and 13a), in which the data are illuminated from the southeast (135° sun azimuth). The study area is bounded on the southeast by a pair of major northeast-trending positive magnetic anomalies, labeled GM-1 and GM-2 on Figure 12b. A pair of discontinuous northeast-trending positive gravity anomalies are nearly coincident with the previously described positive magnetic anomalies (GG-1 and GG-2, Fig. 13b). GM-1, GM-2, GG-1, and GG-2 may be tracing the southern extension of the Grenville Front because of their similarity to magnetic and gravity anomalies mapped from contour maps along the exposed portion and the inferred southwest extension of the front beneath Paleozoic rocks in Lake Huron (O'Hara and Hinze, 1980; Gibb and others, 1983).

The Grenville Front is a northeast-trending belt of highly deformed, high-grade metamorphic rocks of late Precambrian age, which are separated from less deformed, older Precambrian-age intrusives and lower-grade metamorphics by a steeply dipping reverse or thrust fault. The age of deformation and metamorphism along the front is thought to be around 1.1 Ga (Green and others, 1988). The northwestern boundaries of positive magnetic and gravity anomalies GM-1 and GG-1 are nearly parallel with, and offset southeastward from, the southeastern edge of the Saginaw Bay "Graben," indicating that the trend and location of this edge may have been influenced by the structures along the Grenville Front.

The southwest portion of the Saginaw Bay study area, shown in Figure 2, is bounded by a west-northwest–trending gravity high (i.e., Mid-Michigan Gravity High of Hinze and others, 1971), which may represent a late Precambrian rift that formed shortly before or during the development of structures along the Grenville Front (Baer, 1976; MMGH, Fig. 13b). A broad positive gravity anomaly extends northeastward from the Mid-Michigan Gravity High into Bay County (GH, Fig. 13b). This is the type of anomaly that might be expected to be associated with a graben similar to those postulated to exist along the Mid-Michigan Gravity High. However, the southeast and northwest boundaries of this positive anomaly are parallel to, but offset northwestward from, the postulated boundaries of the Saginaw Bay "Graben," as originally defined by Drake (1976).

### Interpretation of magnetic data in the Saginaw Bay area

The 135° directional second-derivative TOPOIMAGE of magnetic data reveals a series of prominent, northeast- and east-west–trending positive anomalies extending southwestward from Saginaw Bay, labeled M1 through M5 on Figure 14a and b. The edges of the "graben" correlate with the flanks of several of these anomalies. Both the Pinconning oil field and Quanicassee River lineaments occur along the northwest boundaries of northeast-trending positive residual magnetic anomalies (M1 and M2, Figs.

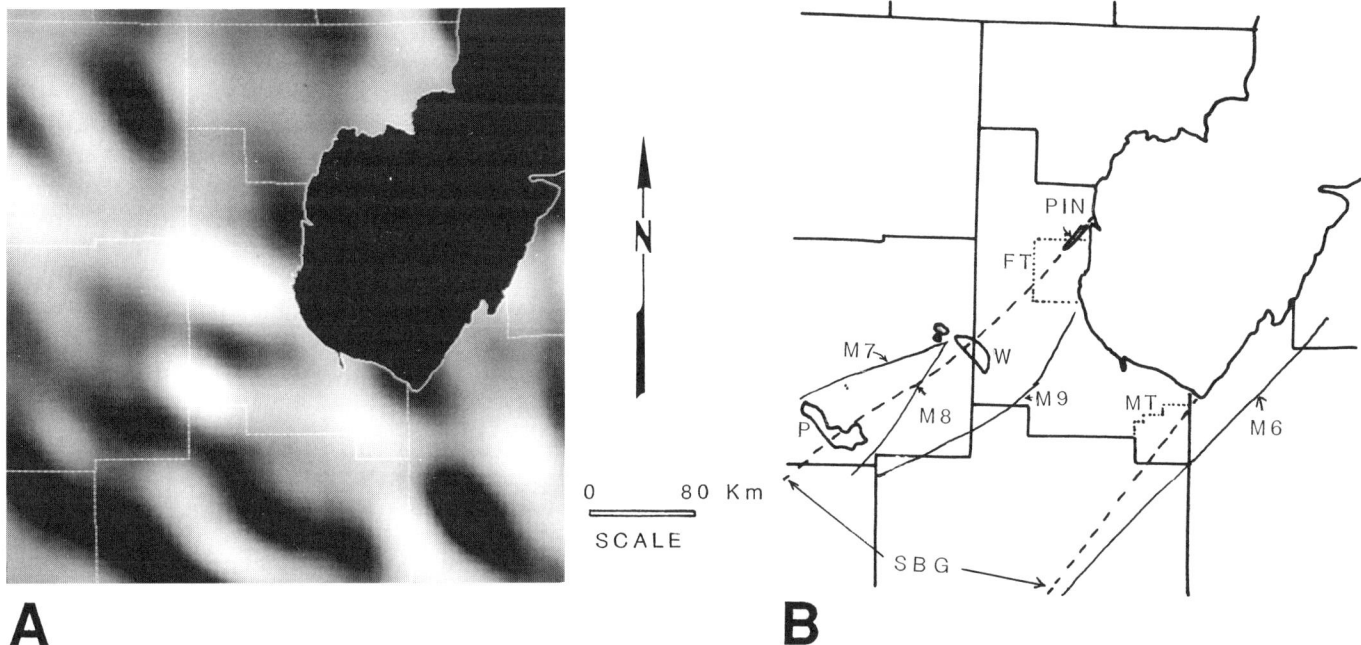

Figure 15. A. 45° directional, second-derivative TOPOIMAGE of magnetic intensity covering the Saginaw Bay study area. B. Interpretation of image on Figure 15A showing major anomalies along the edges of the Saginaw Bay "Graben." See Figure 14B caption for list abbreviations.

14a, b). This correspondence may indicate that basement structures and/or paleo-relief across basement contacts controls the subsurface structure in the Paleozoic section associated with these lineaments. Southwest of Fraser Township (T16N, R4E), a nearly east-west–trending positive anomaly (M3) cuts across the northwest edge of the "graben."

Farther southwest of M3, the northwest "graben" edge is nearly coincident with the northwestern flanks of northeast-trending positive anomalies (M4). Where the northwestern flank of M4 intersects the Porter and Williams oil fields, detailed structure maps by Helmboldt (1968) and Gunn (1988) show northeast-trending, northwest-dipping monoclinal structures or faults. This association may indicate basement control of these Paleozoic structures along M4. Southwest of Merritt Township (T13N, R6E), the southeast edge of the "graben" is nearly coincident with the southeastern flank of a northeast-trending positive anomaly (M5). The flank of M5 is nearly coincident with a northeast-trending bedrock elevation lineament mapped by Rieck (1989), and northeast-trending monoclinal structures mapped at the Mississippian Coldwater Shale, Sunbury Shale, and Berea Sandstone levels by Cohee and others (1951) and Fisher (1980a, b; Figs. 10a, b, c). These associations imply a possible basement influence on the geomorphologic and structural configuration of the near-surface rock units, along the southeast edge of the "graben."

The 45° directional second-derivative TOPOIMAGE of magnetic intensity shows little obvious evidence of northeast-trending lineaments or anomalies along the edges of the "graben," as might be expected with this illumination azimuth. However,

prominent northwest-trending positive anomalies are truncated or apparently offset along a northeast-trending lineament (M6, Figs. 15a, b), which is nearly parallel to but offset about 3 km southeast from the southeast boundary of the "graben." Anomaly M6 may be associated with a fault zone within the Precambrian basement. Northwest-trending positive anomalies are offset or truncated along at least three northeast-trending lineaments (M7, M8, and M9, Fig. 11b). These lineaments may also be due to faults within the Precambrian basement.

The directional second-derivative TOPOIMAGE of magnetic data shows a "U-shaped" group of positive anomalies extending southwestward from Saginaw Bay (Fig. 3c). These anomalies may be caused by intrusive complexes in the basement, emplaced along old northeast-trending zones of weakness near the locations of the "graben" edges. The northwestern flanks of the northwest anomalies and the southeastern flanks of the southeast anomalies occur near the edges of the "graben" hypothesized by Drake (1976).

### Interpretation of gravity data in the Saginaw Bay area

The 135° directional second-derivative TOPOIMAGE of the gravity data shows some striking northeast-trending lineaments, two of which are nearly parallel to the western and southeastern edges of Saginaw Bay (G1 and G2, Figs. 16a, b). These lineaments form the northwest and southeast boundaries of a negative residual gravity anomaly that extends from Saginaw Bay southwestward into central Saginaw County (GN, Figs. 16a, b). The southeast boundary of the negative gravity anomaly is parallel to, and offset to the southeast of, the "graben" edge

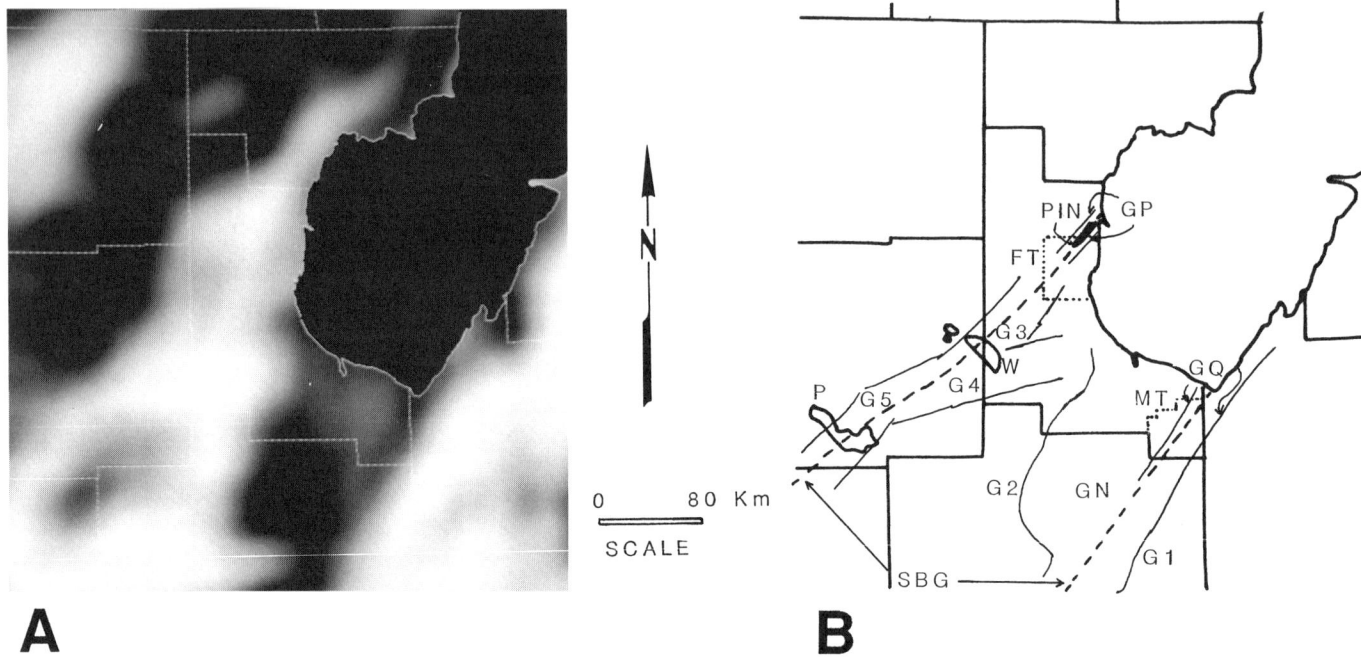

Figure 16. A. 135° directional second-derivative TOPOIMAGE of gravity intensity covering the Saginaw Bay study area. B. Interpretation of image on Figure 16A showing major anomalies along the edges of the Saginaw Bay "Graben." See Figure 14B caption for explanation of abbreviations.

defined by Drake (1976). The northwest edge of the negative gravity anomaly occurs in the middle of the "graben." The northwest edge of the "graben" occurs along a series of prominent northeast-trending residual positive gravity anomalies (G3, G4, and G5, Figs. 16a, b). An exception is along the Pinconning oil field, which occurs in a subtle, northeast-trending low (GP, Fig. 16b). The Quanicassee River lineament at the southeast edge of the graben also occurs along a northeast-trending residual gravity low (GQ, Fig. 16b).

These associations indicate that the subsurface Paleozoic structures responsible for the Pinconning oil field and Quanicassee River lineaments correlate with similar basement features. Southwest of Fraser and Merritt Townships in Bay County, however, the northwest and southeast edges of the "graben" do not correlate with same type of gravity anomaly, implying existence of different basement features in those areas (Figs. 16a, b).

The 45° directional second-derivative TOPOIMAGE of gravity data shows little obvious evidence for northeast-trending anomalies parallel to the "graben" extending southwest of Saginaw Bay, as may be expected. However, prominent northwest-trending positive residual gravity anomalies are truncated or apparently offset along or near the edges of the graben along lineaments of G6 and G7 (Figs. 17a, b). G6 and G7 may be associated with northeast-trending basement faults nearly parallel to, and in some places coincident with, the edges of the graben.

The second-derivative gravity image displays two rather prominent northeast-trending lineaments that extend southwestward from Saginaw Bay into Gratiot and Saginaw Counties (Fig. 3d, G8 and G9). These lineaments occur along the northwest boundaries of prominent positive residual anomalies, which are parallel to but offset from the boundaries of the "graben" (Fig. 3d). G8 and G9 may be caused by deep-seated density contrasts within the Precambrian basement.

## SUMMARY AND CONCLUSIONS

Previous satellite and geophysical contour-map studies in the region around Saginaw Bay, Michigan, led to the hypothesis that a northeast-trending graben occurs between a southeast-bounding fault along the old trace of the Quanicassee River in southeastern Bay County, and a northwest-boundary fault along and southwest of the Pinconning oil field in northeastern Bay County. A study of old well logs contributed to the idea that the suspected graben was tilted down to the northeast.

Follow-up studies of seismic data, gravity, magnetic, and digital-terrain images, as well as examination of well-log data generated from exploration for hydrocarbon reserves in the Ordovician has led to a revision of the simple graben hypothesis of Drake (1976). Seismic and well-log data collected as a result of hydrocarbon exploration along the northwest and southeast edges of the "graben" verified the existence of folding that may be related to deep-seated faulting both in the area southwest of the Pinconning field, and along the Quanicassee River in Bay County. However, this oil exploration data indicated down-to-the-northwest dip or faulting on both edges of the "graben" in Fraser and Merritt Townships of Bay County (Figs 4, 5, and 6). If the structural feature identified by Drake (1976) is a simple graben expressed by present-day structural configuration, then

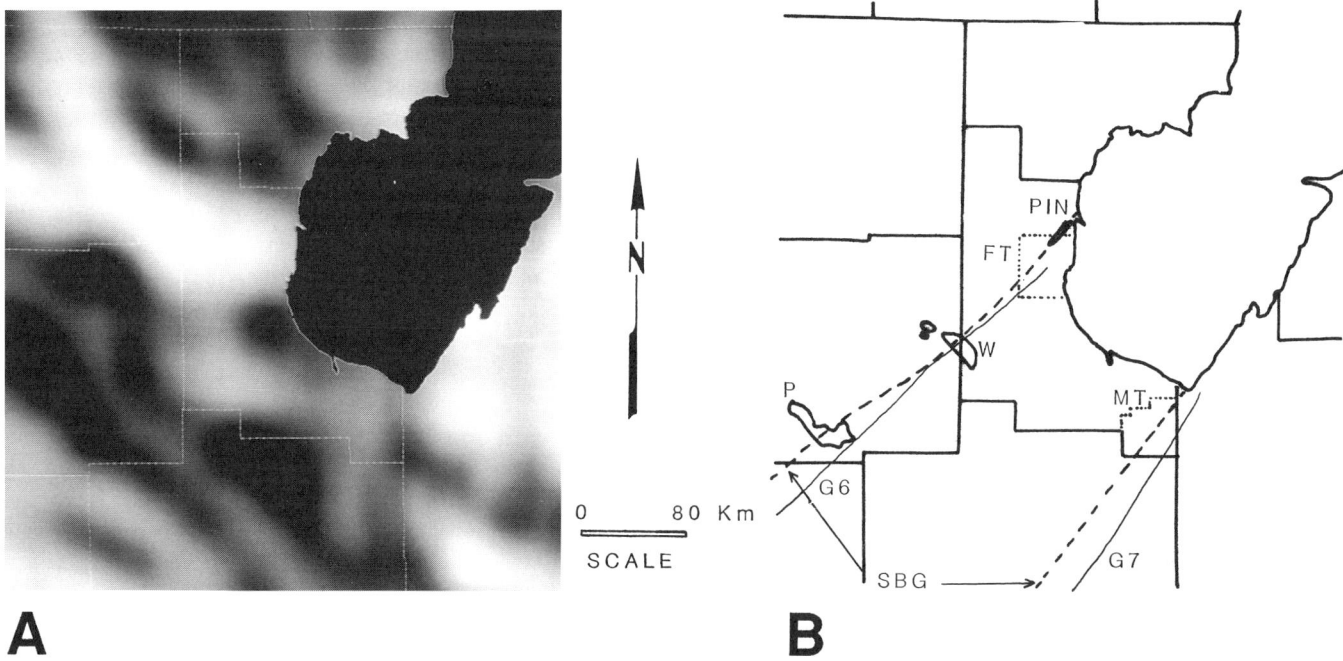

Figure 17. A. 45° directional second-derivative TOPOIMAGE of gravity intensity covering the Saginaw Bay study area. B. Interpretation of image on Figure 17A showing major anomnalies along the edges of the Saginaw Bay "Graben." See Figure 14B caption for explanation of abbreviations.

down-to-the-southeast faulting or structural dip should have been expected at the northwest edge of the graben in Fraser Township.

Structure data from well logs in the Saginaw Bay region do not show a discrete, simple, northeast-trending graben structure extending southwestward from Saginaw Bay (Fig. 7). Instead, both edges of the "graben" correlate with monoclinal structures or hinge lines that dip northwest. This association is corroborated by comparison of the "graben" edges with regional structure maps of the Mississippian Coldwater Shale, Sunbury Shale, and Berea Sandstone (Fig. 10a, b, c).

Data from hydrocarbon exploration wells drilled into the Ordovician showed that the combined thickness of the Silurian, Upper Ordovician, and Middle Ordovician Trenton and Black River Groups reached a maximum within the "graben" in apparent support of the original hypothesis (Drake, 1976) that the Saginaw Bay region was an ancient depocenter. However, the thickness of the Silurian and Ordovician rocks in southeast Gladwin County, just outside the northwest edge of the "graben," is nearly as great as that within the "graben" (Figs. 8 and 9). Isopach maps of Pennsylvanian, Mississippian, and Devonian units show northeast-trending thickness variations along the edges of the "graben" (Fig. 11a, b, c). Regional isopach and facies maps of the Middle to Lower Ordovician "Goodwell," Foster, and "Umlor" units of Brady and DeHaas (1988) show rapid thinning and facies changes along a northeast-trending zone extending southwest of the southeastern boundary of Saginaw Bay, sub-parallel to the southeast edge of the "graben." These thickness variations show that northeast-trending structural features such as monoclines or faults along parts of the "graben" edges controlled sedimentation patterns during portions of several geologic periods.

Analysis of a color, shaded-relief image of digital terrain (elevation) data illuminated from the southeast (135° sun angle), revealed a prominent, northeast-trending, topographic low extending southwest of Saginaw Bay (FIg. 3b). This low is parallel to, but not coincident with, the "graben" hypothesized by Drake (1976). However, subtle northeast topographic lineaments, consisting of stream valleys and scarps, detectable on the digital-terrain image, were found to coincide with the edges of the "graben." These lineaments are hypothesized to be the surface expressions of subsurface folds or faults with small displacements, not necessarily bounding a graben-like structure.

Study of directional, second-derivative TOPOIMAGES of magnetic data covering the Saginaw Bay region (Figs. 3c, 14, and 15) showed that the edges of the "graben" follow the flanks of positive, northeast-trending magnetic anomalies. The data further showed that northwest-trending positive and negative magnetic anomalies are truncated and displaced along, or very near, the "graben" edges.

The directional, second-derivative images of gravity data (Figs. 3c, 14, and 15) showed that northeast-trending positive and negative anomalies occurred along the "graben" edges, and that northwest-trending positive and negative gravity anomalies are truncated or displaced along lineaments parallel to the "graben" edges. The "graben," as defined by Drake (1976), cannot be linked to a discrete or simple type of magnetic or gravity anomaly in the same way that, for example, the axis of the North American Mid-Continent Rift System correlates with a series of prominent Bouguer gravity anomalies (Van Schmus and Hinze, 1985). Rather, it appears that the edges of the "graben" correlate with magnetic and gravity anomalies of differing sources. This is most clearly demonstrated by examination of the 135° directional second-derivative gravity TOPOIMAGE (Figs. 16a, b). The extreme northeast sections of the onshore edges of the "graben," in Fraser and Merritt Townships of Bay County, both correlate with negative residual gravity anomalies. However, if the northwest edge of the "graben" is followed southwestward from Fraser Township, it correlates with the edges and crests of positive residual gravity anomalies. In contrast, the southeast edge of the "graben" correlates with the middle of a negative residual gravity anomaly southwest of Merritt Township.

These images of geophysical data, in conjunction with digital terrain, structure, and isopach data indicate that northeast-trending, possibly basement-influenced, structures controlled sedimentation patterns intermittently during Paleozoic time southwest of and around Saginaw Bay along large segments of the edges of what Drake (1976) postulated to be a graben. The southeast edge of the "graben" may have had a more profound influence on structural and depositional patterns throughout the history of the Michigan Basin than the northwest edge, because of its closer proximity to the Grenville Front.

In conclusion, the existence of a discrete, simple, northeast-trending graben, including Saginaw Bay and extending southwestward from it within the limits originally hypothesized by Drake (1976), is not completely supported by well-log information and images of geophysical data. Rather, interpretation and integration of the Landsat, digital terrain, geophysical, and well-log data indicate the presence of northeast-trending, basement-influenced, monoclinal or anticlinal structures within the Paleozoic section; these structures extend southwest from the northwest and southeast sides of Saginaw Bay. The structure in the Paleozoic section along the southeast side of Saginaw Bay, extending southwestward, may have been influenced by Precambrian basement faults or contacts that developed during the formation of the Grenville Front. Intermittent movement along these northeast-trending structures during Paleozoic time controlled sedimentation patterns, as illustrated by northeast-trending thickness variations of Pennsylvanian, Mississippian, Devonian, and Ordovician units extending southwestward from SAginaw Bay.

## ACKNOWLEDGMENTS

The authors thank William Hinze at Purdue University for providing the aeromagnetic data used for this study. In addition, we thank Greg Pezda for drafting many of the figures.

## REFERENCES CITED

Baer, A. J., 1976, The Grenville Province in Helikian times; A possible model of evolution: Royal Society of London Philosophical Transactions, series A, v. 280, p. 499–515.

Brady, R. B., and DeHaas, R., 1988, The "deep" (pre-Glenwood) formations of the Michigan Basin, Parts 1 through 10: Michigan's Oil and Gas News, February through December monthly issues in 1988.

Cohee, G. V., Macha, C., and Holk, M., 1951, Thickness and lithology of Upper Devonian and Carboniferous rocks in Michigan: U.S. Geological Survey Oil and Gas Investigation Chert OC-41.

Drake, B., 1976, Saginaw Bay Graben and its implications for the origin of the Michigan Basin and Pleistocene glaciation [abs.]: Midwestern Regional American Geophysical Union Meeting, Ann Arbor, Michigan.

Drake, B., and Vincent, R. K., 1975, Geologic Interpretation of LANDSAT 1 imagery of the greater part of the Michigan Basin, *in* Proceedings, 10th International Symposium on Remote Sensing of the Environment, Ann Arbor, Michigan: Environmental Research Institute of Michigan, p. 933–947.

Fisher, J. H., 1980a, Structure map on Sunbury Shale: Dow Chemical Co., U.S. Department of Energy Report no. FE2346-80, plate 1, scale 1:1,000,000.

——, 1980b, Structure map on Berea Sandstone, Dow Chemical Co., U.S. Department of Energy Report no. F2346-80, plate 2, scale 1:1,000,000.

Gardner, W. C., 1974, Middle Devonian stratigraphy and depositional environments in the Michigan Basin: Lansing, Michigan Basin Geological Society Special Paper 1, 138 p.

Gibb, R. A., Thomas, M. D., LaPointe, P. L., and Mukhopadhyay, M., 1983, Geophysics of proposed Proterozoic sutures in Canada: Precambrian Research, v. 19, p. 349–384.

Green, A. G., and 9 others, 1988, Crustal structure of the Grenville Front and adjacent terranes: Geology, v. 16, p. 788–792.

Gunn, G. R., 1988, Stratigraphy, reservoir characteristics of Berea Sandstone in Michigan fields: Oil and Gas Journal, March 7, p. 57–60.

Helmboldt, D., 1968, The Porter field, *in* Michigan Basin Geological Society Oil and Gas Fields Symposium: East Lansing, Michigan Basin Geological Society, p. 137–142.

Hinze, W., Kellogg, R. L., and Merritt, D. W., 1971, Gravity and magnetic anomaly maps of the Southern Peninsula of Michigan: Michigan Department of Natural Resources, Geological Survey Division Report of Investigation 14, 14 p.

Landes, K. K., 1970, Petroleum geology of the United States: New York, Wiley-Interscience, 571 p.

O'Hara, N. W., and Hinze, W., 1980, Regional basement geology of Lake Huron: Geological Society of America Bulletin, part 1, v. 91, p. 348–358.

Rieck, R. L., 1989, Regionalization of southern Michigan's bedrock surface, *in* Symposium Abstracts with Programs; Michigan, its geology and geologic resources: Michigan Department of Natural Resources Geologic Survey Division and Michigan State University Department of Geology, p. 21.

Strutz, T. A., 1978, A pre-Pennsylvanian paleogeologic study of Michigan [M.S. thesis]: East Lansing, Michigan State University, 65 p.

Van Schmus, W. R., and Hinze, W. J., 1985, The Mid-Continent rift system: Annual Reviews of Earth and Planetary Science, v. 13, p. 345–383.

Vincent, R. K., and Coupland, D. C., 1980, Petroleum exploration with LANDSAT in Bay County, Michigan; An interim case study, *in* Proceedings, 14th International Symposium on Remote Sensing of the Environment: Ann Arbor, Michigan, Environmental Research Institute of Michigan, p. 379–387.

Vincent, R. K., and Drake, B., 1976, Potential use of LANDSAT data for petroleum exploration in the Michigan Basin, *in* Proceedings, 15th Annual Conference of the Ontario Petroleum Institute, Chatham, Ontario, Canada: Ontario Petroleum Institute, Inc., Technical Paper 8, p. 1–23.

Vincent, R. K., and Etzler, P. J., 1982, Magnetic and gravity image processing for petroleum and mineral exploration [abs.]: Society of Exploration Geophysicists, 52nd International Meeting, 1982, Dallas, Texas.

MANUSCRIPT ACCEPTED BY THE SOCIETY JUNE 1, 1990

# Index

[Italic page numbers indicate major references]

*Abathomphalus mayaroensis*, 473, 509
A-1 Carbonate, 98
Acadian orogeny, 187
Acadian tectophase 2, 189
acid rain, 129
agglomerate, 19, 23
aggregate, 129, 215
Albion-Scipio field, 73, 74, 79, 80, 81, 83
algae, 94
  red, 76
Alger County, 58
Algonquin Arch, 139, 142, 182, 196
Algonquin River, 214
Allegheny Basin, 141, 142, 143
Allen Pass, 146
Alpena County, Michigan, 159, 209
alteration, 23
Amabel Formation, 118, 119
Amasa Formation, 35, *36*
Amasa Oval, 35, 37
Amherstburg Formation, 9
Amoco, Schiller 1-10 well, 58
amphibolite, 17, 18, 19
amygdaloid, *42*
*Ancyrodella*
  *gigas*, 163
  *lobata*, 164
*Ancyrognathus*
  *ancyrognathoideus*, 162, 163
  *ncyrodella alata*, 162
  *rotundiloba*, 162
  *rugosa*, 162
  *soluta*, 162
  *symmetricus*, 164
*Ancyrognathus triangularis* Zone, 156, 163, 164, 169, 182, 186, 189
andesites, 38, 39
anhydrite, 57, *82*, 94, 140, 142, 145, 146, 150, 151, 215
anomalies
  gravity, 55
  magnetic, 236
anticlines, 40, 73
Antler orogeny, 187
Antler trough, 186
Antrim basin, *195*
Antrim County, 209, 211
Antrim Formation, 7, 8, 9, 10
Antrim Shale, 156, 157, *158*, *162*, *167*, 182, *187*, *199*, 206, *209*
Anvil Member, 34, 35
Appalachian Basin, 6, 111, 119, 130, *184*
Archean, *15*
Arenac County, 206
argillite, 16
Arkansas Novaculite, 186
asbestos bearing, 19
ash, volcanic, 19
ashbed flow, 44
*Asterocalamites*, 164, 185

asymmetric Zone, 156
Atlantic mine, 45
Attawapiskat Formation, 145
Au Train Formation, 64

*Bactrites warthini*, 162
Bailey Formation, 118, 119
Bakken Formation, 156
Baltic lodes, 44
Banner Lake area, *17*
Baraga County, 29
Baraga Group, 23, *24*, 35, 36
barite, 82, 83
barrier, supratidal, 95
basalts, 17, 19, 38
  olivine, 38
  pillow, 19
basins
  deep starved, 198
  lacustrine, 55
  margins, *167*
  subsidence, 1, *148*
  transition, *187*
Bass Islands Formation, 145
Bass Islands Group, 140, *142*, 150
Battle Creek trough, 145
Bay City, 144
Bay County, Michigan, 213, 225, 228
Bayport Limestone, *215*
Beaver Island, 58
Bedford mud delta, progradation, *196*
Bedford Shale, 156, 158, 165, *166*, 182, 187, *195*, 206, 209, *211*
bedrock, 22, 32, 42, 156, 168, 221
beds
  anhydrite, 145, 215
  dolostone, 211
  Red Rock, 212
  sedimentary, 38
  shaly, 161
Bell Formation, 7, 8, 9
belts
  greenstone, 16
  volcanic, *16*
Berea sands, eastward regression, *197*
Berea Sandstone, 9, 156, 158, 165, *166*, 182, 183, 187, *196*, 209, *211*, 235
Berthelet Member, 162, 167, 169
Bessemer, Michigan, 32
Bijiki Iron Formation, 29
bioherms, 145
biomicrites, 74, 75
biosparites, 74, 75
biostratigraphy, *155*, *158*, *167*
Bird Iron-bearing Member, 36
*Bispathodus*
  *aculeatus anteposicornis*, 166
  *stabilis*, 157
Black River Formation, 9
  diagenetic history, *73*
  dolomitization, 7
Black River Group, 225, 239

blastoids, 161
Blocher Member, 184, 189
Bluffton Reef, Indiana, 120
Bois Blanc Formation, 9, *73*
bornite, 23, 45
Bouguer gravity high, 189
boundstone, 94
Bowling Green fault, 73
brachiopods, 76, 84, 94
Branch County, 212, 213
*Branmehla*
  *culminidirectus*, 166
  *fissilis*, 166
Brazos shale, 61
breccia, 23, 37
  flow, 17, 24
  tuff, 23
  volcanic, 16
Briar Member, 32
brines, 145, 151, 214
  depth, *147*
  level, *146*
  migration, 83
bromine, 14
Brown Lime, 215
Bruce Peninsula exposures, Ontario, 120
Bruggers Formation, 64
bryozoans, 94, 161
*Buchiola* sp., 162
Buckland, Ohio, 117
buildups
  clinothem, *114*
  discrete, 128
  eastern Iowa, 120
  ecologic, *114*
burrows, 210

*Caballos novaculite*, 186
Cain Formation, 92, 97
*Calamites*, 214
calcite, 39, 40, *82*
  marine, 84
calcium chloride, 14
Calhoun County, 212, 213, 214
caliche, 123
Callahan Mining Corporation, 22
*Callixylon*, 185, 210, 211
Calumet Conglomerate, 46
Calumet mine, 45
Canada reefs, eastern, 117
Canadian Shield, 206, 213, 214
Canol Shale, 169
carbon, 37
  graphitic, 24
carbonate, 23, 27, 34, 61, 83, 89, 117, 140, 141, 206, 216, 217
  banks, *118*, 128, 148
  muds, 146
  platform, 92, *187*, 198
Carey pits, 129
Carp Lake mine, 45
Carp River, 26

Cataract Formation, 10
cathodoluminescence, 78, 81
Catskill complex, 184, 185
Catskill facies, 198
Cayugan Series, 111, 125
Cedarville Dolomite, 127
Celina, Ohio, 127
Celina complex, 117
cement
  dolomite, 78, 81, 86
  sulfate, 83
Central mine, 41, 44
Central province, 73
*Ceratoikiscum*
  *planistellare*, 163
  *spinosiarcuatum*, 163
Cerium, 24
Chagrain Shale, 182, 186, 194
chalcocite, 23, 45, 46
chalcopyrite, 23, 24, 45
Chamberlin, T. C., 107
channel, inflow, 146
Charlevoix County, 57, 58, 209, 211
Chatham sag, 142, 151, 182, 195
Chattanooga Shale, 156, 210
Cheboygan County, 209
Cheney Limestone Company, 215
chert, 19, 29, 32, 34, 36, 37, 75, 186, 216
chert carbonate, 36
Chesterian Series, 206
chlorite, 29, 37, 40, 46, 214
chloritization, 23
Chocolay Group, *23*
*Chonetes*, 167
Cincinnati Arch, 118, 126, 127, 130
clastics, 16, 55, 145, 150, 217
clays, *150*, 168, 214
Cleveland Iron Mining Company, 26
Cleveland Member, 166, 185, 191, 195
Cliff fissure, 44
Cliffs Shaft mine, 24, 31
clinothem buildups, *114*
Clinton Formation, 9, 10
Clinton Pass, 146
coal, 5, 6, 164
  metamorphism, 6
Colby mine, 32
Coldwater Formation, 8
Coldwater Lime, 212
Coldwater Red Rock, 212
Coldwater Shale, 167, 206, 209, *212*, 235
Coldwater-Marshall contact, 213
Collingwood Shale, 75
Compeau Creek, 19
Compeau Creek Gneiss, *19*
compression, 189
concentrates, 15
conductivity, thermal, 5
conglomerate, 16, 18, 38, 40, 42, 44, 93, 94, 141
conodonts, *156*, 168, 187, 206
Consolidated Copper Company, 22
Copper Harbor, 42
Copper Harbor Conglomerate, 38, 45, *55*
Copper Range Company, 45

copper, 14, 38, 47
  mineralization, 45
  mines, 15
  native, 14, 38, *40*
copper oxide, 42
copper sulfide, 23, 38, 46
Copps Formation, 34
corals, 94, 161
Corrigan McKinney and Company, 22
crinoid, 76, 78, 84, 94, 161
crosscuts, 22
Crystal Falls Iron District, *35*
  geology, *36*
  production, *37*
crystals
  gypsum, 216
  halite, 96
  millerite, 167
Cumings, Edgar R., 107
Cup Lake area, *17*
cuprite, 42
Curry Member, 32
Cuyahoga County, 211

dacite flows, 39
Dart, Christian 1-16 well, 225
Dart, Christian 1-16A well, 225
Dart, Edwards 7-36 well, 57
data
  digital terrain, *235*
  geophysical, *235*
  well-log, *235*
Dead River, 26
Dead River Pluton, *19*
Dead River Storage Basin, 19
debate, great reef-evaporite, *108*
Deer Lake, 19
Deer Lake Peridotite, *19*, 23
Deer Lake serpentinite, 23
deformation, compressional, 84
Delaware County, 212
Delphi Reef, Indiana, 120
deposition, evaporite, *95*
deposits
  braided stream, 55
  carbonate, 89, 216
  fluvial-deltaic, 5, 6
  hydrocarbon, 84
  oil, 80
  sulfide, 38
  supratidal, 95
detritus, 128
Detroit mine, 142
Detroit River Formation, 9
Detroit River Group, 140, 143, 151
Detroit River Sour Zone, 228
Detroit salt mine, 150
Devonian, 1, *209*
  Late, *181*, *185*, *189*, 198
  Middle, *161*
  Upper, *155*, *161*, *162*, *165*
diagenesis, 96
  history, *73*
  reef, *95*, *122*
*Diaphorostoma pugnus*, 162
Dickinson County, 31, 32
Dickinson Group, *18*
dikes, 15, 18, 22, 29, 34, 35

dolomite, 23, 57, 58, 61, 67, *77*, 83, 84, 129, 140, 142, 145, *149*
  cap, *79*, *85*
  cement, 78, 81, 86
  diagenetic, 84
  formation, 84
  fracture, *80*, *85*
  oolitic, 145
  regional, *77*, *85*
dolomitization, *77*, 89, *96*, 122, *149*
  regional, 86
dolomudstones, 146
dolostone, 167, 211, 212, 215, 216
Dow-Doe #110 well, 157
downwarp, 139, 140
drifts, 22
  glacial, 41, 167, 221
Dundee Formation, 9, *228*
Dundee Limestone, 225, 228, 229, 235
Dunkirk Shale, 185
Dunn Creek State, *37*

Eagle Harbor, 41
Early *hassi* Zone, 156
Early *praesulcata* Zone, 166
East Branch Arkose, *18*
eastern Canada reefs, 117
Eastern Interior seaway, 182, 185, 189, 198
Eau Claire Formation, 7, *58*, 61, 68
economics, reef, *128*
ejecta, volcanic, 18
Elfelian Lucas Formation, 151
Elk Point evaporites, 151
Ellsworth area, *161*, 183
Ellsworth delta, *190*
  development, *191*
  progradition, *191*, *194*
  retreat, *195*
Ellsworth Shale, 156, 158, *161*, *164*, 168, 182, *187*, 194, 206, 209, *211*
Elmhurst quarry, *168*
*Elvina* sp., 61
Emperior Volcanic Complex, 24
*Endosporites lacunosus*, 166
Engadine Dolomite, 90
Engadine Group, 119
*Entactinosphaera*
  *eostrongyla*, 163
  *fredericki*, 163
*Entomoze prolifica*, 164
*eosteinhornensis*, 126
*eosteinhornensis* Zone, 126
erosion, 1, 3, 6, 68, 114, 213
Essexville oil field, 235
Eureka Iron Company, 27
European series terms, 111
evaporites, 85, 89, *123*, 128, *139*, 148, *149*, 206
  A-1, 118, 119, 124
  B, 148
  deposition, 95
  formation, 141
  post-Salina, *151*
  precipitation, 92, 96, 140
  sabkha, 124
  Salina, 95, *139*
Everton Dolomite, 64

*expansa* Zone, 191
exposure, subaerial, 89

facies
   back-barrier, 89
   greenstone, 23
Falmouth Field, 57
*falsiovalis* Zone, 161, 162
fan complex, alluvial, 55
fault zones, 83
faulting, 37
faults, 34, 83, 222
   bedding, 34
   Bowling Green, 73
   Keweenaw, 40
   Lucas, 73
   Monroe, 73
   Sunday Lake, 34
   transverse, 34, 40
Felch Mountain District, *31*
   geology, *32*
   production, *32*
feldspar, 40, 46, 57
feldspar clasts, 57
felsites, 38
Fence River Formation, 37
fields
   Albion-Scipio, 73, 74, 79, 80, 81, 83
   gas, 73
   Northville, 74, 79, 80, 81, 82, 83
   oil, 73, 77, 81, 83, 222, 225, 235, 236, 238
   Porter oil, 235
   Stoney Point, 74, 79, 80, 81
Findlay Arch, 111, 118, 120, 123, 126, 127, 129, 130, 131, 142, 182, 194, 195
flagstones, 128
   construction, 214
flow breccias, 24
flows
   ashbed, 44
   dacite, 39
   lava, 16, 24, 38, 42
   mafic, 16
   mud, 19
   volcanic, 22
fluorite, 83
fluxes, 129
*Foerstia*, 210
folding, 37
Fort Wayne Bank, 118, 119, 125, 145, 146
Fort Wilkins State Park, 42
Foster Formation, 61, 64
fractionation, isotopic, 84
Framennian, later, *191*
Franconia Formation, 10, *61*
Fraser Township, *225*, 237
Fraserdale Arch, 139
Freda Formation, 38, *40*
Freda Sandstone, 55
Frontenac Arch, 139, 144

gabbro, 19
galena, 23
Galesville Formation, 61
Galesville Sandstone, *61*, 68

garnet, 214
gas fields, 73
   reef-based, 129
gas, 211, 214, 217, 225
   leases, 225
   natural, 53, 57, 215
   reserve, 89
Genesee Group, 189
Geneseo Shale, 184, 210
geology, economic, *13*, *203*
geothermal gradient, 1, *4*
   subsurface, 3
glaciation, 169
Gladwin County, 64, 209
glauconite, 57, 58
Glenwood Formation, 9, 10, 53, 64, *67*, 68, 75
gneisses, 15, 18
Goat Island Dolomite, 119
goethite, 29
goethite-hematite, 37
Gogebic County, 32
Gogebic Iron Range, 24, *32*
   geology, *34*
   production, *35*
gold, 22, 24
   native, 23
gold mines, geology, *15, 22, 23, 28, 32, 34, 36, 38, 45, 67*
Gondwana basins, Africa, 5
Goodrich Quartzite, 24
Goose Lake Member, *29*
Gower Formation, 121
Grand Marais, Michigan, 58
granites, 15, 17
granodiorite, 15, 19
gravel, 14, 44
graywacke, 16, 18, 19
Great Lakes reef province, 112
Green Bay, Wisconsin, 40
Greenfield contact, lower, 125
Greenfield Dolomite, 126
greenockite, 45
greenschist, 17
greenstone, 16, 23, 37, 38
Grenville Front, 236, 240
Grenville province, 73
grindstone, 213
Groveland mine, 31, 32
grunerite, 32, 36
Guelph Dolomite, 119, 127
Guelph Formation, 118
Gwinn Iron District, 29
gypsum, 14, *215*, 216

halite, 140, 141, 142, 143, 144, *147*, 149, 151
   crystals, 96
   precipitation, 140, 146
Hall, James, 104
Hancock, Michigan, 40
hanging wall, 46
heat, *6*
   flow, 1, 2, 3
Hecla Conglomerate, 45
Hecla mine, 45
hematite, 28, 29, 32, 36, 39, 40
Hemlock Formation, 35, *36*
Hillsdale County, 79, 212, 213, 214

hornblende, 18
hornblende diorite, 22
Horseshoe Curve section, 166
Howell anticline, 73, 83
Hudson's Bay Basin, 145
Humboldt Iron Mine, 22
Hunt Energy, Martin 1-15 well, 64
Huron County, 58, 212, 213, 214, 215
Huron Member, 157
Huron River, 214
hydrocarbon, 89, *101*
   deposits, 84
   production, *129*
   reservoirs, 83
*Hymenozonotriletes lepidophytus*, 165
*Hyperammina*, 162

*Icriodus*
   *latericrescens latericrescens*, 161
   *symmetricus*, 161, 162
Illinois, northeastern, *168*
Illinois Basin, 6, 94, 112, 118, 119, 120, 127, 130, 184, 206
illite, 214
image processing techniques, *221, 229*
inclusions, oil, 96
Indiana Basin, 141
Indiana group, 119, 121, 126
Iosco County, 214
Iowa group, 121
iron, 14, *24*
   ferrous, 77
   formation, 18, *23*, 29, 31, 34, 36, 37
   magnetic, 32
   mines, 15
   ore, 14, 15, *24*, 34, 47
   pig, 27
   resources, 24
   silicates, 28, 34
   siliceous, 31
iron oxides, 32
Iron River District, *35*
   geology, *36*
   production, *37*
Ironwood area, *16*
Ironwood Iron Formation, *34*
Isabella County, 215
Ishpeming, Michigan, 24, 29
Isle Royale, 41
Isle Royale lode, 44
Isle Royale mine, 45

Jackson County, 81, 214
Jackson mine, 26
Jackson Mining Company, 26
Jacobsville Sandstone, *55*
*jamieae* Zone, 156, 186
Jelinek-Ferris well, 9
Joliet Formation, 119
Jordan Sandstone, 64
Jordon River Formation, *159*, 189
Jude's Quarry, 214

Kankakee Arch, 127, 145, 150, 206
kaolinite, 214
Karch Quarry, 127
Kawkawlin oil field, 235

Kearsarge lode, 44
Kenneth Limestone Member, 121
Kenogami River Formation, 145
Kent County, 206, 214, 215
Kentucky Lockport reefs, 112
Kettle Point Formation, 156, 159, *168*, 183, 197, 210
Keweenaw fault, 40
Keweenaw Peninsula, 38, 40, 41, 44
Keweenawan rift zone, 73
Kinderhookian Series, *209*
Kitchi Schist, *19*
*Klapperina disparilis* Zone, 161, 162
Knowles, Wisconsin, 127, 128
Knox Sandstone, 64
*Koenenites cooperi*, 162
Kokomo Sea, 139, 146
Kona Dolomite, 23, 24

Lachine Member , 163, *164*, 187, *190*
LaHar 1-7 well, 228
Lake *expansa* Zone, 165
Lake Gogebic, 17
Lake Mine syncline, 40
Lake Shore Traps, 45
Lake Superior basin, 38
Lake Superior district, 41
Lake Superior Iron Company, 26
Lanthanum, 24
Late *expansa* Zone, 166, 169
Late *hassi* Zone, 156, 186
Late *praesulcata* Zone, 197
Late *rhenana* Zone, 186
latite, 19
laumontite, 39, 40
lava, 38
    basaltic, 23
    dacitic, 23
    flows, 16, 24, 38, 42
    mafic, 18
leaching, 28, 89, 96, 123
Leatham Formation, 165, 166, 186
LeClaire, Iowa, 120
LeClaire Facies, 121
LeClaire Member, 121
*Leiorhynchus*, 164
Lenewee County, 58
*Lepidodendron*, 214
leucoxene, 214
Lighthouse Point Member, *19*, 27
Limberlost Dolomite, 126
lime, agricultural, 129
limestone, 67, 76, 78, 85, 129, 161, 167, 211, 212, *215*
    high-calcium, 129
Lindwurm Member, 162, *167*, 169
*linguiformis* Zone, 164, 169, 182, 190
*Lingula*, 168, 210
Liston Creek Limestone Member, 119
Livingston County, 58, 213
Lockport contact, upper, 125
Lockport Dolomite, 127
Lockport Formation, 93, 97
Lockport Group, 111, 119, 130
Lockport-Greenfield contact, 126
Long Rapids Formation, 185
Loptain method, 3
Loretto Member, 32
Louisiana Limestone, 165, 166

Louisville Limestone, 112, 119, 126, 128
Louisville-Mississinewa Contact, 121
Lowenstam, Heinz, 108
Lower *crenulata* Zone, 158, 167, 198
Lower *expansa* Zone, 156, 158
Lower *gigas* Zone, 163, 168, 169, 190
Lower *marginifera* Zone, 157, 158, 164, 168, 169, 191, 194
Lower Marshall, *213*
Lower *Masrginifera* Zone, 191, 199
Lower Member, *19*
Lower *Praesulcata* Zone, 165, 166, 182, 191, 196, 199
Lower *rhomboidea* Zone, 164, 169
Lower Sandstone, 46
Lowermost *asymmetricus* Zone, 161
Lucas fault, 73
Ludington, Michigan, 150
Lyons quarry, 168

Mackinac Formation, 151
magnesium, 14, 78, 123
magnetite, 28, 29, 32, 34, 36, 84
manganese ore, 43
Manitoulin Island, Ontario, 122
Mansfield Iron-bearing Slate Member, 35, 36
marble, 22
marcasite, 167
*Marginifera* Zone, 182, 191
margins, basin, *167*
Marine Reef, 130
Marquette County, 22, 24, 27
Marquette Greenstone Belt, *18*, 22
Marquette Iron Company, 26
Marquette Iron Range, 23, *24*, 28
    geology, *31*
    production, *31*
Marquette Quadrangle, 19
Marquette River, 214
Marshall Sandstone, 206, *213*
Marshall Zone, *213*
Mason County, 77
Mass anticline, 40
Mass-Greenland area, 38, 39
Massive Sand, 64
Mather B mine, 26
maturity data, organic, 2, 7
Maumee Algal Stromatolite, 90, 93
Maumee Reef, 116, 120, 123, 125
McClure Oil, Beaver Island #1 well, 57
McClure Oil, Beaver Island #2 well, 57
McClure Oil, Sparks and others 1-8 well, 55
Mecosta County, 215
Menominee Group, *23*
Menominee Iron-bearing District, *31*
    geology, *32*
    production, *32*
Menominee Range, 37
Merritt Township, graben, *225*
*Mesotaxis*
    *asymmetrica*, 162, 163
    *dengleri*, 162
metamorphism, 15
    coal, 6
    high-grade, 32
    regional, 19, 23

Metz 1-15 wells, 228
mica, white, 18
Michigamme Formation, 22, 32, 35
Michigan, *13*, *47*
    copper production, *49*
    iron-ore production, *48*
    northern peninsula, *55*
    southern peninsula, *55*, *73*
Michigan Basin
    biostratigraphy, *155*
    brine depth, *147*
    brine level, *146*
    central, *221*
    diagenetic history, 73
    evaporite thicknesses, *148*
    evaporites, *139*
    formation stratigraphy, 3, *53*
    formations, *53*
    fossil record, *156*
    geological evaluation, *221*
    geophysical evaluation, *221*
    history (Late Devonian), *181*
    Mississippian System, *203*
    organic maturation model, 2
    organic maturity data, 2, 7
    paleogeography, *155*, *185*
    pinnacle reefs, *89*
    reef history (Silurian), *101*
    reefs, *101*
    sediments, 3
    thermal evolution, 3
    thermal maturity, *1*, 6
Michigan copper district, *38*
Michigan Formation, 7, 29, 206, *214*
Michigan paleoriver, 211
Michigan Stray, 215
micofossils, 206
micrites, 75
microcline, 18
Mid-continent Gravity High, 38
Mid-Continent Rift System (MRS), 38, 55, 239
Mid-Michigan Gravity Anomaly, 222
Mid-Michigan Gravity High, 235
Middle *triangularis* Zone, 168
Middle *expansa* Zoen, 169
Middle *praesulcata* Zone, 158, 165, 166, 197, 199
Middle *varcus* Subzone, 182, 187, 198
Middlesex Shale, 184
*Mihlina strigosa*, 157
millerite crystals, 167
Milwaukee Formation, 162, *167*, 189
mines, *35*, *37*
    Atlantic, 45
    Calumet, 45
    Carp Lake, 45
    Central, 41, 44
    Cliffs Shaft, 24, 31
    Colby, 32
    copper, 15
    Detroit, 142
    gold, *15*, 22, 23, 28, 32, 34, 36, 38, *45*, *67*
    Groveland, 31, 32
    Hecla, 45
    Humboldt Iron, 22
    iron, 15
    Isle Royale, 45

Jackson, 26
Mather B, 26
Minnesota, 41, 44
Nonesuch, 45, 46
Phoenix, 44
Quincy, 44, 45
Ropes Gold, *22*
salt, 150
Taylor, 29, 31
White Pine, 45
White Pine Extension, 45, *46*
mineralization
 copper, 45
 post-dolomite, *81*
minerals
 ferromagnesium, 214
 metallic, *13*
 nonfuel, *14*
 nonmetallic, 14
 opaque, 39
mining, prehistoric copper, 41
Minnesota mine, 41, 44
minnesotaite, 28, 34
Missaukee County, 57, 215
Mississippian, *198, 203, 209*
 Lower, *167*
Mobil Oil, Messmore #1 well, 58
Mobil-Jelinck #27907 well, 2
Moccasin Springs Formation, 119, 120, 121
models
 organic maturation, *2*
 reef growth timing, *92*
 rift, 55
Molluscan bed, 162
molybdenite, 23
Mona Schist, *19*, 27
monazite, 24
monocline, 17
Monroe fault, 73
Montcalm County, 215
Montpelier, Indiana, 120
Montpelier Reef, 121
Moose River Basin, *145*, 151, 185
Mt. Simon Sandstone, *57*
mud, 206
 basinal, 148
 black, 187, *189, 190*
 carbonate, 146
 cores, 117
 flows, 19
 hemipelagic, 198
 marine, 168
 prodeltaic, *190*
mudstone beds, 161
 calcareous, *162*
mudstones, 5, 55, 75, 94, 140, 141, 145, 167
 line, 89
Munising Formation, 61
Munising Group, 61, 140
muscovite, 18, 46

Napoleon, *213*
Napoleon Sandstone, 213
National Associates Petroleum Co. #1, Wolgamott well, *161*
Nazbro Quarry, 127
Nealy Creek Member, *19*

Negaunee, Michigan, 26
Negaunee Iron Formation, 24, 27, *28*
Negaunee Quadrangle, *19*
neodymium, 24
New Albany Shale, 156, 184, 189, 210
New Richmond Sandstone, 64
New York series terms, 111
Newago County, 206
Newburg Sand, 130
Niagara Escarpment, 101, 105, 130
Niagara Formation, 9, 10
Niagara Group, 113, 118, 119, 121, 124
Niagaran Guelph Formation, 90
Niagaran Series, 111, 112,
Niagaran-Cayugan unconformity, 109, 112, 121, *125*, 127
Nonesuch Formation, 38, *39, 45, 55*
 geology, *45*
 production, *46*
Nonesuch mine, 45, 46
Nonesuch Mining Company, 45
Norrie Member, 34, 35
North American craton, 185
North Point Member, *167*
Northern Central Basin, biostratigraphy, *158*
Northern Complex, Michigan, 15
Northern Peninsula, Michigan, 13
 surveys (linear), 25
Northville field, 74, 79, 80, 81, 82, 83
Norwood area, *159*, 183
Norwood Member, *163*, 169, 187, *189*

Oak Bluff Formation, *38*
Oceana County, 58, 77
Ohio Basin, 141, 143
Ohio Shale, 156, 185, 191, 210
oil, 89, 129, 211, 214, 215, 225
 crude, 1
 deposits, 80
 generation, 1
 inclusions, 96
 leases, 225
 reserve, 89
 source beds, 2
oil fields, 73, 77, 81, 83
 Essexville, 235
 Kawkawlin, 235
 Pinconning, 222, 225, 236, 238
 Saginaw, 235
oil wells, 129
 *See also* wells
Old Red Sandstone, 145
Ontario, southwestern, *168*
Ontario paleoriver, 211, 214
Ontonagon boulder, 42
Ontonagon County, 40, 41, 44
Ontonagon River, 42
oolite bars, 128
opaques, 39
Ordovician, 4
 Early, 8
 Middle, *53*
Ore Host Rock, *23*
Oronto Group, *55*
Osceola lode, 44
ostracodes, 216
Ottawa County, 213, 214

Ouachita trough, 186
overburden, 1, *4*
 thickness, 3, 4, 5
oxidation, 28, 34, 37, 58
*Oxinoxis*, 162, 167

packstone, 74, 75, 76
Paint River Group, 23, *24*, 35, *37*
*Palaeoscenidium cladophorum*, 163
paleogeography, *185*
*Paleoneilo*, 167
paleotemperatures, 5
Palhis Formation, 34
Palisades complex, 120
*Palmatolepis*
 *foliacea*, 163
 *gigas*, 163
 *glabra distorta*, 157
 *glabra lepta*, 164
 *glabra prima*, 164
 *gracilis sigmoidalis*, 157
 *hassi*, 163
 *minuta loba*, 164
 *minuta minuta*, 164
 *perlobata helmsi*, 157
 *perlobata sigmoidea*, 157
 *poolei*, 164
 *punctata*, 163
 *quadrantinodosa*, 168
 *quadrantinodosalobata*, 164
 *regularis*, 164
 *subperlobata*, 164
 *transitans*, 162, 163
 *unicornis*, 164
*Palmatolepis linguiformis* Zone, 190
*Palmatolepis triangularis* Zone, 190
Palmer area, 24
Pan Am, Crasey #1 well, 58
Parting Shale, *46*
Partridge Point, 158, *159*, 161, *164*, 183
Paxton Member, *163*, 182, 187, *190*, 199
Peace Member, 34
peat, 14
Peebles Dolomite, 119, 126
pegmatite, 18
Pence Member, 34, 35
Penokean orogeny, 23
Penokean province, 73
*Pentamerus oblongus*, 119, 126
peridotite, *19*, 22
 serpentinized, 22
periods
 discovery, *104*
 enlightenment, *107*
 stratigraphic integration, *111*
permeability, 89
Peterson Group, 32
petroleum, 1
phenocrysts, 19
Phoenix mine, 44
phosphorus, 37
phyllite, 23
Pinconning oil field, 222, 225, 236, 238
Pinconning South prospect, *225*
Pipe Creek Jr. Reef, 113, 114, 123, 129

pisolites, 123
plagioclase, 18, 46
*Plamatolepis*, 163
plutons, 15
Plymouth Member, 34, 35
Pocono Formation, 166
pods, 18
Point Au Gres Limestone, 215
*Polygnathus*, 163
  *brevilaminus*, 157
  *cristatus*, 161
  *decorosus*, 164
  *dengleri*, 161
  *dubius*, 161, 162, 163
  *glaber*, 164
  *ordinatus*, 161
  *pennatus*, 161
Porcupine Mountains, 39, 38, 40, *45*, 46
porosity, 74, 89, *96*
  intercrystalline, 80
porphyry
  granodiorite, 22
  quartz-feldspar, 38
Port Byron, Illinois, 120
Portage Lake District, 45
Portage Lake Volcanics, *38*, 44
Porter oil field, 235
*postera* Zone, 183
potassium feldspar, *81*
*Praecrdium* sp. (*Paneka* sp.), 162
Prairie du Chien Formation, 2, 10, *61*, 68
Prairie du Chien Group, *61*, 228
Prairie du Chienwell, 228
pre-Bedford Shale, paleogeology, *195*
Pre-Mt. Simon Clastics, 55
pre-Sunbury Shale, paleogeology, *197*
Precambrian, 55
Precambrian Formation, 7
precipitation, evaporite, 92, 96, 140
Presque Isle basin, 46
Presque Isle County, 58
production
  copper, *49*
  hydrocarbon, *129*
  Iron River District, *37*
  iron-ore, *31*, *48*
  Menominee Iron-bearing District, *32*
  native copper, *44*
  Nonesuch Formation, *46*
  salt, 14
products, stone, *128*
Pross 1-12 wells, 228
Proterozoic, 57
  early, *23*
  middle, *38*, *53*
*Protosalvinia*, 156, 158, 159, 164, 168, 169
*Protosalvinia (Foerstia)*, 156, 185, 191
*Psammosphaera*, 162
*punctata* Zone, 163, 168, 184, 189, 199
*punctata*, 169
Puritan Quartz Monzonite, 16
pyrite, 17, *23*, 37, 45, *81*, 144, 167, 210

pyritization, 23
pyroclastics, 16, 19
pyrrhotite, *81*

Quanicassee River, 228
Quanicassee River lineaments, 236, 238
Quanicassee River prospect, *228*
quartz, 18, 19, 22, 23, 39, 40, 46, 57, 61
quartzite, 16, 18, 23, 36, 58
Queensland, 6
Quincy mine, 44, 45

Racine Formation, 119, 121
Racine units, 122
radioactivity, 195
Ramsay Formation, *16*
Randville Dolomite, 32
Ray's Quarry, 214
rebound, isostatic, *190*
reclamations, native copper, 15
Red Rock, beds, 212
Red Rock unit, 167
red beds, 55, 145, 206
reefing
  A-1, 120
  A-2, 120
  Guelph, 120
reefs, *118*
  barrier, 89
  coralgal, 93
  diagenesis, *95*, *122*
  discovery period, *104*
  eastern Canada, 117
  economics, *128*
  enlightenment period, *107*
  generation-6 Kenneth, 122
  generations, *120*, 121
  Great Lakes province, 112
  growth, *92*, *94*, 112, 148, *149*
  historical phases, *103*
  history, *101*
  Illinois Basin, 120, 127
  Kentucky Lockport, 112
  Michigan Basin, *101*, 120, 123, 127
  numbers, *127*
  organic, 94
  patch, 140
  pinnacle, *89*, *92*, *93*, 120, 122, 146
  reef-bearing rocks, 111, *124*, 126
  Silurian, *101*, *103*, *104*, *107*, *111*
  sizes, *127*
  spongiostromatid, 127
  stratigraphic integraion period, *111*
  Wabash Valley, 117
  water depth, *122*
  See also specific reefs
regression, marine, 68
remagnetization, 84
reservoir fluids, 97
reservoir occurrence, *96*
*Retispora lepidophyta*, 158, 165, 166, 169
*rhenana* Zone, 156
Rhinestreet Shale, 185
*Rhipidomella missouriensis*, 165
*rhomboidea* Zone, 157, 169

rhyodacite, 16, 19
rhyolite, 16, 19, 38
rhythmites, 194
Richfield Formation, 228
Richmond Formation, 9
rift, crustal, 38
rifting, continental, 67
Riverton Iron Formation, 24, 35, *37*
Rockford complex, 127
Rockford Limestone, 167
Rockford Quarry Complex, 117
rocks
  carbonate, 158
  country, 16, 23
  crystalline, 32, 55
  evaporite-bearing, 111, *124*, 126
  felsic, 15, 19, 40
  host, 15, 47
  igneous, 47
  mafic, 19, 38
  metamorphic, 47
  parent, 23
  plutonic, 18
  reef-bearing, 111, *124*, 126
  sedimentary, 18, 37, 47, 53, *55*, *57*, 84, 203
  volcanic, 15, 16, 18, 22, 23, 32, 39, 40
Ropes Gold Mine, *22*
Roscommon County, 209
Ruff Formation, 90, 93, 141
Ruhr Valley, Germany, 6

Saginaw Bay, Michigan, *221*
  gravity data, *237*
  magnetic data, *236*
Saginaw Bay Graben, 228
  hypothesis, *222*, *228*, *238*
Saginaw County, 237
Saginaw Formation, 7, 8, 206, 213
Saginaw oil field, 235
St. Clair Limestone, 118, 119, 120, 122
St. Lawrence Formation, 61
St. Peter Sandstone, 61, *64*, 68
Ste. Marie's Falls Ship Canal, 28
Salamonie Dolomite, 119, 120, 122, 126
Salina A-1, 9
Salina A-2, 9
Salina C Formation, 7, 9
Salina deposition, 140
Salina evaporites, 92, 95, *139*
Salina Group, 90, 111, 113, 114, 119, 121, 122, 124, *140*, *149*
  A-carbonates, 121
  depositional history, *142*
  units, *140*, *149*
salinity, 121, 140, 214
salt, 14, 125, *143*
  cycles, 142, 148
  production, 14
salt mine, Detroit, 150
sand, 14, 39, 44, 61, 206, 211
  industrial, 14
sandstone, 38, 39, 40, 45, 46, 53, 55, *57*, 67, 145, 206, 211, 214, 215, 216, 217
Sanilac County, 58, 212, 213

Sappington Member, 165, 166
Sauk Sequence, upper boundary, *67*
Saverton Shale, 165
*Scaphignathus velifer*, 157
Schacht shaft, 46
scheelite, 23
Schiller 1-10 well, Amoco, 58
schist, 22
   amphibole, 19
   biotite, 18
   felsic, 16
   laminated, 19
   quartz-mica, 18
   volcanic, 22, 23
*Schmidtognathus wittekindti*, 162
sea level, change, *95*
sediment lithology, 3
   vertical variations, 2
sedimentation, 184, 189, 194
   subtidal marine, 74
sedimentology, *203*
sediments, 15, 23, 38, *187*, 213, 217
   interflow, 38
   thermal evolution, *3*
seepage, subsurface, *146*
sericitization, 23
serpentinite, 23
serpentization, 19
shale, 5, 19, 40, 45, 46, 55, 57, 58,
   62, 67, 75, 140, 141, 142, 150,
   151, 156, 167, 184, 185, 197,
   206
Sheared Rhyolite Tuff Member, *19*
Shell, Whyte 1-33 well, 235
Shrock, Robert R., 107
Siamo Slate, 29
Sibley, Badour 1-11 well, 228
siderite, 28, 29, 37
silica, 19, 31, 37, 39, 57
silicates, iron, 28, 34
silicification, *23*
sills, 15, 23, 29
siltstone, 16, 38, 39, 40, 46, 55, 57,
   58, 145, 167, 211, 214
Silurian, 4, *101*
   Late, *89*
silver, 22, 23, 45
   native, 40
*Siphonodella*
   *praesulcata*, 158, 165, 166
   *sulcata*, 165
*Siphonodella crenulata* Zone, 165
*Siphonodella sulcata* Zone, 148, 165
Six Mile Lake Amphibolite, *18*
Skunk Creek Member, 18
slate, 23, 34, 36
   clasts, 18
   magnetic, 37
   quartz, 34
   sericitic, 19
Slaven Chert, 186
Solberg Schist, *18*
*Spathognathodus costatus*, 165
sphalerite, 23, 83, 167
spongiostromatid reefs, *127*
sporopollen, 2
   coloration indices, 2, 7, 8
Squaw Bay Limestone, 156, 158, 159,
   *161*, 162, 163, *168*, 182, 189

Stambaugh, Michigan, 35
Stambaugh Formation, *37*
stilpnomelane, 28, 29, 34, 36
stone products, *128*
   crushed, 129
   foundation, 128
Stoney Point field, 74, 79, 80, 81
strait entrance, *144*
strata, interreef, 128
stratigraphy, 3, *53*, *55*, 94, *127*,
   *203*
stromatolites, 94, 128, 216
stromatoporoids, 94, 161
*Styliolina fissurella*, 162, 163
stylolites, 83
subsidence, 3, 6, 90, 140, 144, 182,
   184, 189, 209
   basin, 1, *148*
Sudbury sag, 144, 150
sulfate cements, 83
sulfides, 38, 45
   volcanogenic, 16
Sunbury black muds, *198*
Sunbury Shale, 165, 166, *167*, 183,
   *197*, 209, *212*, 235
Sunday Lake Fault, 34
supratidal islands, 94
surface temperature, 3
syenite, porphyritic, 22
sylvinite, *143*, 148, 149
sylvite, 143
synclines, 32, 40, 167
*Syringopora*, 117
*Syringothyris hannibalensis*, 165

talc, 23
talc-carbonate, 19
*Tasmanites*, 163, 164, 167, 168, 210
Taverse Group, 1, *158*
Taylor mine, 29, 31
tectonics, 83
tectonism, *127*
tenorite, 42
*Tentaculites*, 167
Terre Haute, Indiana, 129
Terre Haute Bank, 118, 127
tetrahedrite, 23
thermal alteration indices, 2, 8, 9
thermal blanket, 4, 5, 6
thermal contraction, 1
thermal evolution, 3
thermal maturity, *1*, *6*
thorium, 24
Thornton quarry, 168
Thornton Reef, 113, 123, 129
Three Forks Formation, 165, 166
Thunder Bay Limestone, 155, 159,
   *161*, 182, 189
time-temperature index, 3
*Tolypammina*, 162
tonalite, 15, 19
TOPIMAGE, *229*
*Tornoceras (T.)*
   *arcuatum*, 162
   *uniangulare*, 162
tourmaline, 23, 214
*trachytera* Zone, 182
Traders Member, 32
Transcontinental arch, 150

*transitans* Zone, 161, 162, 167, 168,
   184, 189, 198
Traverse Formation, 9, 225
Traverse Group, 156, 161, 182, 189,
   209
Trempealeau Formation, 10, *61*, 68
trenching, 22
Trenton Formation, 2, 8, *73*
   diagenetic history, *73*
   dolomitization, *77*
Trenton Group, *229*, 239
*triangularis* Zone, 190, 199
trilobite debris, 76
Triple Gyp, 215
troughs, 35
tuff, 16, 18, 19, 23, 24
   water-laid, 19
Tuscola County, 213, 228
Tyler Formation, 34
Tymochtee Dolomite, 119

Umlor Formation, 61
Umlor unit, 239
unconformities
   grand postreef, *125*
   reef-evaporite, 125
   reef-top, 126
Undifferentiated Greenstone, *19*
Union Oil Co., #1 Smith well, 159
Upper *asymmetrica* Zone, 163, 169,
   189
Upper *crepida* Zone, 164, 168, 169
Upper *expansa* Zone, 169, 199
Upper *gigas* Zone, 168, 169, 186
Upper *hermanni-cristatus* Zone, 161
Upper *marginifera*, Zone, 164, 191
Upper Marshall, *213*
Upper *Mesotaxis asymmetrica* Zone,
   186
Upper *praesulcata* Zone, 165, 166,
   169, 199
Upper *rhomboidea* Zone, 164, 168,
   190, 199
Upper Shale, *46*
Upper *trachytera* Zone, 156, 157, 169
Upper *varcus* Subzone, 161, 184
Upson, Wisconsin, 32
uranium, 24
Utica Formation, 7, 9, 10
Utica Shale, 75, 80, 85

Valders, Wisconsin, 120
Van Buskirk Gneiss, 16
*varcus* Zone, 167
Verde Antique marble, 22
viscosity, 31
visicles, 38
vitrinite, 2, 7
   anisotropy, 6
   reflectance index, 7, 8, 9
volcanics, 15
Vulcan Iron Formation, 31, *32*

Wabash Formation, 119
Wabash platform, 92, 95, 96
Wabash prong, 145
Wabash Valley, 122, 127
Wabash Valley reefs, 117
wackestone, 74, 75, 76, 94

Wakefield area, *16*, 34
Waldron Shale, 117, 126
Wallace Stone Company, 215
Washtenaw County, 79
waters
    acid mine, 37
    diagenetic, 84
    inflow, 149
    marine, *85*
    meteoric, *85*
Watersmeet, 17
Waubakee Formation, 130
Wauseca Pyritic Member, *37*
wave base, *122*
waves, inland-sea, 117
*Webbinelloidea*, 162
wells, 7
    Amoco, Schiller 1-10, 58
    Dart, Christian 1-16, 225
    Dart, Christian 1-16A, 225
    Dart, Edwards 7-36, 57
    Dow-Doe #110, 157
    Hunt Energy, Martin 1-15, 64
    Jelinek-Ferris, 9
    LaHar 1-7, 228
    McClure Oil, Beaver Island #1, 57
    McClure Oil, Beaver Island #2, 57
    McClure Oil, Sparks and others 1-8, 55
    Metz 1-15, 228
    Mobil Oil, Messmore #1, 58
    Mobil-Jelinck #27907, 2
    National Associates Petroleum Co. #1 Wolgamott, *161*
    Pan Am, Crasey #1, 58
    Prairie du Chien, 228
    Pross 1-12, 228
    Shell, Whyte 1-33, 235
    Sibley, Badour 1-11, 228
    Union Oil Co., #1 Smith, 159
West Kiernan Sill, 36
West Milgrove, Ohio, 120
Whiskers Creek Gneiss, 16
Whiskey Creek Formation, 159
White Pine area, 46
White Pine Copper Company, 45
White Pine Extension mine, 45, *46*
White Pine mine, 45
Williston basin, 156
Wisconsin, eastern, *167*
Wisconsin dome, 150
Wisconsin Highlands, 206, 214
Wisconsin River, 214
wood, carbonized, 210
Woodford Shale, 186
Wyandotte, 27

Yale Member, 34, 35
Yorktown Reef, 128

zircon, pink, 214
*Zoophycos*, 161, 162, 167, 169, 189